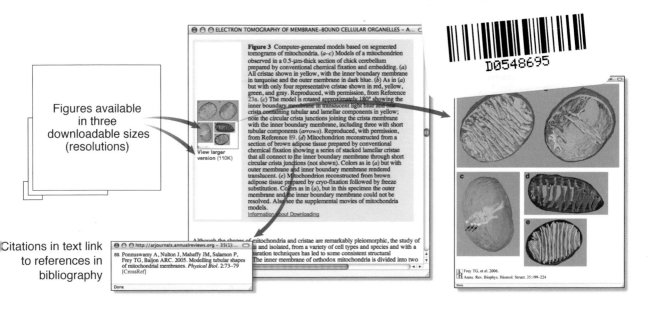

Figures available in three downloadable sizes (resolutions)

Citations in text link to references in bibliography

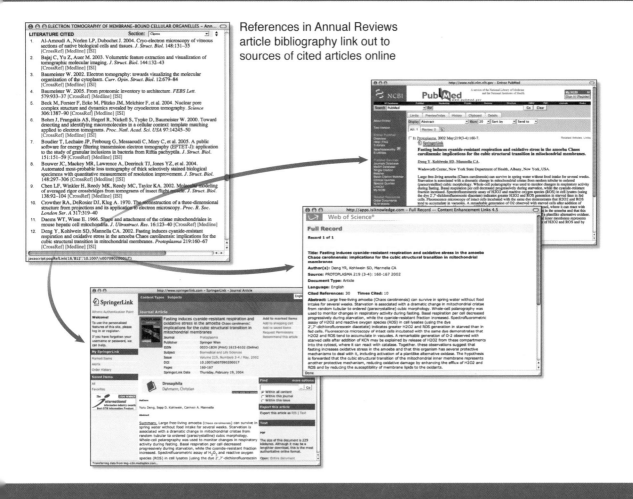

References in Annual Reviews article bibliography link out to sources of cited articles online

Annual Review of Biophysics

Annual Review of Biophysics

Volume 38, 2009

Douglas C. Rees, *Editor*
California Institute of Technology

Ken A. Dill, *Associate Editor*
University of California, San Francisco

Michael P. Sheetz, *Associate Editor*
Columbia University

James R. Williamson, *Associate Editor*
The Scripps Research Institute

www.annualreviews.org • science@annualreviews.org • 650-493-4400

Annual Reviews
4139 El Camino Way • P.O. Box 10139 • Palo Alto, California 94303-0139

Annual Reviews
Palo Alto, California, USA

International Standard Serial Number: 1936-122X
International Standard Book Number: 978-0-8243-1838-3
Library of Congress Catalog Card Number: 79-188446

TYPESET BY APTARA
PRINTED AND BOUND BY FRIESENS CORPORATION, ALTONA, MANITOBA, CANADA

Contents

Annual Review of
Biophysics
Volume 38, 2009

Index

Errata

An online log of corrections to *Annual Review of Biophysics* articles may be found at
http://biophys.annualreviews.org/errata.shtml

Related Articles

Protein Translocation Across the Bacterial Cytoplasmic Membrane
Arnold J.M. Driessen and Nico Nouwen

From the ***Annual Review of Cell and Developmental Biology***, Volume 24 (2008)

Microtubule Dynamics in Cell Division: Exploring Living Cells
with Polarized Light Microscopy
Shinya Inoué

Systems Approaches to Identifying Gene Regulatory Networks in Plants
Terri A. Long, Siobhan M. Brady, and Philip N. Benfey

Sister Chromatid Cohesion: A Simple Concept with a Complex Reality
*Itay Onn, Jill M. Heidinger-Pauli, Vincent Guacci, Elçin Ünal,
and Douglas E. Koshland*

Disulfide-Linked Protein Folding Pathways
Bharath S. Mamathambika and James C. Bardwell

Unconventional Mechanisms of Protein Transport to the Cell Surface
of Eukaryotic Cells
Walter Nickel and Matthias Seedorf

Prelude to a Division
Needhi Bhalla and Abby F. Dernburg

Evolution of Coloration Patterns
Meredith E. Protas and Nipam H. Patel

Cell Polarity Signaling in *Arabidopsis*
Zhenbiao Yang

From the ***Annual Review of Genetics***, Volume 42 (2008)

How *Saccharomyces* Responds to Nutrients
Shadia Zaman, Soyeon Im Lippman, Xin Zhao, and James R. Broach

Myxococcus—From Single-Cell Polarity to Complex Multicellular Patterns
Dale Kaiser

Rhomboid Proteases and Their Biological Functions
Matthew Freeman

The Organization of the Bacterial Genome
Eduardo P.C. Rocha

Individuality in Bacteria
Carla J. Davidson and Michael G. Surette

Determination of the Cleavage Plane in Early *C. elegans* Embryos
Matilde Galli and Sander van den Heuvel

The Dynamics of Photosynthesis
Stephan Eberhard, Giovanni Finazzi, and Francis-André Wollman

Sunney I. Chan

A Physical Chemist's Expedition to Explore the World of Membrane Proteins

Sunney I. Chan

Noyes Laboratory of Chemical Physics 127-72, California Institute of Technology, Pasadena, California 91125; and Institute of Chemistry, Academia Sinica, Nankang, Taipei 115, Taiwan; email: sunneychan@yahoo.com, chans@its.caltech.edu

Annu. Rev. Biophys. 2009. 38:1–27

The *Annual Review of Biophysics* is online at biophys.annualreviews.org

This article's doi:
10.1146/annurev.biophys.050708.133713

1936-122X/09/0609-0001$20.00

Key Words

NMR of base-stacking interactions in nucleic acids in solution, dynamic structure of bilayer membranes, membrane biophysics, metal cofactor structure and function, redox linkage and proton pumping in cytochrome c oxidase, methane hydroxylation by the particulate methane monooxygenase, early kinetic events in protein folding

Abstract

Despite growing up amid humble surroundings, I ended up receiving an excellent education at the University of California at Berkeley and postdoctoral training at Harvard. My academic career at Caltech was shaped by serendipity, inspirational colleagues, and a stimulating research environment, as well as smart, motivated students and postdocs who were willing to join my search for molecular understanding of complex biological systems. From chemical physics I allowed my research to evolve, beginning with the application of NMR to investigate the base stacking of nucleic acid bases in solution, the dynamic structure of membranes, and culminating with the use of various forms of spectroscopy to elucidate the structure and function of membrane proteins and the early kinetic events in protein folding. The journey was a biased random walk driven by my own intellectual curiosity and instincts and by the pace at which I learned biochemistry from my students and postdocs, my colleagues, and the literature and through osmosis during seminars and scientific meetings.

Contents

MY HUMBLE BEGINNINGS

I was born in San Francisco and brought up in relatively humble surroundings. My parents were both immigrants from Southern China. My father came to the United States as a teenager in the 1920s, and my mother emigrated in her early twenties in 1936. Neither of my parents had any formal education. Both worked in a factory manufacturing denim for Levi Strauss & Co. The working hours were long, as was typical of sweatshops during those days.

Coming from a Chinatown ghetto community, I grew up culturally disadvantaged, if not deprived. I was certainly not destined to be a university professor. There was not a drop of academic blood in my body. Moreover, there were no role models, scholars, scientists, or professionals for me to emulate. I was the first in my family to attend university, receive a college degree, and obtain a PhD. How I ever got to where I ended up makes for an interesting saga. It was mostly serendipity, and not by design or program!

After the Second World War ended in 1945, my parents decided that I should move to Hong Kong to receive a Chinese education. After all, I was growing up to behave like a "hollow bamboo," without any knowledge or appreciation of my Chinese heritage or culture. So in the spring of 1949, I was thrown into a Chinese middle school in Hong Kong. I had just turned 12, and by age I was assigned to the first year of middle school (the equivalent of seventh grade in the United States). Everything was taught in the native language, including the subject "English." Not surprisingly, I flunked out at the end of the term. It was a socially humiliating experience, if not a psychological trauma that took many years to shake off.

Ultimately, I ended up in an English-speaking school run by the Irish Jesuits. It was here that my interests in academics were kindled and I developed a fascination for mathematics and science. A number of excellent math and science teachers had aroused my interest in these subjects with their inspiring style of teaching.

In the fall of 1953, I returned to California and entered the University of San Francisco (USF) with the intention of joining the priesthood and becoming a science teacher. At the age of 16 going on 17, I was not up to the challenge of a rigid lifestyle and I soon gave up the idea of becoming a Jesuit. Instead of continuing my studies at USF, I transferred to U.C. Berkeley with the help of the chairman of the science faculty at USF.

BERKELEY DAYS: FROM UNIT OPERATIONS TO QUARTIC OSCILLATORS

U.C. Berkeley was a difficult school for me. I was not prepared for the rigorous curriculum. Berkeley was also intimidating in other ways. Because I was brought up in a relatively sheltered environment throughout my adolescence, it took some effort to build up the self-confidence to survive in this seemingly "unsupportive" environment. I had to find myself.

Two professors at U.C. Berkeley were inspirational to me, both inside and outside the classroom. Professor George Pimentel's lucid lectures on chemical equilibrium did much to stimulate my interest in the complex chemistry of solutions. More importantly, he maintained an interest in my overall development as a young man and steered me toward cultivating intellectual outlets other than just chemistry. Professor Andrew Acrivos taught me applied mathematics, chemical engineering thermodynamics, and kinetics, and he did me a greater favor by introducing me to the process of self-study and independent research. Thus, by the time I became a senior, I was prepared to learn things on my own from sources outside the classroom. During my senior year, I began auditing courses in physics and chemistry that were not part of the formal curriculum.

After my graduation from U.C. Berkeley with a BS in chemical engineering in 1957, my original plans were to do graduate studies at the University of Minnesota. For health reasons, I ultimately stayed at Berkeley to work toward a PhD in physical chemistry. Professor Kenneth Pitzer, the then dean of the College of Chemistry, arranged for my admission and late registration. Even so, by the time I was allowed to

register and discuss research with professors in the College, essentially all the vacancies in the research groups I was interested in were filled. I eventually signed up to work in the area of EPR for Professor Rollie Myers, who happened to be on sabbatical leave in Switzerland at the time. Professor William Gwinn was willing to take me on as a graduate student for one year until Dr. Myers returned. A year later, Dr. Gwinn decided to keep me on as his own student, as I was well on my way toward completing my thesis research. Dr. Myers was not aware of this arrangement and outcome until many years later when he was on sabbatical with me at Caltech.

My PhD thesis was on the microwave spectrum of oxetane (or trimethylene oxide) and its molecular structure. The work was concerned with the nature of the potential function describing the out-of-plane vibration of the four-membered ring. At that time the planarity of this heterocycle was controversial. According to the microwave spectrum, the molecule was essentially planar, but the ring was bent according to far-infrared spectroscopy. My thesis work was to reconcile the microwave and far-infrared data. It was possible to fit both sets of data in terms of a potential function consisting of the quartic potential with a small central barrier, with the zero-point vibrational level above the top of the barrier (15). In other words, the four-membered ring of oxetane was essentially planar. The remarkable feature of the fit was how well the far-infrared data were accounted for by the quartic oscillator, including the isotope shifts of the $\Delta n = 1$ transition frequencies when the three CH_2 groups on the periphery of the four-membered ring were replaced by CD_2. Despite the excellent agreement between theory and experiment, it took almost a decade before the story was finally accepted by the scientific community.

ONWARD TO HARVARD

I was set on an industrial career after obtaining my PhD. In fact, by the early spring of 1960, I had lined up attractive positions at MIT Lincoln Laboratories, Lockheed, General Electric (to work with Bruno Zimm), and IBM (to work with Enrico Clementi). My PhD mentor, however, had a different plan for me. Upon his insistence, I applied for an NSF Postdoctoral Fellowship to work with Professor Norman Ramsey in the department of physics at Harvard University. High-resolution solution NMR was developing at the time, and Professor Dudley Herschbach, who was a beginning assistant professor of chemistry at Berkeley, suggested that a good way for me to enter the field with my background in microwave spectroscopy of gases was to study the NMR of small molecules in molecular beams. At Harvard, I would learn the physics of NMR rigorously from Professors Ramsey, Edward Purcell, and R.V. Pound. Thus, Ramsey seemed like the perfect choice. Nonetheless, the NSF application was just an exercise for me to keep peace with my professor. When the word came on March 15, 1960, that I had won an NSF Fellowship, I was taken aback. When I accepted to go to Harvard in April, Dr. Gwinn was pleased, but my father was disappointed that I took the $4500 fellowship instead of the much-higher-paying positions at General Electric and IBM that I was seriously considering. He showed his disappointment by reminding me, "Son, I don't have a PhD, but I make much more money than that."

NMR IN MOLECULAR BEAMS

I arrived in Cambridge, MA, before Labor Day 1960. The Ramsey laboratory had just assembled a molecular beam machine with the universal detector. The mass spectrometer with electron-bombardment ionization to detect the molecular beam allowed molecular beam magnetic resonance studies to be extended to molecules beyond molecular hydrogen. As I was the only chemist in the group, it was incumbent upon me to prepare other molecules for these experiments. Toward this end, I prepared HF, HCl, HBr, HI, and HCN, and the proton spin-rotation coupling constants of these molecules were subsequently determined by Jim Pinkerton, a PhD student in the Ramsey laboratory.

For my own research, I embarked on measurements of the spin-rotation coupling of ^{15}N in molecular $^{15}N_2$ and the rotational magnetic moment of this molecule (13). Because the ^{15}N nuclear moment and the rotational magnetic moment were one order of magnitude smaller than the corresponding magnetic moments in H_2, narrower slit widths had to be used in collimating the molecular beam and detecting magnetic resonance, and to make up for the loss in sensitivity of detection of the magnetic resonance signal, the standard single slit was replaced by multiple slits. The ^{15}N spin-rotation coupling constant and the rotational magnetic moment of $^{15}N_2$ culminated in the determination of the paramagnetic contribution to the ^{15}N nuclear magnetic shielding and the absolute shielding scale for the ^{15}N chemical shift, as well as the paramagnetic contribution to the magnetic susceptibility in molecular N_2.

It soon became obvious that the molecular beam magnetic resonance method could be extended to molecular systems with larger magnetic moments, e.g., molecules with unpaired electrons. The first system I tried was NO, a diatomic with an unpaired electron in a π-molecular orbital. This electron spin is coupled via spin-orbital coupling to the orbital angular momentum along the N−O bond axis, leading to $^2\pi_{1/2}$ and $^2\pi_{3/2}$ Λ-doublet states separated by 121.1 cm^{-1}. NO in the $^2\pi_{1/2}$ state is nonmagnetic, but owing to rapid end-over-end rotation of the molecule, a large rotational magnetic moment is produced (12). Professor J.H. Van Fleck happened to be in the laboratory the afternoon this discovery was made, and by the next morning he had explained the observation using perturbation theory to mix the $^2\pi_{1/2}$ and $^2\pi_{3/2}$ Λ-doublets by electron spin-rotation coupling and the Zeeman interaction of the electron spin. To a novice scientist such as myself, this was an impressive demonstration of real raw intellectual power. The following week I picked up a copy of Van Fleck's book *Theory of Electric and Magnetic Susceptibility* at the Harvard Coop bookstore in Harvard Square and read it from cover to cover. Subsequently, I had many scientific discussions with Dr. van Fleck, and these encounters left a deep and lasting impression on me, reinforcing that I had made the right decision to do my postdoctoral study at Harvard.

The second free radical I tried to study on the molecular beam apparatus was the methyl radical, which was produced by pyrolysis of tetra-methyl lead in a hot source constructed from a quartz tube. I was able to deflect the $CH_3 \cdot$ in both the dispersing and refocusing inhomogeneous magnetic field regions, but the electron spin resonance elicited in the homogeneous magnetic field region of the molecular beam apparatus was spread out because of the large coupling of the electron spin with the large rotational magnetic fields produced by end-over-end rotation of the radical.

BEGINNING MY ACADEMIC CAREER AT THE UNIVERSITY OF CALIFORNIA AT RIVERSIDE

During the spring of 1961, I was offered the position of assistant professor at the University of California at Riverside (UCR). The graduate program in chemistry at UCR had just started, and I thought I could contribute to building a high-quality program. It was here that my original interest in science teaching was finally beginning to take shape. I also began to develop an interest in the mentoring of graduate students. My graduate training at Berkeley and my postdoctoral experience at Harvard had convinced me that research training was a great way to encourage young people to develop their potential to become original and independent research scientists. Thus, my career as an academic scholar and teacher was beginning to gel. I was about to turn 25.

Although I had finally settled on my career goals, my research plans were far from crystallized. I did not want to continue in the NMR of molecular beams. This type of spectroscopy, like microwave spectroscopy, was maturing and becoming less likely to generate new concepts or breakthroughs. To train graduate students, I felt that the research they worked on had to be "discovery" driven. In curiosity-driven sci-

ence, the young person could be trained to define the scientific question, develop the research plan, formulate the hypotheses to be tested, and design the experiments to test them. So, during my first year at Riverside, I tried a number of areas that were new to me, including the solution structure of transition metal complexes in solution by NMR (introduced to me by Professor Donald Sawyer), NMR of purine and pyrimidine bases (introduced to me by Professor George Helmkamp), and EPR of concentrated alkali metal ammonia solutions. Magnetic resonance in solution appeared to be a developing field, and it was capturing my intellectual attention.

I happened to be at the right place at the right time. It was my good fortune that a Varian DP 60 NMR spectrometer was sitting idle in the Citrus Experimental Station and that it would become the tool of my research. All I had to do was get the spectrometer fired up and running. Before long, I was the faculty member in charge of the spectrometer, the NMR technician, and the NMR operator for the department. I worked together with Richard Kula and Donald Sawyer on the NMR of metal complexes of ethylenediaminetetraacetic acid (EDTA), methyliminodiacetic acid (MIDA), and nitrilotriacetic acid (NTA) (19), and with Marty Schweizer and George Helmkamp on the NMR of purines (87). Oscar Paez started to prepare concentrated alkali metal liquid ammonia solutions, and we determined the microwave skin depth of these metal solutions by EPR to estimate the electrical conductivity (11). Robert Iwamasa started a project on the selective hydration of alkali and alkaline earth metal ions in dilute solutions in acetonitrile.

During this period, I maintained an interest in spin-rotation coupling and the paramagnetic contribution to nuclear magnetic shielding. While at Harvard, I had many conversations with Professor John Baldeschwieler on ^{19}F NMR relaxation. It was apparent that spin-rotation coupling was an important relaxation process for ^{19}F. It turned out that this was the case for ^{13}C and ^{15}N as well.

MOVING ACROSS TOWN TO CALTECH

During the spring of 1961, Professor G.W. Robinson invited me to present a seminar at Caltech. Much to my surprise, I was invited to join the Caltech chemistry faculty the following week. It didn't take me long to decline the offer. As my research program was just about to gel, I did not feel that it was time for me to make a move. I was also reluctant to move to such a high-power place. I wasn't sure that I was up to the challenge. Besides, it could mean a dramatic change in lifestyle for me, more than what I was willing to make. When Professors Harden McConnell and Jack Roberts subsequently asked me to reconsider, and a similar offer was extended to me from the chemistry department at Yale, I decided that perhaps I should consider the long-term impact of a stimulating research environment and high-quality graduate students on the development of my research career. After some soul searching, I finally made the decision to move to Pasadena.

Caltech is an amazing place. As the Caltech logo now says, "There is only one Caltech." Yes, it is a unique place to do science. My colleagues are world class and supportive. The graduate students are first rate. All the way around, there is quality. It is simply a classy institution. I have been associated with Caltech for well over 45 years. Not for a single moment have I ever regretted the decision I made in 1961.

MY FIRST RESEARCH PROGRAM IN BIOPHYSICS: BASE-STACKING OF NUCLEIC ACIDS IN AQUEOUS SOLUTION

Soon after I arrived in Pasadena, I began to consolidate my research program to focus on the application of NMR to address the structure of biological molecules and their interactions in solution. As I was leaving UCR, I was already studying the interactions between nucleic acid bases, nucleosides, and nucleotides in water solution. From studies of the colligative properties of these molecules in aqueous medium, it was evident that these molecules associate in water. The issue was whether they associate via horizontal hydrogen-bonding or vertical base-stacking via hydrophobic and Van der Waal interactions. I was exploiting the ring current shifts of the nucleic acid bases to distinguish between the two modes of association. Together with Dr. Paul O.P. Ts'o and George Helmkamp, Marty Schweizer and I studied the NMR chemical shifts of the ring protons of purine and 6-methylpurine as a function of concentration (22). Upfield shifts were observed, consistent with base stacking, and the extent of the concentration shifts was consistent with the degree of association implicated by osmotic coefficient measurements (100). This conclusion was received with some skepticism at the time. Dr. Y. Kyogoku, working with Professors Richard Lord and Alex Rich at MIT, had independently shown by infrared spectroscopy that derivatives of A and T, and G and C, formed horizontal complementary base pairs in $CHCl_3$, just as predicted by Watson and Crick for double-helical DNA (52, 53). It soon became clear that the difference in solvent, water in the case of the NMR experiments and $CHCl_3$ in the case of the infrared experiments, accounted for the differences in the solution behavior of these molecules. Thus, nucleic acid bases do stack in aqueous buffer. The same vertical forces must be operative in stabilizing the secondary structures of DNA and RNA molecules to overcome in part the significant entropic penalty required to form complementary base pairs in the double helix.

Following up on this work, James Nelson studied the conformational properties of the dinucleotide ApA in aqueous solution (21); Benedict Bangerter, the dinucleotides ApC, CpA, ApU and UpA (3); and Paulus Kroon, George Kreishman, and James Nelson, the conformational properties of ApApA, ApApApA, and ApApApApA (50). Ben Bangerter and Heinrich Peter also discovered that purine could insert into the ApA stacks and form weak short-lived intercalated complexes with ApC, CpA, and ApU, further demonstrating the importance of base-stacking interactions (14). Ben was also able to use polyU as a template to form a triple helix from two polyU strands and a vertical stack of adenosine, which was stabilized by Watson-Crick and Hoogsteen base-pairing in addition to the vertical interactions along the adenosine stack (2).

PHASING OUT THE CHEMICAL PHYSICS PROGRAM

Part of the consolidation of my research program included the gradual phase out of my interests in spin-rotation coupling and NMR relaxation. One of the first graduate students to join the Chan laboratory at Caltech was Alan Dubin. He looked for evidence of spin-internal-rotation coupling, the coupling of a nuclear spin to the rotational magnetic field generated by rapid internal rotation of a rotor in molecules containing $-CH_3$ or $-CF_3$ tops. We selected benzotrifluoride for this work. In this molecule, the barrier to internal rotation of the $-CF_3$ group is sixfold; sixfold barriers are typically very small, a few calories per mole. ^{19}F was chosen because the rotational magnetic fields per unit angular momentum experienced by the three ^{19}F nuclei in the $-CF_3$ top were expected to be large. Alan, together with Tom Burke, built a pulse NMR machine to measure the ^{19}F NMR longitudinal relaxation rates ($1/T_1$) in benzotrifluoride and hexafluorobutyne-2 (10, 33). The ^{19}F T_1s were short; moreover, part of the NMR relaxation was observed to be independent of the viscosity of the medium, as expected for spin-internal-rotation coupling. Fluctuations in the rotational magnetic field due to internal rotation should be inertially controlled. As predicted, Alan quenched the spin-internal-rotation coupling in the case of benzotrifluoride by introducing a substituent at one of the *ortho* positions of the aromatic ring, converting the barrier from sixfold to threefold (33). Threefold barriers to internal rotation are typically 1000 times higher. Additional confirmation of this mechanism of NMR relaxation came from a study of the pressure dependence of the NMR relaxation rates by Professor Jiri Jonas and his coworkers at the University of Illinois, Urbana-Champaign (9). Later, Charles Schmidt established the importance of ^{13}C spin-internal-rotation coupling in the ^{13}C NMR relaxation of the methyl top in toluene (84).

As I began to embark on the new field of biophysics, it was not obvious to me that my background in chemical physics would serve me well in my future endeavors. Moreover, given that I have had no formal training in biochemistry or biology, I was entering the field with a major handicap. Molecular biophysics was a relatively virgin field, and not knowing much about it was a blessing in disguise. I learned the biochemistry as my interests and understanding developed, and I had no choice but to focus. The more serious hurdle was finding out what was known and what was not known, as well as identifying the scientific questions that needed to be solved and the tools that would get me to the heart of the matter. Here, my physical background and experience in molecular spectroscopy, particularly NMR and EPR, proved to be invaluable. My training in physics armed me with the insights to formulate and analyze complex problems rigorously from the outset. I was also in a better position to develop reasonable conceptual models and hypotheses to test in the laboratory.

FORAYS INTO MEMBRANE BIOPHYSICS

Dynamic Structure of Lipid Bilayer Membranes

In 1968, Michael Sheetz joined my laboratory. He was interested in developing biophysical methods to probe the structure and dynamics of biological membranes consisting of both phospholipids and membrane proteins. The impetus of the work actually came from Professor S.J. Singer of the University of California at San Diego, who was eager to seek more direct physical evidence for the fluid mosaic model of biological membranes (93). Mike initiated a collaborative effort: The Singer laboratory prepared red blood cell ghosts and Mike examined

T_1: longitudinal spin relaxation time

T$_2$: transverse spin
relaxation time

the proton NMR of the ghost membranes. A Varian HR 220 superconducting NMR spectrometer had just been installed at Caltech, and it seemed like the timely experiment to try.

No discernable features were apparent in the NMR spectrum under ambient temperatures. However, as the temperature was raised to 40°C and above, reproducible resonances appeared characteristic of denatured proteins (41). These resonances originated from spectrin, the major cytoskeletal protein in the membrane of the red cell ghosts (92). Surprisingly, no phospholipid features were evident, even though the lipid bilayer was supposed to be fluid. The obvious conclusion was that the local motions of the lipid molecules were restricted and slow in the membranes; the overall rotation of the ghost was too slow to average out any residue magnetic dipolar interactions among the proton spins.

To address this issue, Mike prepared single-walled phospholipid bilayer vesicles, fractionated the vesicles according to size, and recorded their proton NMR spectra. The proton NMR of the smallest bilayer vesicles, which were 250–300 nm in diameter, was extremely sharp, resembling the spectrum of a small hydrocarbon droplet in a micelle of the same diameter. Moreover, the line widths were not influenced by the viscosity of the medium in which the vesicles were suspended. Thus, the overall tumbling of the vesicle was not a controlling factor in limiting the line widths of the lipid resonances in the small sonicated vesicle. We surmised that the observed fluidity arose from disorder in the packing of the phospholipid molecules to fit the extreme curvature imposed by the bilayer structure (91). Although this explanation of the unusual fluidity of small lipid vesicles seemed intuitively obvious, the curvature effect remained controversial for many years.

An outcome of the curvature effect was that the two monolayer leaflets of a small bilayer vesicle were intrinsically nonequivalent owing to differences in the magnitude and sign of the curvature experienced by the inside- and outside-facing lipid molecules. This asymmetry was subsequently demonstrated by Joseph

Schuh, Utpal Banerjee, and Luciano Mueller (85) and Kenneth Eigenberg (34) in proton NMR experiments at 500 MHz that revealed different chemical shifts for the choline methyl groups of the lecithin molecules occupying the two halves of the bilayer vesicle.

During the 1970s, the Chan laboratory devoted considerable effort toward comparing the properties of single-walled bilayer vesicles with those of multilamellar dispersions. As opposed to single-walled bilayer vesicles, the multilamellar structures corresponded to the global thermodynamic free-energy minimum of the phospholipid-water dispersion. In contrast, single-walled bilayer vesicles were metastable and they tended to fuse with one another to form larger structures. This vesicle fusion was catalyzed by local defects in the bilayer structure. Rüdiger Lawaczeck, Jean-Luc Girardet, and Masatsune Kainosho introduced structural defects into the bilayer membranes by sonicating phospholipid dispersion below the phase transition temperature of the lecithin dispersions (56). These structural defects also mediated transmembrane solute and ion transport (57), reminiscent of the structural fluctuations found in lipid structures containing coexisting gel and fluid domains. Arthur Lau also discovered that amphiphatic peptides like alamethicin could mediate the rapid fusion of small bilayer vesicles prepared by sonication above the thermal transition (55). ^{1}H NMR experiments by Arthur (54) and ^{31}P and ^{2}H-NMR experiments by Utpal Banerjee, Robert Birge, and Raphael Zidovetzki (1) showed that these peptides were surface active and could serve as nucleating centers to lower the activation energy for vesicle fusion.

To compare the dynamic structure of the bilayer vesicle with that of the multilamellar phospholipid dispersion, the Chan laboratory appealed to NMR line width and relaxation measurements, exploiting the different sensitivity of the T$_1$ and T$_2$ to molecular motions of different timescales. Unlike the high-resolution liquid-like spectrum observed for the small lipid vesicle, in lecithin multilamellar dispersions, the choline methyl resonance, the resonances of the methylene protons, and the terminal methyl protons of the acyl chains were significantly broader in the ^{1}H NMR spectrum. In addition, the spectrum exhibited intensity anomalies expected for a powder spectrum of a system with motional restriction (23). To estimate the degree of motional restriction, Gerry Feigenson and Charlie Seiter recorded the early part of the free induction decay (FID) in the pulsed ^{1}H NMR of the lecithin dispersion (17).

Charlie Seiter interpreted the decay in terms of the magnetic dipolar interactions among the methylene protons near the top of the acyl chains that were not averaged by molecular motions, including contributions from both the geminal dipolar coupling as well as intergeminal interactions. To account for the observed motional averaging of the ^{1}H–^{1}H dipolar interactions, he suggested rapid lipid or chain motion about the normal to the plane of the bilayer membrane together with off-axis motion within a restricted range of ~40° within the timescale of the NMR measurement (89). Although this analysis was relatively crude, it was an important beginning because it introduced the boundary conditions under which the NMR of membranes needed to be interpreted.

In due course, the ^{1}H NMR experiment was superseded by ^{2}H NMR, in which the orientational order of the acyl chains was determined by the extent of motional averaging of the ^{2}H electric quadrupolar interactions of the C–D bonds within each of the C–D$_2$ segments (88). The ^{2}H NMR method became the method of choice because it allowed direct determination of the order parameters of the various methylene segments along the acyl chains, providing a much more precise picture of the flexibility gradient along the hydrocarbon chains.

Nils Petersen elaborated on the motional model of Seiter by introducing the possibility of *trans-gauche* isomerization of the methylene segments along the acyl chains, in addition to the restricted fluctuations of the lipid chains about the director of the membrane (79). In principle, many types of motions could contribute to the reduction of the orientational order of the individual segments of the acyl chains

in a partially disordered lipid system, including *trans-gauche* isomerization, the formation of kinks via β-coupled *gauche$^+$-trans-gauche$^-$* isomerizations, diffusion of the kinks up and down the acyl chains, dynamic tilting of the hydrocarbon chains, and the entire phospholipid molecule about the membrane director, as well as elastic waves arising from undulations of a collection of phospholipid molecules within the thin bilayer film (74). The formation of ($\pm$) *gauche* bonds should be facile and highly probable, particularly toward the tails of the hydrocarbon chains. The chain fluctuations were expected to be slower, but these motions could involve a single lipid molecule undergoing restricted motion within its accessible volume, as well as a cooperative domain of many molecules undergoing highly damped surface undulations within the lipid film. The timescales of these motions extended from relatively rapid individual molecular fluctuations to slow collective motions of larger amplitudes. Any disorder in the molecular packing of the phospholipid molecules would reduce the cooperativity and increase the amplitude of the restricted motion and the frequency of the fluctuations. In contrast, cholesterol introduced considerable molecular ordering into the motional state of lecithin bilayer vesicles (49).

In those days it was popular to attribute the reduction of the orientational order of the hydrocarbon chains solely to *trans-gauche* isomerization. ^{13}C T_1s and ^{2}H NMR order parameters and T_1s were interpreted in this manner (58, 98). Gerry Feigenson measured ^{1}H T_1s of lecithin multilamellar dispersions over a range of temperatures and NMR frequencies and concluded that there were at least two types of motions involved: one correlation time of the order of 10^{-9} s and the other of the order of $\geq 10^{-7}$ s (40). He attributed the faster motion to kink diffusion and the slower motion to chain fluctuation. Masatsune Kainosho, Paulus Kroon, Rudiger Lawazcek, and Nils Petersen (47) and the group headed by Mel Klein at U.C. Berkeley also measured the T_1s and T_2s of the protons of the headgroups and the hydrocarbon chains in small sonicated lecithin bilayer

CcO: cytochrome *c* oxidase

vesicles (44). From the temperature dependence and the NMR frequency dependence of T_1s, it was concluded that the formation and diffusion of kinks occurred on the timescale of 10^{-10} to 10^{-9} s, but the chain fluctuations occurred in the vicinity of 10^{-8} s (74). As expected, the correlation length of a cooperative domain was substantially shorter in the disordered bilayer vesicle.

In these systems, the observed intrinsic T_2s were always substantially shorter (typically by a factor of 10) than the observed T_1s, reflecting the more complex power spectrum of the fluctuating magnetic dipolar fields brought about by the complex motions highlighted above. Thus, the dynamic structure of a bilayer membrane is highly complex and the motions are characterized by a hierarchy of motions of different amplitudes and timescales. The Chan group exploited the sensitivity of multiple-frequency NMR measurements to different timescales to sample these motions during the 1970s (8).

Lipid-Protein Interactions and Lipid-Mediated Protein-Protein Forces

According to the fluid mosaic model of Singer & Nicolson (93), the structure of a biological membrane consisted of a fluid lipid bilayer with membrane proteins embedded and anchored vectorially in the thin film. For the embedded proteins to be thermodynamically stable, the protein must be folded with the hydrophobic residues exposed to the lipid molecules, and there must be matching of the hydrophobic hydrocarbon chains to the hydrophobic surface of the proteins (75, 77, 78). Any hydrophobic mismatch could be alleviated by adjusting the number of *gauche* bonds toward the tails of the hydrocarbon chains of the phospholipids, tilting the phospholipids and the hydrocarbon chains about the normal to the bilayer plane, and redistributing the otherwise heterogeneous distribution of lipids in the cell membrane to bring lipids with the appropriate hydrocarbon chain lengths to form the boundary lipids

at the interface of the proteins (78). Thus, any hydrophobic mismatch would necessarily perturb the lipid bilayer and lead to protein-protein forces. Owicki, McConnell, and colleagues were the first to treat this problem theoretically (72, 73). Tim Pearson and Jay Edelman extended this work and showed that these lipid-protein interactions did not readily damp out because of the two-dimensional structure of the bilayer film (77). Correlation lengths were long, of the order of 10 nm. Accordingly, at the appropriate protein concentrations in the membrane, when the perturbed lipid domains overlapped, there could be lipid-mediated protein-protein forces and the protein pair radial distribution function would be modified from the statistically random distribution.

Tim Pearson was the first to look for experimental evidence in support of this prediction. Working with Barbara Lewis and Don Engelman, he examined the protein distribution in recombinants of bacteriorhodopsin in bilayer dispersions formed from lecithins with different acyl chain lengths by freeze-fracture electron microscopy (78). Indeed, the pair protein distribution function differed from the statistically random distribution and varied in a manner predictable by the hydrophobic mismatch at the lipid-protein interface. Both attractive and repulsion interactions were implicated depending on the nature of the mismatch. Subsequently, Tim Pearson extended the study to protein-lipid recombinants of cytochrome *c* oxidase (CcO) and reported evidence for a similar lipid-mediated protein-protein force (76).

Later, Paula Watnick, Phoebe Dea, and Luciano Mueller studied the effects of the transmembrane peptide gramicidin A on the dynamic properties of the bilayer membrane (65, 104). Different deuterated lecithins of varying chain lengths were examined by ^{2}H NMR. As expected, gramicidin A had no noticeable effects on the ^{2}H T_1s. However, there were dramatic effects on the homogeneous T_2s of the ^{2}H (65). These results were consistent with the expected effects of the incorporated peptide on the elastic constants and the lateral cooperativity of the bilayer structure (104). The peptide reduced the overall effective correlation length of the cooperative domain.

THE ERA OF MEMBRANE PROTEINS

Cytochrome *c* Oxidase: A Redox-Linked Proton Pump

Our early attempts to study lipid-protein interactions and to obtain evidence for the boundary lipid were carried out using CcO from bovine heart. This work was carried out in collaboration with Professor Tsoo E. King and Drs. Chang-An Yu and Linda Yu. Chang-An and Linda Yu (110) provided the laboratory with high-quality protein, and my postdoc Dr. Ming-Chu Hsu and graduate student Valerie Hu prepared lipid combinants of the CcO for NMR studies. Unfortunately, we were unsuccessful in obtaining reproducible-quality ^{1}H NMR data. Moreover, the NMR signals were broad, presumably because of the difficulty in obtaining a sufficiently homogeneous system and because of chemical exchange effects arising from sampling of the phospholipids over domains of different sizes and compositions. After about one year, we decided to move on to study the protein itself.

At the time, CcO was not molecularly characterized. The subunit composition was still unclear. The protein was thought to contain two iron heme As and two copper ions (4). The hemes were called cytochrome *a* and cytochrome a_3, and the copper ions were called Cu_A and Cu_B. It was established that the enzyme participated in dioxygen reduction and was thus the terminal enzyme in the respiratory chain of mitochondria. But little was known about the electron transport and the dioxygen chemistry mediated by the enzyme, not to mention the role played by the metal cofactors in mediating this chemistry. Märten Wikström had obtained the first evidence that CcO might be a proton pump (106), but the issue was highly debated. In any case, the general consensus then was that CcO was a highly complex molecular machine.

A molecule of this complexity was certainly beyond the training and expertise of a physical chemist. In fact, an eminent membrane biologist thought that there was no room for a physical chemist in this field and he advised me to apply my knowledge of physical methods to work on better-defined systems. On the other hand, several laboratories, including those of Helmut Beinert and Tsoo King, among others (43, 110), were beginning to isolate the enzyme from mitochondria and purify the protein to homogeneity. The methods of molecular biology were also becoming available at the time, and my instincts suggested that it was only a matter of time before this membrane protein was going to be cloned and overproduced. With the field at this stage of development, it was a perfect opportunity to train graduate students and postdocs. They could learn to define the issues that needed to be addressed, come up with the methods of the solution, and contribute to molecularizing the problem, which was the goal of my jumping into the quagmire to begin with.

The first attempts by the Chan laboratory were directed toward characterizing the metal centers of CcO. National synchrotron radiation facilities had just been established, and X-ray absorption spectroscopy seemed like a good way to identify and characterize the Fe and Cu atoms in the enzyme. Dr. Bob Shulman of Bell Laboratories contributed part of his beam time at the Stanford Synchrotron Radiation Laboratory to allow Valerie Hu and me to perform these experiments. Strictly speaking, we had to obtain Dr. Bill Blumberg's blessings before undertaking the Cu Kα-edge experiments, since the turf was divided up into Fe and Cu X-ray absorption between the Shulman and Blumberg groups at Bell Laboratories, respectively. Fortunately, I was able to forestall a political crisis because Dr. George Brown of Bell Laboratories was helping Valerie and me with the experiments.

The interpretation of the Fe Kα-edge of CcO was straightforward. However, the Cu Kα-edge suggested that one of the two copper ions was Cu^{1+} when the enzyme was fully oxidized, or the ligand environment of this copper ion was highly covalent (45). This observation suggested that cysteine sulfurs were associated with this copper cofactor. During those times, the enzyme was thought to contain two iron and two copper ions, and it wasn't until 20 years later that Caughey and coworkers (109) discov- ered that the correct metal content was two iron and three copper ions. It is now known that Cu_A is really a mixed valent Cu(I)Cu(II) species bridged by two thiolates in the oxidized enzyme (**Figure 1a**) (46, 101, 102). Thus, the Cu Kα-edge was giving us the right story, though we

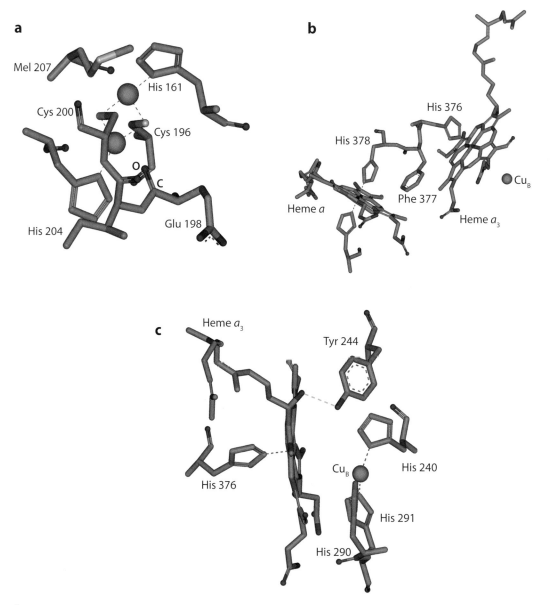

Figure 1

Ligand structures of (*a*) Cu_A, (*b*) cytochrome *a* and cytochrome a_3, and (*c*) cytochrome a_3 and Cu_B in bovine cytochrome *c* oxidase. Taken from Reference 101.

could not distinguish between a reduced copper and a highly covalent Cu(II) during those earlier times. It turned out the answer was both.

In the meantime, Gary Brudvig, Tom Stevens, and David Bocian joined the CcO project. They decided to use NO to probe the dioxygen-binding site of the enzyme. At the time, the dioxygen reduction site was known to consist of cytochrome a_3 and Cu_B, but this site was EPR silent in both the fully oxidized and reduced forms of the enzyme. In the fully reduced form, both cytochrome a_3 Fe^{2+} and Cu_B^{1+} were diamagnetic, but in the fully oxidized state, the high-spin cytochrome a_3 Fe^{3+} and Cu_B^{2+} were thought to be exchange coupled to yield an $S = 2$ electronic state, which did not possess a Kramer's doublet to facilitate EPR observations. On the other hand, NO was often used as a probe of dioxygen-binding sites. With its unpaired electron spin, the NO could be used to convert an integral spin system into an electronic state with a Kramer's doublet. Armed with this strategy, Gary, Tom, and David treated the reduced and the oxidized oxidases with NO to explore the structure of the dioxygen reduction site (95). EPR signals were elicited for all the cofactors in one or the other of these experiments. In a neat experiment, in which they treated the azide adduct of the oxidized enzyme with NO, they demonstrated that the reaction of NO with the azide adduct led to disproportionation of the azide adduct to give N_2O and N_2, followed by reduction of the cytochrome a_3 and binding of a second NO to give an exchange-coupled cytochrome a_3^{2+}-NO, Cu_B^{2+} triplet species. The resulting zero-field splitting suggested a NO-Cu_B^{2+} interspin distance of 3.5 Å or a Fe Cu distance of ~5 Å for the dioxygen reduction site. This result indicated that the dioxygen-binding site on cytochrome a_3 was located on the same side of the porphyrin plane as Cu_B, providing the first direct evidence that the binuclear site could be bridged by a peroxide intermediate during the initial stages of dioxygen activation (**Figure 1***b,c*).

Tom Stevens and Craig Martin moved the project forward another notch by obtaining information on the ligands of the various metal cofactors. The prevailing view at the time was that this information was needed before the redox potentials and electron transfer rates mediated by the metal cofactors of the enzyme could be understood. With the assumption that these ligands must be highly conserved, Tom and Craig exploited amino acid auxotrophs to incorporate selected amino acids with isotopic labels into the CcO to reveal the identity of the ligands by the nuclear superhyperfine interactions in the EPR that could be elicited from each of the four metal cofactors, either by conventional EPR experiments or electron nuclear double resonance (ENDOR) measurements of the various EPR signals. Thus, ^{15}N ring-labeled histidines and ^{2}H- and ^{13}C-labeled cysteines were used to obtain yeast CcO when appropriate auxotrophs of *Saccharomyces cerevisiae* were cultured. In this manner, Tom and Craig systematically identified the histidines of cytochrome a, cytochrome a_3, Cu_A and Cu_B, and the cysteines of Cu_A (6, 63, 96, 97) (**Figure 1**). That cysteines were associated with Cu_A was particularly significant, and judging from the extended X-ray absorption fine structure (EXAFS) in X-ray absorption experiments, thiolates were implicated as the bridging ligand at the binuclear site by other investigators. Because there were only two conserved thiolates in the subunits containing the Fe and Cu cofactors (subunits I and II), and these cysteines were in subunit II, the Chan group had insisted that Cu_A was located in subunit II and that the two cysteines were associated with this site. This controversy over the cysteines persisted for many years until the crystal structure of CcO was finally obtained. The two thiolates provided the bridging sulfurs for the mixed valent Cu(I)Cu(II) Cu_A structure (101).

The Chan laboratory has also contributed to the understanding of the mechanism of dioxygen activation of CcO, which is paramount to the enzyme's function as a proton pump. The binding of dioxygen to CcO had been well studied by many groups ever since the enzyme was identified and purified. Britton Chance championed the method of flash photolysis at low temperature to study the formation of the dioxygen adduct and the subsequent formation of the peroxide intermediate (Compound C) (27). Malmström and coworkers (32) studied the transfer of the "third" electron to the Compound C by incrementally increasing the temperature, monitoring the electron transfer process by low-temperature UV-visible spectroscopy. David Blair and Steve Witt (7, 25) followed the reduction of Compound C by rapid quench 4 K EPR. Surprisingly, two dioxygen intermediates at the three-electron level of reduction were identified in the low-temperature experiments, with the first intermediate undergoing a bond-breaking step to yield the oxyferryl species. These findings confirmed that the O–O bond in dioxygen was cleaved at the three-electron level of reduction, as in the case of cytochrome P_{450}, rather than after the succession of two two-electron reductions as proposed by Malmström and coworkers (32). The formation of the peroxide intermediate and the oxyferryl during dioxygen reduction provides for the two high-potential intermediates that are needed to accomplish the redox linkages during the proton pumping mediated by the enzyme (20, 105).

During the 1980s, the evidence for CcO as a proton pump became increasingly compelling (105, 106). Certain properties of the proton pump were also becoming evident. From thermodynamic reasoning, the transfer of the first two electrons from cytochrome c to the dioxygen reduction site via cytochrome a and Cu_A could not be directly linked to the pumping of protons, unless the protein had been energized in an earlier redox linkage step and the transfer of these electrons to the oxidized enzyme was to dissipate any residue conformational excitation of the protein to eject the protons that have already been moved across the osmotic barrier. There is simply insufficient driving force to move a proton uphill across the osmotic barrier otherwise (20). Redox-linked proton pumping in CcO is a highly complex process consisting of electron input, electron transfer, proton uptake, redox linkage, and proton ejection steps (86). However, without redox linkage there can be no transfer of redox energy from

the electron degrees of freedom to the protein degrees of freedom, and therefore no proton pumping (66, 67). The putative redox linkage could occur only during the second half of the turnover cycle when the final two electrons are passed to the activated dioxygen, i.e., when the peroxide and the oxyferryl intermediates are presumably formed (67). The redox potentials of these dioxygen intermediates are of the order of 1 volt, which is approximately 700–800 mV higher than the intrinsic redox potentials of the cytochrome a_3 and Cu_B without the dioxygen bound (105).

To achieve this redox linkage, the flow of both electrons and protons must be gated. Electron transfer without linkage corresponded to an electron leak without transfer of redox energy to the protein. Without proton gating, protons cannot be effectively pumped because of proton backflow (molecular slip). David Blair and Jeff Gelles (5) introduced the concept of electron gating and emphasized its importance for the robust functioning of a redox-linked proton pump. Siegfried Musser (66) emphasized the importance of proton gating. He also developed the ideas of redox linkage in some detail by embodying the rules in several examples with varying driving forces for proton pumping (**Figure 2**). Molecular models were also highlighted to illustrate these principles (67). At this juncture, the rules remain untested. Although the structure of CcO has been available for almost two decades (46, 102), the experimental testing of these ideas has been slow. We require a detailed understanding of the structural changes in the enzyme when the dioxygen-binding site has become activated and when one of the low-potential centers has been reduced.

In principle the electron transfers linked to proton pumping could originate from either cytochrome a or Cu_A. In recent times, the electron flow within the enzyme has been thought to follow the canonical pathway for every step regardless of the state of activation of the dioxygen reduction site: cytochrome $c \rightarrow Cu_A \rightarrow$ cytochrome $a \rightarrow$ dioxygen-binding site. This need not be the case. In fact, no hard data bear on the transfers of the final two electrons, the two electrons linked to redox linkage. The Chan laboratory has suggested that the electron flow is bifurcated, with the first two electrons following the canonical pathway, and the second two electrons involved in redox linkage occurring through a different pathway bypassing cytochrome *a* (**Figure 2**).

As noted above, the direct transfer of the latter two electrons would occur with high driving force, and in fact, if the driving drive is sufficiently high, the rate could be governed by the Marcus theory of electron transfer on the right side of the inverted region in the plot of k_{et} versus driving force (62). Thus, the Marcus theory could be exploited to gate the electron flow. If the reduction of Cu_A is linked to a conformational change within the protein that significantly raises the reduction potential of the site (**Figure 2**), then the putative electron transfer from Cu_A to the dioxygen intermediates anchored at the binuclear center could be rendered more facile following the redox linkage. In this scenario, the reduction potential of Cu_A would become higher than that of cytochrome *a*, forestalling the electron transfer from Cu_A to cytochrome *a*. In the absence of redox linkage, the electron equilibration between Cu_A and cytochrome *a* would be rapid, of the order of 20,000 s (64), and possibly even faster (113). Aside from the redox potential difference between these two redox centers, the kinetics of the electron transfer between Cu_A and cytochrome *a* could also be tuned by subtle conformational changes in the enzyme.

It goes without saying that there must also be allosteric coupling between the Cu_A and the activated binuclear site in this model for the enzyme to distinguish between proton-pumping and nonproton-pumping events. The Cu_A site, which consists of a Cu(I)Cu(II) bridged by two thiolates, is electron rich and is poised to act as a conformational trigger upon reduction by an additional electron (67). We expect conformational changes at the dioxygen reduction site as well in order to optimize the biological energy transduction (**Figure 2**). The idea is that following redox linkage, the electron transfer from Cu_A to the dioxygen intermediate(s) will

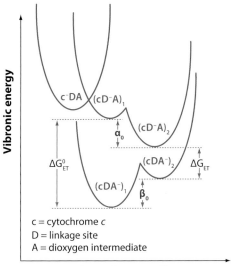

Coupled reaction path (proton pumping occurs):

$$c^-DA \longrightarrow (cD^-A)_1 \longrightarrow (cD^-A)_2 \longrightarrow (cDA^-)_2 \longrightarrow (cDA^-)_1$$

Uncoupled reaction path (proton pumping does not occur):

$$c^-DA \longrightarrow (cD^-A)_1 \longrightarrow (cDA^-)_1$$

Figure 2

Schematics of redox linkage occurring during proton pumping by cytochrome *c* oxidase (CcO). In this model, both Cu_A and the dioxygen-activated cytochrome a_3 and Cu_B are involved in the redox linkage. The primary electron donor that initiates the process is cytochrome *c*, which is denoted by c. D denotes the electron donor for the electron transfer coupled to the proton pump (i.e., the input linkage site), and A represents the peroxide or oxyferryl dioxygen intermediate (electron acceptor) (i.e., the output linkage site). Proton-pumping and nonproton-pumping conformational states of the CcO complex are highlighted by the subscripts $_2$ and $_1$, respectively. These symbols are used in unison [e.g., $(c^-DA)_1$, $(cD^-A)_1$, $(cD^-A)_2$, $(cDA^-)_2$] to denote a specific redox and conformational state of the CcO complex. Plotted along the ordinate is the total vibronic energy of the enzyme complex, the electron donating cytochrome *c*, and the protons involved in the dioxygen chemistry and the proton-pumping process. The abscissa denotes the protein nuclear coordinates. α_0 and β_0 denote the amounts of redox energy transferred to the protein to excite the vibrational degrees of freedom during the redox linkage from the input and output redox linkage sites, respectively. ΔG_{ET} denotes the driving force for electron transfer. Figure adapted from Reference 66.

occur with a driving force on the left side of the inverted region to allow the process to be completed efficiently and irreversibly.

CcO can pump up to 1 H^+/e^- (105), and this H^+/e^- ratio decreases with increasing

MALDI-TOF MS: matrix-assisted laser desorption ionization mass spectrometry

protonmotive force. The enzyme seems to behave as if there is a built-in clutch to switch the proton-pumping machinery between productive and nonproductive cycles (**Figure 2**). In this manner, variable H^+/e^- ratios could be obtained depending on the external conditions under which the pump must operate.

Membrane Proteins that Mediate Ion and Solute Transport

Aside from CcO, the Chan group has maintained a strong interest in other membrane transporters that mediate ion transport as well as the transport of other solutes. Our interest in ion transporters actually started with the structural studies on nonactin, valinomycin, and alamethicin. This work was inspired by Professor Max Delbrück, whose interests in the biophysics of membranes did much to inspire me to initiate research on the structure and dynamics of the bilayer membrane and the mechanism of ion transport across membranes.

During the late 1960s Jim Prestegard worked on the structure of nonactin and tried to determine the molecular basis of the ion selectivity for K^+. Jim carried out ion-binding studies and concluded that the difference in the free energy of hydration and complexation of the naked cation by the nonactin provided the driving force for the selectivity (80, 81). The same result was obtained in the case of valinomycin. However, it was generally assumed that the selectivity was determined by the ability of the polar C=O and ether oxygen groups on the transporter to adapt and accommodate the naked cation. The size of the cation contributed to the selectivity to the extent that it determined the ability of the macrocyclic ligand to adapt and bind to the cation to attain the optimal free energy of complexation. Ion selectivity was actually determined by the difference in this free energy of complexation of the naked cation and the free energy of hydration of the same cation in the bulk solvent.

When Joseph Falke joined the Chan laboratory, he wanted to work on the anion trans-

porter in the red cell membrane. He asked me to come up with a biophysical method to probe the Cl^-/HCO_3^- exchange, and I suggested ^{35}Cl and ^{37}Cl NMR. Joe Falke exploited this technique to study the binding of Cl^- to band 3 (38). He prepared ghost membranes with both the inside-facing and outside-facing sides of the ghosts exposed to a Cl^- solution, and ghost membranes with either the inside- or the outside-facing side exposed. Joe was able to recruit all the sites to one side only by using a reversible transport inhibitor (37, 39). By using the arginine-specific reagent glyoxal to modify the transporter covalently, he argued that the transport site was an arginine, which was alternatively exposed to the inside-facing and outside-facing sides during the transport cycle. Thus, the anion transport occurs via an alternating transport mechanism, or the ping-pong mechanism (36). Following up on Joe Falke's work, Kathy Kane tried to identify this arginine transport site on the primary sequence via ^{14}C radioactive labeling. The band 3 was cleaved by specific proteases after chemical modification, and attempts were made to separate the transmembrane peptides by reverse phase hydrophobic column chromatography and to identify the glyoxal-labeled peptide by scintillation counting. The experiment failed. These latter studies would have been facilitated by modern matrix-assisted laser desorption ionization mass spectrometry (MALDI-TOF MS), but unfortunately for the project, these were times well before the advent of modern mass spectroscopy.

Following his success with the band 3 system, Joe Falke moved on to the glucose transporter in the red cell ghost. There were many fewer copies of the glucose transporter in the red cell (10^4 copies of the glucose transporter versus 10^6 copies of the anion transporter), but glucose binding and transport were specific to this transporter. Thus, a substrate-binding assay was developed based on changes in the spectroscopy of the substrate when it became bound to the transporter. Joe and Jin Wang exploited ^1H NMR nuclear Overhauser enhancement as an assay for glucose binding (103).

Particulate Methane Monooxygenase: A Membrane-Bound Enzyme that Oxidizes Methane to Methanol

During the early 1990s, Caltech was developing a program in C–H activation with the goal of designing a catalyst for the facile conversion of methane to methanol. As part of this effort, Professor Mary Lidstrom, a microbial geneticist working in the Division of Engineering and Applied Science at Caltech, and I assembled a team to try to understand how this chemistry was carried out by methanotrophic bacteria, which consumed methane produced in anaerobic sediments. These bacteria possess enzymes that catalyze the conversion of methane to methanol using dioxygen as a cosubstrate at ambient pressures and temperatures with high efficiency. The manner by which this conversion is accomplished in bacteria is of great interest to industrial chemists, as the controlled oxidation of methane to methanol is a difficult process in the laboratory, requiring expensive catalysts operating at high temperatures and pressures.

There are two of these enzymes: the membrane-bound monooxygenase (pMMO) and the soluble methane monooxygenase (sMMO), which are found in the cytoplasm of certain strains of bacteria. During the 1980s Professor John Lipscomb of the University of Minnesota isolated and purified sMMO to homogeneity (61). Accordingly, Mary Lidstrom and I decided to focus on the pMMO. Preliminary evidence for the existence of a membrane-bound methane monooxygenase was first reported by the laboratory of Professor Howard Dalton at the University of Warwick (82, 94). However, the pMMO had resisted both initial identification and subsequent isolation and purification for biochemical and biophysical characterization because of its instability outside the lipid bilayer and its tendency to lose essential metal cofactors during isolation and purification.

Mary Lidstrom was to focus on the molecular biology, and the Chan laboratory on the isolation, purification, and characterization of the pMMO from the membranes. Drew Schimke,

a postdoc, was recruited to initiate this effort. One of his first observations was that the pMMO was an abundant protein in the membranes of *Methylococcus capsulatus* (Bath) when this bacterium was cultured under high copper concentrations in the growth media. Accordingly, some of the properties of the enzyme could be inferred from physical characterization of the membranes obtained under these conditions.

Soon thereafter, Hoa Nguyen joined the pMMO project as a PhD student. Together with Drew, he carried out a detailed study of the effects of copper concentration on the methane monooxygenase activity of the membranes isolated from cells of *M. capsulatus*, and correlated the specific activity with the level of copper that was determined to be associated with these membranes. They concluded that the pMMO was indeed produced in high levels in the membranes when the bacterium was cultured at high copper concentration in the growth media, and that it was a multicopper protein containing as many as 15 copper ions per 100 kDa (71). From a quantitative proteomic analysis of bacterial cells obtained when the bacterium was grown under copper stress (48), we now know that copper ions are essential for the expression of the pMMO gene and growth of the organism. Not only do copper ions serve as a transcriptional activator of the pMMO gene, but they also enhance the expression of all the genes involved in cellular metabolism, including lipid biosynthesis. In addition, pMMO is a membrane protein containing high levels of copper ions, and these coppers are required for protein function (16).

Toward studying the enzyme in greater depth, Hoa succeeded in developing protocols to isolate and purify the enzyme to homogeneity despite numerous false starts (69). From sodium dodecyl sulfate gels of the subunits, Hoa obtained the N-terminal sequences of the subunits, information that was needed to fish out and clone the pMMO genes (90). Hoa also began to characterize both the pMMO-enriched membranes and the purified protein by EPR, X-ray Cu Kα-edge absorption spectroscopy,

pMMO: particulate methane monooxygenase

sMMO: soluble methane monooxygenase

and SQUID magnetization measurements (69, 70). It was evident from these experiments that pMMO contained many copper ions and that some of these copper ions were sequestered into clusters.

In a groundbreaking experiment carried out in collaboration with Heinz Floss of the University of Washington and Philip Williams of the Tritium NMR Facility at the Lawrence Berkeley Laboratory, we showed that the hydroxylation of cryptically chiral ethane proceeded with total retention of configuration at the carbon center oxidized (107). This result suggested concerted oxenoid insertion across the C–H bond of the alkane, a reaction pathway distinct from the radical mechanism suggested for the nonheme diiron cluster in sMMO (61). Mei Zhu and Sean Elliott carried out further studies to clarify the regiospecificity and stereochemistry of the hydroxylation mediated by the pMMO, extending the substrates to include propane, butane, pentane, propene, and 1-butene (35). In propane, butane, and pentane, only the secondary alcohol was produced. Steve Yu synthesized deuterated chiral butanes and confirmed that the transfer of the O-atom to the secondary carbon occurs with total retention of configuration (112), in accord with a concerted oxo-insertion mechanism.

I retired from Caltech and moved to Academia Sinica in Taiwan in the fall of 1997. Despite my administrative responsibilities, initially as Director of the Institute of Chemistry, and subsequently as Vice President of Academia Sinica, I maintained an active research program. The conversion of methane to methanol by pMMO has been one of the research targets of the Chan laboratory in Taiwan during the past decade. During this period, Steve Yu, Kelvin Chen, Charlie Chen, and others developed methods to obtain large quantities of high-quality pMMO that were judged to be homogenous according to subunit composition, complement of metal cofactors, and molecular mass (28, 111). Progress in our work on pMMO was also facilitated by major technological advances in the study of proteins since

our work on CcO some 30 years earlier, including high-throughput genome sequencing, bioinformatics, MALDI-TOF MS and ESI MS (electrospray ionization mass spectrometry) for proteomics and peptide mass fingerprinting, and ATR FTIR (attenuated total reflectance Fourier transform infrared spectroscopy) for secondary structure determination of proteins in membranes. Our progress toward understanding pMMO structure and function is just another example of how modern technology can fuel the advances in basic science.

In-depth studies have now revealed how the enzyme functions. It was necessary to carry out the structural and biophysical/biochemical studies on the enzyme with all the cofactors intact. In our work, a preparation of the enzyme with the full complement of 15 Cu ions was always used (26). At the active site of this enzyme is a trinuclear copper cluster (24, 26, 68) that, upon activation by molecular oxygen, mediates the transfer of a singlet oxene to the methane at the active site. A recent crystal structure of pMMO from *M. capsulatus* did not reveal a trinuclear copper center in the structure (60), but the protein preparation on which the crystal structure was based did not possess enzyme activity. The lesson we can all learn from this example is that protein isolation and purification must be carried out with adequate controls to ensure the biochemical integrity before we embark on biophysical and structural measurements. Those of us who conduct biophysical measurements are often less circumspect about the quality of the sample than they should be. Fortunately, in the case of pMMO, it was possible to reconcile the biochemical/biophysical data with the crystal structure by introducing the missing metallic cofactors back into the protein scaffold (24, 26, 68) (**Figure 3**).

Using a triad of copper ions to harness a singlet oxene offers a new mechanism for the controlled oxidation of an organic substrate (**Figure 4**). Analysis by density functional theory of various mechanistic scenarios by Peter Chen confirmed that the proposed direct singlet oxene insertion mediated by the trinuclear

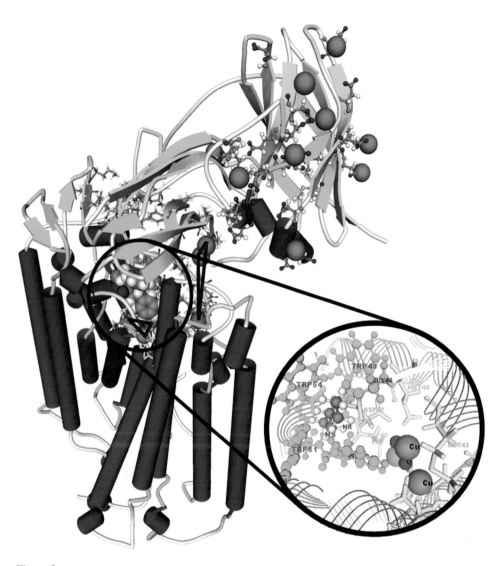

Figure 3

Structure of the particulate methane monooxygenase (pMMO) from *Methylococcus capsulatus* (Bath) with the active site, consisting of both the three-copper cluster and the hydrocarbon-binding pocket (*inset*). Figure taken from Reference 26.

copper cluster offered the most facile pathway for the alkane hydroxylation (29), and this mechanism yielded catalytic turnover rates and $^1H/^2H$ kinetic isotope effect values in close agreement with experiment.

The Chan laboratory is now exploiting the new chemistry toward developing a cheap and efficient catalyst for the controlled oxidation of methane and other small alkanes under ambient conditions of temperature and pressure. A model trinuclear copper cluster has been designed and synthesized to provide further support for the ideas derived from the enzyme studies. Our model trinuclear copper cluster can mediate facile oxo-insertion across C–C and C–H bonds in small organic substrates (30). The same chemistry seems to be operative here as in pMMO.

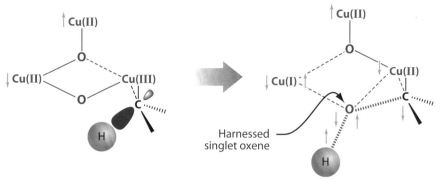

Figure 4

The singlet oxene mechanism for oxene insertion across a C–H bond from a dioxygen-activated trinuclear copper cluster to methane in pMMO. The harnessed singlet oxene is highlighted in the transition state. Figure adapted from Reference 29.

RETURN TO A SIMPLER ISSUE: EARLY KINETIC EVENTS IN PROTEIN FOLDING

To close out my scientific career, I decided to return to a simpler issue as well. Before the protein folding funnel of Peter Wolynes, José Onuchic, and David Thirumalai (108), I had begun to ask "Why do proteins fold so rapidly in water?" Small globular proteins fold into well-defined, three-dimensional structures with biological function quickly, usually within a few seconds or less. During the early 1990s, before I took early retirement from Caltech, I initiated another project that was subsequently carried over to Academia Sinica for continuation.

This work was premised on the hypothesis that the kinetic events that took place during the early stages of protein folding consisted of the formation of structural elements, and that the formation of the molten globule involved the hydrophobic collapse of secondary structures that were already preformed. To test this idea, Ron Rock and Kirk Hansen developed the photolabile caged-peptide strategy to measure the kinetics of the formation of different structural motifs (α-helices, β-sheets, hairpins). The method exploited a photolabile caged compound based on 3′-methoxybenzoin to cyclize the N terminus of the peptide to a side chain to disrupt the secondary structure (18, 42, 83). To trigger the refolding of the peptide, the photolinker was cleaved rapidly (10^{-10} s) by a light pulse with good yield (quantum yield 0.6–0.7), and the refolding of the uncaged peptide was monitored by photoacoustic calorimetry. The method allowed us to observe the refolding of a structural element at times as early as 10^{-9} s and over a time span as long as 10^{-6} s. Another advantage was that the refolding process could be observed under ambient temperature and physiological pH without the interference from denaturants.

Using this approach, Rita Chen, Joseph Huang, Nicole Kuo, Howard Jan, Marappan Velusamy, Chung-Tien Lee, Wunshain Fann, Jaroslava Miksovska, and Randy Larsen measured the folding/refolding kinetics of specific structural motifs, including α-helices, hairpins, and β-sheets (31). These experiments revealed that these different structural motifs form quickly but, more surprisingly, according to a hierarchy of well-defined timescales, that is, a different time window for each type of structural motif. Moreover, the method was sufficiently sensitive, enabling them to observe subtle details in the protein folding (51).

Protein folding begins with a series of local folding processes driven by different interactions, and they are separated in time according to a highly ordered hierarchy. For example, when the folding process is driven by the local turn structure formation, the folding rate is much faster than that driven by interstrand hydrophobic interaction. Thus, α-helices are formed first, followed by hairpins and β-sheets with strong turn-promoting sequences, which are in turn followed by hairpins and β-sheets that refold via nucleation by hydrophobic interactions. At the outset, local structural preference is derived from sequence steric hindrance, so the conformational space that needs to be sampled in the search is significantly reduced. These events would then be followed by further hydrophobic collapse of the remaining hydrophobic residues to form the molten globule and by subsequent annealing of the protein toward its global minimum. The global tertiary collapse needed to form the molten globule is expected to be slower than the refolding of hairpins via hydrophobic nucleation because the remaining hydrophobic residues must traverse over longer distances to displace the water molecules in the tertiary collapse. In the case of multidomain proteins, chaperones might be required to keep the domains apart spatially during this global hydrophobic collapse. On the other hand, for a protein with a relatively smooth free-energy landscape, the limit often referred to as minimal frustration, there would practically be no molten globule formed, and the structure would collapse to the native state once the topological fold is identified. Finally, the subsequent annealing of the protein is even slower yet, and a number of folding intermediates might be involved. The overall process is schematically illustrated in **Figure 5**, together with timescales expected for the various steps.

The early formation of the secondary structure or structural elements has been advocated by Baldwin, Englander, and others long ago. Their celebrated ideas were motivated mostly on thermodynamic grounds. In the end, folding pathways are dictated by kinetic considerations. In other words, protein folding occurs by kinetic channeling. As a kinetically controlled process, the protein follows a pathway of steepest descent in free energy within the protein folding funnel without sampling all points in conformational space in search of the na-

tive state. In this manner, Levinthal's paradox is avoided (59).

Because the information regarding the propensity of a polypeptide toward forming different structural motifs is built into the primary sequence of a protein, it follows that the details relating to protein folding are also encoded in the primary sequence of the polypeptide. This is a perfect example of how evolution has led to modularity of structural elements to allow a protein to fold in an orderly fashion in the otherwise highly complex free-energy landscape, as suggested by Michael Deem (99).

EPILOGUE

I have had an exciting career. The journey that ultimately led me to the world of membrane proteins was hardly programmed. It evolved serendipitously (logically though) from the chemical physics of simpler systems, from the dynamic structure of the lipid bilayer to lipid-protein interactions, to membrane biophysics of ion transport, and culminated in the structure and function of several membrane protein systems. I have taken advantage of my training in physical chemistry and modern spectroscopy and applied this knowledge to probe and solve interesting problems in modern biochemistry. As the systems became more complex, I learned how to manage the research undertaking, how to define a myriad of complex biological problems from scratch without knowing much about them, and how to formulate research strategies to approach the various issues. The complexity of these problems dictates that progress could only evolve, with the next generation of experiments building on the outcome of earlier ones. A problem like pMMO is so complex that it transcends generations of coworkers. In academia, and as a research mentor, I worked with students and postdocs as they joined my research group. With limited research resources, it is not possible to build a large team and solve problems in a short amount of time as is done in industry. However, with a more deliberate pace, it is possible to take advantage of emerging technologies as they

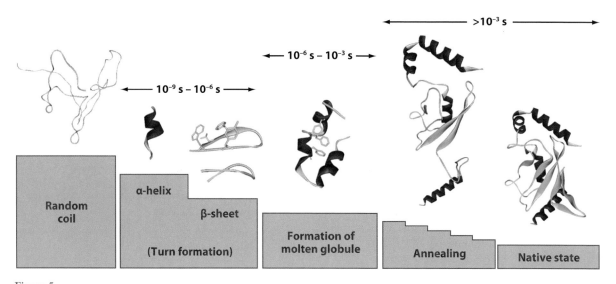

10^{-9} s – 10^{-6} s

10^{-6} s – 10^{-3} s

>10^{-3} s

Random coil

α-helix

β-sheet

(Turn formation)

Formation of molten globule

Annealing

Native state

Figure 5

Kinetic hierarchy of the events in the folding of a single-domain protein, starting with the formation of the secondary structure to the formation of the molten globule, followed by thermal annealing of partially folded intermediates to reach the native structure.

develop. I have learned that technologies fuel the development of basic science, and a certain key technology can often make the difference.

The bulk of the membrane research accomplished at Caltech was supported by the National Institute of General Medical Sciences of NIH. Because my work on membranes evolved logically from model membranes to proteins in the cell membrane, I was able to sustain NIH funding for this project for more than 30 years. The level of support was adequate to cover most research expenditures, with the exception of upgrades of research instrumentation or the purchases of major pieces of equipment. After I moved to Taiwan, my research program was funded by Academia Sinica. Although the level of support was not as high as the NIH support I had at Caltech, Academia Sinica provided funds for upgrades of research facilities and the acquisition of state-of-the-art instrumentation as justified by the research. This capability greatly accelerated the progress of the scientific work I conducted in Taiwan.

I have learned much from my graduate students, postdocs, and colleagues. In addition, many people have opened doors for me and have contributed to the shaping of my scien-

tific career. Many of my colleagues at Caltech have been inspirational. As many as 200 graduate students, postdocs, and visitors (I have lost count) have passed through my laboratory. Unfortunately, this brief chapter does not allow me to mention each and every one.

Many of my former students and postdocs have gone on to do new science and even open new areas of research. I am proud of them. But that's the way it should be. In retrospect, I like to think that my style of training students works. The most important part of doing science is asking questions, and I taught my students to ask questions and how to ask questions. Without the questions, good or bad ones, there is no problem to solve. Discoveries in science are made by asking questions, from following up on leads, making observations in experiments, formulating hypotheses, and testing them by designing new experiments. The whole process from hypothesis to truth is often short-circuited by good intuition, which is, of course, based on knowledge or sound grounding in fundamentals, knowing what is reasonable and unreasonable in order to avert blind alleys.

Finally, I like to think that no matter how complicated biological problems may appear to be, there is a simple answer to the outcome when all is said and done. In my view, nature is intrinsically simple, and it is fun to discover how the behavior of a complex biological system is explicable in terms of the laws of physics and chemistry as we now understand them to be.

DISCLOSURE STATEMENT

The author is not aware of any biases that might be perceived as affecting the objectivity of this review.

ACKNOWLEDGMENTS

I am indebted to Drs. Steve S.-F. Yu and Joseph J.-T. Huang of the Institute of Chemistry, Academia Sinica, for advice and assistance with the preparation of the manuscript.

LITERATURE CITED

1. Banerjee U, Zidovetzki R, Birge R, Chan SI. 1985. Interaction of alamethicin with lecithin bilayers: a ^{31}P and ^{2}H NMR study. *Biochemistry* 24:7621–27
2. Bangerter BW, Chan SI. 1968. A proton magnetic resonance study of the interaction of adenosine with polyuridylic acid: evidence for both adenine-uracil base-stacking and base-pairing. *Proc. Natl. Acad. Sci. USA* 60:1144–51
3. Bangerter BW, Chan SI. 1969. Proton magnetic resonance studies of ribose dinucleoside monophosphates in aqueous solution. The nature of the base-stacking interaction in adenylyl (3′ 5′) cytidine and cytidylyl (3′ 5′) adenosine. *J. Am. Chem. Soc.* 91:3910–21
4. Beinert H, Griffiths DE, Wharton DC, Sands RH. 1962. Properties of the copper associated with cytochrome oxidase as studied by paramagnetic resonance spectroscopy. *J. Biol. Chem.* 237:2337–46
5. Blair DF, Gelles J, Chan SI. 1986. Redox-linked proton translocation in cytochrome oxidase: the importance of gating electron flow. *Biophys. J.* 50:713–33
6. Blair DF, Martin CT, Gelles J, Wang H, Brudvig GW, et al. 1983. The metal centers of cytochrome *c* oxidase: structures and interactions. *Chem. Scr.* 21:43–53
7. Blair DF, Witt SN, Chan SI. 1985. The mechanism of cytochrome *c* oxidase-catalyzed dioxygen reduction at low temperatures. Evidence for two intermediates at the three-electron level and entropic promotion of the bond-breaking step. *J. Am. Chem. Soc.* 107:7389–99
8. Bocian DF, Chan SI. 1978. NMR studies of membrane structure and dynamics. *Annu. Rev. Phys. Chem.* 29:307–35
9. Bull TE, Barthel JS, Jonas J. 1971. The effect of pressure on the F-19 spin-rotation interactions in benzotrifluoride in the liquid state. *J. Chem. Phys.* 54:3663–66
10. Burke TE, Chan SI. 1970. Nuclear spin relaxation in the presence of internal rotation. Nuclear-spin-internal-rotational coupling in benzotrifluoride and hexafluorobutyne-2. *J. Magn. Reson.* 2:120–40
11. Chan SI, Austin JA, Paez OA. 1969. Electron spin resonance studies of concentrated alkali metal-ammonia solutions. *Proc. Int. Union Pure Appl. Chem. Conf.*, pp. 425–38. New York: Cornell Univ. Press
12. Chan SI, Baker MR. 1961. Large rotational magnetic moment of NO. *Bull. Am. Phys. Soc.* 6:271
13. Chan SI, Baker MR, Ramsey NF. 1964. Molecular-beam magnetic resonance studies of the nitrogen molecule. *Phys. Rev.* 136:A1224–28
14. Chan SI, Bangerter BW, Peter HH. 1966. Purine binding to dinucleotides: evidence for base stacking and insertion. *Proc. Natl. Acad. Sci. USA* 55:720–27
15. Chan SI, Borgers TR, Russell JW, Strauss HL, Gwinn WD. 1966. Trimethylene oxide. Far-infrared spectrum and double-minimum vibration. *J. Chem. Phys.* 44:1103–11
16. Chan SI, Chen KHC, Yu SSF, Chen CL, Kuo SSJ. 2004. Toward delineating the structure and function of the particulate methane monooxygenase (pMMO) from methanotrophic bacteria. *Biochemistry* 43:4421–30

17. Chan SI, Feigenson GW, Seiter CHA. 1971. Nuclear relaxation studies of lecithin bilayers. *Nature* 231:110–12
18. Chan SI, Huang JJT, Larsen RW, Rock RS, Hansen KC. 2005. Early kinetic events in protein folding: the development and applications of caged peptides. In *Dynamic Studies in Biology: Phototriggers, Photoswitches and Caged Biomolecules*, ed. M Goeldner, R Givens, pp. 479–94. Weinheim, Ger.: Wiley-VCH Verlag GmbH
19. Chan SI, Kula RJ, Sawyer DT. 1964. The proton magnetic resonance spectra and structures of ethylene-diaminetetraacetic acid, methyliminodiacetic acid, and nitrilotriacetic acid chelates of molybdenum. *J. Am. Chem. Soc.* 86:377–79
20. Chan SI, Li PM. 1990. Cytochrome *c* oxidase: understanding nature's design of a proton pump. *Biochemistry* 29:1–12
21. Chan SI, Nelson JH. 1969. Proton magnetic resonance studies of ribose dinucleoside monophosphates in aqueous solution. The nature of the base-stacking interaction in adenylyl (3′ 5′) adenosine. *J. Am. Chem. Soc.* 91:168–83
22. Chan SI, Schweizer MP, Ts'o POP, Helmkamp GK. 1964. Interaction and association of bases and nucleosides in aqueous solutions. A nuclear magnetic resonance study of the self-association of purine and 6-methylpurine. *J. Am. Chem. Soc.* 86:4182–88
23. Chan SI, Seiter CHA, Feigenson GW. 1972. Anisotropic and restricted molecular motion in lecithin bilayers. *Biochem. Biophys. Res. Commun.* 46:1488–92
24. Chan SI, Wang VCC, Lai JCH, Yu SSF, Chen PPY, et al. 2007. Redox potentiometric studies of the particulate methane monooxygenase: support for a trinuclear copper cluster active site. *Angew. Chem. Int. Ed.* 46:1992–94
25. Chan SI, Witt SN, Blair DF. 1988. The dioxygen chemistry of cytochrome *c* oxidase. *Chem. Scr.* 28a:51–56
26. Chan SI, Yu SSF. 2008. Controlled oxidation of hydrocarbons by the membrane-bound methane monooxygenase: the case for a tricopper cluster. *Acc. Chem. Res.* 41:969–79
27. Chance B, Saronio C, Leigh JS. 1975. Functional intermediates in the reaction of membrane-bound cytochrome oxidase with oxygen. *J. Biol. Chem.* 250:9226–37
28. Chen KHC, Chen CL, Tseng CF, Yu SSF, Ke SC, et al. 2004. The copper clusters in the particulate methane monooxygenase (pMMO) from *Methylococcus capsulatus* (Bath). *J. Chin. Chem. Soc.* 51:1081–98
29. Chen PPY, Chan SI. 2006. Theoretical modeling of the hydroxylation of methane as mediated by the particulate methane monooxygenase. *J. Inorg. Biochem.* 100(4):801–9
30. Chen PPY, Yang RBG, Lee JCM, Chan SI. 2007. Facile O-atom insertion into C–C and C–H bonds by a trinuclear copper complex designed to harness a "singlet oxene." *Proc. Natl. Acad. Sci. USA* 104:14570–75
31. Chen RP, Huang JJ, Chen HL, Jan H, Velusamy M, et al. 2004. Measuring the refolding of β-sheets with different turn sequences on a nanosecond timescale. *Proc. Natl. Acad. Sci. USA* 101:7305–10
32. Clore GM, Andreasson LE, Karlsson B, Aasa R, Malmström BG. 1980. Characterization of the intermediates in the reaction of mixed-valence state soluble cytochrome oxidase with oxygen at low temperatures by optical and electro-paramagnetic-resonance spectroscopy. *Biochem. J.* 185:155–67
33. Dubin AS, Chan SI. 1967. Nuclear spin internal-rotation coupling. *J. Chem. Phys.* 46:4533–35
34. Eigenberg KE, Chan SI. 1980. The effect of surface curvature on the head-group structure and phase transition properties of phospholipid bilayer vesicles. *Biochim. Biophys. Acta* 599:330–35
35. Elliott SJ, Zhu M, Nguyen HHT, Yip JHK, Chan SI. 1997. The regio- and stereoselectivity of particulate methane monooxygenase from *Methylococcus capsulatus* (Bath). *J. Am. Chem. Soc.* 119:9949–55
36. Falke JJ, Chan SI. 1985. Evidence that anion transport by band 3 proceeds via a ping-pong mechanism involving a single transport site. A ^{35}Cl NMR study. *J. Biol. Chem.* 260:9537–44
37. Falke JJ, Chan SI. 1986. Molecular mechanisms of band 3 inhibitors: transport site inhibitors. *Biochemistry* 25:7888–94
38. Falke JJ, Pace RJ, Chan SI. 1984. Chloride binding to the anion transport binding sites of band 3: a ^{35}Cl NMR study. *J. Biol. Chem.* 259:6472–80
39. Falke JJ, Pace RJ, Chan SI. 1984. Direct observation of the transmembrane recruitment of band 3 transport sites by competitive inhibitors: a ^{35}Cl NMR study. *J. Biol. Chem.* 259:6481–94
40. Feigenson GW, Chan SI. 1974 . Nuclear magnetic relaxation behavior of lecithin multilayers. *J. Am. Chem. Soc.* 96:1312–19

41. Glaser M, Simpkins H, Singer SJ, Sheetz M, Chan SI. 1970. On the interactions of lipids and proteins in the red blood cell membrane. *Proc. Natl. Acad. Sci. USA* 65:721–28

42. Hansen KC, Rock RS, Larsen RW, Chan SI. 2000. A method for photoinitiating protein folding in a non-denaturing environment. *J. Am. Chem. Soc.* 122:11567–68

43. Hartzell CR, Beinert H. 1974. Components of cytochrome *c* oxidase detectable by EPR spectroscopy. *Biochim. Biophys. Acta* 368:318–38

44. Horwitz AF, Horsley WJ, Klein MP. 1972. Magnetic resonance studies on membrane and model membrane systems: proton magnetic relaxation rates in sonicated lecithin dispersions. *Proc. Natl. Acad. Sci. USA* 69:590–93

45. Hu VW, Chan SI, Brown GS. 1977. X-ray absorption edge studies on oxidized and reduced cytochrome *c* oxidase. *Proc. Natl. Acad. Sci. USA* 74:3821–25

46. Iwata S, Ostermeier C, Ludwig B, Michel H. 1995. Structure at 2.8 Å resolution of cytochrome *c* oxidase from *Paracoccus denitrificans*. *Nature* 376:660–69

47. Kainosho M, Kroon PA, Lawaczeck R, Petersen NO, Chan SI. 1978. Chain length dependence of the ^{1}H NMR relaxation rates in bilayer vesicles. *Chem. Phys. Lipids* 21:59–68

48. Kao WC, Chen YR, Yi EC, Lee H, Tian Q, et al. 2004. Quantitative proteomic analysis of metabolic regulation by copper in *M. capsulatus* (Bath). *J. Biol. Chem.* 279:51554–60

49. Kroon PA, Kainosho M, Chan SI. 1975. The state of molecular motion of cholesterol in lecithin bilayers. *Nature* 256:582–84

50. Kroon PA, Kreishman GP, Nelson JH, Chan SI. 1974. The effects of chain length on the secondary structure of oligoadenylates. *Biopolymers* 13:2571–92

51. Kuo NNW, Huang JJT, Miksovska J, Chen RPY, Larsen RW, et al. 2005. Effects of turn mutation on the structure, stability, and folding kinetics of a β-sheet peptide. *J. Am. Chem. Soc.* 127:16945–54

52. Kyogoku Y, Lord RC, Rich A. 1966. Hydrogen bonding specificity of nucleic acid purines and pyrimidines in solution. *Science* 154:518–20

53. Kyogoku Y, Lord RC, Rich A. 1967. An infrared study of hydrogen bonding between adenine and uracil derivatives in chloroform solution. *J. Am. Chem. Soc.* 89:496–504

54. Lau ALY, Chan SI. 1974. Nuclear magnetic resonance studies of the interaction of alamethicin with lecithin bilayers. *Biochemistry* 13:4942–48

55. Lau ALY, Chan SI. 1975. Alamethicin-mediated fusion of lecithin vesicles. *Proc. Natl. Acad. Sci. USA* 72:2170–74

56. Lawaczeck R, Kainosho M, Chan SI. 1976. The formation and annealing of structural defects in lipid bilayer vesicles. *Biochim. Biophys. Acta* 433:313–30

57. Lawaczeck R, Kainosho M, Girardet JL, Chan SI. 1975. Effects of structural defects in sonicated phospholipid vesicles on fusion and ion permeability. *Nature* 256:584–86

58. Levine YK, Birdsall NJM, Lee AG, Metcalfe JC. 1972. Carbon-13 nuclear magnetic resonance relaxation measurements of synthetic lecithins and the effect of spin-labeled lipids. *Biochemistry* 11:1416–21

59. Levinthal C. 1968. Are there pathways for protein folding? *J. Chem. Phys.* 85:44–45

60. Lieberman RL, Rosenzwig AC. 2005. Crystal structure of a membrane-bound metalloenzyme that catalyzes the biological oxidation of methane. *Nature* 434:177–82

61. Lipscomb JD. 1994. Biochemistry of the soluble methane monooxygenase. *Annu. Rev. Microbiol.* 48:371–99

62. Marcus RA, Sutin N. 1985. Electron transfers in chemistry and biology. *Biochim. Biophys. Acta* 811:265–322

63. Martin CT, Scholes CP, Chan SI. 1988. On the nature of cysteine coordination to Cu_A in cytochrome *c* oxidase. *J. Biol. Chem.* 263:8420–29

64. Morgan JE, Li PM, Jang DJ, El-Sayed MA, Chan SI. 1989. Electron transfer between cytochrome *a* and copper A in cytochrome *c* oxidase: a perturbed equilibrium study. *Biochemistry* 28:6975–83

65. Mueller L, Chan SI. 1983. Two-dimensional deuterium NMR of lipid membranes. *J. Chem. Phys.* 78:4341–48

66. Musser SM, Chan SI. 1995. Understanding the cytochrome *c* oxidase proton pump: thermodynamics of redox linkage. *Biophys. J.* 68:2543–55

67. Musser SM, Stowell MHB, Chan SI. 1995. Cytochrome *c* oxidase: chemistry of a molecular machine. *Adv. Enzymol. Relat. Areas Mol. Biol.* 71:79–208

68. Ng KY, Tu LC, Wang YS, Chan SI, Yu SS. 2008. Probing the hydrophobic pocket of the active site in the particulate methane monooxygenase (pMMO) from *Methylococcus capsulatus* (Bath) by variable stereo-selective alkane hydroxylation and olefin epoxidation. *Chembiochem* 9:1116–23

69. Nguyen HHT, Elliott SJ, Yip JHK, Chan SI. 1998. The particulate methane monooxygenase from *Methylococcus capsulatus* (Bath) is a novel copper-containing three-subunit enzyme: isolation and characterization. *J. Biol. Chem.* 273:7957–66

70. Nguyen HHT, Nakagawa KH, Hedman B, Elliott SJ, Lidstrom ME, et al. 1996. X-ray absorption and EPR studies on the copper ions associated with the particulate methane monooxygenase from *Methylococcus capsulatus* (Bath). Cu(I) ions and their implications. *J. Am. Chem. Soc.* 118:12766–76

71. Nguyen HHT, Shiemke AK, Jacobs SJ, Hales BJ, Lidstrom ME, et al. 1994. The nature of the copper ions in the membranes containing the particulate methane monooxygenase from *Methylococcus capsulatus* (Bath). *J. Biol. Chem.* 269:14995–5005

72. Owicki JC, McConnell HM. 1979. Theory of protein-lipid and protein-protein interactions in bilayer membranes. *Proc. Natl. Acad. Sci. USA* 76:4750–54

73. Owicki JC, Springgate MW, McConnell HM. 1978. Theoretical study of protein-lipid interactions in bilayer membranes. *Proc. Natl. Acad. Sci. USA* 75:1616–19

74. Pace RJ, Chan SI. 1982. Molecular motions in lipid bilayers. II. Magnetic resonance of multilamellar and vesicle systems. *J. Chem. Phys.* 76:4228–40

75. Pearson LT, Chan SI. 1982. Effects of lipid-mediated interactions on protein pair distribution functions. *Biophys. J.* 37:141–42

76. Pearson LT, Chan SI. 1987. Pair distribution functions of cytochrome *c* oxidase in lipid bilayers: evidence for a lipid-mediated repulsion between protein particles. *Chem. Scr.* 27B:203–9

77. Pearson LT, Edelman J, Chan SI. 1984. Statistical mechanics of lipid membranes: protein correlation functions and lipid ordering. *Biophys. J.* 45:863–71

78. Pearson LT, Lewis BA, Engleman DM, Chan SI. 1983. Pair distribution functions of bacteriorhodopsin and rhodopsin in model bilayers. *Biophys. J.* 43:167–74

79. Petersen NO, Chan SI. 1977. More on the motional state of lipid bilayer membranes. The interpretation of order parameters obtained from nuclear magnetic resonance experiments. *Biochemistry* 16:2657–67

80. Prestegard JH, Chan SI. 1969. Proton magnetic resonance studies of the cation-binding properties of nonactin. The K^+-nonactin complex. *Biochemistry* 8: 3921–27

81. Prestegard JH, Chan SI. 1970. Proton magnetic resonance studies of the cation-binding properties on nonactin. Comparison of the sodium ion, potassium ion, and cesium ion complexes. *J. Am. Chem. Soc.* 92:4440–46

82. Prior SD, Dalton H. 1985. The effect of copper ions in membrane content and methane monooxygenase activity in methanol-grown cells of *Methylococcus capsulatus* (Bath). *J. Gen. Microbiol.* 131:155–63

83. Rock RS, Chan SI. 1998. Preparation of a water-soluble "cage" based on 3′,5′-dimethoxybenzoin. *J. Am. Chem. Soc.* 120:10766–67

84. Schmidt CF Jr, Chan SI. 1971. Nuclear spin-lattice relaxation of the methyl carbon-13 in toluene. *J. Magn. Reson.* 5:151–54

85. Schuh J, Banerjee U, Mueller L, Chan SI. 1982. The phospholipid packing arrangement in small bilayer vesicles as revealed by proton magnetic resonance studies at 500 MHz. *Biochim. Biophys. Acta* 687:219–25

86. Schultz BE, Chan SI. 2001. Structures and proton-pumping strategies of mitochondrial respiratory enzymes. *Annu. Rev. Biophys. Biomol. Struct.* 30:23–65

87. Schweizer MP, Chan SI, Helmkamp GK, Ts'o POP. 1964. An experimental assignment of the proton magnetic resonance spectrum of purine. *J. Am. Chem. Soc.* 86:696–700

88. Seelig J, Seelig A. 1974. Dynamic structure of fatty acyl chains in a phospholipid bilayer measured by deuterium magnetic resonance. *Biochemistry* 13:4839–45

89. Seiter CHA, Chan SI. 1973. Molecular motion in lipid bilayers: a nuclear magnetic resonance linewidth study. *J. Am. Chem. Soc.* 95:7541–53

90. Semrau J, Chistoserdov A, Lebron J, Costello A, Davagnino J, et al. 1995. Particulate methane monooxygenase genes in methanotrophs. *J. Bacteriol.* 177:3071–79

91. Sheetz MP, Chan SI. 1972. Effect of sonication on the structure of lecithin bilayers. *Biochemistry* 11:4573–81

92. Sheetz MP, Chan SI. 1972. Proton magnetic resonance studies of whole human erythrocyte membranes. *Biochemistry* 11:548–55

93. Singer SJ, Nicolson GL. 1972. The fluid mosaic model of the structure of cell membranes. *Science* 175:720–31

94. Smith DDS, Dalton H. 1989. Solubilization of methane monooxygenase from *Methylococcus capsulatus* Bath. *Eur. J. Biochem.* 182:667–72

95. Stevens TH, Brudvig GW, Bocian DF, Chan SI. 1979. The structure of the cytochrome a_3-Cu_{a3} couple in cytochrome *c* oxidase as revealed by nitric oxide binding studies. *Proc. Natl. Acad. Sci. USA* 76:3320–24

96. Stevens TH, Chan SI. 1981. Histidine is the axial ligand to cytochrome a_3 in cytochrome *c* oxidase. *J. Biol. Chem.* 256:1069–71

97. Stevens TH, Martin CT, Wang H, Brudvig GW, Scholes CP, et al. 1982. The nature of Cu_a in cytochrome *c* oxidase. *J. Biol. Chem.* 257:12106–13

98. Stockton G, Polnaszek CF, Tulloch AP, Hasan F, Smith ICP. 1976. Molecular motion and order in single-bilayer vesicles with multilamellar dispersions of egg lecithin and lecithin-cholesterol mixtures. A deuterium nuclear magnetic resonance study of specifically labeled lipids. *Biochemistry* 15:954–66

99. Sun J, Deem MW. 2007. Spontaneous emergence of modularity in a model of evolving individuals. *Phys. Rev. Lett.* 99:228107(4)

100. Ts'o POP, Chan SI. 1964. Interaction and association of bases and nucleosides in aqueous solutions. Association of 6-methylpurine and 5-bromouridine and treatment of multiple equilibria. *J. Am. Chem. Soc.* 86:4176–81

101. Tsukihara T, Aoyama H, Yamashita E, Tomizaki T, Yamaguchi H, et al. 1995. Structures of metal sites of oxidized bovine heart cytochrome *c* oxidase at 2.8 Å. *Science* 269:1069–74

102. Tsukihara T, Aoyama H, Yamashita E, Tomizaki T, Yamaguchi H, et al. 1996. The whole structure of the 13-subunit oxidized cytochrome *c* oxidase at 2.8 Å. *Science* 272:1136–44

103. Wang JF, Falke JJ, Chan SI. 1986. A proton NMR study of the mechanism of the erythrocyte glucose transporter. *Proc. Natl. Acad. Sci. USA* 83:3277–81

104. Watnick PI, Chan SI, Dea P. 1990. Hydrophobic mismatch in gramicidin A'/lecithin systems. *Biochemistry* 29:6215–21

105. Wikström M. 1989. Identification of the electron transfers in cytochrome oxidase that are coupled to proton-pumping. *Nature* 338:776–78

106. Wikström MK. 1977. Proton pump coupled to cytochrome *c* oxidase in mitochondria. *Nature* 266:271–73

107. Wilkinson B, Zhu M, Priestley ND, Nguyen HHT, Morimoto H, et al. 1996. A concerted mechanism for ethane hydroxylation by the particulate methane monooxygenase from *Methylococcus capsulatus* (Bath). *J. Am. Chem. Soc.* 118:921–22

108. Wolynes PG, Onuchic JN, Thirumalai D. 1995. Navigating the folding routes. *Science* 267:1619–20

109. Yoshikawa S, Tera T, Takahashi Y, Caughey WS. 1988. Crystalline cytochrome *c* oxidase of bovine heart mitochondrial membranes: composition and X-ray diffraction studies. *Proc. Nat. Acad. Sci. USA* 85:1354–58

110. Yu C, Yu L, King TE. 1975. Studies on cytochrome oxidase. Interactions of the cytochrome oxidase protein with phospholipids and cytochrome *c*. *J. Biol. Chem.* 250:1383–92

111. Yu SSF, Chen KHC, Tseng MYH, Wang YS, Tseng CF, et al. 2003. Production of high quality pMMO in high yields from *Methylococcus capsulatus* (Bath) with a hollow-fiber membrane bioreactor. *J. Bacteriol.* 185:5915–24

112. Yu SSF, Wu LY, Chen KHC, Luo WI, Huang DS, et al. 2003. The stereospecific hydroxylation of [2,2-$^2H_2$]butane and chiral dideuteriobutanes by the particulate methane monooxygenase from *Methylococcus capsulatus* (Bath). *J. Biol. Chem.* 278:40658–69

113. Zaslavsky D, Sadoski RC, Wang K, Durham B, Gennis RB, et al. 1998. Single electron reduction of cytochrome *c* oxidase compound F: resolution of partial steps by transient spectroscopy. *Biochemistry* 37:14910–16

Crystallizing Membrane Proteins for Structure Determination: Use of Lipidic Mesophases

Martin Caffrey

Membrane Structural and Functional Biology Group, University of Limerick, Limerick, Ireland; email: martin.caffrey@ul.ie

Annu. Rev. Biophys. 2009. 38:29–51

First published online as a Review in Advance on December 16, 2008

The *Annual Review of Biophysics* is online at biophys.annualreviews.org

This article's doi: 10.1146/annurev.biophys.050708.133655

Key Words

crystallization mechanism, cubic phase, macromolecular crystallography, robot, structure-function, X-ray methods

Abstract

The principal route to determine the structure and the function and interactions of membrane proteins is via macromolecular crystallography. For macromolecular crystallography to be successful, structure-quality crystals of the target protein must be forthcoming, and crystallogenesis represents a major challenge. Several techniques are employed to crystallize membrane proteins, and the bulk of these techniques make direct use of solubilized protein-surfactant complexes by the more traditional, so-called *in surfo* methods. An alternative *in meso* approach, which employs a bicontinuous lipidic mesophase, has emerged as a method with considerable promise in part because it involves reconstitution of the solubilized protein back into a stabilizing and organizing lipid bilayer reservoir as a prelude to crystallogenesis. A hypothesis for how the method works at the molecular level and experimental evidence in support of the proposal are reviewed here. The latest advances, successes, and challenges associated with the method are described.

Contents

INTRODUCTION

If a picture is worth a thousand words then a structure is worth a million.

In the mid-1990s, a mild galvanic jolt passed through the membrane structural biology community with the observation that three-dimensional (3D) structure-grade crystals of bacteriorhodopsin (bR) grew in a lipidic cubic mesophase (33, 51). Initially, it was highly anticipated that the method would be the panacea, lowering one of the major barriers to high-resolution structure with the insights into the activity, function, and interaction that molecular structure offers. However, the immediate yield was not as expected, and the so-called *in meso* method was regarded by many then as a novelty that worked with a restricted set of membrane proteins.

Despite this obvious setback, the Caffrey group persevered with the method for several

reasons. First, we had spent the previous decade and beyond working with lipids and lipid phase behavior and had acquired a comfort level dealing with and handling these refractory states of matter. Second, the *in meso* method involved the cubic phase, which for some time had fascinated the author because of its intricate bicontinuous structure, its ability as a liquid crystal to diffract X-rays at small angles (40), and its proposed involvement as an intermediate in fat digestion (50) and membrane fusion (55). Third, while the method had been dismissed by many as relevant to a limited set of membrane proteins, the author was of the view that nature is a little more general than that. The sense was that there is enough variety at the level of membrane constitution, and in the biophysics and physical chemistry that prevail in that nanoliter volume in which crystallogenesis typically occurs, that more than bR and its homologs would yield to the *in meso* method. And so our group persevered, buoyed by a funding agency prepared to take a calculated risk and convinced that we had the knowledge and experience to exploit the methodology and, at the very least, to give it a thorough testing.

The path toward an exploration of the method was neither direct nor simple, however. At the time, the group faced two impediments: We were not in the business of producing membrane proteins nor indeed were we macromolecular crystallographers. The author had gained considerable experience with membrane proteins of natural abundance from graduate work with milk lipid synthesis enzymes (12) and with the calcium-ATPase from sarcoplasmic reticulum (11). However, in the intervening decade and a half the author had focused almost exclusively on understanding lipid phase behavior while a revolution of sorts had taken place in molecular biology. These deficiencies were compensated for by engaging with enthusiastic and generous collaborators. At the same time, as a group we set about learning and developing the relevant recombinant DNA technology and membrane protein production, purification, crystallization, and structure determination

---

bR: bacteriorhodopsin

Mesophase: a liquid crystalline state of matter with order intermediate between that of a crystalline solid and an isotropic liquid

***In meso*:** a method for crystallizing membrane proteins by way of a bicontinuous cubic or sponge mesophase

skills for in-house use, all of which have paid dividends.

The review that follows begins with an overview of a hypothesis for how the *in meso* crystallization of membrane proteins takes place at the molecular level and the phase science that defines the physicochemical bases for such events. The rest of the review deals with practical aspects of the method and evidence in support of the hypothesis and ends with an overview of the structures solved and proteins crystallized by way of the *in meso* method.

HYPOTHESIS

A proposal has been advanced for how *in meso* crystallogenesis takes place at the molecular level (7, 8, 48) (**Figure 1**). It begins with an isolated biological membrane that is treated with detergent to solubilize the target protein. The

Bicontinuous: a mesophase consisting of a single, curved bilayer that permeates 3D space and that separates two continuous, interpenetrating but noncontacting channels filled with aqueous medium

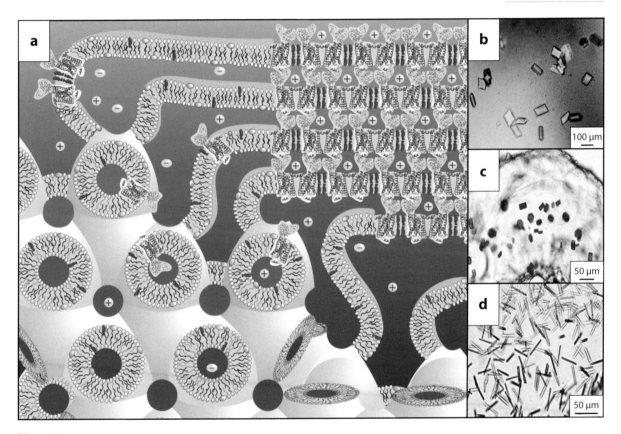

Figure 1

In meso crystallization model and crystals. (*a*) Schematic representation of the events proposed to take place during the crystallization of an integral membrane protein from the lipidic cubic mesophase. The process begins with the protein reconstituted into the highly curved bilayers of the bicontinuous cubic phase (*bottom left quadrant*). Added precipitants shift the equilibrium away from stability in the cubic membrane. This leads to phase separation, wherein protein molecules diffuse from the continuous bilayered reservoir of the cubic phase by way of a sheet-like or lamellar portal (*upper left quadrant*) to lock into the lattice of the advancing crystal face (*upper right quadrant*). Salt (positive and negative signs) facilitates crystallization, in part, by charge screening. Cocrystallization of the protein with native or added lipid (cholesterol) is shown in this illustration. As much as possible, the dimensions of the lipid (*light yellow oval with tail*), detergent (*pink oval with tail*), native membrane or added lipid (*purple*), protein (*blue*; β2AR-T4L; PDB code 2RH1), and bilayer and aqueous channels (*dark blue*) have been drawn to scale. The lipid bilayer is approximately 40 Å thick. (Panel *a* taken from Reference 9.) Crystals of (*b*) BtuB, (*c*) bacteriorhodopsin, and (*d*) light-harvesting complex II growing *in meso*. (Panels *b–d* taken from References 8 and 19.)

protein-detergent complex is purified by standard wet biochemical methods. Homogenizing with a monoacylglycerol (MAG) effects a uniform reconstitution of the purified protein into the bilayer of the cubic phase. The latter is bicontinuous in the sense that both the aqueous and bilayer compartments are continuous in 3D space (**Figure 1**). Upon reconstitution, the protein retains its native conformation and activity and has partial or complete mobility within the plane of the cubic phase bilayer. A precipitant is added to the mesophase, which triggers phase separation. Under conditions leading to crystallization, one of the separated phases is enriched in protein, which nucleates and develops into a bulk crystal. The hypothesis includes a local lamellar phase that acts as a medium in which nucleation and crystal growth occur. This phase also serves as a conduit or portal for proteins on their way from the cubic phase reservoir to the growing face of the crystal. Initially at least, the proteins leave the lamellar conduit and ratchet into the growing crystal face to generate a layered-type (Type I) (42) packing of protein molecules within the crystal (**Figure 1**). Given that proteins are reconstituted into the bilayer of the cubic phase with no preferred orientation and the 3D continuity of the mesophase, it is possible for the resulting crystals to be polar or nonpolar. These correspond to situations in which adjacent proteins in a layer have their long-axis director oriented in the same or in the opposite directions.

The proposal for how nucleation and crystal growth occur *in meso* relies absolutely on the 3D continuity of the mesophase. Under the assumption that the sample exists as a monodomain, continuity ensures that the mesophase acts essentially as an infinite reservoir from which all protein molecules in the sample can end up in a bulk crystal. Neither the lamellar nor the inverted hexagonal phases, both of which are accessible in lipidic systems, have 3D continuity and alone are unlikely to support membrane protein crystallogenesis by the *in meso* method.

Because of the proposed need for the diffusion of proteins in the bilayer and of the precipitant component in the aqueous channels of the mesophase, the expectation is that crystal growth rates might be tardy *in meso*. However, crystals have been seen to grow within 1 h, which suggests that the slowness associated with restricted diffusion can be compensated for by a reduction in dimensionality. The latter is a result of the protein being confined to a lipid bilayer with its long axis oriented perpendicular to the membrane plane. Thus, the number of orientations that must be sampled to effect nucleation and crystal growth is few *in meso* compared with its *in surfo* counterpart, in which all 3D space is accessible.

If indeed crystal growth takes place in a nanoporous mesophase, then it likely occurs in a convection-free environment. This is analogous to growth under conditions of microgravity, which offers the advantage of a stable zone of depletion around the growing crystal and thus a slower and more orderly growth (41). Settling of crystals and subsequent growth into one another are also avoided under these conditions, as is the likelihood that impurities are wafted in from the surrounding solution to poison the face of the crystal and limit growth. For all these reasons *in meso* crystallogenesis is similar to crystallization in space with the prospect of producing high-quality, structure-grade crystals.

UNDERLYING PHASE BEHAVIOR

In meso crystallogenesis takes place in a lipid-based liquid crystal, or mesophase. The many components present and the conditions that prevail during crystallogenesis can affect phase behavior and, by extension, the outcome of the crystallization trial. Accordingly, the crystal grower should understand the phase behavior of the relevant lipid/water system that forms the basis of the *in meso* method. To date, monoolein is the lipid most commonly used and its temperature-aqueous composition phase diagram (5, 52) (**Figure 2**) is described briefly to provide the necessary phase behavior background and to provide the basis for further discussion.

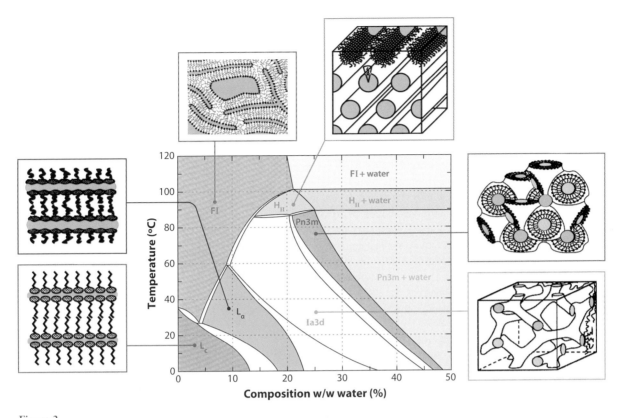

Figure 2

Temperature-composition phase diagram of the monoolein/water system determined under conditions of use in the heating and cooling directions from 20°C. A schematic representation of the various phase states is included in which colored zones represent water. The liquid crystalline phases below ∼17°C are metastable (52). Figure reproduced from Reference 19. Abbreviations: FI, fluid isotropic phase; H_{II}, inverted hexagonal phase; L_{α}, lamellar liquid crystalline phase; L_c, lamellar crystal phase.

In the dry state, the hydrating lipid monoolein goes from a solid lamellar crystal to a liquid, so-called fluid isotropic phase at about 37°C (i.e., it melts). Instead of effecting solidification, recooling the melt often leads to an undercooled liquid that can persist for extended periods at room temperature. The addition of water to the system gives rise to a number of lyotropic (in this case, water-induced) mesophases, the identity of which depends on temperature. Thus, at elevated temperatures the inverted hexagonal phase forms. This gives way to two different cubic phases upon lowering temperature: The cubic-Ia3d phase forms at lower hydration levels than the cubic-Pn3m phase does. The latter can exist in equilibrium with excess water (in actuality it is a solution containing monomers and micelles of monoolein) as a two-phase system over a wide temperature range. At intermediate hydration levels and temperatures the lamellar liquid crystalline phase forms. The equilibrium phase diagram for the monoolein/water system shows that the mesophases are no longer stable below ∼17°C (52). However, as was observed with the dry lipid, long-lived undercooling of the lamellar and cubic mesophases is possible and is commonly observed (52).

METHOD DEVELOPMENT

Manual and Robotic Approaches

Crystallogenesis by the *in meso* method begins with the formation of the lipidic mesophase into which the target membrane protein is

Lyotrope: describes a liquid that, when combined with a lipid or detergent amphiphile, forms a liquid crystal or mesophase

reconstituted. This is done by mixing to homogeneity approximately equal volumes of lipid, in either solid or liquid form, and protein-detergent solution. A disk-shaped bolus of the mesophase surrounded by precipitant is then sandwiched between parallel glass plates in a sealed well, and crystal growth is allowed to progress at a fixed temperature (14, 23). Components of the precipitant diffuse inward along radial lines toward the center of the disk. Crystallization presumably occurs in response to the varied gradients of precipitant components that develop in the porous mesophase and to the microstructural and compositional changes these gradients induce in the supporting mesophase (9). Periodically, wells are examined by microscope to evaluate how crystallization is progressing.

One of the major challenges experienced with the *in meso* method was the creation and handling of the mesophase into which the protein is reconstituted and from which crystals grow. The cubic phase is viscous, analogous to thick toothpaste, and sticky. It is not easy to handle. Described below is the development of the tools for manipulating such materials and for performing crystallization trials with them. The emphasis was placed on using materials that were commercially available and inexpensive.

The mesophase preparation and handling challenge was overcome by employing a device, called the lipid-mixer, developed in the laboratory for lipid phase diagram work (13). It consists of two 100 µL gas-tight Hamilton syringes joined by a narrow-bore, low dead-volume coupling needle. Lipid is placed in one syringe, and protein solution is placed in the other. The two are connected via the coupler, and mixing is effected by moving the contents of the two syringes gently back and forth through the coupler. The homogenous mesophase into which the protein has reconstituted is transferred to one of the syringes, the coupler is replaced by a standard needle, and the loaded syringe is used to dispense the mesophase. This device is simple, easy to assemble mostly using commercially available parts, and robust. It can handle and dispense the highly viscous mesophases typically encountered because movement happens by positive displacement. Further, when used in conjunction with a commercially available repeating dispenser, it functions to dispense the mesophase reproducibly in small volumes of the type typically used in crystallization trials. A two- to threefold reduction in the amount of mesophase dispensed per well (up to 70 nL) was realized when the number of teeth in the ratchet of the dispenser was increased from 50 to 120 (15).

The next challenge was to produce a plate on which conditions that support crystal growth could be screened efficiently. After much experimentation we ended up using standard 3×1 inch^2 glass microscope slides as the base (14). Wells were created on the slide by covering it with double-stick tape perforated with uniformly spaced, 7-mm-diameter holes. The tape was chosen for its water resistance. The plates were sealed with standard glass microscope coverslips. Each coverslip sealed a 3×3 square grid of wells and the slide contained a total of 27 wells.

The glass sandwich plates proved to be robust and functional, and they nicely matched the manual dispensing of mesophase as described. The mesophase itself is optically isotropic and nonbirefringent, and when sealed between two pieces of glass, the plates offer unprecedented optical quality and clarity. Thus, crystals in the micrometer-size range can be detected with ease, especially when plates are examined with both normal and polarized light.

The 27-well plates used in conjunction with the mixer/dispensing device are employed when conditions are to be optimized or for labeling crystals in situ with heavy atoms. However, the combination is not suited to high-throughput screening, in which hundreds, and possibly thousands, of conditions must be evaluated. To respond to this need, two robots have been developed; the first robot automates the filling of crystallization plates and the second robot is used to view wells and to record images for subsequent inspection by the crystal grower. The first robot, also called the *in meso* robot, is a commercial liquid-handling device modified for use with viscous mesophase samples (23). It includes two robotic arms. The first arm was built to work with a Hamilton syringe preloaded with mesophase. It places the tip of the syringe needle a fixed distance above the base of the well on a glass crystallization plate into which a 20 or 50 nL bolus of mesophase is dispensed. Accuracy in tip positioning is key to successful delivery. The second arm, which has eight tips, places 800 nL precipitant solutions on top of the mesophase boluses.

The *in meso* robot is used in combination with a barcoded, 96-well version of the glass sandwich plate introduced above (23). The robot operates such that six plates can be filled and sealed in one hour. Loaded *in meso* plates are placed in a commercial, combined incubator-imaging system (the second robot) custom-made to accommodate the low-profile (1.5-mm-high) sandwich plates. The imager, which houses 1000 glass and 500 standard commercial crystallization plates, records images of wells on a defined schedule in normal and polarized light. Images, recorded at different positions across the 0.14-mm-thick sample, can be viewed individually and as a combined edge-enhanced image for improved visualization. With this system, crystals in the micrometer-size range can be detected. Such sensitivity is key to finding leads upon which optimization is based.

The mechanical reproducibility and fine step sizes available with the *in meso* robot enabled us to further reduce the amount of valuable protein and lipid used in a given crystallization trial. Thus, successful crystallization trials have been performed with just 530 picoliters of mesophase corresponding to 210 picoliters of protein solution (2–4 ng protein) and 320 picoliters of lipid (16). This attests to the efficiency of the *in meso* method for membrane protein crystallization.

Screening with and Compatibility of Precipitants

The *in meso* method relies on spontaneous bicontinuous mesophase formation when a lipid and a protein solution are homogenized under suitable conditions of chemistry, composition, and temperature. To this typically cubic mesophase, a precipitant solution is added with the intent to induce nucleation and crystal growth. However, most commercial precipitants contain additives and solvents with the potential to destabilize the lipidic mesophase and are not compatible for direct use with the *in meso* method. In 2001, several commercial screens were surveyed for compatibility, using small-angle X-ray scattering (SAXS) as the diagnostic, and in some kits up to half of the conditions destabilized the cubic phase (20). Since that time, many more screens have appeared on the market. A study aimed at evaluating the compatibility of these new screens with the *in meso* method is in progress (D. Aragao, B. Sun, D. Li, J. Tan, N. Höfer & M. Caffrey, unpublished observations). In total, 19 screen kits have been examined. Measurements were performed under conditions similar to those used for standard crystallogenesis. By eliminating replicates and conditions that destabilize the cubic mesophase from within the original set of ~1500 screens, a panel of unique precipitants has been identified that makes efficient use of expensive protein and lipid as well as of imaging space and time.

Screening with Lipids

Lipids play prominent roles in *in meso* crystallogenesis. They create the hosting mesophase, and they are included with the host as important additives.

Host lipid and rational design. The default lipid for the bulk of the *in meso* crystallogenesis studies performed to date is monoolein. However, there is no good reason why monoolein should be the preferred lipid for all membrane proteins. The latter come from an array of biomembrane types with varying properties that include hydrophobic thickness, intrinsic curvature, lipid makeup, and compositional asymmetry. Thus, it seems reasonable that screening for crystallizability based on the identity of the lipid creating the hosting mesophase would be worthwhile. For this, MAGs with differing acyl chain characteristics such as length and olefinic bond position must be available. A lipid synthesis and purification program is in place in the author's laboratory to serve this need.

The MAGs that have been used in successful structure determination studies based on *in meso*–grown crystals have had chains lengths of 16 and 18 carbon atoms. A proposal was advanced that a shorter-chain lipid producing a thinner bilayer would facilitate and possibly enhance crystallization. A 14-carbon MAG was chosen as the lipid with which to test the proposal (43). To be compatible with the *in meso* method, a *cis* olefinic bond had to be placed in the acyl chain. Its position was arrived at by applying rational design principles to a collection of temperature-composition phase diagrams for homologous MAGs. The target lipid, 2,3-dihydroxypropyl-(7Z)-tetradec-7-enoate (7.7 MAG), was identified and synthesized, and its phase properties were characterized by SAXS. As designed, this short-chain lipid formed the requisite cubic mesophase at room temperature. In support of the hypothesis, it produced crystals of three disparate integral membrane proteins by the *in meso* method. These included bR, cytochrome *caa*₃ oxidase, and the bacterial outer membrane cobalamin (vitamin B_{12}) transporter, BtuB. The last protein is notable in that it was the first β-barrel protein to be crystallized by the *in meso* method. Other short-chain MAGs that are successful in the crystallogenesis of *Pseudomonas aeruginosa* membrane protein targets have been produced (D. Li, J. Tan, D. Aragao & M. Caffrey, unpublished observations). The means by which these lipids optimize crystallogenesis has not been established, but it surely reflects a preferential partitioning of the protein between the crystal and the hosting mesophase. Thus, bilayer thickness, or some other mesophase property, must not match the requirements of the protein such that it prefers to exist in the more ordered environment of a crystal.

Reference has been made to the sensitivity of phase microstructure to lipid identity. Support for this statement is based on SAXS measurements performed on the cubic phase prepared with a homologous series of MAGs (2–5, 8). The data show expected behavior in that as chain length decreases so too does the thickness of the lipid layer that creates the apolar fabric of the cubic phase, when evaluated at a single temperature. Less intuitive perhaps is the finding that the aqueous channel diameter drops as chain length increases. This is consistent with a flattening and attenuating curvature at the polar/apolar interface with the shorter-chain lipids.

Although lipid identity can be used to tailor phase microstructure, it is possible that the desired microstructure might not be accessible with a single lipid species in the temperature range of interest. In this case, it is possible to fine-tune by using mixtures of MAGs with different acyl chain lengths for which the mole ratio is adjusted to set the microstructure at the desired intermediate value.

The microstructure of the mesophase can be engineered over relatively wide limits by manipulating temperature and/or lipid identity and composition. However, the two metrics of the cubic phase—the polar and apolar compartment dimensions—are not independently adjustable and indeed are tightly coupled (8). Nonetheless, this feature of tunability is a valuable tool available to the crystal grower in search of a suitable lipid matrix in which to grow crystals. Thus, proteins with extramembranal domains that come in a variety of sizes can be accommodated as can those that originate from native membranes with different hydrophobic thicknesses (46).

The original *in meso* method does not work reliably at low temperatures, at which proteins are generally more stable, because the hosting lipid, monoolein, becomes solid. The need exists therefore for a lipid that forms the cubic phase and that supports crystal growth at low temperatures. As with the 7.7 MAG in the example above, a database of phase diagrams was mined and used to design such a lipid (45). SAXS showed that the new lipid, 7.9 MAG, exhibited expected phase behavior. Further, it produced diffraction-quality membrane protein crystals by the *in meso* method at 6°C. These results demonstrate that lipidic materials, like their protein counterparts, are amenable to rational design. The same approach, as used in these design studies, should find application in extending the range of membrane proteins amenable to crystallization by the *in meso* method.

Added lipid. It is possible that the hosting cubic phase created by MAG alone, which itself is a most uncommon membrane component, will limit the type of membrane proteins crystallizable by the *in meso* method. With a view to expand the range of applicability of the method and to make the hosting cubic phase more familiar to its guest protein, the degree to which the reference cubic-Pn3m phase formed by hydrated monoolein can be modified by other lipid types was quantified using SAXS (18). These included phosphatidylcholine (PC), phosphatidylethanolamine (PE), phosphatidylserine (PS), cardiolipin, lyso-PC, 2-monoolein, and cholesterol. The study showed that all lipids were accommodated in the cubic phase of (1-)monoolein to some extent without altering phase identity. The positional isomer, 2-monoolein, was tolerated to the highest level. The least well tolerated were the anionic lipids, followed by lyso-PC. The others were accommodated to the extent of 20–25 mol%. These data are for use in the rational design of cubic phase crystallization matrices with lipid profiles that match the needs of a greater range of membrane proteins. A case in point is the recent structure determination of the GPCR chimeras, from which diffraction-quality crystals were obtained only when the system contained added cholesterol (24, 30, 31a). The benefits of these added lipids range from altering the microstructure, dynamics, and chemistry of the mesophase, to favoring lamellar phase formation for nucleation and growth, to stabilizing the protein as does a ligand, cofactor, or substrate. Although a lipid additive may destabilize the cubic phase, the latter can be recovered upon incubation with certain precipitant solutions (9, 10, 43).

Evaluating Crystallization

A problem in the area of crystallogenesis derives from the need to establish whether the crystal observed is composed of protein or is instead a crystal of salt, detergent, or lipid. The ultimate proof is to harvest the crystal and test its diffraction. However, this puts the crystal grower at a considerable disadvantage in terms of lost time because months may have been needed to grow the crystal. Negative protein-free controls can always be run, but these also require time and are rarely done in parallel unless one is at the stage of optimization. Protein-staining dyes have been used to selectively label protein crystals, but usually these work best with crystals grown by the *in surfo* methods. Under *in meso* conditions staining of the lipid mesophase by the lipophilic dye causes background problems. It is also possible to covalently tag the protein with a chromophore, but this requires separate labeling and purification and can alter the target in unpredictable ways. Increasingly the community is turning to the use of spectroscopic methods to identify protein crystals (39). These methods rely on the target containing enough tryptophan and/or tyrosine to give the crystal a characteristic absorbance and/or fluorescence in the UV region. The glass base plates used for *in meso* crystallogenesis have reasonable UV transmittance and are being evaluated for suitability in absorbance and fluorescence imaging. However, the instrumentation to make such measurements in high-throughput fashion is still expensive.

Harvesting Crystals Grown *In Meso*

A frequently asked question concerning the *in meso* method is, How are crystals harvested? As noted, the cubic phase is viscous and sticky, but with the right tools (13) it can be handled with ease. In a similar vein, harvesting crystals grown *in meso* is relatively straightforward—with the right tools (10). These consist of needles, spatulas, and loops of various sizes and shapes that are used to open up the bulk mesophase and to expose a crystal or to move a crystal to the surface. Once exposed, it can be picked up with a standard cryo- or litho-loop and immediately flash-cooled, where the cubic phase acts as a cryo-protectant. Under certain conditions, the lipid from the mesophase can solidify and typically it produces diffraction rings, from closely packed acyl chains, in the vicinity of 4 Å. Occasionally, it is possible to vitrify the mesophase, in which case it generates diffuse scatter, again in the 4 Å range. To minimize interference with diffraction data from the protein crystal, the less mesophase that accompanies the crystal into the harvesting loop, the better. Because crystals grown *in meso* are often small, harvesting is best done with the aid of a reasonably high-power polarizing microscope.

Detergents (38), oils (10), and enzymatic hydrolysis (47) have been used to release protein crystals from the lipidic mesophase for harvesting. As with all manipulations, however, the possibility is that diffraction quality is compromised, and the full range of alternative methods should be explored. In our hands, the least damaging and most successful involves direct harvesting from the cubic phase.

All of the above assumes that the mesophase in which crystals have formed is accessible. When trials are performed in glass sandwich plates, accessibility is a real issue. This has been dealt with in part by using glass plates strictly for screening and then performing trials in commercial plates that can be easily opened and from which crystals can be easily harvested. However, we have found that using a different crystallization environment invariably requires an optimization screen at the very least. With certain proteins, structure-yielding crystals could only be grown in glass plates. In such circumstances a glass-cutting tool was used to remove the cover slip from the well (10).

It is also possible to perform *in meso* crystallization in glass or quartz X-ray capillaries (45). In this case, there is no need to isolate the crystal for separate mounting, and diffraction data can be recorded directly.

Working with Microcrystals

The *in meso* method typically produces small crystals. Accordingly, diffraction data collection requires use of a naturally collimated synchrotron X-ray source and benefits from so-called micro- or minibeams. In such cases, the beam size matches that of the crystal such that background scatter, and diffraction from the mesophase carried along with the crystal, is reduced to a minimum. This was used advantageously in the original bR work (51) and in recent GPCR studies (24, 30, 31a). One of the lingering challenges in this area concerns finding and positioning in the X-ray beam crystals buried in an opaque, cryo-cooled mesophase sample. Under active investigation at the various synchrotron sources worldwide is the development of an automated diffraction-based protocol for working with such refractory samples.

It is interesting to enquire what might limit the size of the crystal given that the cubic phase is composed of a single lipid bilayer and thus, presumably a single protein reservoir. While it is true that just one continuous bilayer permeates a given domain within the cubic phase, the macroscopic sample itself is usually composed of multiple microdomains. Evidence in support of this statement comes from the observation that the cubic phase typically produces a powder diffraction pattern (6). However, it is not uncommon for the same phase to produce a spotty diffraction pattern, the hallmark of large crystallites or domains within the sample (17). It is not clear what form the lipid/water dispersion takes at the surface of a domain, nor indeed the nature of the communication between domains. If movement of protein were confined to a given domain, then its size and protein payload would ultimately limit that of the crystal. Back-of-the-envelope calculations show that a spherical microdomain of cubic phase (prepared with a 10 mg protein per millimeter of solution) with a diameter of 100 μm is required to produce a single 20-μm-diameter (spherical) crystal. Thus, relatively large domains and/or interdomain diffusion are needed to produce reasonably sized crystals. Currently, ways to increase domain size and interdomain diffusion with a view to producing larger, diffraction-quality crystals are being investigated.

A Role for Detergents

The solutions used to spontaneously form the protein-enriched cubic phase usually contain significant amounts of detergents that were employed initially to purify and to solubilize the membrane protein. By virtue of their amphiphilic and surface active natures, detergents have the potential to affect the phase properties of the *in meso* system and, by extension, the outcome of the crystallization process. Accordingly, studies have been performed to quantify the effects that commonly used detergents have on the phase behavior of hydrated monoolein (1, 35, 44). Phase identity and microstructure were characterized by SAXS measurements on samples prepared to mimic *in meso* crystallization conditions. The results show that the cubic phase is relatively insensitive to small amounts of alkyl (hexyl, octyl, nonyl, decyl) glucosides, dodecyl maltoside, alkyl (dodecyl, hexadecyl) fos-cholines, lauryldimethylamine-oxide (LDAO), sodium dodecyl sulfate (SDS), and Cymal. However, at higher levels the detergents trigger a transition to the lamellar phase and, where studied, do so in a temperature-, lipid-, and salt concentration-dependent manner (44).

These data have important implications for *in meso* crystallization. First, a small amount of detergent may facilitate crystallogenesis by favoring formation of lamellar domains in which nucleation and crystal growth are proposed to take place. Second, proteins with a high concentration of detergent can give rise to the bulk lamellar phase upon homogenization with lipid as a preliminary to crystallization. If the precipitant solutions used have a high concentration of salt, the lamellar-to-cubic phase transition can be reversed for successful crystal growth, as has been demonstrated (44). Third, detergents

can increase mesophase domain size, thereby supporting the production of larger crystals.

The Sponge Phase

In the presence of certain precipitant components such as Jeffamine, butanediol, and polyethylene glycol (PEG), the aqueous channels of the cubic phase enlarge and its lattice parameter, as monitored by SAXS, rises (19). Such additives are proposed to act by interacting with the lipid head group and to increase the cross-sectional area per molecule at the aqueous/apolar interface of the mesophase. In parallel, the bilayer of the emerging, so-called sponge phase becomes more flexible and the regular 3D periodicity of the original cubic phase is lost. Despite dramatic microstructural and textural changes, the mesophase remains bicontinuous and this is key to the ability of the sponge phase to support *in meso* crystallogenesis. With the enlarged aqueous channels and the less highly curved lipid bilayer, the prospect is that the sponge phase will offer advantages for the crystallization of proteins with large cross-sectional areas in the membrane plane and/or with large extramembranal domains. Because it is considerably less viscous than the cubic phase and because it flows, specialty tools may not be required for its handling. The fact that the sponge phase works with several membrane proteins (19, 24, 25, 49, 57) presents a rational case for including mesophase-swelling additives in screens for *in meso* crystallogenesis. Their use will contribute to broadening the range of membrane proteins that yield structures.

VALIDATING THE PROPOSAL

Homogenous Reconstitution

The *in meso* method begins with what is assumed to be a uniform reconstitution of the protein into the lipid bilayer of the cubic phase (**Figure 1**). The protein is combined with MAG typically in a ratio that should produce the cubic phase provided the detergent concentration of the protein solution is not too high. When this was done with BtuB (25), OpcA (22), and gramicidin D (35) for example, the cubic-Pn3m phase was produced as evidenced by SAXS. The lattice parameter of the cubic phase was similar to the value observed with control, protein-free samples. Upon addition of precipitant solution to trigger nucleation and crystal growth in the case of BtuB, the cubic phase swelled and formed the sponge phase (25). It is from this swollen, bicontinuous mesophase that the crystals of BtuB were harvested. These data are consistent with the view that the protein reconstitutes into the cubic phase in a way that is homogenous and that does not perturb the original mesophase.

That reconstitution is uniform throughout the cubic mesophase is obvious when working with highly colored proteins such as bR (33), the photosynthetic reaction center (31), and the light-harvesting complex II (LHII) (M. Caffrey, unpublished observations). After the lipid and protein solution is homogenized, an optically clear mesophase is produced that, to the naked eye, is uniformly colored.

The electronic fluorescence properties of the gramicidin molecule directly reconstituted into the lipid bilayer of the cubic phase suggest that it resides in an apolar environment (35). Thus, the yield and wavelength of maximum intensity of the fluorescence from the tryptophans in gramicidin were increased and blueshifted, respectively, compared with tryptophan in aqueous solution.

Quenching of intrinsic tryptophan fluorescence by a lipid with a dibrominated acyl chain (bromo-MAG) has been used to demonstrate reconstitution of BtuB (25), OpcA (22), and gramicidin (35) in the lipidic mesophase (**Figure 3c**). Respectively, these have 13, 4, and 4 tryptophans, of which 12, 3, and 4 should be accessible to quenching by bromo-MAG, provided the target is reconstituted into the cubic phase bilayer. The extent of quenching observed (>80%) is consistent with this expectation (**Figure 3c**) and supports the view that the targets are reconstituted prior to crystallization.

Some additional evidence in support of a bilayer location derives from the fact that the quenching behavior of gramicidin was sensitive to acyl chain identity of the accompanying, nonquenching MAG (35). Because the chains are confined to the bilayer interior, some property(ies) of the bilayer itself changes with the different MAGs. This is sensed by gramicidin presumably only when it is associated with that same lipid bilayer. In so doing, it responds differently to the quenching effect of the brominated lipid, which has a distinct character imprinted on it by the different nonquenching MAGs. One of the properties that changes with MAG identity is bilayer thickness. This, in turn, defines the relative positions of the apolar/polar interface across the membrane that will affect the fluorescence behavior of the tryptophans that sample such an environment.

A final piece of evidence for bilayer location hinges on the logic that gramicidin is so apolar that it is unfavorable for it to reside anywhere else within the confines of the mesophase. SAXS data show that the cubic phase can accommodate gramicidin up to a point. Beyond that limit, it triggers a transformation to the inverted hexagonal phase (35). This presumably reflects a change in the energetics associated with mismatch between the peptide and the lipid/water interface, which is a result of a gramicidin that is bilayer bound.

Conformation

Spectroscopic measurements were made to examine the conformational state of three membrane proteins and gramicidin reconstituted in the cubic phase. The UV-visible spectra of the BtuB (25) and OpcA (22) preparations in micellar form and *in meso* were, within experimental error, the same regardless of the protein dispersion state. In the case of LHII a similar observation held, with the exception of a slight change in bacteriochlorophyll absorption in the 780- to 900-nm range (19). Circular dichroism spectra showed that the gross secondary structure of both BtuB (25) and OpcA (22) was insensitive to whether they were in a micellar or a bilayer environment. Together these data suggest that cubic phase reconstitution does not dramatically

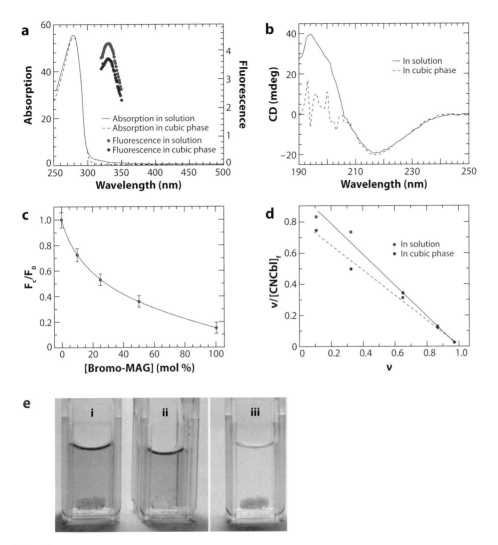

Figure 3

Spectrophotometric and visual properties of BtuB in detergent solution and in the cubic phase. (*a*) UV-visible absorption and fluorescence emission spectra of apo-BtuB in detergent solution and in the cubic-Pn3m phase. (*b*) Circular dichroism spectra of apo-BtuB in detergent solution and in the cubic phase. The region of the cubic phase spectrum below ~208 nm is not reliable because of strong background absorption by the lipid. (*c*) Quenching of apo-BtuB intrinsic fluorescence by bromo-MAG in the cubic phase of hydrated monoolein. Fluorescence intensity (F_c) was normalized to the value recorded in the absence of quenching lipid (F_0). (*d*) Scatchard analysis for the binding of CNCbl to apo-BtuB in micellar solution (*blue circles*) and in the cubic phase of hydrated monoolein (*red circles*). The corresponding K_d values are 1.02 and 1.24 nM. (*e*) Photograph of a bolus of cubic phase with (*i* and *iii*) and without (*ii*) reconstituted apo-BtuB equilibrated for 6 days at 20°C with a solution of 67 μM CNCbl (*pink*). In (*iii*), the bathing CNCbl solution was replaced with CNCbl-free buffer just before the photograph was taken to make the labeling of the bolus more obvious. The bolus of cubic phase can be seen as an elliptically shaped object at the bottom of each cuvette. Taken from Reference 25.

alter the conformation of crystallogenesis targets consistent with the hypothesis.

Functional Activity

It is assumed that proteins reconstituted prior to crystallization retain functionality *in meso*. In the case of BtuB, this was examined by measuring substrate (cyanocobalamin, CNCbl) binding to the protein reconstituted into the cubic phase (25). Protein-free control samples exhibited no binding, whereas test *in meso* BtuB-containing samples showed convincing evidence of substrate uptake (**Figure 3d,e**). Binding was shown by quenching of intrinsic fluorescence of aromatic residues by CNCbl and by direct ligand binding (**Figure 3e**) to be tight with a K_d value of ~1 nM (**Figure 3d**). Similar K_d values have been reported for the native membrane-bound and micellarized form of the protein. Sialic acid binding to OpcA was identical *in meso* and in detergent solution (22). Taken together, these data support the view that the protein reconstitutes into the bilayer of the cubic phase in an active form prior to *in meso* crystallization.

Transport

Crystallization, regardless of how it happens, requires transport (i.e., the movement of the crystallant from the bulk medium up and into the face of the crystal). If transport is impeded or does not happen, crystallogenesis will suffer. *In meso*, mobility must take place both in the bilayer and in the aqueous channels of the mesophase. When working with colored proteins, such as bR, mobility under *in meso* conditions is noticeable with a simple light microscope. As crystals form, flecks of dark purple appear surrounded by a zone of colorless mesophase, whose extent away from the crystal expands with time and crystal growth. This is evidence that the protein is moving from the mesophase reservoir, presumably in the lipid bilayer, to the crystal. Additional and more quantitative evidence that mobility in the bilayer is required for *in meso* crystallogenesis comes from

recent fluorescence recovery after photobleaching measurements performed with labeled bR and a GPCR chimera (21). In this case, diffusion and a high fractional recovery of fluorescence in the bleached area correlated with crystallizability.

The diffusion of gramicidin, LHII, and Sudan Red, a highly lipophilic dye, has been used to characterize the transport properties of the cubic phase (26). To this end, a bolus of diffusant-loaded cubic phase, the source, was placed in direct contact with a bolus of diffusant-free cubic phase, the sink, at a sharp interface. Transfer of diffusant between the two boluses and subsequent diffusion in the sink were monitored by UV-visible spectroscopy and were shown to occur. In the case of the Sudan Red transport could be seen with the naked eye (**Figure 4**). These data show that the cubic phase supports transport and, because at least two of these diffusants are highly apolar, that diffusion is most likely taking place within the lipid bilayer. The results also highlight the fusogenic nature of the cubic phase, suggesting that the bilayer of one bolus can become continuous with the bilayer of the other bolus with which it makes contact. This means that the bilayer composition of a given bolus can, within limits, be adjusted at will, which has implications for seeding, cocrystallization, and complex formation by a stepwise approach to *in meso* crystallogenesis.

In the context of *in meso* crystallogenesis the cubic phase is viewed as a porous molecular sponge consisting of two interpenetrating nanochannels filled with an aqueous medium and coated by a common lipid bilayer. In the preceding paragraphs it was shown that proteins move within the membrane—a requirement for crystallogenesis. Mobility within the aqueous channels is also a prerequisite for crystal growth, at the very least to enable precipitant components to access the interior of the bolus and to trigger nucleation and crystal growth.

Several studies have been performed that support such transport, and for reasons of experimental simplicity, most were done by following release of water-soluble diffusants from a bolus of preloaded cubic phase (27, 28, 36). The studies show that the diffusion rate was dependent on the size of diffusant molecules as expected, given that the channels within the mesophase have a diameter of approximately 50 Å. Remarkably, transport was observed with apo-ferritin, whose size (∼100 Å diameter) far exceeds that of the aqueous channel, suggesting a molecular breathing or peristalsis type of facilitated diffusion (27). Exquisite control over the rate of movement within the aqueous channels was achieved by adjusting (*a*) channel dimensions, (*b*) the partitioning of the diffusant on or into the lipid bilayer, (*c*) the electrostatic interaction strength, and (*d*) histidine-tag displacement. Thus, although the mesophase channels are small and confined—just 15 water molecules wide—they enable simple and well-behaved transport.

In support of this, ultrafast hydration dynamics studies revealed that the channels include a water core with bulk-like dynamics and orientational relaxation properties consistent with transport (32). In contrast, the water at the aqueous/bilayer interface is dynamically rigid. The latter is surrounded by a hydrogen-bonded network of water with dynamic relaxations intermediate between those of the interfacial and core water. Taken together these data support the view that the cubic phase behaves as a nanoporous molecular sponge into and out of which water-soluble substances of a wide range of sizes can diffuse, which is integral to the *in meso* crystallization model.

Lamellar Phase Involvement

The hypothesis posits that the protein migrates from the bulk mesophase reservoir to the face of the crystal by way of a lamellar conduit (7–9). Using a submicrometer-sized X-ray beam, Cherezov & Caffrey (17) investigated the interface between a growing membrane protein crystal and the bulk cubic phase with micrometer spatial resolution. Characteristic diffraction from the lamellar phase was observed at the interface, which supports the proposal that the protein uses a lamellar portal on its way from

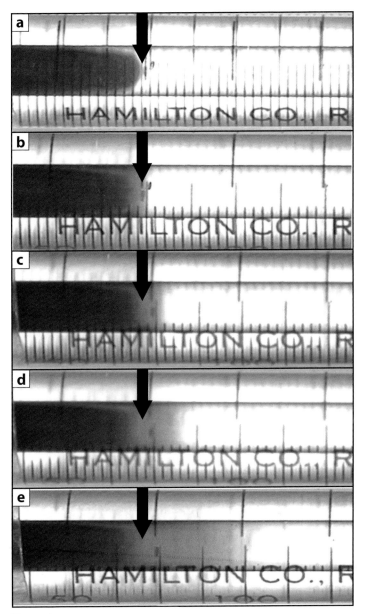

Figure 4

Evidence for mobility within the bilayer of the cubic phase and for the fusogenic nature of the mesophase. Two boluses of cubic phase were brought into contact with one another (*black arrow*) inside the barrel of a 0.1 mL Hamilton syringe. One bolus was doped with the lipophilic dye, Sudan Red (*left of arrow*). The other bolus (*right of arrow*) was prepared without dye. Individual panels show progress in the diffusion of the dye from the donor (*left*) to the acceptor bolus (*right*) following contact at elapsed times of (*a*) 0, (*b*) 0.1, (*c*) 0.2, (*d*) 1.1, and (*e*) 5.2 days.

the bulk mesophase up and into the face of the crystal.

There are two reports based on microscopy that address the *in meso* growth of membrane protein crystals by way of a lamellar conduit. The first of these involved freeze-fracture electron microscopic (EM) examinations of acetylcholine receptor-α-bungarotoxin complex microcrystals grown from within a lipid mesophase (49). EM images showed highly ordered domains of the complex next to lipid lamellae, consistent with the working hypothesis. In the second study, atomic force microscopy was used to demonstrate the existence of a lamellar conduit between the bR crystal and the bulk cubic phase (53).

In meso crystallization is predicted to produce Type I crystals (**Figures 1** and **5**). Here, proteins are arranged in planar sheets that stack one atop the other. Protein-protein interactions within the plane of a given layer can be extensive in Type I crystals. Type II crystals (42) are commonly encountered when grown by *in surfo* methods. In this case, a torus of detergent coats the protein where it contacted the apolar region of the biomembrane from which it came originally. As a result, direct contact between the apolar parts of the protein is much less likely in Type II crystals, and packing density and diffracting power can be low. To date, all membrane proteins that have been crystallized by the *in meso* method have given rise to Type I crystals (54) (**Figure 5**) (**http://www.mpdb.ul.ie/**), consistent with the

hypothesis. However, nonlamellar-type packing could be observed at some point with *in meso*–grown crystals. This might come about by a polymorphic transition in the solid state (9). Presumably, the Type I crystal will form first to be replaced by a more stable polymorph in which the proteins are no longer arranged in distinct lamellae.

STRUCTURES AND CRYSTALS

The Membrane Protein Data Bank (MPDB; **http://www.mpdb.ul.ie/**) (54) is a convenient online resource for perusing and analyzing statistics on published membrane protein structures. It has been used to establish that of over 500 published X-ray structures for integral membrane proteins about 10% (54 structures) are based on crystals grown *in meso*. The method made its first appearance in 1997 with the structure determination of bR (51). A decade and some eight distinctly different membrane protein types later came its most recent successes, the human GPCR chimeras of the β_2-adrenergic and the A_{2A}-adenosine class. The method would therefore appear to be general, having been used to solve structures of prokaryotic and eukaryotic proteins; proteins that are monomeric, homo- and heteromultimeric, chromophore-containing, and chromophore-free; and α-helical and β-barrel proteins (**Figure 5**). To date, all structures have been solved by molecular replacement. The highest published resolution

---

Figure 5

In meso crystal packing arrangement and molecular structures of membrane proteins. The packing arrangement is shown in columns 1–3 within each panel. Expanded views of individual proteins or oligomers are shown in columns 4 and 5. The views in columns 1 and 5 represent a projection along the stacking axis. The views in columns 2–4 are from within the plane of the stacked lamellae along the two other unit cell axes. In columns 1–3 black outlines are projections of the unit cell. Images in column 4 are from the Orientations of Proteins in Membranes (OPM) database (**http://opm.phar.umich.edu/**), in which the red and blue horizontal lines define the hydrophobic thickness of the protein. In column 4 (*d, e*) the horizontal red and blue lines correspond to the extracellular and periplasmic surfaces, respectively. In column 4 (*a–c, f–i*) the red and blue lines correspond to the extracellular and intracellular surfaces, respectively. Proteins are identified by name, source organism, resolution, and Protein Data Bank (PDB; **http://www.rcsb.org/pdb/home/home.do**) accession number. (In the pdf version of this review, links to the PDB and the OPM are provided by clicking on the accession number and on the images in column 4, respectively.) Currently, *in meso* records in the Membrane Protein Data Bank (**http://www.mpdb.ul.ie/**) number 33 for bR, 5 for sensory rhodopsin, 4 for the photosynthetic reaction center, 3 for the sensory rhodopsin/transducer complex, 3 for halorhodopsin, 3 for GPCR-T4 chimera, and 1 each for BtuB, OpcA, and LHII. The examples included in this figure represent the highest resolution available for each of the nine protein types (54).

a   Bacteriorhodopsin (*H. salinarum*), 1.43 Å, <u>1M0K</u>

b   Halorhodopsin (*H. salinarum*), 1.70 Å, <u>2JAF</u>

c   Sensory rhodopsin II/transducer (*N. pharaonis*), 1.93 Å, <u>1H2S</u>

d   Adhesin OpcA (*N. meningitidis*), 1.95 Å, <u>2VDF</u>

e   Cobalamin transporter (*E. coli*), 1.95 Å, <u>2GUF</u>

f   Light harvesting complex II (*Rps. acidophila*), 1.95 Å, <u>2FKW</u>

g   Sensory rhodopsin II (*Anabaena sp.*), 2.00 Å, <u>1XIO</u>

h   Photosynthetic reaction center (*R. sphaeroides*), 2.20 Å, <u>2GNU</u>

i   β₂-adrenergic GPCR-lysozyme (*H. sapiens*-phage), 2.40 Å, <u>2RH1</u>

attributable to the method is 1.43 Å for the K and M1 intermediates of bR.

The structures that have resulted from the *in meso* method have made important contributions to our understanding of how membrane proteins function at the molecular level. Some of the highlights include the structural basis for the photocycle and proton pumping mechanism in bR and related photosensitive proteins (34). By solving the structure of a sensory rhodopsin II-transducer binding domain complex, Gordeliy et al. (29) revealed key molecular details of how signal communication happens in phototaxis and chemotaxis. The structures of LHII (19), BtuB (25), and OpcA (22) highlighted the dense packing and high diffracting power of crystals grown *in meso*. New pigment (LHII), surfactant (LHII), and lipid (BtuB) binding features also emerged from these studies. In the case of OpcA, the putative sialic acid binding site in the core of this β-barrel was revealed (22). Molecular dynamics based on *in meso* and *in surfo* structures of OpcA made a convincing case that mobility in the loops extending from the barrel plays a role in securing by induced fit the corresponding host proteoglycan as a prelude to infection that leads to meningitis (22, 37). These studies highlight the benefit of using different crystal forms with which to interpret crystallographic structure data. Most recently, the high-resolution structure of the first two nonrhodopsin GPCRs was determined using crystals grown *in meso* (24, 31a). Although great effort, which included mutating, truncating, ligand (and possibly sterol) binding, and making a chimera of the receptor, was needed to stabilize the protein for crystallization, the end result was the first close-up view of how these representatives of an important group of signal-transducing proteins operate molecularly. The work has been extended to show that specific cholesterol binding is likely to play a role in stabilizing the receptor and that sterol binding may be exploitable pharmacologically (30).

In addition to solved structures, the method has been used to grow crystals of the following proteins: cytochrome *caa₃* oxidase from *Thermus thermophilus* (56), cytochrome oxidase from *Rhodobacter sphaeroides* (O. Slattery, S. Ferguson-Miller & M. Caffrey, unpublished observations), two putative β-barrel proteins from the outer membrane of *Pseudomonas aeruginosa* (D. Li, J. Tan & M. Caffrey, unpublished observations), LHII from *R. sphaeroides* (57), photosynthetic reaction center from *Blastochloris viridis* (57), aquaporin from *Plasmodium falciparum* and *Spinacia oleracea* (57), Complex II (57) and a short integral membrane peptide from *Bacillus subtilis* (N. Höfer & M. Caffrey, unpublished observations), and the photosynthetic RC-LH1 core complex from *Bacillus viridis* (57). Together with the solved structures above, these data attest to the wide-ranging applicability of and potential for the *in meso* method.

CONCLUDING REMARKS

In meso crystallogenesis is a relatively new and incompletely tested method for crystallizing membrane proteins. It makes use of a bicontinuous lipid mesophase and it works with a wide range of membrane protein types. The method offers the advantage that the protein is removed from the potentially harmful environment of a detergent micelle and is placed into a more familiar and hospitable lipid bilayer in which crystallogenesis takes place. Crystallization is proposed to happen as a result of the development in the porous mesophase of a series of overlapping concentration gradients of precipitant components. These trigger local bilayer composition and microstructure transitions that involve the lamellar phase. Proteins partition preferentially into the latter and, under the right conditions, nucleate and produce crystals. Growth of the crystal is fed from a bicontinuous reservoir of protein to which it is tethered by a lamellar portal. Screening for crystallization by the *in meso* method involves the full range of conditions examined by more traditional methods. In addition, optimizations that include different lipids and detergents are likely to play important roles. Although the method is proven, there remain many challenges, not the least of which involve making the method a routine option for every grower of membrane protein crystals. Harvesting is still not straightforward, nor is the handling of these typically small crystals for optimum diffraction data collection, especially under cryo-conditions. The fact that two representatives of a high profile receptor family have yielded to the method means that it will receive more attention. With this attention will come additional successes and an improvement in one of the more challenging steps along the tortuous and bumpy road that has variously been referred to as extending from "bug to drug."

SUMMARY POINTS

1. The *in meso* method for crystallizing membrane proteins is general, having been used to solve structures of prokaryotic and eukaryotic proteins; proteins that are monomeric, homo- and heteromultimeric, chromophore-containing, and chromophore-free; and α-helical and β-barrel proteins. It accounts for ~10% of the published X-ray structures of integral membrane proteins.

2. The method makes use of a bicontinuous lipid mesophase. It offers the advantage that the protein is removed from the potentially harmful environment of a detergent micelle and is placed into a more familiar, hospitable, stabilizing, and organizing lipid bilayer in which crystallogenesis takes place.

3. A mechanism for how the method works at the molecular level has been proposed, and considerable experimental evidence is available in support of the model. Overlapping concentration gradients of precipitant components that develop in the porous mesophase are proposed to trigger local bilayer composition and microstructure transitions that involve the lamellar phase. Proteins partition preferentially into and stabilize the latter and, under the right conditions, nucleate and produce crystals. Growth of the crystal is fed from a bicontinuous reservoir of protein to which it is tethered by a lamellar portal.

4. With this understanding of how the method works at the molecular level a more rational approach to membrane protein crystallization is now possible and is being implemented.

5. The method itself has presented some formidable experimental challenges that derive from the viscous and sticky nature of the mesophase into which the protein is reconstituted as a prelude to crystallization. These challenges have been overcome, and the method can now be used for robot-based high-throughput screening using extraordinarily small quantities of protein and lipid.

6. Future challenges for the *in meso* method include making it a user-friendly, routine approach for crystallizing membrane proteins, developing convenient high-throughput assays of functionality and stability of proteins *in meso*, extending it to large and small proteins and to complexes, and providing reliable screens that include lipids, detergents, and spongifying agents.

FUTURE ISSUES

1. The *in meso* method works with a wide range of membrane protein types. The next challenge will be to extend the methodology to large proteins and complexes and to small proteins and peptides.

2. A robust, reliable, and possibly automatic method for harvesting crystals from the cubic and sponge phases in glass sandwich plates is needed.

3. It is phase science that dictates *in meso* crystallogenesis. Accordingly, a major effort must be devoted to establishing, understanding, and exploiting the phase behavior of the multicomponent lipid-based systems that work, and indeed those that do not work, in *in meso* crystallization.

4. For the *in meso* approach to become routine and standard in crystal-growing laboratories, the community must become familiar and comfortable with the theory behind and the phase science underlying the method. Further, the materials and the tools with which to perform *in meso* crystallogenesis must be simple and made more generally and inexpensively available.

5. Precipitants and screen additives (including the full range of detergents) that are compatible with the *in meso* method must be identified and a minimal set of conditions developed and made commercially available with which to perform reliable initial screens on new protein targets. With time, screens specific to different protein types, such as β-barrel and α-helical proteins, should emerge.

6. Methods must be developed for convenient, inexpensive, and high-throughput pre-screening of *in meso* crystallizability, for quantifying in high-throughput fashion the functional activity and stability of membrane proteins reconstituted *in meso* as a prelude to crystallogenesis, for growing bigger crystals *in meso* (likely to include the use of detergents and other additives), for convenient *in meso* seeding and heavy-atom labeling, and for automatically finding and centering micrometer-sized crystals in opaque cryo-cooled mesophase samples at synchrotron facilities.

7. The utility of the *in meso* method for crystallizing cell-free expressed as well as relatively impure membrane protein preparations and inclusion bodies must be evaluated.

8. The use of different lipids to create the hosting mesophase and to serve as additives must become a routine screening feature of *in meso* crystallogenesis. This must be supplemented by a lipid synthesis, purification, and characterization program. With time, a better understanding of why and how particular lipids and other components do or do not work should emerge, leading to a more rational approach to *in meso* crystallogenesis.

DISCLOSURE STATEMENT

The author is not aware of any biases that might be perceived as affecting the objectivity of this review.

ACKNOWLEDGMENTS

There are many who contributed to this work and most are from my own group, both past and present members. To all I extend my warmest thanks and appreciation. This work was supported in part by grants from Science Foundation Ireland (07/IN.1/B1836), the National Institutes of Health (GM75915), and the National Science Foundation (IIS-0308078).

LITERATURE CITED

1. Ai X, Caffrey M. 2000. Membrane protein crystallization in lipidic mesophases. Detergent effects. *Biophys. J.* 79:394–405
2. Briggs J. 1994. *The phase behavior of hydrated monoacylglycerols and the design of an X-ray compatible scanning calorimeter.* PhD thesis. The Ohio State Univ. 393 pp.
3. Briggs J, Caffrey M. 1994. The temperature-composition phase diagram and mesophase structure characterization of monopentadecenoin in water. *Biophys. J.* 67:1594–602
4. Briggs J, Caffrey M. 1994. The temperature-composition phase diagram of mono-myristolein in water: equilibrium and metastability aspects. *Biophys. J.* 66:573–87
5. Briggs J, Chung H, Caffrey M. 1996. The temperature-composition phase diagram and mesophase structure characterization of the monoolein/water system. *J. Phys. II France* 6:723–51
6. Caffrey M. 1987. Kinetics and mechanism of transitions involving the lamellar, cubic, inverted hexagonal and isotropic phase of hydrated monoacylglycerides. *Biochemistry* 26:6349–63
7. Caffrey M. 2000. A lipid's eye view of membrane protein crystallization in mesophases. *Curr. Opin. Struct. Biol.* 10:486–97
8. Caffrey M. 2003. Membrane protein crystallization. *J. Struct. Biol.* 142:108–32
9. **Caffrey M. 2009. On the mechanism of membrane protein crystallization in lipidic mesophases. *Crystal Growth Des.* 8:4244–54**
10. **Caffrey M, Cherezov V. 2009. Crystallizing membrane proteins using lipidic mesophases. *Nat. Prot.* In press**
11. Caffrey M, Feigenson GW. 1981. Fluorescence quenching in model membranes: the relationship between Ca^{2+}-ATPase enzyme activity and the affinity of the protein for phosphatidylcholines with different acyl chain characteristics. *Biochemistry* 20:1949–61
12. Caffrey M, Infante J, Kinsella JE. 1975. Isoenzymes of an acyltransferase from rabbit mammary gland: evidence from biphasic substrate saturation kinetics. *FEBS Lett.* 52:116–20
13. **Cheng A, Hummel B, Qiu H, Caffrey M. 1998. A simple mechanical mixer for small viscous lipid-containing samples. *Chem. Phys. Lipids* 95:11–21**
14. Cherezov V, Caffrey M. 2003. Nano-volume plates with excellent optical properties for fast, inexpensive crystallization screening of membrane proteins. *J. Appl. Crystallogr.* 36:1372–77
15. Cherezov V, Caffrey M. 2005. A simple and inexpensive nanoliter-volume dispenser for highly viscous materials used in membrane protein crystallization. *J. Appl. Crystallogr.* 38:398–400
16. Cherezov V, Caffrey M. 2006. Picoliter-scale crystallization of membrane proteins. *J. Appl. Crystallogr.* 39:604–9
17. **Cherezov V, Caffrey M. 2007. Membrane protein crystallization in lipidic mesophases. A mechanism study using X-ray microdiffraction. *Faraday Disc.* 136:188–205**
18. Cherezov V, Clogston J, Misquitta Y, Abdel Gawad W, Caffrey M. 2002. Membrane protein crystallization *in meso*. Lipid type-tailoring of the cubic phase. *Biophys. J.* 83:3393–407
19. **Cherezov V, Clogston J, Papiz M, Caffrey M. 2006. Room to move. Crystallizing membrane proteins in swollen lipidic mesophases. *J. Mol. Biol.* 357:1605–18**
20. Cherezov V, Fersi H, Caffrey M. 2001. Crystallization screens: compatibility with the lipidic cubic phase for *in meso* crystallization of membrane proteins. *Biophys. J.* 81:225–42
21. Cherezov V, Liu JL, Griffith M, Hanson MA, Stevens RC. 2009. LCP-FRAP assay for prescreening membrane proteins for *in meso* crystallization. *Crystal Growth Des.* 8:4307–15
22. **Cherezov V, Liu W, Derrick J, Luan B, Aksimentiev A, et al. 2008. *In meso* crystal structure and docking simulations suggest an alternative proteoglycan binding site in the OpcA outer membrane adhesin. *Proteins* 71:24–34**
23. **Cherezov V, Peddi A, Muthusubramaniam L, Zheng YF, Caffrey M. 2004. A robotic system for crystallizing membrane and soluble proteins in lipidic mesophases. *Acta Crystallogr. D* 60:1795–807**
24. **Cherezov V, Rosenbaum DM, Hanson MA, Rasmussen SGF, Thian FS, et al. 2007. High-resolution crystal structure of an engineered human β2-adrenergic G protein coupled receptor. *Science* 318:1258–65**

9. By deciphering the underlying molecular mechanism, clearer insights emerge for practicing a rational approach to crystallizing membrane protein *in meso*.

10. Recipes and step-by-step instructions for performing *in meso* crystallization of membrane proteins.

13. How to build your own lipid-mixing, protein-reconstitution device.

17. SAXS evidence for a lamellar conduit between the crystal and the bulk cubic phase.

19. *In meso* crystallization in the sponge phase method is introduced.

22. *In meso* crystal structure reveals a new binding site conformation in an adhesin protein.

23. How to build and operate an *in meso* robot.

24. First engineered human GPCR chimera to have its structure solved using *in meso*–grown crystals.

25. First β-barrel and noncolored protein to have its structure solved using *in meso*–grown crystals.

25. **Cherezov V, Yamashita E, Liu W, Zhalnina M, Cramer WA, Caffrey M. 2006.** *In meso* **structure of the cobalamin transporter, BtuB, at 1.95 Å resolution.** *J. Mol. Biol.* **364:716–34**

26. Clogston J. 2005. *Applications of the lipidic cubic phase: from controlled release and uptake to in meso crystallization of membrane proteins.* PhD thesis. The Ohio State Univ. 352 pp.

27. Clogston J, Caffrey M. 2005. Controlling release from the cubic phase. Amino acids, peptides, proteins and nucleic acids. *J. Control. Release* 107:97–111

28. Clogston J, Graciun G, Hart DJ, Caffrey M. 2005. Controlling release from the lipidic cubic phase by selective alkylation. *J. Control. Release* 102:441–61

29. Gordeliy VI, Labahn J, Moukhametzianov R, Efremov R, Granzin J, et al. 2002. Molecular basis of transmembrane signalling by sensory rhodopsin II-transducer complex. *Nature* 419:484–87

30. Hanson MA, Cherezov V, Griffith MT, Roth CB, Jaakola V-P, et al. 2008. A specific cholesterol binding site is established by the 2.8 Å structure of the human β_2-adrenergic receptor. *Structure* 16:897–905

31. Hochkoeppler A, Landau EM, Venturoli G, Zannoni D, Feick R, Luisi PL. 1995. Photochemistry of a photosynthetic reaction center immobilized in lipidic cubic phases. *Biotechnol. Bioeng.* 46:93–98

31a. Jaakola V-P, Griffith MT, Hanson MA, Cherezov V, Chien EYT, et al. The 2.6 Å crystal structure of a human A_{2A} adenosine receptor bound to an antagonist. *Science* 322:1211–17

32. Kim J, Lu W, Qiu W, Wang L, Caffrey M, Zhong D. 2006. Ultrafast hydration dynamics in the lipidic cubic phase: discrete water structures in nanochannels. *J. Phys. Chem. B* 110:21994–2000

33. First report that crystals of a membrane protein grow *in meso*.

33. **Landau EM, Rosenbusch JP. 1996. Lipidic cubic phases: a novel concept for the crystallization of membrane proteins.** *Proc. Natl. Acad. Sci. USA* **93:14532–35**

34. Lanyi JK. 2004. X-ray diffraction of bacteriorhodopsin photocycle intermediates. *Mol. Membr. Biol.* 21:143–50

35. Liu W, Caffrey M. 2005. Gramicidin structure and disposition in highly curved membranes. *J. Struct. Biol.* 150:23–40

36. Liu W, Caffrey M. 2006. Interactions of tryptophan, tryptophan peptides and tryptophan alkyl esters at curved membrane interfaces. *Biochemistry* 45:11713–26

37. Luan B, Caffrey M, Aksimentiev A. 2007. Structural refinement of the OpcA adhesin using molecular dynamics. *Biophys. J.* 93:3058–69

38. Luecke H, Schobert B, Richter H-T, Cartailler J-P, Lanyi JK. 1999. Structure of bacteriorhodopsin at 1.55 angstrom resolution. *J. Mol. Biol.* 291:899–911

39. Lunde CS, Rouhani S, Remis JP, Ruzin SE, Ernst JA, Glaeser RM. 2005. UV microscopy at 280 nm is effective in screening for the growth of protein microcrystals. *J. Appl. Crystallogr.* 38:1031–34

40. Luzzati V. 1997. Biological significance of lipid polymorphism. The cubic phases. *Curr. Opin. Struct. Biol.* 5:661–68

41. McPherson A. 1999. *Crystallization of Biological Macromolecules.* Cold Spring Harbor, NY: Cold Spring Harbor Lab. Press

42. Michel H. 1983. Crystallization of membrane proteins. *Trends Biochem. Sci.* 8:56–59

43. Misquitta LV, Misquitta Y, Cherezov V, Slattery O, Mohan JM, et al. 2004. Membrane protein crystallization in lipidic mesophases with tailored bilayers. *Structure* 12:2113–24

44. Misquitta Y, Caffrey M. 2003. Detergents destabilize the cubic phase of monoolein. Implications for membrane protein crystallization. *Biophys. J.* 85:3084–96

45. Misquitta Y, Cherezov V, Havas F, Patterson S, Mohan J, et al. 2004. Rational design of lipid for membrane protein crystallization. *J. Struct. Biol.* 148:169–75

46. Munro S. 1998. Localization of proteins to the Golgi apparatus. *Trends Cell Biol.* 8:11–15

47. Nollert P, Landau EM. 1998. Enzymic release of crystals from lipidic cubic phases. *Biochem. Soc. Trans.* 26:709–13

48. Nollert P, Qiu H, Caffrey M, Rosenbusch JP, Landau EM. 2001. Molecular mechanism for the crystallization of bacteriorhodopsin in lipidic cubic phases. *FEBS Lett.* 504:179–86

49. Paas Y, Cartaud J, Recouvreur M, Grailhe R, Dufresne V, et al. 2003. Electron microscopic evidence for nucleation and growth of 3D acetylcholine receptor microcrystals in structured lipid-detergent matrices. *Proc. Natl. Acad. Sci. USA* 100:11309–14

50. Patton JS, Carey MC. 1979. Watching fat digestion. *Science* 204:145–48

51. Pebay-Peyroula E, Rummel G, Rosenbusch JP, Landau EM. 1997. X-ray structure of bacteriorhodopsin at 2.5 angstroms from microcrystals grown in lipidic cubic phases. *Science* 277:1607–8

52. Qiu H, Caffrey M. 2000. Phase diagram of the monoolein/water system: metastability and equilibrium aspects. *Biomaterials* 21:223–34

53. Qutub Y, Reviakine I, Maxwell C, Navarro J, Landau EM, Vekilov PG. 2004. Crystallization of transmembrane proteins *in cubo*: mechanisms of crystal growth and defect formation. *J. Mol. Biol.* 343:1243–54

54. Raman P, Cherezov V, Caffrey M. 2006. The Membrane Protein Data Bank. *Cell. Mol. Life Sci.* 63:36–51

55. Siegel DP, Cherezov V, Greathouse DV, Koeppe RE II, Killian JA, Caffrey M. 2006. Transmembrane peptides stabilize inverted cubic phases in a biphasic length-dependent manner: implications for protein-induced membrane fusion. *Biophys. J.* 90:200–11

56. Slattery O, Caffrey M, Soulimane T. 2008. Crystallization and preliminary X-ray diffraction analysis of *caa₃*-cytochrome *c* oxidase from *Thermus thermophilus*. *Biochim. Biophys. Acta* 1777:S74

57. Wöhri AB, Johansson LC, Wadsten-Hindrichsen P, Wahlgren WY, Fischer G, et al. 2008. A lipidic-sponge phase screen for membrane protein crystallization. *Structure* 16:1003–9

RELATED RESOURCES

http://blanco.biomol.uci.edu/Membrane_Proteins_xtal.html
Lipid Data Bank (**http://www.caffreylab.ul.ie/**)
LIPID MAPS (**http://www.lipidmaps.org/**)
Membrane Protein Data Bank (**http://www.mpdb.ul.ie/**)
Orientations of Proteins in Membranes (**http://opm.phar.umich.edu/**)

Advances in Imaging Secondary Ion Mass Spectrometry for Biological Samples

Steven G. Boxer,[1] Mary L. Kraft,[2] and Peter K. Weber[3]

[1] Department of Chemistry, Stanford University, Stanford, California 94305; email: sboxer@stanford.edu

[2] Department of Chemical and Biomolecular Engineering, University of Illinois, Urbana-Champaign, Urbana, Illinois 61801; email: mlkraft@illinois.edu

[3] Glenn T. Seaborg Institute, Lawrence Livermore National Laboratory, Livermore, California 94551; email: weber21@llnl.gov

Annu. Rev. Biophys. 2009. 38:53–74

First published online as a Review in Advance on December 16, 2008

The *Annual Review of Biophysics* is online at biophys.annualreviews.org

This article's doi: 10.1146/annurev.biophys.050708.133634

Key Words

membrane organization, chemical composition imaging, NanoSIMS, ToF-SIMS, dynamic SIMS

Abstract

Imaging mass spectrometry combines the power of mass spectrometry to identify complex molecules based on mass with sample imaging. Recent advances in secondary ion mass spectrometry have improved sensitivity and spatial resolution, so that these methods have the potential to bridge between high-resolution structures obtained by X-ray crystallography and cyro-electron microscopy and ultrastructure visualized by conventional light microscopy. Following background information on the method and instrumentation, we address the key issue of sample preparation. Because mass spectrometry is performed in high vacuum, it is essential to preserve the lateral organization of the sample while removing bulk water, and this has been a major barrier for applications to biological systems. Recent applications of imaging mass spectrometry to cell biology, microbial communities, and biosynthetic pathways are summarized briefly, and studies of biological membrane organization are described in greater depth.

Contents

INTRODUCTION

Structural biology has seen huge advances during the past 20 years due to the confluence of efficient methods for protein overexpression and purification, synchrotron X-ray sources, and array detectors. As a result, we now have the three-dimensional structures of many proteins, including membrane proteins and large assemblies such as the ribosome. These structures provide the basic level of organization of biological systems, often at atomic resolution, on the length scale up to roughly 10 nm. The development of new methods for obtaining information on the organization and dynamics of assemblies on a longer length scale, such as the lateral organization of proteins and lipids in a biological membrane, is an important frontier in structural biology. Many types of imaging methods and microscopies seek to fill this need. This review focuses on secondary ion mass spectrometry (SIMS), still in its infancy as applied to biological systems. This approach to biological imaging has the potential to bridge between atomic-level structures and conventional light microscopy by providing direct compositional information at the 50 nm to several micron length scale. Comparisons with other state-of-the-art imaging methods are briefly discussed in the concluding section.

Conventional mass spectrometry has been transformed from a method largely limited to analytical chemistry and geochemistry into an indispensable tool for the characterization of biological molecules and even assemblies. This is the result of revolutionary advances in matrix-assisted laser desorption/ionization (MALDI) (14, 15, 70, 91, 104) and electrospray ionization (48, 114) methods. Both methods deliver large molecules into the gas phase, where they are ionized so that their mass-to-charge ratio can be measured with high precision. Further processing into fragments (tandem mass spectrometry) can provide high-resolution information on molecular level structure, and by clever use of isotopes, even the kinetics of assembly of large complexes can be characterized (105). These once exotic methods have become essential, even routine, analytical tools for all laboratories investigating the structure of biological molecules. The information content of mass spectrometry is extraordinary, leading to unambiguous identification of molecules by their mass. The promise of imaging mass spectrometry is to combine this level of chemical identification with spatial information.

There are two broad classes of imaging mass spectrometry, depending upon the method used to scan the sample and generate an image. The first class is based on MALDI and takes advantage of the laser used to desorb molecules within a matrix from a surface as the laser is scanned across the sample. Typically, pulsed lasers are used to ablate the sample, and the resulting ions or fragments are detected by using a time-of-flight (ToF) mass spectrometer. Because the spot size of the laser is, at best, given by the diffraction limit, lateral resolution less than 1 μm is difficult. Typically, a much larger spot size (>10 microns) is used to produce sufficient secondary ion signal intensity for molecular imaging. This method is routinely applied to organ-scale biomolecule imaging. MALDI imaging of biological samples has recently been reviewed and is not covered further here (14, 71).

SIMS: secondary ion mass spectrometry

MALDI: matrix-assisted laser desorption/ionization

ToF: time of flight

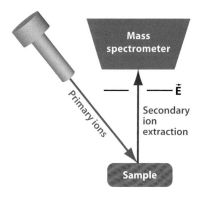

Figure 1

Simplified schematic of a secondary ion mass spectrometry (SIMS) instrument. The common components of all SIMS instruments are a primary ion source, a sample chamber with secondary ion extraction optics, and a mass spectrometer. Mass spectrometers include time-of-flight, magnetic sector, and quadrapoles.

The second class of imaging mass spectrometry, SIMS, is based on the use of an accelerated primary ion beam that bombards the surface and generates secondary ions (**Figure 1**). As described below, lateral resolution better than 100 nm is possible with specialized instruments. SIMS has the potential to image the distributions of specific species within complex biological samples and to measure the amount of each species within a specified region on the sample. Thus, the potential exists for a true analysis of the sample, although many challenges must be resolved before this long-term goal becomes a reality.

There are two fundamentally different approaches to SIMS analysis of biological samples based on the instrument design: ToF-SIMS and dynamic SIMS using a magnetic sector mass spectrometer. For ToF-SIMS, the primary ion beam is pulsed and the resulting secondary ions are detected by a ToF mass spectrometer. This approach allows the full mass spectrum to be monitored during the analysis. The goal of ToF-SIMS is the ejection and detection of molecular species. In contrast, dynamic SIMS is performed with a continuous primary ion beam and a preselected set of ions is detected with a magnetic sector mass spectrometer. Here we use the term "dynamic SIMS" as others have used it. However, ToF-SIMS instruments can be operated in a dynamic mode, meaning that the primary ion beam erodes away the sample surface, exposing fresh sample. This ToF-SIMS approach is typically not used for biological samples because molecular information is lost. A term less used, but potentially more apt, than dynamic SIMS is "magnetic sector SIMS." Typically, in dynamic SIMS, reactive primary ions are used to enhance secondary ion yields. Under continuous ion bombardment, molecular bonds are broken and only monatomic and small molecular ions are produced and detected. In this approach, molecule-specific elemental or isotopic tags are used to locate and quantify molecules of interest.

SIMS is widely used for studying hard materials in materials science, geology, and cosmochemistry. SIMS instruments are commonly found in materials science departments, and they are integrated into semiconductor fabrication for quality control and analysis. By comparison, the application of SIMS to biological sample analysis has been limited, with only a small number of laboratories putting significant effort into biologically related problems. Broadly speaking, the major reasons that SIMS has not historically been more widely adopted for biological sample analysis are (*a*) insufficient sensitivity and spatial resolution, (*b*) challenges of molecular identification, and (*c*) challenges of biological sample preservation for high-vacuum analysis. However, the invention of new ion sources for ToF-SIMS and the development of a high-resolution dynamic SIMS instrument, the NanoSIMS 50 from CAMECA Instruments (16), have raised the potential for SIMS to become a valuable imaging technique in the biosciences. The state-of-the-art of SIMS imaging of biological samples is the subject of this review, with an emphasis on dynamic SIMS; excellent reviews of ToF-SIMS have recently been published (53, 69). We emphasize references from the literature on biological imaging; several reviews on the underlying technology are cited but not discussed in detail.

NanoSIMS: brand name of a high-spatial resolution dynamic SIMS instrument from CAMECA

INSTRUMENTATION

Figure 1 shows a generic scheme for a SIMS instrument. The major components are a primary ion source, a sample chamber, and a secondary ion mass spectrometer. Typically, the primary ion beam column is oriented obliquely to the sample surface, and secondary ions are extracted for analysis by an electrostatic field normal to the surface. This configuration allows the primary ion beam focusing to be independent of the secondary ion beam focusing. CAMECA fundamentally changed this configuration with the introduction of the NanoSIMS 50 by bringing the primary ion beam in normal to the sample surface, coaxial with the secondary ion beam (**Figure 2**). This change enables the primary focusing lens to be brought closer to the sample, thereby reducing focusing aberrations (51, 100). Imaging is performed by scanning the focused primary beam across the sample and digital reconstruction of a map of different masses corresponding to different parent molecules. The best lateral resolution achieved by these instruments is on the order of 50 nm with significant reduction in primary beam current, while 100 nm can be achieved routinely with a beam current of $\sim$2 pA Cs^+. Similar lateral resolution was previously achieved in SIMS with liquid metal sources (63). The NanoSIMS is significant because it achieves this lateral resolution with reactive primary ions, which enhances sensitivity, while maintaining high mass resolving power, which enhances specificity.

ToF-SIMS

For ToF-SIMS, many different primary ion beams have been developed, and this is an active area of research and development. The primary ion beam is accelerated to high energy and focused onto the region of the sample whose composition is being imaged. The interaction of the primary beam with the sample depends on the energy, current, and nature of whatever is accelerated and, in many cases, the environment, or matrix, in which the molecules of interest are embedded (116). A large body of empirical data has been collected (73, 101) and some models are available to guide new developments (32) because each primary ion source involves a substantial engineering and optimization effort. Each primary ion beam offers different advantages for the generation of secondary ions. For biological samples, the emphasis in recent years has been on larger projectiles such as gold clusters as large as several hundred gold atoms (66), Bi_3^+, SF_5^+, or C_{60}^+, because these primary ions open the potential for higher yields of large molecules. Large secondary ion fragments are desirable because biological samples typically contain complex mixtures and larger fragments have greater chemical information than smaller ones do. At the same time, the amounts of each component in the sample area are often low, so high sensitivity is also at a premium. Unfortunately, most of the species ejected from the surface are neutral and therefore not detected by the mass spectrometer. This factor is the primary limitation on detecting and imaging low-abundance molecules. Post-ionization of the neutrals is possible, although further fragmentation of molecules is a primary challenge for this approach (49, 115). A great deal of instrumentation development has been devoted to focusing and rastering the primary ion beam and optimizing secondary ion collection optics, and although beyond the scope of this review, this greatly affects the sensitivity of the measurement (13, 16).

For ToF-SIMS, the primary ion beam is pulsed to enable ToF detection. The secondary ions are accelerated by an electric field to the same kinetic energy; thus the velocity of any individual ion depends on its mass-to-charge ratio (heavier particles move more slowly). By measuring the time for an ion to reach a detector, this mass-to-charge ratio can be obtained with high precision and a wide range of masses can be monitored. The ability to distinguish between adjacent masses is characterized by mass resolving power, which is defined as the nominal mass divided by the difference in mass between the two species ($M/\Delta M$). Because ToF-SIMS ion detection is based on the time of

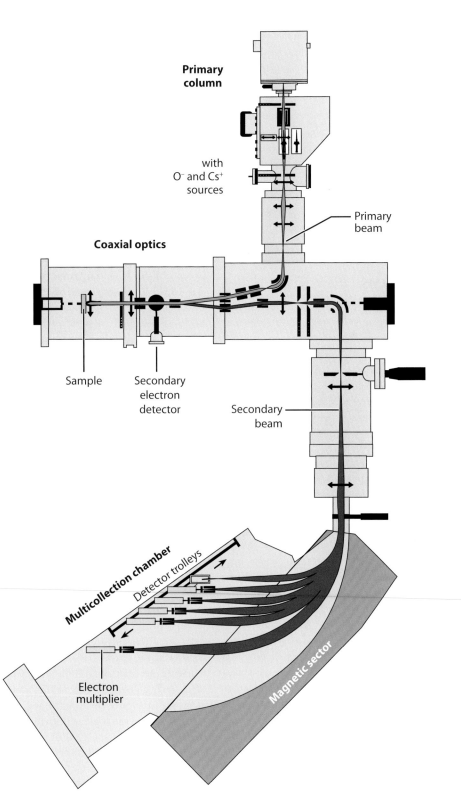

Primary column

with O⁻ and Cs⁺ sources

Primary beam

Coaxial optics

Sample

Secondary electron detector

Secondary beam

Multicollection chamber

Detector trolleys

Electron multiplier

Magnetic sector

Figure 2

Schematic of the NanoSIMS 50 from CAMECA showing the major components for biological imaging. The instrument uses either an oxygen or cesium primary ion source to generate positive or negative secondary ions, respectively. The coaxial optics enable copropagation of the primary and secondary ions, which places the primary focusing optics closer to and normal to the sample, thereby reducing the spot size on the sample. Simultaneous ion detection of up to seven species is performed on electron multipliers in the multicollection chamber. This figure is adapted from Reference 16.

flight to the detector, instrument mass resolving power is determined by the ratio of the duration of secondary ion generation (primary ion pulse length) and the length of the secondary ion path. This relationship results in a trade-off between spatial resolution at the sample and mass resolving power because the primary ion beam pulses must be compressed to increase mass resolution, which degrades the lateral resolution of the primary beam. Limitations on maximum count rates, duty cycle, and the ratio of analyzed to sputtered material can also place limits on ToF-SIMS analysis speed and sensitivity.

Dynamic SIMS

In dynamic SIMS instruments, a continuous primary ion beam generates a continuous flow of secondary ions. Typically, an oxygen primary beam (O^-, O_2^-, or O_2^+) is used to generate positive secondary ions, and a cesium primary beam (Cs^+) is used to generate negative secondary ions. These reactive primary ions implant into the sample, increasing the probability of producing positive or negative secondary ions, respectively. The development of a microcesium source by CAMECA reduced the primary spot size to less than 1 micron for dynamic SIMS instruments. The standard oxygen source, the duoplasmatron, however, has not been improved since the development of the CAMECA f-series instruments, leaving room for the development of brighter sources of electronegative primary ions (42). Positive ion imaging can be useful for probing metals in living systems (89, 109) and for tracking metabolic pathways (e.g., for Ca^{2+} and Zn^{2+}) (4, 9, 18). Negative secondary ions are formed as the fragments of organic molecules, including all common classes of biological macromolecules. For biological samples the most important atomic secondary ions, including those introduced as atom or isotopic labels, are $^1H^-$, D^-, $^{12}C^-$, $^{13}C^-$, $^{16}O^-$, $^{18}O^-$, $^{19}F^-$, $^{31}P^-$, and $^{32}S^-$; molecular secondary ions include $^{12}CH^-$, $^{13}CH^-$, $^{12}CD^-$, $^{13}CD^-$, $^{12}CH_2^-$, $^{12}C^{14}N^-$, $^{13}C^{14}N^-$, $^{12}C^{15}N^-$, and $^{13}C^{15}N^-$ ($^{13}C^{14}N^-$, 27.0064 amu, and $^{12}C^{15}N^-$, 27.0001 amu, can be distinguished). For species with the highest ionization probability, such as O^-, F^-, S^-, and CN^-, as many as 1 in 20 atoms in the sample can be detected.

In dynamic SIMS, atomic, diatomic, and larger molecular secondary ions are generated, generally in decreasing quantities. These secondary ions are separated and analyzed continuously with a modern version of the bending magnets and electric fields introduced by J.J. Thompson in the original development of the mass spectrometer. The mass spectrometers focus in both mass and energy to achieve high mass resolving power ($M/\Delta M$). Note that mass resolving power is not directly comparable between ToF-SIMS and dynamic SIMS because ToF-SIMS produces mass spectra peaks over which counts are integrated, whereas dynamic SIMS is designed to perform ion detection at a fixed dispersion on a flattop peak. These mass spectrometers produce flattop mass peaks, which enable quantification down to 1 in 1000 precision for highly abundant species. The details of the separation system depend on the target application of the instrument and determine the ultimate mass resolution. Standard dynamic SIMS instruments (e.g., the CAMECA f series) use small radius magnets, and the transmission (the fraction of collected secondary ions making it to the detector) falls off rapidly with increase in mass resolution. Large radius SIMS instruments are used to maximize transmission at high mass resolving power (full transmission at $\sim$5000 $M/\Delta M$); these instruments are typically dedicated to geochronology and high-precision isotopic analysis. The NanoSIMS achieves relatively high transmission at high mass resolution (full transmission at $\sim$2000 $M/\Delta M$) by optimizing transmission with a narrow energy window at the entrance slit to the mass spectrometer. An example of high mass resolution of secondary negative ions on the NanoSIMS is shown in **Figure 3**. For imaging, the ions are detected with high sensitivity and little background with some type of electron multiplier. Count rates are limited to less than 1 million counts per second to prevent premature aging of detectors. These instruments are

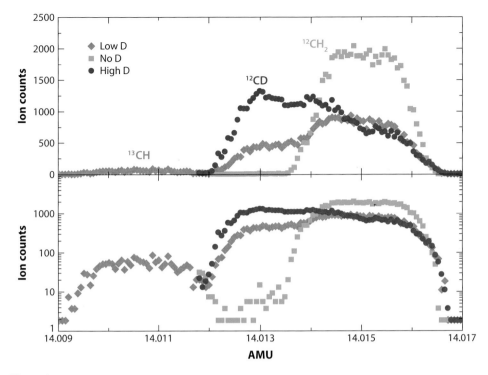

Figure 3

Mass spectra of ^{12}CD, ^{13}CH, and $^{12}CH_2$ at high mass resolving power for three different D/H ratios in spotted lipid samples, shown on linear (*upper panel*) and logarithmic scales (*lower panel*). The sample with no added D shows the peak shape with this tuning. The samples with added D show the ability to discriminate the ^{12}CD from the $^{12}CH_2$ peak by counting with the mass setting on the left side of the peak. The slope on the ^{12}CD peak top for the high D/H sample is caused by the sample being sputtered away during the mass scan. Nominal mass resolution of these scans is 8000 $M/\Delta M$.

typically equipped with Faraday cups to enable higher secondary ion count rates, but the response time for Faraday cups is too slow for imaging.

The NanoSIMS instrument can detect five or seven secondary ions in parallel, depending on the model, so that a precise map of several different fragments detected exactly at the same location can be generated. Unlike ToF-SIMS, only preselected ions can be detected. Because biological samples are complicated mixtures and the palate of secondary ions available is limited, isotopic and/or elemental labeling that selectively discriminates the molecule(s) of interest is essential, and this has been used in most experiments reported to date. In sectioned samples, the number of ions in a given region can also be used to generate image contrast; for

example, DNA produces high count rates of CN^- and P^- (60, 90).

Quantification is a central challenge for all SIMS methods. For dynamic SIMS, the species of interest are normalized to a major element in the sample of known concentration or reported as isotopic ratios. Standards that closely match the major element composition of the unknowns are analyzed under the same conditions as the unknown to control for differences in ion yield between the species of interest and the normalizing species. Differences in ion yield can be substantial (113). The same methods can be applied to ToF-SIMS molecular imaging, but at this time, quantification is not the central thrust of the work on biological samples and remains an important direction for the next generation of experiments.

SAMPLE PREPARATION

Because imaging mass spectrometry is performed in ultrahigh vacuum, sample preparation methods that preserve the biologically relevant organization being probed are essential. As described in detail below, evaluation of sample quality with other methods prior to SIMS analysis is prudent. Regions of the sample can be examined with comparable lateral resolution by near-field methods such as atomic force microscopy (AFM) and electron microscopy, and larger regions can be compared with optical measurements. Many of the same issues arise in electron microscopy. Although some imaging mass spectrometry instruments have a provision for a cold stage, in most cases it is necessary to remove water and minimize salts. This is best achieved by rapid freezing, e.g., using liquid ethane or propane, which prevents crystallization of water, and by keeping the sample frozen while subliming off the vitreous ice. In addition to removing water, it is essential that the molecules of interest not sublime off the surface during freeze-drying or in the high-vacuum environment of the mass spectrometer, and that contamination of the surface by pump-oil or the laboratory environment should be minimized. Just as background fluorescence from a variety of sources compromises high-sensitivity fluorescence measurements, environmental contamination can be a serious problem in mass spectrometry.

Even with careful control and relatively simple samples (lipid bilayers on solid supports), we find that samples are often poorly preserved. This is best seen on a micron scale either by adding a fluorescent dye (e.g., a dye-labeled lipid for lipid monolayer or bilayer samples) or by imaging ellipsometry. The latter does not require a dye and can measure small thickness variations that may arise if the sample separates from the substrate. Prior assessment of regions that should be avoided is essential, and regions identified as interesting to probe at a later time in the imaging mass spectrometer must be found again, so combinations of imaging methods require the placement of some sort of landmark on the surface so the same region can be located with high precision. In the case of lipid bilayers on solid supports, this is readily achieved by membrane patterning (38), in which a grid pattern spacing on the order of tens of microns provides visual landmarks that can be mapped with optical techniques, such as fluorescence microscopy or imaging ellipsometry, and used to locate the same region on the surface for analysis by multiple complementary techniques, including AFM and imaging mass spectrometry.

Methods of sample preparation for cells and tissues need to be evaluated according to the experimental goals. Mapping diffusible ions requires fast freezing methods (19). Outside of this extreme case, little work has been done to determine the trade-offs between ease of sample preparation and sample integrity. Fast freezing and low temperature dehydration are generally regarded as the gold standard (19, 35, 41, 89); other methods have been used such as standard resin embedding and ultramicrotomy (60, 61), chemical fixation and air drying (65, 87), and even no treatment for stable structures such as bacterial spores (33).

NanoSIMS IMAGING OF METABOLIC PATHWAYS AND MICROBIAL COLONIES

To date only a handful of laboratories have reported results using a NanoSIMS on biological samples. This is partly because there are fewer than 20 instruments in the world and most are largely dedicated to studying hard materials. In addition, there is a significant hurdle for new entrants into the field because the instruments are complex and require extensive training to acquire meaningful data, and methods of data analysis and sample preparation are not standardized. Much of the published literature is found in relatively specialized journals that would not be routinely seen by biochemists or structural biologists, and the few higher-profile papers are the result of a great deal of effort. Much of this work is still in the method development and capability demonstration stage.

Approximately one-quarter of the publications fall in the category of studies of metabolic pathways (4, 17, 25, 40, 56, 60, 89, 90, 95). Several studies have demonstrated the value of NanoSIMS for tracer studies, including ^{13}C-labeled free fatty acid transport across cell membranes (56), thyroid uptake of iodine (17, 25), and cellular uptake and distribution of an ^{15}N-labeled peptide vector (95). Lechene et al. (60) provide an extensive exposition of the use of stable isotopes for cell biology. A recent study probes the origin of the CN$^-$ ion based on the formation of a ^{13}C^{15}N cluster from adjacent ^{13}C- and ^{15}N-labeled proteins (62). This work follows an early study of the origin of the CN$^-$ molecular ion that demonstrated that it could come directly from directly bonded atoms or from atom recombination (72). This phenomenon may potentially be exploited to obtain proximity information on a length scale much smaller than the primary beam spot size.

Applications of NanoSIMS in microbiology have expanded rapidly. NanoSIMS has been used to image the distribution of isotopic and elemental tracers at the scale of individual bacterial cells—which can be submicron in size—to infer microbial metabolism (2, 6, 20, 50, 60, 61, 65, 78, 87, 110). This application is valuable for microbiologists because microbial metabolism has typically been studied in pure cultures or inferred from genomic data, even though most microbes cannot be cultured, pure cultures may not reflect metabolism in mixed cultures, and genomic data reflect capability, not actual processes. Orphan et al. (80) were the first to use SIMS to monitor microbial metabolism in a natural consortium. They used fluorescent probes hybridized to ribosomal RNA to identify microbes in a relatively simple, spherically symmetric consortium amenable to analysis by SIMS with micron-scale lateral resolution. To determine microbial metabolism, the fluorescent probes were correlated to natural abundance isotopic analysis with a large-radius SIMS instrument. Natural abundance isotopic analysis provided insight into microbial metabolism because of the large natural fractionation in carbon isotopes by the methanogenic bacteria in

the consortium. For most NanoSIMS studies, stable isotope-labeled substrates are provided to the microbial consortium to determine which microbes metabolize the substrate. The higher spatial resolution of the NanoSIMS allows individual cells and subcellular regions to be resolved, and simultaneous ion detection facilitates imaging of stable isotope analysis and elementally tagged molecular probes within single micron-scale cells (6, 65, 78).

NanoSIMS has also been used to characterize microbially mediated geochemical processes using isotopic tracers (27), natural abundance stable isotopes (28), and organic nitrogen detection and mapping in mineral aggregates (77). Other biologically related geochemistry studies include elemental mapping and tracer studies on reef-building corals (20, 75), statoliths (118), coccoliths (36, 92), diatoms (2), and fossil organic mapping (79).

IMAGING MASS SPECTROMETRY OF BIOLOGICAL MEMBRANES

Although the lipid bilayer is the universal structural element of biological membranes, relatively little is known about the lateral organization of lipids and membrane-associated proteins. Biological membranes are highly dynamic as lipids generally diffuse in the plane of the membrane, as do many membrane-associated proteins. Specific associations among lipids such as sphingomyelin, cholesterol, and membrane-anchored proteins, often called lipid rafts, are hypothesized to form an organized entity with collective function. These microdomains are believed to play a central role in organizing this fluid system, enabling the cell membrane to carry out essential cellular processes, including protein recruitment and signal transduction (43, 117). Yet direct visualization of these microdomains has proved to be difficult, the precise compositions and other physical characteristics of these domains have not been established, and thus the existence of rafts is controversial (24).

There has been extensive work on lateral phase separation using model membranes, such

POPC: 1-palmitoyl-
2-oleoylphosphati-
dylcholine

as monolayers at the air-water interface or on hydrophobic supports, and bilayers, either on solid supports or in giant vesicles. Domains are often visualized by the partitioning of fluorescent probes between phases. Although many dye-labeled lipids are available, their physical properties can be different from those of native lipids, which is not surprising because the dyes are often covalently attached via important functional groups and they introduce unnatural charge and functionality. Imaging mass spectrometry has the potential to make a major contribution to this field by directly imaging the lateral distributions of components within the membrane without changing the chemical structures of the components of interest, and therefore the delicate interactions that are essential for function are not perturbed. ToF-SIMS has been used to directly detect and image various membrane components in synthetic model membrane systems, biological tissues, and individual cell membranes on the length scale of 200 nm–1 μm. Dynamic SIMS performed with the NanoSIMS has been combined with isotopic labeling of specific components to extract comparable information from supported lipid bilayers on a scale of 100 nm. In this section, we review this work.

ToF-SIMS

The majority of SIMS analysis of lipid membranes has been performed using ToF-SIMS. Because large molecular fragments often have distinctive masses, labeling is, in principle, not required for component identification. Winograd and coworkers (74, 102) have pioneered the ToF-SIMS analysis of domain formation in Langmuir-Blodgett lipid monolayers deposited onto self-assembled monolayers of alkane thiols on gold. To date, data have been reported at the 1 μm length scale; although a primary ion beam diameter of 100–200 nm can be achieved, the lateral resolution of these studies was limited by the pixel size. This body of work was discussed in a recent review of ToF-SIMS imaging of lipid membranes (54), so we only briefly summarize key results here.

ToF-SIMS analysis of biological and synthetic lipid membranes offers the advantages that unlabeled model or native cell membrane samples can be characterized, and the resulting mass spectra contain a wealth of information on the molecules present even if they were unidentified at the time of analysis. A prime example of the strengths of ToF-SIMS is demonstrated in a recent report characterizing an unlabeled lipid monolayer composed of sphingomyelin (m/z 731 and m/z 264), cholesterol (m/z 385 and m/z 369), and a partially unsaturated lipid [1-palmitoyl-2-oleoylphosphatidylcholine (POPC), m/z 760 and m/z 224] with 1 μm lateral resolution (74). The distributions of the characteristic secondary ions revealed domains enriched with sphingomyelin and cholesterol but deficient in POPC (74). ToF-SIMS has also been used to image phase separation within lipid monolayers that model lung surfactants (44) and contain a surfactant protein (10, 12) with 200 nm–1 μm lateral resolution.

The ability to directly image the lateral distributions of unlabeled molecular components within a sample renders ToF-SIMS an ideal method to image lipid distribution in actual cell membranes. There are several reports of imaging the distributions of membrane components within biological tissues and individual cells with a lateral resolution of one to a few microns (1, 21, 83, 85, 93, 94, 99, 106, 107). The most impressive example of the potential of this method is a report from Ostrowski et al. (82), who demonstrate a decreased abundance of phosphatidylcholine and an increase in an aminoethylphosphonolipid detected at the plasma membrane sites of fusion between *Tetrahymena* cells. If this result can be generalized, it can provide a unique opportunity to analyze the composition of specific regions of membranes that are important in other cellular processes.

The primary issues for ToF-SIMS analysis of biological samples are sensitivity, spatial resolution, specificity, and quantification. Unfortunately, the high-mass fragment ions used for component identification often have low yields. The further development of polyatomic primary ion sources is a promising approach to decrease molecular fragmentation and increase the yields of molecular ions and high-mass molecular fragment ions, thereby simplifying component identification and increasing the working lateral resolution (29, 81). Molecular fragment identification can be challenging for complex samples, and in some cases, high-mass molecular fragment ions are not uniquely characteristic of a single membrane component (102), so selective deuterium labeling may also be required for component identification (8). The central challenge for quantification is determining the relationship between secondary ion signal intensity and component concentration. This relationship is subject to matrix effects that result in differences in ion yields. For example, the yields of ToF-SIMS secondary ions can vary because of changes in the lipid packing density in the monolayer (11, 73) and other differences in the local membrane environment (84, 88). These complications can be mitigated by using relative sensitivity factors measured on calibration samples (74). Furthermore, the yields of large molecules are often low and may be different for different molecules (e.g., lipids versus proteins, phospholipids versus cholesterol), limiting sensitivity and compromising quantification. Though beyond the scope of this review, multivariate analysis techniques that glean component identity from the abundant low-mass fragment ions that are otherwise unexploited can also improve component identification (108). ToF-SIMS is especially powerful for detecting lateral variations that give rise to visible contrast in the component-specific ion image, although care must be taken in translating this change in the signal intensity into variations in the abundance of a specific species, as opposed to changes in the local environment that alter the secondary ion yield.

Dynamic SIMS

In contrast to ToF-SIMS, only tiny molecular fragments such as atomic and diatomic secondary ions are detected in dynamic SIMS analysis. Therefore the incorporation of distinct stable isotopes or unique atoms (e.g., F) into the components of interest is required for component identification. Isotopically labeled molecules and the corresponding unlabeled molecules have identical chemical structures, and much of what we know about biochemical pathways is derived from studies that use stable and radioactive isotopes as tracers. As described in the section on instrumentation, while ToF-SIMS generates an entire mass spectrum, dynamic SIMS performed with the NanoSIMS only allows the parallel detection of five or seven different m/z ratios (depending on the NanoSIMS model). The trade-off is that a higher spatial resolution and greater sensitivity to small mol fractions of each component can be achieved by this approach than by ToF-SIMS. Taken together, dynamic SIMS performed with the NanoSIMS is advantageous when the components of interest can be isotopically labeled and when high lateral resolution and sensitivity are the primary goals. To date, a direct comparison of ToF-SIMS and NanoSIMS images for identically prepared samples has not been reported. This would provide a useful assessment of the strengths of each approach.

Model lipid membranes are an ideal test bed for dynamic SIMS because they have been characterized by many surface-sensitive methods and isotopic labels can be easily incorporated. Our joint published work to date focuses on supported lipid bilayers, one of the simplest model systems that captures essential features of biological membranes (31, 57, 58). At the time we began this work, it was not clear that the sensitivity would be sufficient to detect a single bilayer using the NanoSIMS, let alone to attempt quantitative analysis with high spatial resolution. Furthermore, there was no precedent for preparing supported bilayers for high vacuum analysis, and our initial efforts often led to surfaces that, when imaged by adding a small amount of a fluorescent dye-labeled lipid, were of poor quality.

Supported lipid bilayers are prepared by exposing glass or silica surfaces to a suspension of lipid vesicles. This self-assembly process does

DLPC: 1,2-
dilauroylphosphati-
dylcholine

DSPC: 1,2-
distearoylphosphati-
dylcholine

not occur on many materials, and by placing such materials on substrates as barriers, bilayer assembly can be directed only to those regions where the glass or silica is exposed. This is called membrane patterning, and many methods for varying and manipulating the composition of lipid and membrane-associated protein components in these patterned surfaces have been developed (37). Although supported bilayers can be formed on the native SiO$_2$ that forms spontaneously on Si in air, bilayers on native oxide are not particularly stable. By growing a thicker but still very thin SiO$_2$ layer on the Si substrates, stable bilayers can routinely be formed, and these SiO$_2$ layers can be patterned with other materials (we currently use Cr) so that bilayer regions on the surface can be relocated by different imaging methods such as fluorescence, ellipsometry, AFM, and ultimately the NanoSIMS (which destroys the sample).

We use 5 × 5 mm Si wafers, which are held firmly and grounded in our NanoSIMS sample holders. The oxide thickness on these wafers is important because charge buildup on the surface during exposure to the Cs$^+$ beam can reduce the ion yield and resolution, so a compromise between SiO$_2$/bilayer stability and charge dissipation through the grounded Si wafer must be sought. Although a comprehensive analysis has not been performed, it appears that a roughly 10-nm-thick SiO$_2$ layer is a good compromise, which is consistent with the mean implantation depth of 20 nm for the 16 kV Cs$^+$ primary ions.

This thin oxide layer has a second consequence for fluorescence imaging. Because the underlying Si substrate is an excellent mirror, incoming light is reflected off the surface and interferes, giving a standing wave pattern above the substrate with a null at the surface, a maximum at one-quarter of the wavelength of light, a null again at one-half the wavelength of light and so on [this is the basis of the interferometric technique fluorescence inference contrast microscopy (59)]. Because the SiO$_2$ layer is very thin, fluorescence from molecules in a bilayer that are only a few more nanometers

from the Si mirror is very weak. After freeze-drying, the molecules are likely even closer to the mirror, and we also find that what little fluorescence there is rapidly photobleaches in air. Thus, although a thin SiO$_2$ layer is optimal for the NanoSIMS measurement, it makes conventional epifluorescence microscopy more difficult.

Preserving the bilayer and its lateral organization on the surface by freeze-drying is a tricky process. Supported bilayers are not stable when exposed to air, so the sample must be under water at all times, but the amount of water must be kept to a minimum for rapid freezing (the 1-mm-thick Si wafer also presents a substantial thermal mass). The sample is rapidly immersed in liquid ethane with a minimal perturbation of the surface and then transferred under liquid nitrogen into a freeze-drying chamber where the water is removed at low temperature (we use an oil-free scroll pump to minimize contamination). Once freeze-dried, the surfaces appear to be stable for a long time, e.g., domains visualized by fluorescence microscopy (necessarily microns or larger) are preserved pre- and postfreeze-drying and remain for long periods. Although lipid mobility is a hallmark of biological membranes, once freeze-dried this mobility ceases. Also, because the sample is frozen in a small fraction of a second, a typical lipid molecule with a diffusion coefficient of approximately 1 μm^2 s^{-1} will at most exchange positions with only a few neighbors, and this is much less than the lateral resolution of the NanoSIMS measurement due to the diameter of the primary ion beam.

As a step toward the ultimate goal of high-resolution membrane composition analysis, we have demonstrated that small domains within a phase-separated lipid bilayer could be imaged with 100-nm lateral resolution and that the lipid composition within small regions of the membrane could be quantified by NanoSIMS imaging (58). Homogeneous bilayers composed of 1,2-dilauroylphosphatidylcholine-^{15}N (^{15}N-DLPC), 1,2-distearoylphosphatidylcholine-^{13}C$_{18}$ (^{13}C$_{18}$-DSPC), and 0.5 mol% of a fluorescent lipid that enabled the bilayer to be visualized by fluorescence microscopy during sample preparation were formed by vesicle fusion on chrome-patterned silicon substrates at a temperature at which both lipids were in the fluid state. Pure DSPC has a gel-to-liquid phase transition temperature of 55°C, and pure DLPC has a gel-to-liquid phase transition temperature of –1°C. The freely mixed supported bilayers were cooled to induce phase separation due to this difference in their phase behavior associated with acyl chain lengths. The bilayers were flash-frozen and freeze-dried to remove the water without perturbing the bilayer's organization, and the locations and geometries of the gel and fluid phases within the freeze-dried lipid bilayer were imaged by AFM (**Figure 4d**). The height difference between these two phases was consistent with measurements made on hydrated, phase-separated bilayers composed of DSPC and DLPC (67).

The ^{13}C^{1}H$^-$ and ^{12}C^{15}N$^-$ secondary ion signals were used to visualize the ^{13}C$_{18}$-DSPC and ^{15}N-DLPC distributions, respectively, within the lipid membrane. The lipid-specific secondary ion images showed that domains enriched with ^{13}C$_{18}$-DSPC, as evidenced by a locally elevated ^{13}C^{1}H$^-$ signal and decreased ^{12}C^{15}N$^-$ signal, were dispersed within a ^{15}N-DLPC–rich bilayer (**Figure 4**). The positions and geometries of the ^{13}C$_{18}$-DSPC–enriched domains visualized by the NanoSIMS were nearly identical to the gel phase domains imaged by AFM. A few features in the AFM image did not produce lipid-specific secondary ions (**Figure 4**); the height difference between these objects and the bilayer (>5 nm) verified that these objects were unlabeled debris, not lipid domains. This side-by-side comparison is particularly important because AFM and NanoSIMS images contain different information and each enhances the value of the other.

Quantitative information on the lipid composition within small regions of the membrane was obtained by calibrating the lipid-specific secondary ion signal intensities with standard samples. To establish the calibration curves, two sets of homogeneous supported lipid bilayer samples that each systematically varied in the ^{13}C$_{18}$-DSPC or ^{15}N-DLPC content were made. Using the NanoSIMS, measurements were made on each calibration sample within the two sets. The normalized ^{13}C^{1}H$^-$ or ^{12}C^{15}N$^-$ signal intensities (^{13}C^{1}H$^-$/^{12}C$^-$ or ^{12}C^{15}N$^-$/^{12}C$^-$) were determined at several regions in each calibration sample, and the standard deviation relative to the average ^{13}C^{1}H$^-$ or ^{12}C^{15}N$^-$ signal intensity measured on each sample was below 3% and 5%, respectively. Calibration curves were constructed from this data, and the normalized ^{13}C^{1}H$^-$ and ^{12}C^{15}N$^-$ signal intensities had excellent linear correlations with the isotopic enrichment in the bilayer, which suggests that matrix effects were negligible in these samples. The normalized ^{12}C^{15}N$^-$ signal intensity was sensitive to the thickness of the SiO$_2$ layer of the substrate that supported the lipid membrane, so the SiO$_2$ thickness used for these experiments varied <1 nm. The magnitude of matrix effects within bilayer samples that contain more diverse membrane compositions is currently being investigated.

Using these calibration curves, the gel and fluid phase compositions were evaluated by converting the component-specific secondary ion signal intensities collected from small regions of the bilayer into mol% concentrations (**Figure 5**). Most of the gel phase domains consisted of a ~9:1 molar ratio of ^{13}C$_{18}$-DSPC to ^{15}N-DLPC, consistent with the phase diagram for this mixture. A statistically significant elevation in the amount of ^{15}N-DLPC was detected at a localized area in one gel phase domain. Inspection of the AFM image acquired at this location revealed that the elevated ^{15}N-DLPC concentration within the domain corresponded to a small (<200-nm-diameter) depression in the thickness of the domain, and this could be interpreted as a fluid phase subdomain trapped within the gel phase (**Figure 5**). In the bilayer regions that corresponded to the fluid phase, the molar ratio of ^{15}N-DLPC to ^{13}C$_{18}$-DSPC was greater than 19:1, again consistent with the

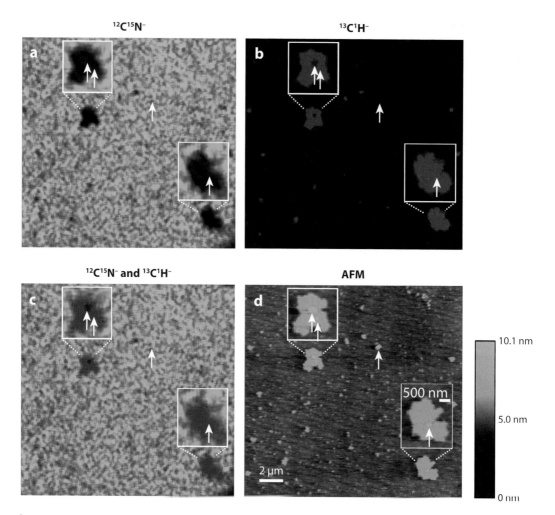

Figure 4

NanoSIMS analysis of a freeze-dried, phase-separated supported lipid bilayer composed of 1,2-dilauroylphosphatidylcholine-^{15}N (^{15}N-DLPC) and 1,2-distearoylphosphatidylcholine-^{13}C$_{18}$ (^{13}C$_{18}$-DSPC) (*a–c*) and the atomic force microscopy (AFM) image of the topography of the same bilayer location (*d*). The analysis was performed with the NanoSIMS 50 using a cesium primary ion beam with a diameter of 100 nm and a pixel size of 100 × 100 nm. (*a*) The normalized ^{12}C^{15}N$^-$ secondary ion signal intensity (*green*) reveals the distribution of ^{15}N-DLPC in this area, and (*b*) the normalized ^{13}C^{1}H$^-$ secondary ion signal (*red*) shows the distribution of ^{13}C$_{18}$-DSPC in the membrane. (*c*) Overlay of the two lipid-specific ion signals. A comparison of the AFM image (*d*) of the membrane topography that was acquired at the same sample location prior to SIMS analysis to the lipid-specific secondary ion images (*a–c*) shows that the sizes and shapes of the ^{13}C$_{18}$-DSPC–enriched lipid domains visualized by SIMS are nearly identical to the gel phase domains imaged by AFM. Arrows indicate objects in the AFM image that are unlabeled debris, not labeled domains, and their locations in the SIMS images.

phase diagram. For 300 × 300 nm^2 bilayer regions, the uncertainty in the lipid composition was under 10%, although uncertainties as large as 20% were occasionally measured. The uncertainty in the estimation of lipid composition is based on counting statistics, as described in detail in the supplementary information of our previous report (58).

This work validates the ability to image the distribution of membrane components on the submicron length scale that is relevant to organization within biological membranes.

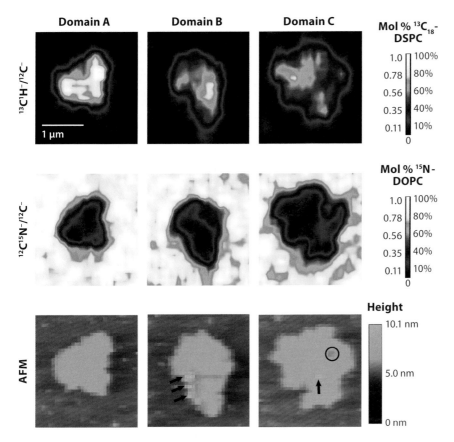

Figure 5

Quantitative analysis of membrane domains as described in **Figure 4**. The normalized $^{13}C^{1}H^{-}$ and $^{12}C^{15}N^{-}$ secondary ion signal intensities were calibrated using sets of homogeneous lipid bilayers that systematically varied in isotopic composition. The thermal false color scales represent the abundance of 1,2-distearoylphosphatidylcholine-$^{13}C_{18}$ ($^{13}C_{18}$-DSPC) and 1,2-dilauroylphosphatidylcholine-^{15}N (^{15}N-DLPC) within the bilayer. AFM images acquired at the same locations before SIMS analysis reveal topography. Lower concentrations of both lipids were detected in the locations where debris was identified by AFM (*arrows*). The lipid composition within the gel phase was typically ~9:1, $^{13}C_{18}$-DSPC to ^{15}N-DLPC (domain A), but unusually high amounts of ^{15}N-DLPC were occasionally detected within the gel phase domains (domains B and C). AFM imaging indicated a small (<200-nm-diameter) depression that might indicate that a small fluid phase domain (*circle*) was entrapped within the gel phase (domain C). The presence of this fluid phase subdomain is confirmed by the NanoSIMS image, which reveals a locally elevated concentration of ^{15}N-DLPC at this location. SIMS images were acquired with the NanoSIMS 50. Pixel size is 100 × 100 nm, and the images are smoothed over three pixels.

NanoSIMS analysis of actual cell membranes has not yet been attempted. The obvious challenges will be to selectively incorporate distinct stable isotopes into the components of interest and to isolate the cell membrane intact for analysis. Methods for achieving this have been reported, mostly for fluorescence imaging (23, 86).

COMPARISON WITH OTHER ADVANCED IMAGING METHODS

There are many approaches to imaging biological systems, depending on the nature of the samples, in particular whether they are alive, fixed, or otherwise preserved and/or sectioned, the length scale, sensitivity, and type of information that can be extracted. We close this review with a brief comparison of imaging mass spectrometry to a few methods that have seen great recent advances. It is our strong belief that combinations of methods are the best approach; that is, these are not competing, but complementary. An excellent example is provided by the NanoSIMS images of phase-separated supported bilayers compared with AFM images in **Figure 4**. AFM is a well-established method that can provide atomic resolution topographical images of the surface of hard and soft objects under ideal conditions, but not chemical information [although a recent example with a functionalized AFM tip shows the potential for molecule-specific information (111)]. The NanoSIMS data contain little topographic information but provide unambiguous identification of isotopic composition that translates into chemical identification. The spatial resolution (both lateral and vertical) is much poorer than with AFM, but the molecular information content is much greater; thus side-by-side measurements on the same sample are much more powerful than either alone. Another comparative example probing the architecture of pathogen cell walls has been reported recently (22).

Cryo-electron microscopy is a powerful tool for imaging complex assemblies such as the ribosome, microtubules, and nuclear pores, to name a few spectacular examples (30, 103). Advances in electron microscopes and image processing open the possibility of nearly atomic-level resolution for noncrystalline samples. At the same time, advances in image reconstruction and thin sectioning combined with state-of-the-art electron microscopy make electron tomography a reality (39, 68, 96). Each of these approaches, like mass spectrometry imaging, depends upon careful sample preservation. Molecule-specific information can be obtained by affinity labels that are typically conjugated to large metal clusters, which is a disadvantage compared with the intrinsically multiplexed mass-specific information of mass spectrometry. As with imaging mass spectrometry, these techniques are still limited to a few laboratories, as the instrumentation

and expertise are specialized, and results emerge slowly. The potential for side-by-side comparisons of imaging mass spectrometry and electron tomography is particularly attractive.

Commercial light microscopes come equipped with extraordinary optics, detection systems, and software, so that advanced methods of data acquisition and analysis have become routine. Traditional approaches to sample staining with dyes or dye-labeled antibodies have been supplemented by genetically encoded labels based on green fluorescent protein (34). The advantage of these approaches, especially using green fluorescent protein, is that measurements can be performed on living samples and single molecules can be detected due to the extreme sensitivity of fluorescence detection (76). The obvious limitations for imaging the organization of complex systems are the need for specific labels (one only sees what is labeled with a fluorescent tag), the label may interfere with an important interaction, and the diffraction limit (on the order of several hundred nanometers under ideal conditions). The last limitation has been challenged by a series of increasingly impressive methods. Most of these approaches involve some sort of structured illumination. Near-field scanning optical microscopy scans a subdiffraction-sized aperture over the sample (45). Structured illumination near nonbiological surfaces can be achieved by evanescent wave approaches, in which the light field decays exponentially from a surface with a higher refractive index than the material being probed (e.g., total internal reflection fluorescence microscopy) (3), or by interference contrast, in which a standing wave is created near a reflective surface (e.g., fluorescence interference contrast microscopy) (59). Most impressive and potentially most general are methods that create structured illumination using the microscope optics or exploit the photophysical processes of the fluorescent molecules used for imaging to obtain resolution much below the diffraction limit. In particular the works of Hell and coworkers (46, 47, 55, 97, 98, 112) have demonstrated resolution better than 100 nm.

These state-of-the-art methods are not yet routine, and in most cases the images take considerable time to obtain, but they can be applied to intact three-dimensional objects. Finally, a new class of super-resolution methods based on the location of individual molecules and complex photophysics have recently been developed that avoid structured illumination but typically require multiple labels (5, 7, 52). These methods are evolving rapidly and offer spatial resolution that is comparable to imaging mass spectrometry. While the sensitivity is at the level of individual molecules, multiplexed detection and identification are limited to what is labeled.

Although fluorescence microscopy is widely used because the instrumentation and labels are so readily available, infrared microscopy can provide spectral signatures of specific types of molecules based on the characteristic vibrational spectra of the molecules in the sample. In this sense, infrared-based methods parallel imaging mass spectrometry in that they give chemically specific information. The advantage is that these methods, like fluorescence microscopy, can be applied to live cells. The disadvantages are the diffraction limit (on the order of several microns for typical vibrational frequencies), sensitivity (infrared transitions are typically detected much less sensitively than electronic transitions are), and background absorption, e.g., from bulk water. Raman microscopy can be used to avoid the last, but it is quite insensitive. Coherent anti-Stokes Raman scattering can improve the sensitivity by many orders of magnitude, and this is currently being developed as a microscopy method in several laboratories (26, 64).

Each of these advanced imaging methods offers specific advantages and none provides a complete story by itself. For planar samples like model membranes, imaging mass spectrometry combined with AFM takes optimum advantage of both. For complex topologies or where fast reorganization is important for function, these methods are much less useful, although it may prove possible to trap intermediates by freezing and obtain planar images by thin sectioning. Ultimately, mass spectrometry offers the highest level of information because it is mass specific, so the future of this approach appears to be bright.

DISCLOSURE STATEMENT

The authors are not aware of any biases that might be perceived as affecting the objectivity of this review.

ACKNOWLEDGMENTS

We thank Ian Hutcheon, Nick Winograd, Francois Hillion, and Francois Horreard for many useful discussions. M.L.K. was supported by an NIH NRSA fellowship while at Stanford and holds a Career Award at the Scientific Interface from the Burroughs Wellcome Fund. This work is supported by grants from the NSF Biophysics program and NIH GM06930 (S.G.B.) and from the LLNL Lab Directed Research and Development Program and DOE Genomes to Life Program (P.K.W.) and was performed under the auspices of the U.S. Department of Energy under contract DE-AC52-07NA27344. We are grateful to the Stanford Nanofabrication Facility for fabrication and the NSF MRSEC CPIMA for analysis (ellipsometry and AFM).

LITERATURE CITED

1. Altelaar AFM, Klinkert I, de Lange RPJ, Adan RAH, Heeren RMA, Piersma SR. 2006. Gold-enhanced biomolecular surface imaging of cells and tissue by SIMS and MALDI mass spectrometry. *Anal. Chem.* 78:734–42

2. Audinot JN, Guignard C, Migeon HN, Hoffmann L. 2006. Study of the mechanism of diatom cell division by means of Si-29 isotope tracing. *Appl. Surf. Sci.* 252:6813–15

3. Axelrod D. 2001. Total internal reflection fluorescence microscopy in cell biology. *Traffic* 2:764–74

4. Azari F, Vali H, Guerquin-Kern J-L, Wu T-D, Croisy A, et al. 2008. Intracellular precipitation of hydroxyapatite mineral and implications for pathologic calcification. *J. Struct. Biol.* 162:468–79

5. Bates M, Huang B, Dempsey GT, Zhuang X. 2007. Multicolor super-resolution imaging with photo-switchable fluorescent probes. *Science* 317:1749–53

6. Behrens S, Losekann T, Pett-Ridge J, Weber PK, Ng WO, et al. 2008. Linking microbial phylogeny to metabolic activity at the single-cell level by using enhanced element labeling-catalyzed reporter deposition fluorescence in situ hybridization (EL-FISH) and NanoSIMS. *Appl. Environ. Microbiol.* 74:3143–50

7. Betzig E, Patterson GH, Sougrat R, Lindwasser OW, Olenych S, et al. 2006. Imaging intracellular fluorescent proteins at nanometer resolution. *Science* 313:1642–45

8. Biesinger MC, Miller DJ, Harbottle RR, Possmayer F, McIntyre NS, Petersen NO. 2006. Imaging lipid distributions in model monolayers by ToF-SIMS with selectively deuterated components and principal components analysis. *Appl. Surf. Sci.* 252:6957–65

9. Bordat C, Guerquin-Kern J-L, Lieberherr M, Cournot G. 2004. Direct visualization of intracellular calcium in rat osteoblasts by energy-filtering transmission electron microscopy. *Histochem. Cell Biol.* 121:31–38

10. Bourdos N, Kollmer F, Benninghoven A, Ross M, Sieber M, Galla H-J. 2000. Analysis of lung surfactant model systems with time-of-flight secondary ion mass spectrometry. *Biophys. J.* 79:357–69

11. Bourdos N, Kollmer F, Benninghoven A, Sieber M, Galla H-J. 2000. Imaging of domain structures in a one-component lipid monolayer by time-of-flight secondary ion mass spectrometry. *Langmuir* 16:1481–84

12. Breitenstein D, Batenburg JJ, Hagenhoff B, Galla H-J. 2006. Lipid specificity of surfactant protein B studied by time-of-flight secondary ion mass spectrometry. *Biophys. J.* 91:1347–56

13. Brunelle A, Touboul D, Laprévote O. 2005. Biological tissue imaging with time-of-flight secondary ion mass spectrometry and cluster ion sources. *J. Mass Spectrom.* 40:985–99

14. Burnum KE, Frappier SL, Caprioli RM. 2008. Matrix-assisted laser desorption/ionization imaging mass spectrometry for the investigation of proteins and peptides. *Annu. Rev. Anal. Chem.* 1:689–705

15. Caldwell RL, Caprioli RM. 2005. Tissue profiling by mass spectrometry: a review of methodology and applications. *Mol. Cell Proteomics* 4:394–401

16. CAMECA. 2008. *NanoSIMS 50/50L. SIMS microprobe for ultra fine feature analysis.* **http://www.cameca.fr/html/product_nanosims.html**

17. Champion C, Elbast M, Wu TD, Colas-Linhart N. 2007. Thyroid cell irradiation by radioiodines: a new Monte Carlo electron track-structure code. *Braz. Arch. Biol. Technol.* 50:135–44

18. Chandra S. 2004. Subcellular SIMS imaging of isotopically labeled amino acids in cryogenically prepared cells. *Appl. Surf. Sci.* 231–232:462–66

19. Chandra S, Morrison GH. 1992. Sample preparation of animal tissues and cell cultures for secondary ion mass spectrometry (SIMS) microscopy. *Biol. Cell* 74:31–42

20. Clode PL, Stern RA, Marshall AT. 2007. Subcellular imaging of isotopically labeled carbon compounds in a biological sample by ion microprobe (NanoSIMS). *Microsc. Res. Tech.* 70:220–29

21. Colliver TL, Brummel CL, Pachloski ML, Swanek FD, Ewing AG, Winograd N. 1997. Atomic and molecular imaging at the single-cell level with TOF-SIMS. *Anal. Chem.* 69:2225–31

22. Dague E, Delcorte A, Latge J-P, Dufrene YF. 2008. Combined use of atomic force microscopy, X-ray photoelectron spectroscopy, and secondary ion mass spectrometry for cell surface analysis. *Langmuir* 24:2955–59

23. Danelon C, Perez JB, Santschi C, Brugger J, Vogel H. 2006. Cell membranes suspended across nanoaperture arrays. *Langmuir* 22:22–25

24. Edidin M. 2003. The state of lipid rafts: from model membranes to cells. *Annu. Rev. Biophys. Biomol. Struct.* 32:257–83

25. Elbast M, Wu TD, Guiraud-Vitaux F, Guerquin-Kern JL, Petiet A, et al. 2007. Kinetics of intracolloidal iodine in thyroid of iodine-deficient or equilibrated newborn rats. Direct imaging using secondary ion mass spectrometry. *Cell Mol. Biol.* 53(Suppl.):OL1018–24

26. Evans CL, Xie XS. 2008. Coherent anti-stokes Raman scattering microscopy: chemically selective imaging for biology and medicine. *Annu. Rev. Anal. Chem.* 1:883–909

27. Fayek M, Utsunomiya S, Pfiffner SM, White DC, Riciputi LR, et al. 2005. The application of HRTEM techniques and NanoSIMS to chemically and isotopically characterize *Geobacter sulfurreducens* surfaces. *Can. Mineral.* 43:1631–41

28. Fike DA, Gammon CL, Ziebis W, Orphan VJ. 2008. Micron-scale mapping of sulfur cycling across the oxycline of a cyanobacterial mat: a paired NanoSIMS and CARD-FISH approach. *ISME J.* 2:749–59

29. Fletcher JS, Lockyer NP, Vickerman JC. 2006. C60, Buckminsterfullerene: its impact on biological ToF-SIMS analysis. *Surf. Interface Anal.* 38:1393–400

30. Frank J. 2006. *Three-Dimensional Electron Microscopy of Macromolecular Assemblies*. New York: Oxford Univ. Press

31. Galli Marxer C, Kraft ML, Weber PK, Hutcheon ID, Boxer SG. 2005. Supported membrane composition analysis by secondary ion mass spectrometry with high lateral resolution. *Biophys. J.* 88:2965–75

32. Garrison BJ, Postawa Z. 2008. Computational view of surface based organic mass spectrometry. *Mass Spectrom. Rev.* 27:289–351

33. Ghosal S, Fallon SJ, Leighton T, Wheeler KE, Hutcheon ID, Weber PK. 2008. Imaging and 3D elemental characterization of intact bacterial spores with high-resolution secondary ion mass spectrometry (NanoSIMS) depth profile analysis. *Anal. Chem.* 80:5986–92

34. Giepmans BNG, Adams SR, Ellisman MH, Tsien RY. 2006. The fluorescent toolbox for assessing protein location and function. *Science* 312:217–24

35. Grignon N. 2007. Using SIMS and MIMS in biological materials. In *Electron Microscopy: Methods and Protocols*, ed. J Juo, pp. 569–91. Totowa, NJ: Humana Press Inc.

36. Grovenor CRM, Smart KE, Kilburn MR, Shore B, Dilworth JR, et al. 2006. Specimen preparation for NanoSIMS analysis of biological materials. *Appl. Surf. Sci.* 252:6917–24

37. Groves JT, Boxer SG. 2002. Micropattern formation in supported lipid membranes. *Acc. Chem. Res.* 35:149–57

38. Groves JT, Ulman N, Boxer SG. 1997. Micropatterning fluid lipid bilayers on solid supports. *Science* 275:651–53

39. Gruska M, Medalia O, Baumeister W, Leis A. 2008. Electron tomography of vitreous sections from cultured mammalian cells. *J. Struct. Biol.* 161:384–92

40. Guerquin-Kern J-L, Hillion F, Madelmont J-C, Labarre P, Papon J, Croisy A. 2004. Ultra-structural cell distribution of the melanoma marker iodobenzamide: improved potentiality of SIMS imaging in life sciences. *Biomed. Eng. Online* 3:10. **http://www.biomedical-engineering-online.com/content/3/1/10**

41. Guerquin-Kern J-L, Wu T-D, Quintana C, Croisy A. 2005. Progress in analytical imaging of the cell by dynamic secondary ion mass spectrometry (SIMS microscopy). *Biochim. Biophys. Acta* 1724:228–38

42. Guillermier C, Lechene CP, Hill J, Hillion F. 2003. Vacuum bench for the characterization of thermoionization ion sources. *Rev. Sci. Instrum.* 74:3312–16

43. Hancock JF. 2006. Lipid rafts: contentious only from simplistic standpoints. *Nat. Rev. Mol. Cell Biol.* 7:456–62

44. Harbottle RR, Nag K, McIntyre NS, Possmayer F, Petersen NO. 2003. Molecular organization revealed by time-of-flight secondary ion mass spectrometry of a clinically used extracted pulmonary surfactant. *Langmuir* 19:3698–704

45. Hecht B, Sick B, Wild UP, Deckert V, Zenobi R, et al. 2000. Scanning near-field optical microscopy with aperture probes: fundamentals and applications. *J. Chem. Phys.* 112:7761–74

46. Hell SW. 2007. Far-field optical nanoscopy. *Science* 316:1153–58

47. Hell SW, Wichmann J. 1994. Breaking the diffraction resolution limit by stimulated emission. *Opt. Lett.* 19:780–82

48. Hendrickson CL, Emmett MR. 1999. Electrospray ionization Fourier transform ion cyclotron resonance mass spectrometry. *Annu. Rev. Phys. Chem.* 50:517–36

49. Henkel T, Tizard J, Blagburn DJ, Lyon IC. 2007. Interstellar dust laser explorer: a new instrument for elemental and isotopic analysis and imaging of interstellar and interplanetary dust. *Rev. Sc. Instrum.* 78:055107

50. Herrmann AM, Clode PL, Fletcher IR, Nunan N, Stockdale EA, et al. 2007. A novel method for the study of the biophysical interface in soils using nano-scale secondary ion mass spectrometry. *Rapid Commun. Mass Spectrom.* 21:29–34

51. Hillion F, Daigne B, Girard F, Slodzian G. 1993. A new high performance instrument: the CAMECA NanoSIMS 50. In *Secondary Ion Mass Spectrometry: SIMS IX*, ed. A Benninghoven, Y Nihei, R Shimizu, HW Werner, pp. 254–57. New York: Wiley

52. Huang B, Wang W, Bates M, Zhuang X. 2008. Three-dimensional super-resolution imaging by stochastic optical reconstruction microscopy. *Science* 319:810–13

53. Jacoby M. 2006. Bioimaging with mass spectrometry. *Chem. Eng. News* 84:55–56

54. Johansson B. 2006. ToF-SIMS imaging of lipids in cell membranes. *Surf. Interface Anal.* 38:1401–12

55. Klar TA, Jakobs S, Dyba M, Egner A, Hell SW. 2000. Fluorescence microscopy with diffraction resolution limit broken by stimulated emission. *Proc. Nat. Acad. Sci. USA* 97:8206–10

56. Kleinfeld AM, Kampf JP, Lechene C. 2004. Transport of 13C-oleate in adipocytes measured using multi imaging mass spectrometry. *J. Am. Soc. Mass Spectrom.* 15:1572–80

57. Kraft ML, Fishel SF, Galli Marxer C, Weber PK, Hutcheon ID, Boxer SG. 2006. Quantitative analysis of supported membrane composition using the NanoSIMS. *Appl. Surf. Sci.* 252:6950–56

58. Kraft ML, Weber PK, Longo ML, Hutcheon ID, Boxer SG. 2006. Phase-separation of lipid membranes analyzed with high-resolution secondary ion mass spectrometry. *Science* 313:1948–51

59. Lambacher A, Fromherz P. 2002. Luminescence of dye molecules on oxidized silicon and fluorescence interference contrast microscopy of biomembranes. *J. Opt. Soc. Am. B* 19:1435–53

60. Lechene C, Hillion F, McMahon G, Benson D, Kleinfeld AM, et al. 2006. High-resolution quantitative imaging of mammalian and bacterial cells using stable isotope mass spectrometry. *J. Biol.* 5:20.1–20.30. **http://jbiol.com/content/5/6/20**

61. Lechene CP, Luyten Y, McMahon G, Distel DL. 2007. Quantitative imaging of nitrogen fixation within animal cells. *Science* 317:1563–66

62. Legent G, Delaune A, Norris V, Delcorte A, Gibouin D, et al. 2008. Method for macromolecular colocalization using atomic recombination in dynamic SIMS. *J. Phys. Chem. B* 112:5534–46

63. Levi-Setti R. 1988. Structural and microanalytical imaging of biological materials by scanning microscopy with heavy-ion probes. *Annu. Rev. Biophys. Biophys. Chem.* 17:325–47

64. Li L, Cheng J-X. 2008. Label-free coherent anti-stokes Raman scattering imaging of coexisting lipid domains in single bilayers. *J. Phys. Chem. B* 112:1576–79

65. Li T, Wu TD, Mazeas L, Toffin L, Guerquin-Kern JL, et al. 2008. Simultaneous analysis of microbial identity and function using NanoSIMS. *Environ. Microbiol.* 10:580–88

66. Li Z, Verkhoturov SV, Locklear JE, Schweikert EA. 2008. SIMS with C60+ and Au4004+ projectiles: depth and nature of secondary ion emission from multilayer assemblies. *Int. J. Mass Spectrom.* 269:112–117

67. Lin WC, Blanchette CD, Ratto TV, Longo ML. 2006. Lipid asymmetry in DLPC/DSPC-supported lipid bilayers: a combined AFM and fluorescence microscopy study. *Biophys. J.* 90:228–37

68. Liu J, Bartesaghi A, Borgnia MJ, Sapiro G, Subramaniam S. 2008. Molecular architecture of native HIV-1 gp120 trimers. *Nature* 455:109–13

69. Lockyer NP, Vickerman JC. 2004. Progress in cellular analysis using ToF-SIMS. *Appl. Surf. Sci.* 212–232:377–84

70. Mann M, Hendrickson RC, Pandey A. 2001. Analysis of proteins and proteomes by mass spectrometry. *Biochemistry* 70:437–73

71. McDonnell LA, Heeren RMA. 2007. Imaging mass spectrometry. *Mass Spectrom. Rev.* 26:606–43

72. McMahon G, Saint-Cyr HF, Lechene C, Unkefer CJ. 2006. CN-secondary ions form by recombination as demonstrated using multi-isotope mass spectrometry of 13C- and 15N-labeled polyglycine. *J. Am. Soc. Mass Spectrom.* 17:1181–87

73. McQuaw CM, Sostarecz AG, Zheng L, Ewing AG, Winograd N. 2005. Lateral heterogeneity of dipalmitoylphosphatidylethanolamine-cholesterol Langmuir-Blodgett films investigated with imaging mass spectrometry and atomic force microscopy. *Langmuir* 21:807–13

74. McQuaw CM, Zheng L, Ewing AG, Winograd N. 2007. Localization of sphingomyelin in cholesterol domains by imaging mass spectrometry. *Langmuir* 23:5645–50

75. Meibom A, Cuif JP, Houlbreque F, Mostefaoui S, Dauphin Y, et al. 2008. Compositional variations at ultrastructure length scales in coral skeleton. *Geochim. Cosmochim. Acta* 72:1555–69

76. Moerner WE. 2007. New directions in single-molecule imaging and analysis. *Proc. Natl. Acad. Sci. USA* 104:12596–602

77. Moreau JW, Weber PK, Martin MC, Gilbert B, Hutcheon ID, Banfield JF. 2007. Extracellular proteins limit the dispersal of biogenic nanoparticles. *Science* 316:1600–3

78. Musat N, Halm H, Winterholler B, Hoppe P, Peduzzi S, et al. 2008. A single-cell view on the ecophysiology of anaerobic phototrophic bacteria. *Proc. Natl. Acad. Sci. USA* 105:17861–66

79. Oehler DZ, Robert F, Mostefaoui S, Meibom A, Selo M, McKay DS. 2006. Chemical mapping of proterozoic organic matter at submicron spatial resolution. *Astrobiology* 6:838–50

80. Orphan VJ, House CH, Hinrichs K-U, McKeegan KD, DeLong EF. 2001. Methane-consuming Archaea revealed by directly coupled isotopic and phylogenetic analysis. *Science* 293:484–87

81. Ostrowski SG, Szakal C, Kozole J, Roddy TP, Xu J, et al. 2005. Secondary ion MS imaging of lipids in picoliter vials with a buckminsterfullerene ion source. *Anal. Chem.* 77:6190–96

82. Ostrowski SG, Van Bell CT, Winograd N, Ewing AG. 2004. Mass spectrometric imaging of highly curved membranes during *Tetrahymena* mating. *Science* 305:71–73

83. Pachloski ML, Cannon DM, Ewing AG, Winograd N. 1998. Static time-of-flight secondary ion mass spectrometry imaging of freeze-fractured frozen-hydrated biological membranes. *Rapid Commun. Mass Spectrom.* 12:1232–35

84. Pachloski ML, Cannon DM, Ewing AG, Winograd N. 1999. Imaging of exposed headgroups and tailgroups of phospholipid membranes by mass spectrometry. *J. Am. Chem. Soc.* 121:4716–17

85. Parry S, Winograd N. 2005. High-resolution TOF-SIMS imaging of eukaryotic cells preserved in a trehalose matrix. *Anal. Chem.* 77:7950–57

86. Perez JB, Martinez KL, Segura JM, Vogel H. 2006. Supported cell-membrane sheets for functional fluorescence imaging of membrane proteins. *Adv. Funct. Mater.* 16:306–12

87. Popa R, Weber PK, Pett-Ridge J, Finzi JA, Fallon SJ, et al. 2007. Carbon and nitrogen fixation and metabolite exchange in and between individual cells of *Anabaena oscillarioides*. *ISME J.* 1:354–60

88. Prinz C, Malm J, Höök F, Sjovall P. 2007. Structural effects in the analysis of supported lipid bilayers by time-of-flight secondary ion mass spectrometry. *Langmuir* 23:8035–41

89. Quintana C, Bellefqih S, Lawal JY, Guerquin-Kern JL, Wu TD, et al. 2006. Study of the localization of iron, ferritin, and hemosiderin in Alzheimer's disease hippocampus by analytical microscopy at the subcellular level. *J. Struct. Biol.* 153:42–54

90. Quintana C, Wu T-D, Delatour B, Dhenain M, Guerquin-Kern J-L, Croisy A. 2007. Morphological and chemical studies of pathological human and mice brain at the subcellular level: correlation between light, electron, and NanoSIMS microscopies. *Microsc. Res. Tech.* 70:281–95

91. Reyzer ML, Caprioli RM. 2005. MALDI mass spectrometry for direct tissue analysis: a new tool for biomarker discovery. *J. Proteome Res.* 4:1138–42

92. Rickaby REM, Belshaw N, Kilburn M, Taylor A, Grovenor C, Brownlee C. 2004. Submicron-scale coccolith chemistry revealed by NanoSIMS. *Geochim. Cosmochim. Acta* 68:A215

93. Roddy TP, Cannon DM, Meserole CA, Winograd N, Ewing AG. 2002. Imaging of freeze-fractured cells with in situ florescence and time-of-flight secondary ion mass spectrometry. *Anal. Chem.* 74:4011–19

94. Roddy TP, Cannon DM, Ostrowski SG, Winograd N, Ewing AG. 2002. Identification of cellular sections with imaging mass spectrometry following freeze fracture. *Anal. Chem.* 74:4020–26

95. Romer W, Wu TD, Duchambon P, Amessou M, Carrez D, et al. 2006. Sub-cellular localisation of a 15N-labeled peptide vector using NanoSIMS imaging. *Appl. Surf. Sci.* 252:6925–30

96. Sartori A, Gatz R, Beck F, Rigort A, Baumeister W, Plitzko JM. 2007. Correlative microscopy: bridging the gap between fluorescence light microscopy and cryo-electron tomography. *J. Struct. Biol.* 161:135–45

97. Schermelleh L, Carlton PM, Haase S, Sha L, Winoto L, et al. 2008. Subdiffraction multicolor imaging of the nuclear periphery with 3D structured illumination microscopy. *Science* 320:1332–36

98. Shao L, Isaac B, Uzawa S, Agard DA, Sedat JW, Gustafsson MGL. 2008. I5S: wide-field light microscopy with 100-nm-scale resolution in three dimensions. *Biophys. J.* 94:4971–83

99. Sjovall P, Lausmaa J, Johansson B. 2004. Mass spectrometric imaging of lipids in brain tissue. *Anal. Chem.* 76:4271–78

100. Slodzian G, Daigne B, Girard F, Hillion F. 1993. Ion optics for a high resolution scanning ion microscope and spectrometer: transmission evaluations. In *Secondary Ion Mass Spectrometry: SIMS IX*, ed. A Benninghoven, Y Nihei, R Shimizu, HW Werner, pp. 294–97. New York: Wiley

101. Sostarecz AG, Cannon DM Jr, McQuaw CM, Sun S, Ewing AG, Winograd N. 2004. Influence of molecular environment on the analysis of phospholipids by secondary ion mass spectrometry. *Langmuir* 20:4926–32

102. Sostarecz AG, McQuaw CM, Ewing AG, Winograd N. 2004. Phosphatidylethanolamine-induced cholesterol domains chemically identified with mass spectrometric imaging. *J. Am. Chem. Soc.* 126:13882–83

103. Stahlberg H, Walz T. 2008. Molecular electron microscopy: state of the art and current challenges. *ACS Chem. Biol.* 3:268–81

104. Stoeckli M, Chaurand P, Hallahan DE, Caprioli RM. 2001. Imaging mass spectrometry: a new technology for the analysis of protein expression in mammalian tissues. *Nat. Med.* 7:493–96

105. Sykes MT, Williamson JR. 2009. Assembly of the bacterial 30S ribosomal subunit. *Annu. Rev. Biophys.* 38:197–215

106. Touboul D, Halgand F, Brunelle A, Kersting R, Tallarek E, et al. 2004. Tissue molecular ion imaging by gold cluster ion bombardment. *Anal. Chem.* 76:1550–59

107. Touboul D, Roy S, Germain DP, Chaminade P, Brunelle A, Laprevote O. 2007. MALDI-TOF and cluster-TOF-SIMS imaging of Fabry disease biomarkers. *Int. J. Mass Spectrom.* 260:158–65

108. Tyler BJ, Rayal G, Castner DG. 2007. Multivariate analysis strategies for processing ToF-SIMS images of biomaterials. *Biomaterials* 28:2412–23

109. Usami N, Furusawa Y, Kobayashi K, Lacombe S, Reynaud-Angelin A, et al. 2008. Mammalian cells loaded with platinum-containing molecules are sensitized to fast atomic ions. *Int. J. Rad. Biol.* 84:603–11

110. Verhaeghe E, Fraysse A, Guerquin-Kern J-L, Wu T-D, Devès G, et al. 2007. Microchemical imaging of iodine distribution in the brown alga *Laminaria digitata* suggests a new mechanism for its accumulation. *J. Biol. Inorg. Chem.* 13:257–69

111. Wang H, Obenauer-Kutner L, Lin M, Huang Y, Grace MJ, Lindsay SM. 2008. Imaging glycosylation. *J. Am. Chem. Soc.* 130:8154–55

112. Westphal V, Rizzoli SO, Lauterbach MA, Kamin D, Jahn R, Hell SW. 2008. Video-rate far-field optical nanoscopy dissects synaptic vesicle movement. *Science* 320:246–49

113. Wilson RG, Stevie FA, Magee CW. 1989. *Secondary Ion Mass Spectrometry: A Practical Handbook for Depth Profiling and Bulk Impurity Analysis*. New York: Wiley

114. Wiseman JM, Ifa DR, Zhu Y, Kissinger CB, Manicke NE, et al. 2008. Desorption electrospray ionization mass spectrometry: imaging drugs and metabolites in tissues. *Proc. Natl. Acad. Sci. USA* 105:18120–25

115. Wittig A, Arlinghaus HF, Kriegeskotte C, Moss RL, Appelman K, et al. 2008. Laser postionization secondary neutral mass spectrometry in tissue: A powerful tool for elemental and molecular imaging in the development of targeted drugs. *Mol. Cancer Ther.* 7:1763–71

116. Yu ML. 1982. Matrix effects in the work-function dependence of negative-secondary-ion emission. *Phys. Rev. B* 26:4731–34

117. Zheng H, Chu J, Qiu Y, Loh HH, Law P. 2008. Agonist-selective signaling is determined by the receptor location within the domains. *Proc. Natl. Acad. Sci. USA* 105:9421–26

118. Zumholz K, Hansteen T, Hillion F, Horreard F, Piatkowski U. 2007. Elemental distribution in cephalopod statoliths: NanoSIMS provides new insights into nano-scale structure. *Rev. Fish Biol. Fish.* 17:487–91

Controlling Proteins Through Molecular Springs

Giovanni Zocchi

Department of Physics and Astronomy, University of California Los Angeles, Los Angeles, California 90095-1547; email: zocchi@physics.ucla.edu

Annu. Rev. Biophys. 2009. 38:75–88

First published online as a Review in Advance on December 23, 2008

The *Annual Review of Biophysics* is online at biophys.annualreviews.org

This article's doi:
10.1146/annurev.biophys.050708.133637

1936-122X/09/0609-0075$20.00

Key Words

mechanochemistry, allosteric control, DNA springs, protein engineering, nanodevices

Abstract

We argue that the mechanical control of proteins—the notion of controlling chemical reactions and processes by mechanics—is conceptually interesting. We give a brief review of the main accomplishments so far, leading to our present approach of using DNA molecular springs to exert controlled stresses on proteins. Our focus is on the physical principles that underlie both artificial mechanochemical devices and natural mechanisms of allostery.

Contents

INTRODUCTION AND ENERGY SCALES

Two related properties of globular proteins make their mechanical control interesting and feasible. Proteins are soft. The thermodynamic stability (free energy difference between folded and unfolded states, under folding conditions) of a ~100-amino-acid protein domain is only of the order of $10–20\,kT$ (16) (where T is room temperature; $20\,kT \approx 0.5$ eV, which is much less than the energy of one covalent bond). The native structure can, under opportune conditions, self-assemble from many different conformations, even from the unfolded (random coil) state. The second property has been known for fifty years; its discovery by Anfinsen & Haber (1) helped to launch modern molecular biology. Nonetheless, the implication for the prospect of mechanically controlling the activity of proteins has been considered only very recently. Namely, because of property 1 it is possible to use some covalently linked chain (i.e., another molecule) to exert mechanical stresses on the protein that will substantially alter its conformation and thus its activity level (in the following discussion we think mainly of enzymatic activity). Because of property 2 it is possible not only to inhibit an enzyme (by mechanically forcing the protein into some set of inactive conformations) but also to activate it, by removing a mechanical perturbation that was holding the enzyme in an inactive state. **Figure 1** shows a practical realization: The enzyme guanylate

kinase (GK) is mechanically stressed with the result that the measured activity (speed of the reaction catalyzed by the enzyme) is reduced by a factor of 4. Then for the same sample the mechanical perturbation is removed, and the enzyme activity is completely restored.

But how does one exert controlled mechanical stresses on a protein? Our approach has been to attach a molecular spring to the protein. The spring is made of DNA, and its stiffness (and thus the mechanical stress on the protein) is controlled by the state of hybridization of the DNA. Namely, single-stranded (ss)DNA is a rather flexible polymer, with a persistence length ℓ_p^{ss} of the order of 3 bases or 1 nm. Double-stranded (ds)DNA, on the other hand, is comparatively stiffer, with a persistence length ℓ_p^{ds} of the order of 150 base pairs (bp) or 50 nm. A DNA molecule of length L intermediate between ℓ_p^{ss} and ℓ_p^{ds} will be flexible in the ss form but semirigid in the ds form. In our first experiments we used a 60-base-long DNA strand attached at the two ends to specific locations on the surface of the protein

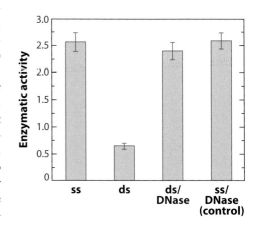

Figure 1

Mechanical control of the enzyme guanylate kinase. In the single-stranded (ss) form the DNA molecular spring is soft and there is no stress on the enzyme. Under stress [chimera in the double-stranded (ds) form] the enzymatic activity is reduced by a factor of 4. When the mechanical stress is removed (ds/DNase), the same sample recovers the original activity. The last column is a control. (Figure adapted from Reference 4.)

(**Figure 2**). In this configuration the semirigid dsDNA is sharply bent, on average roughly in a half-circle; therefore it exerts a significant mechanical stress on the attachment points on the surface of the protein (i.e., the force F is along the line connecting the attachment points) and tends to pull them apart. One can think of a strung bow, in which the protein is the string of the bow. However, one should remember that a static diagram such as **Figure 2** does not convey that the real system is "soft," and thus very large thermal fluctuations occur. The average conformation might be similar to that depicted in **Figure 2** (if the protein is not substantially deformed, which may or may not be the case), except the DNA will fill a roughly cylindrically symmetric shell around the protein (the coupling of the DNA to the protein allows rotational freedom around the attachment points). But in time the system explores many substantially different conformational states: The DNA need not lie in a plane; at times the DNA might be extended and the protein partially unfolded, or conversely the protein folded and the DNA secondary structure partially melted, and so on. Still on average there is a stress on the protein as depicted in **Figure 2**.

Although the detailed dynamics is surely complicated, the fundamental reason why the molecular spring approach works is that the energy scales are right. As a rough estimate of the elastic energy of the DNA spring of **Figure 2**, we may take a continuum linear elasticity model [the so-called worm-like chain (WLC) model] (9, 2) where the energy per unit length of the bent dsDNA is

$$\frac{E}{L} = \frac{1}{2}\frac{B}{R^2}, \qquad 1.$$

where L is the contour length of the DNA, R is the radius of curvature, and B is the bending modulus, which must be related to the persistence length of the polymer by

$$B = kT\ell_p^{ds}. \qquad 2.$$

Equation 1 is probably wrong quantitatively in our case, because it really applies to small bending, whereas the situation in **Figure 2** is one of sharp bending ($R \ll \ell_p^{ds}$), but it gives the

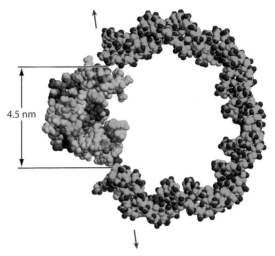

Figure 2

Diagram of a protein-DNA chimera. The guanylate kinase structure is 1S4Q, and the DNA is from the nucleosome structure 1KX5 and is drawn approximately to scale with respect to the protein. The distance between the attachment points of the DNA spring on the protein is 4.5 nm. In the double-stranded form, the DNA spring exerts a stress on the protein that tends to pull the attachment points apart.

right order of magnitude. Putting in the numbers ($\ell_p^{ds} = 50$ nm $\Rightarrow B \approx 50\ kT \times$ nm; $L = 20$ nm, $R = 5$ nm), we get an elastic energy of the order of $E \approx 20\ kT$. The thermodynamic stability of a protein domain is also of the order of 10–$20\ kT$; therefore the molecular spring can substantially perturb the conformation (and thus the activity) of the protein.

In fact, the question is whether the DNA spring in its relaxed (ss) form represents an unacceptable perturbation of the enzyme, which one could envision because of steric/charge/hydrogen-bonding effects (i.e., because of unwanted protein-DNA interactions). In the cases we (3–5) and others (18, 23) have investigated, the experimental answer is no. The reason is that the attachment points of the spring on the surface of the enzyme can be far from the active site, which is why we think of this approach as a form of artificial allostery. In hindsight, enzyme–nucleic acid complexes are common in the cell, as is the coupling of enzymes to other polymers.

In summary, one can in general attach a DNA coil or garland to an enzyme without destroying the enzyme, and the elastic energy of a DNA molecular spring is such that it can substantially perturb the conformation of the enzyme.

MOTIVATION AND RELATED APPROACHES

The allosteric enzyme is the fundamental molecular device in the cell, much as the transistor is the fundamental electronic device in a microprocessor. The concept of allostery was introduced in 1963 by Monod-Changeux-Jacob as a general mechanism of metabolic regulation (15). An allosteric enzyme has a binding site for a regulatory (effector) molecule that is distinct from the active site where the substrates bind. Binding of the effector induces a conformational change in the enzyme that turns the enzyme on (activation) or off (inhibition). The fact that the effector binding site is distinct from the active site is a crucial design element because it allows the activity of the enzyme to be regulated through a molecule that can be chemically completely different from the substrate (15). Thus, an entire metabolic pathway can be efficiently regulated if the end product of that pathway inhibits the first enzyme in the reaction chain, for example. Coming back to the electronics analogy, the allosteric enzyme is a three-terminal device akin to a transistor, with the allosteric site playing the role of the gate. Such devices mainly function as (a) switches,

(b) essential components to construct logic gates, and (c) amplifiers.

The last function is essential for constructing circuitry of any complexity, as the losses in the circuit otherwise wipe out the signals. Compared with electronics, the problem of losses is much more severe for the chemical circuitry of the cell, where the "wiring" occurs mostly through diffusion!

Thus it is interesting to design artificial mechanisms of allostery. The molecular spring approach represents one such mechanism in which a mechanical stress applied away from the active site causes a conformational transition that ultimately affects the active site.

Perhaps the first molecular construction to deliberately use a DNA spring to induce a change of conformation is the molecular beacon, introduced by Tyagi and Kramers in 1996 (19). It consists of a partially self-complementary ssDNA oligomer that forms a hairpin structure consisting of a ds stem and a ss loop. If a complementary strand is hybridized to the loop, the resulting DNA spring forces the stem to open; this conformational change is detected by fluorescence resonance energy transfer, in which the two ends of the beacon are modified with a fluorophore–quencher pair. The beacon is used as a molecular probe.

The next conceptually important development with the DNA springs came from the Scripps Research Institute, where the Ghadiri group (18) built an allosteric control mechanism for an enzyme by using the DNA spring to remove an inhibitor from the enzyme's active site. In their system, the ssDNA arm is attached by one end to the surface of the Cereus neutral protease5 (CNP) (**Figure 3**). The other end of the DNA arm is coupled to an inhibitor. In this configuration, the flexible DNA arm allows the inhibitor to bind to the active site, so that the enzyme is off. Hybridization of the DNA arm with a complementary strand stiffens the DNA spring and removes the inhibitor, activating the enzyme (**Figure 3**). This system was conceived as an amplified molecular probe for oligonucleotide detection; in principle it has the signal amplification advantages of an enzyme-linked

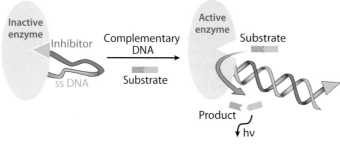

Figure 3

The spring-controlled allosteric enzyme of the Ghadiri laboratory. Hybridization of the complementary strand to the DNA arm stiffens the molecular spring, which removes the inhibitor from the active site, activating the enzyme. (Reproduced from Reference 18.)

assay, but, similar to the beacons, it is a homogeneous assay (i.e., it is not constrained by a surface-bound step).

More fundamentally, this construction also displays an interesting interplay between the elastic energy of the DNA spring, the binding energy of the inhibitor, and the phase space available to the inhibitor in the unbound state. In fact, the result of the DNA tether is that the inhibitor is effectively at a high concentration in the vicinity of the binding site, so one can tolerate a correspondingly low binding affinity and still have the inhibitor bound most of the time. There are many interesting, as yet unanswered, questions one could ask of this system. For instance, what is the optimal length of the DNA arm to have maximum dynamic range of the switch? Let us estimate the minimum binding affinity required for the inhibitor to have effective inhibition. The free energies of the (inhibitor) on and off states of the system can be written as

$$F_{on} = \varepsilon - T S_{loop}$$
$$F_{off} = -T S_{coil},$$

3.

where ε is the binding energy of the inhibitor, S_{loop} is the entropy of the DNA arm with the inhibitor end bound, and S_{coil} is the corresponding entropy with the inhibitor end free. Thus,

$$\Delta F = F_{on} - F_{off} = \varepsilon - T\Delta S, \quad \text{where}$$
$$\Delta S = S_{loop} - S_{coil} < 0.$$

4.

For an ideal chain of N links in three dimensions,

$$\Delta S \approx \frac{3}{2} k \ln N,$$

5.

so that

$$\frac{p(on)}{p(off)} = e^{-\Delta F/kT} \approx N^{-3/2} e^{-\varepsilon/kT}$$

6.

up to factors of the order of 1. Rewriting this as

$$p(on) = \frac{1}{1 + N^{3/2} e^{\varepsilon/kT}},$$

7.

we see that to have $p(on) = 1 - \delta$, $\delta \ll 1$, we need $\delta \approx N^{3/2} e^{-\varepsilon/kT}$. For example, for $\delta = 0.1$, $N = 10$ (corresponding to a 30-base-long DNA arm, the persistence length of ssDNA being ~3 bases), this gives $e^{-\varepsilon/kT} \approx 10^{-5/2} \Rightarrow \varepsilon/kT \approx 7$,

i.e., we expect 90% inhibition with a modest binding energy $\varepsilon \sim 7\ kT$.

In the experiments above, the molecular spring is used to control the binding and unbinding of an inhibitor and thus controls the enzyme. The next important development was to use the molecular spring to control the conformation of the enzyme directly. Our first construction utilized the maltose binding protein (MBP), a transport protein of the bacterial periplasm that is also a classic example of Koshland's induced fit. The overall structure of MBP is similar to that of GK, i.e., two lobes separated by a cleft where the substrate binds. When maltose (or maltotriose) binds to this cleft, the two lobes of the structure rotate toward each other into the closed conformation. We attached the two ends of the 60-base-long DNA molecular spring to opposite locations on the two lobes, reasoning that a mechanical bias toward the open conformation would, by Le Chatelier's principle, lower the binding affinity for the substrate.

In our first construction, the DNA was attached on one side to a Cys residue through a disulfide (covalent) bond and on the other side to a His-tag (through a Ni^{++} complex). This somewhat labile construction indeed showed, under tension (i.e., with the DNA hybridized to the complementary), a modest (35%) reduction in binding affinity for the substrate. We later modified the construction, attaching both ends of the DNA to the protein through covalent bonds, and obtained a larger (60%) reduction in binding affinity under tension (**Figure 4**).

The titration curves in **Figure 4** were constructed by exploiting the difference in Trp fluorescence with maltose bound or not bound. The fluorescence intensity, normalized between 0 and 1, represents the fraction f of proteins with a bound substrate. The binding constant K is extracted in the usual way by fitting to the law of mass action:

$$f = \frac{[M]}{[M] + K^{-1}},$$

8.

where $[M]$ is the equilibrium maltose concentration, which, written in terms of the initial

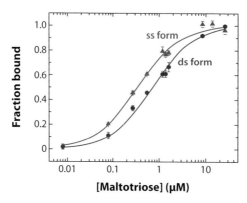

Figure 4

Effect of the mechanical stress on the binding affinity of maltose binding protein for its substrate maltotriose. The titration curves display the fraction of proteins with a bound substrate versus substrate concentration for the chimera in the single-stranded (ss) (*blue*) and double-stranded (ds) forms (*red*). The lines are fits using Equation 9, from which the binding affinity K is obtained. Under stress (*red curve*), K is reduced by 60%. (Figure adapted from Reference 3.)

(total) concentrations of maltose $[M]_0$ and the protein $[P]_0$, becomes

$$[P]_0 f^2 - (K^{-1} + [M]_0 + [P]_0) f + [M]_0 = 0. \quad 9.$$

The lines in **Figure 4** are fits to this form.

To finish the brief history of the topic thus far, the Silverman group independently developed the same concept of controlling the conformation of a macromolecule through a DNA spring, which they initially applied to control the folding-unfolding of an RNA structure (14). In their version, the DNA spring pulls the attachment points together rather than apart, destabilizing the folded structure. They monitor folding-unfolding equilibrium by gel electrophoresis and show that their DNA constraint destabilizes the folded structure by ~10 kT.

Before discussing more examples of protein-DNA chimeras, let us pause to consider what we do know and do not know at present about these molecules. There is little doubt that the mechanism of control is the mechanical stress and not some other binding or steric effect involving the DNA spring. [For the many control

experiments we refer the reader to the original papers (3, 4, 6, 14, 18, 23).] Depending on the protein, other interactions between the DNA spring and the protein may be present, but they are not the origin of the mechanism of control in the cases studied thus far.

We know that mechanical inhibition of activity can be reversible: In the case of MBP, GK, and the ribozyme (23) detaching or digesting the DNA spring restores the original substrate binding affinity or enzymatic activity.

We do not know the typical conformations explored by the protein under stress, or their statistical weights. Even simplifying to only two conformations, native and stressed, we do not know whether the stressed state is akin to the closed versus open conformational change (i.e., a collective displacement of one domain with respect to the others) or whether the protein is partially unfolded. This of course depends on the protein and the molecular spring; at present we do not have the measurements in any specific case. However, for the case of the RNA domain studied by the Silverman group, they show (through gel shift assays) that the mechanical stress provided by their DNA constraint does unfold the RNA's tertiary structure (23).

Related to the considerations in the above paragraph, we do not know whether the enzyme can be completely shut off by the mechanical perturbation provided by the DNA spring, so we do not know the maximum achievable dynamic range of control. The difficulty is that the yield of correctly constructed chimeras in the samples has to be factored in. For example, the apparent fourfold inhibition effect of **Figure 1** is limited by the yield p, which in that case we estimate $p \approx 0.7$. In numbers, if $A(s)$ and $A(r)$ are the activities in the stressed and relaxed states, respectively, the measured reduction in activity $1/\gamma_m$ is

$$\frac{1}{\gamma_m} = \frac{A_m(s)}{A(r)}, \quad 10.$$

where $A_m(s)$ is the measured activity under stress ($\gamma_m = 4$ in the example above).

Introducing the yield p,

$$A_m(s) = p\,A(s) + (1-p)A(r)$$
$$\Rightarrow \frac{1}{\gamma_m} = p\,\frac{A(s)}{A(r)} + (1-p). \qquad 11.$$

Obviously, once $A(s) \ll A(r)$ the measured effect $1/\gamma_m$ is dominated by $(1-p)$, unless $p = 1$, so it is difficult to measure the true effect $1/\gamma = A(s)/A(r)$.

To quantitatively characterize the mechanical perturbation, the most direct approach would be to obtain some measure of the strain (i.e., characterize the stressed conformation, as above) and some measure of the stress. We have recently made progress with the latter. A stress being an energy per unit volume, if one is to measure only one number, the corresponding quantity would be the elastic energy of the system. We discuss some measurements of this quantity below.

Finally, we do not know how anisotropic the response of the protein is. Does it matter where we attach the molecular spring? This is a question that has to do with the energy landscape of the protein; indeed, the molecular spring approach may be a useful tool to explore the structure of this energy surface. For example, a simplified energy landscape for an ideal two-state folder is the golf potential: one deep minimum corresponding to the native state surrounded by a roughly isoenergetic plateau corresponding to the unfolded states. In that case, it essentially would not matter where the mechanical stress is applied—the response would be always the same. Such extreme cooperativity certainly does not apply to a multidomain protein the size of GK, but it could perhaps apply to the individual domains.

SUMMARY OF THE WORK ON ENZYMES

The protein-DNA chimera studied in most detail so far was constructed with the enzyme GK and is shown in **Figure 2**. This enzyme catalyzes the transfer of a phosphate group from ATP to GMP: ATP + GMP → ADP + GDP; binding of the substrates to the cleft between

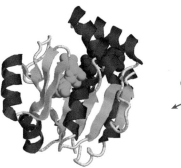 

Closed (active) conformation   Open (inactive) conformation

Figure 5

The conformational change of guanylate kinase upon substrate binding. The structures are 1GKY (*left*) and 1EX6 (*right*). With the molecular spring attached as in **Figure 2**, the mechanical stress (*arrows*) favors the open conformation.

the two lobes induces a conformational change to the closed conformation (**Figure 5**), which is catalytically active. This is an excellent example of induced fit, and indeed it was the substrate specificity of enzymes of this class that inspired Koshland's induced fit theory (13). In the case of GK, the conformational change is driven mainly by GMP binding; the functional significance is that ATP is hydrolyzed only if GMP is also bound. This is known from crystallography studies and is consistent with the behavior of the GK-DNA chimeras, where we find that a mechanical stress favoring the open conformation mainly affects (decreases) the binding affinity for GMP but not ATP.

With the DNA spring attached as in **Figure 2** the mechanical stress favors the open conformation (for which the distance between the attachment points of the spring is larger by ~1 nm; see **Figure 5**). Enzymatic activity is measured either with a luciferase-based assay, which measures the disappearance of ATP, or with a coupled enzymatic reactions assay, which measures the appearance of GDP and ADP (for details, see Reference 6). To attach the DNA spring, the end-modified DNA is coupled to a cross-linker that, on the other side, is reactive toward sulfhydryl (SH) groups. Thus, we modify the protein by site-directed mutagenesis to introduce two Cys residues at the desired locations of spring attachment. With

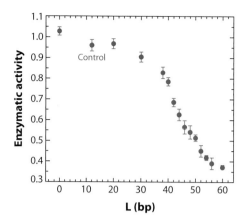

Figure 6

Modulation of the enzymatic activity of guanylate kinase with increasing mechanical stress. L is the length of the complementary DNA added in solution [i.e., L is the length in base pairs (bp) of the hybridized segment of the molecular spring]. The data point L = 12 is a control where the DNA added in solution is a 60mer but only 12 bases are complementary to the DNA of the chimera. (Figure adapted from Reference 6.)

a 60mer DNA spring, we obtain an apparent fourfold inhibition of enzymatic activity under stress (**Figure 1**); as discussed above, the measured effect is limited by the yield of correct chimera constructions in the sample. The mechanical inhibition is reversible (**Figure 1**). The effect can be modulated by hybridizing different lengths of the chimera DNA (**Figure 6**).

A detailed picture of the response of the enzyme can be obtained through titration experiments. In the Michaelis-Menten (MM) description of a one-substrate enzymatic reaction, there is one intermediate state:

$$E + S \underset{k_1, k_{-1}}{\longleftrightarrow} ES \xrightarrow{k_2} E + P, \quad 12.$$

where E is the enzyme, S is the substrate, P is the product, and ES is the intermediate state with the substrate bound to the enzyme. The partition sum of the states of the enzyme contains only two terms (substrate on, substrate off), so the probability that the substrate is on is a Fermi-Dirac distribution that, written in terms of the substrate concentration $[S]$ and

MM constant $K_m = (k_{-1} + k_2)/k_1$, reads:

$$p(on) = \frac{1}{1 + K_m/[S]}. \quad 13.$$

The product accumulates at a rate

$$\frac{d[P]}{dt} = p(on)[E]k_2, \quad 14.$$

where $[E]$ is the (total) enzyme concentration, so the kinetics is described by

$$\frac{d[P]}{dt} = \frac{[E]k_2}{1 + K_m/[S]}. \quad 15.$$

With two substrates G and A (for GMP and ATP, respectively) the partition sum contains four terms (substrates off, A on, G on, A and G on), and the probability that both substrates are bound becomes

$$p(on) = \frac{1}{(1 + K_G/[G])(1 + K_A/[A])}, \quad 16.$$

with the kinetics still given by Equation 14. Equation 16 assumes that binding of one substrate does not influence binding of the other (i.e., the binding energies just add).

In **Figure 7** the titration curves are essentially fitted with these forms [where we have taken certain limits appropriate to the specific experimental conditions and introduced

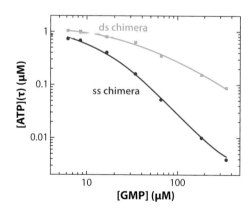

Figure 7

GMP titration experiments. The concentration of ATP remaining after a fixed time τ, $[ATP](\tau)$ is plotted versus the initial GMP concentration, for the single-stranded (ss) and double-stranded (ds) guanylate kinase chimera. The lines are fits to the Michaelis-Menten kinetics for this case. (Figure adapted from Reference 6.)

the yield p of chimeras into the formulas (4)], from which values of K_G, K_A, and k_2 are extracted (to measure all three parameters one needs in practice also ATP titration experiments). The final picture is that the mechanical perturbation applied as in **Figure 2** causes a ~10-fold reduction in K_G, essentially no change in K_A, and a change in k_2 that is less than a factor of 2. This is obviously a more informative result than if all three parameters had changed somewhat. It suggests that the typical stressed conformation is either similar to the open conformation (which by Le Chatelier's principle and the fact that GMP induces the open-to-closed conformational change must bind GMP less strongly than the closed conformation) or is at any rate a state in which the GMP binding site is substantially distorted while the ATP binding site is not. These are qualitative features that could be instructively compared with molecular dynamics simulations of the protein under stress.

The conformation of other macromolecules, not necessarily proteins, can be similarly controlled. The Silverman laboratory demonstrated the mechanical control of the enzymatic activity of a ribozyme in a remarkable study (23) in which they first prepared their own catalytic tools to attach the DNA spring (which they call the DNA constraint; the two terms are used interchangeably in this review) to the ribozyme with high yield by in vitro evolution of a deoxyribozyme (denoted 9FQ4 in Reference 23) that catalyzes formation of an RNA-DNA branch. Using this catalyst they attached two 15-base-long DNA oligomers to two specific locations on the hammerhead ribozyme (**Figure 8**), which catalyzes its own cleavage. This ribozyme consists of two RNA strands: a 43-base enzyme strand and a 20-base substrate strand that contains the cleavage site. The two DNA strands that form the DNA constraint are complementary, and when they are hybridized to each other they force the corresponding attachment points on the ribozyme to move apart (**Figure 8**), disrupting the tertiary structure of the ribozyme, thus leading to loss of activity. This version of the

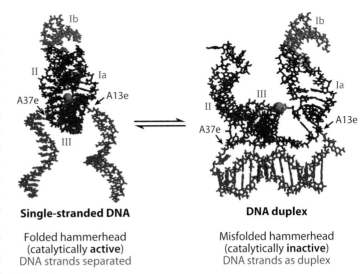

Single-stranded DNA

Folded hammerhead
(catalytically **active**)
DNA strands separated

DNA duplex

Misfolded hammerhead
(catalytically **inactive**)
DNA strands as duplex

Figure 8

Diagram of the mechanical control of the folding equilibrium of the hammerhead ribozyme (Silverman laboratory). The two DNA arms are complementary, and when hybridized they pull apart the attachment points on the ribozyme, causing the RNA tertiary structure to misfold. (Figure reproduced from Reference 23.)

DNA spring is normally in the stiff state and is relaxed by adding a competitor strand in solution. Thus, this configuration achieves allosteric activation by removing the mechanical perturbation that keeps the enzyme in an inactive state.

Figure 9 shows the control of enzymatic activity, in the form of the fraction of cleaved molecules in the course of time, determined by gel electrophoresis. They observe a greater than threefold activation in the presence of a competitor strand. They further show, probing with nucleases that cleave ss regions at specific sites, that in their construct under stress the secondary structure of the ribozyme is intact; therefore the loss of activity is due to deformations of the folded structure. Another interesting result is the behavior with a different pair of attachment points of the DNA constraint further removed from the active site. Silverman and colleagues (23) observe a smaller dynamic range of control (less than twofold activation) for the *trans* form of the ribozyme, in which the two strands constituting the ribozyme are disjoined, but essentially the same effect (as with

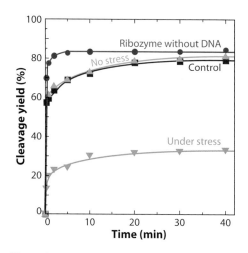

Figure 9

Mechanical control of ribozyme catalysis by the DNA spring of **Figure 8**. The plots show the fraction of molecules cleaved in the course of time t under stress (*inverted orange triangles*) and after relaxing the stress (*upright blue triangles*) using a competitor strand in solution. The red circles show the ribozyme without any DNA attached; the purple squares are a control with two noncomplementary arms attached. (Figure reproduced from Reference 23.)

the previous attachment points) for the *cis* form, in which the two strands are ligated. For the structure that presumably possesses larger cooperativity (the ligated one), the spring attachment points do not matter much, whereas for the lower cooperativity structure they do.

In natural allostery, the binding energy at the allosteric site is used to change either the average conformation (for a review see Reference 12) or the dynamics of the structure (8). With the enzyme protein kinase A (PKA) we tried to use the elastic energy of the molecular spring to induce conformational changes similar to those induced by its natural activator, cAMP (5). PKA is a tetrameric enzyme complex consisting of two regulatory subunits (RS) and two catalytic subunits (CS). The RS binds to the CS through a large surface of contact that includes the active site, which is thus masked in the complex. Binding of the allosteric effector cAMP to the RS causes it to detach from the CS,

activating the enzyme. Part of the corresponding conformational change of the RS consists of two helices tilting away from each other (11). We attached the DNA spring in such a way that under tension it would favor this displacement (5) and observed activation when the complementary DNA was hybridized to the molecular spring.

THE ELASTIC ENERGY OF PROTEIN-DNA CHIMERAS

The simplest characterization of the mechanical perturbation applied by the molecular spring is the elastic energy of the system. We recently measured this quantity through a modified chimera construction in which two separate DNA arms are attached to the protein (**Figure 10**). The hybridized DNA spring contains a nick, but this does not completely or even substantially relax the elastic energy. However, the system can relax the elastic energy by polymerizing (**Figure 10**). The monomer-dimer equilibrium, visualized by gel electrophoresis, provides a sensitive measurement of the elastic energy of the monomer. Namely, the chemical potentials for monomer and dimer can be written as

$$
\begin{cases}
\mu_M = F_{el} + kT \ln X_M \\
\mu_D = kT \ln X_D
\end{cases}, \qquad 17.
$$

where F_{el} is the elastic energy of the monomer, and $kT \ln X$ is the chemical potential associated with the concentration of chimeras: X_M is the mole fraction of monomers, X_D is the mole fraction of dimers. At equilibrium,

$$
2\mu_M - \mu_D = 0, \qquad 18.
$$

so one obtains

$$
F_{el} = \frac{1}{2} kT \ln \frac{X_D}{X_M^2} = \frac{1}{2} kT \ln K_{eq}. \qquad 19.
$$

Evaluating the equilibrium populations of monomers and dimers from the gels, one obtains K_{eq} and thus through Equation 19 a sensitive measurement of F_{el} (**Figure 11**). For example, in the configuration of **Figure 10** and with

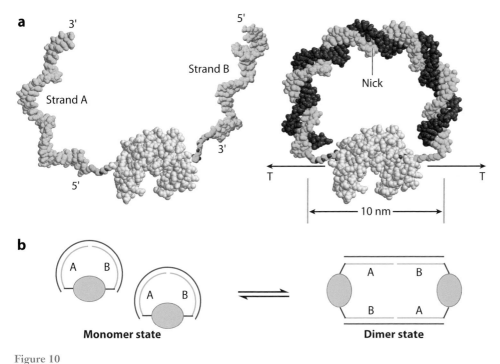

a

3'

5'

Strand B

Strand A

3'

Nick

5'

T   T

10 nm

b

A B

A   B

A   B

B   A

Monomer state   **Dimer state**

Figure 10

Diagram of (*a*) the two-arm protein-DNA chimera and (*b*) the elastic-energy-driven dimerization. (*a*) Two different DNA 30mers are covalently attached to mutated Cys residues on the surface of guanylate kinase; hybridization with a complementary 60mer DNA (*red*) forms a 60-bp DNA molecular spring with a nick in the middle. (*b*) Because of the nick in the molecular spring, monomers can release elastic energy by forming dimers as shown. (Figure adapted from Reference 20.)

$\ell = 60$ we measure $F_{el} = 9.14 \pm 0.10\,kT$. This elastic energy resides partly in the DNA spring and partly in the protein and therefore contains information about the mechanical response of both. For example, **Figure 12** shows that the apparent scatter of the experimental points in **Figure 11** is in fact a periodic modulation of the elastic energy with a period corresponding to one turn (~10 bases) of the DNA helix. Similar measurements for different attachment points of the DNA spring may similarly reveal features of the mechanical response of the protein.

If the mechanical response of the DNA spring was precisely known, the mechanics of the protein could be deduced from measurements such as those in **Figure 11**. However, the mechanics of DNA for sharp bending is at present not completely established. It

seems plausible that the WLC elastic energy (1) should fail (saturate) for sharp bending because of bubble formation in the DNA (22), and some results from cyclization experiments seemed to exhibit this saturation (7); however, these measurements were later reinterpreted to be in agreement with the WLC (10). A recent atomic force microscopy study of the conformations of DNA adsorbed on a surface finds, conversely, that the statistics at short length scales is best described by a linear (rather than quadratic) dependence of the energy on the bending angle (21). Our own measurements (**Figure 11**) show a saturation of the elastic energy with respect to the WLC, but it is not yet clear whether this is due to the DNA or the protein. It is thus an open question how the elastic energy is partitioned between the protein and the DNA spring.

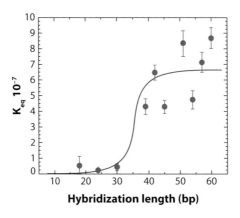

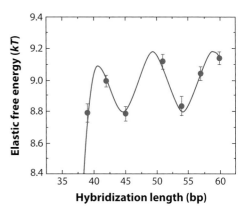

Figure 11

The dimerization equilibrium constant for increasing elastic energy of the monomer; ℓ is the length of the hybridized segment of the DNA spring. For $\ell < 30$ there is no measurable dimerization; geometrically, we do not expect tension in the molecular spring for these hybridization lengths. The increase in the equilibrium constant from $\ell = 39$ to $\ell = 60$ is due to the elastic energy of the spring destabilizing the monomer state. The red line is a "guide to the eye." (Figure reproduced from Reference 20.)

CONCLUDING REMARKS

Two important problems in mechanochemistry remain: the conversion of binding energy into mechanical motion (as in the molecular motors) and conversely the control of chemical reactions by mechanical forces. One interesting approach to investigate the second problem is to use molecular springs to control enzyme conformation. In these polymeric molecular devices, the balance between elastic energy,

Figure 12

Modulation of the elastic energy of the chimera due to the periodicity of the DNA double helix. The same data of **Figure 11** is plotted for $\ell > 39$ in terms of the elastic energy $F_{el}/kT = 1/2 \ln K_{eq}$. The red line is a fit with a function that contains a sinusoidal modulation of period $\lambda = 9.5$ bp. (Figure reproduced from Reference 20.)

binding energy, and configurational entropy can be exploited to drive processes such as the activation or inhibition of an enzyme or polymerization. In particular, the catalytic surface of an enzyme can be disassembled and reassembled mechanically. It is an open question whether mechanical control of enzymes plays an important role in the cell (17). Mechanical control is a general (chemistry-independent) method that in principle lends itself to the modular design of devices. By isolating the elastic energy term, the molecular spring approach appears well suited to explore elastic-energy-driven conformational transitions.

SUMMARY POINTS

1. Proteins are soft molecules that can be reversibly deformed by mechanical stresses. Such mechanically induced conformational transitions modulate the protein's activity.

2. The energy scales are such that DNA molecular springs can be used effectively to control protein conformation.

3. The protein-DNA chimeras display an interesting interplay between entropy and elastic and binding energies.

4. The elastic energy of a spring-loaded molecule can drive a polymerization process.

FUTURE ISSUES

1. Is the molecular spring approach a valuable tool to explore the energy landscape of proteins?

2. Will artificial allostery through the molecular springs bring new understanding of the general features of natural allosteric mechanisms?

3. Can one use protein-DNA chimeras as probes and controllable drugs inside the cell?

4. How widespread is the mechanical control of enzymes in the cell ?

5. What are the stress-strain relations for a self-assembling molecular system?

DISCLOSURE STATEMENT

The author is not aware of any biases that might be perceived as affecting the objectivity of this review.

ACKNOWLEDGMENTS

I thank my students, on whose work this review is based. Work in the Zocchi laboratory was supported by NSF grant DMR-0405632. This material is based on research sponsored by the Defense Microelectronics Activity (DMEA) under agreement numbers H94003-06-2-0607 and H94003-07-2-0702.

LITERATURE CITED

1. Anfinsen CB, Haber E. 1961. Studies on the reduction and re-formation of protein disulfide bonds. *J. Biol. Chem.* 236:1361–63

2. Bustamante C, Marko JF, Siggia ED, Smith S. 1994. Entropic elasticity of lambda-phage DNA. *Science* 265:1599–600

3. Choi B, Zocchi G, Canale S, Wu Y, Chan S, Perry LJ. 2005. Artificial allosteric control of maltose binding protein. *Phys. Rev. Lett.* 94:038103–6

4. Choi B, Zocchi G, Wu Y, Chan S, Perry LJ. 2005. Allosteric control through mechanical tension. *Phys. Rev. Lett.* 95:078102–5

5. Choi B, Zocchi G. 2006. Mimicking cAMP dependent allosteric control of protein kinase A through mechanical tension. *J. Am. Chem. Soc.* 128:8541–48

6. Choi B, Zocchi G. 2007. Guanylate kinase, induced fit, and the allosteric spring probe. *Biophys. J.* 92:1651–58

7. Cloutier TE, Widom J. 2004. Spontaneous sharp bending of double-stranded DNA. *Mol. Cell* 14:355–62

8. Cooper A, Dryden DT. 1984. Allostery without conformational change. A plausible model. *Eur. Biophys. J.* 11:103–9

9. Doi M, Edwards SF. 1985. *Theory of Polymer Dynamics*. New York: Oxford Univ. Press

10. Du Q, Smith C, Shiffeldrim N, Vologodskaia M, Vologodskii A. 2005. Cyclization of short DNA fragments and bending fluctuations of the double helix. *Proc. Natl. Acad. Sci. USA* 102:5397–402

11. Heller WT, Vigil D, Brown S, Blumenthal DK, Taylor SS, Trewhella J. 2004. C subunits binding to the protein kinase A RI alpha dimer induce a large conformational change. *J. Biol. Chem.* 279:19084–90

12. Kern D, Zuiderweg ERP. 2003. The role of dynamics in allosteric regulation. *Curr. Opin. Struct. Biol.* 13:748–57

13. Koshland DE Jr. 1994. The key-lock theory and the induced fit theory. *Angew. Chem. Int. Ed. Engl.* 33:2375–78

14. Miduturu CV, Silverman SK. 2005. DNA constraints allow rational control of macromolecular conformation. *J. Am. Chem. Soc.* 127:10144–45

15. Monod J, Changeux J-P, Jacob F. 1963. Allosteric proteins and cellular control systems. *J. Mol. Biol.* 6:306–29

16. Privalov PL. 1992. Physical basis of the stability of the folded conformations of proteins. In *Protein Folding*, ed. TE Creighton, pp. 83–126. New York: Freeman

17. Purcell TJ, Sweeney HL, Spudich JA. 2005. A force-dependent state controls the coordination of processive myosin V. *Proc. Natl. Acad. Sci. USA* 102:13873–78

18. Saghatelian A, Guckian KM, Thayer DA, Ghadiri MR. 2003. DNA detection and signal amplification via an engineered allosteric enzyme. *J. Am. Chem. Soc.* 125:344–45

19. Tyagi S, Kramer FR. 1996. Molecular beacons: Probes that fluoresce upon hybridization. *Nat. Biotechnol.* 14:303–8

20. Wang A, Zocchi G. 2008. Elastic energy driven polymerization. *Biophys. J.* 96:In press

21. Wiggins PA, et al. 2006. High flexibility of DNA on short length scales probed by atomic force microscopy. *Nat. Nanotech.* 1:137–41

22. Yan J, Marko JF. 2004. Localized single-stranded bubble mechanism for cyclization of short double helix DNA. *Phys. Rev. Lett.* 93:108108

23. Zelin E, Silverman SK. 2007. Allosteric control of ribozyme catalysis by using DNA constraints. *Chembiochem* 8:1907–11

Electron Crystallography as a Technique to Study the Structure on Membrane Proteins in a Lipidic Environment

Stefan Raunser[1] and Thomas Walz[2]

[1] Max Planck Institute of Molecular Physiology, 44227 Dortmund, Germany; email: stefan.raunser@mpi-dortmund.mpg.de

[2] Howard Hughes Medical Institute and Department of Cell Biology, Harvard Medical School, Boston, Massachusetts 02115; email: twalz@hms.harvard.edu

Annu. Rev. Biophys. 2009. 38:89–105

First published online as a Review in Advance on December 23, 2008

The *Annual Review of Biophysics* is online at biophys.annualreviews.org

This article's doi:
10.1146/annurev.biophys.050708.133649

Key Words

membrane protein structure, lipid–protein interactions, two-dimensional crystal

Abstract

The native environment of integral membrane proteins is a lipid bilayer. The structure of a membrane protein is thus ideally studied in a lipidic environment. In the first part of this review we describe some membrane protein structures that revealed the surrounding lipids and provide a brief overview of the techniques that can be used to study membrane proteins in a lipidic environment. In the second part of this review we focus on electron crystallography of two-dimensional crystals as potentially the most suitable technique for such studies. We describe the individual steps involved in the electron crystallographic determination of a membrane protein structure and discuss current challenges that need to be overcome to transform electron crystallography into a technique that can be routinely used to analyze the structure of membrane proteins embedded in a lipid bilayer.

Contents

LIPID-PROTEIN INTERACTIONS

Integral membrane proteins exist in a lipid bilayer, but structural studies on membrane proteins usually focus only on the structure of the membrane protein itself solubilized by detergent. Relatively little information is thus available on how membrane proteins interact with their surrounding lipids. Because lipids can play important roles in the folding, membrane insertion, and even modulating the function of membrane proteins, a better understanding of how membrane proteins interact with lipids is of fundamental importance. The lipids that form a shell around a membrane protein and are in direct contact with it are called annular lipids. A tight interaction of these annular lipids with the hydrophobic surface of the protein is crucial to prevent leakage of solutes across the membrane, which would compromise cell homeostasis. Because membrane proteins present a relatively rough and often mobile surface to the surrounding lipid bilayer, the fatty acyl chains of the lipids have to distort to follow the surface contours of the proteins to form a tight seal (51, 52). The interaction of lipids with proteins, however, involves not only van der Waals interactions between the lipid tails and the hydrophobic, intramembranous surface of the protein but also ionic interactions of the lipid head groups with the hydrophilic, extramembranous surface of the protein.

Several structures of membrane proteins determined by X-ray or electron crystallography also include lipid molecules (reviewed in Reference 39). To crystallize a membrane protein, it has to be purified by chromatographic methods. All loosely bound lipids are usually lost during this process. Most membrane protein structures thus include only tightly bound lipid molecules, such as those associated, for example, with FhuA (19), cytochrome c oxidase (32), and formate dehydrogenase (43). The strong protein association of these lipids suggests that they occupy specific binding sites on the protein surfaces and may be essential for activity and/or stability of the proteins. The removal of tightly bound cardiolipin from cytochrome c oxidase, for example, leads to activity loss, which can be recovered by subsequent reintroduction of lipids (79). In many cases lipids have to be added during purification and/or three-dimensional (3D) crystallization to keep the proteins stable in solution, to prevent dissociation of native lipids bound to the proteins, and/or to stabilize protein-protein interfaces in multimeric membrane proteins, as was the case for the 3D crystallization of light-harvesting complex II (54) and the human β_2-adrenergic G protein–coupled receptor (12, 73) (**Figure 1**).

The majority of lipids interacting with membrane proteins in a biological membrane are not specifically bound lipids but rather annular lipids, which only engage in transient and relatively nonspecific interactions with the protein (59). Because these interactions are weaker than those formed by specifically bound lipids, high-resolution crystal structures of only four membrane proteins resolved annular lipids.

Annular lipids: lipids that surround a membrane protein and are in direct contact with it

Electron crystallography: an EM method, similar to X-ray crystallography, used to study the structure of molecules in a 2D crystal

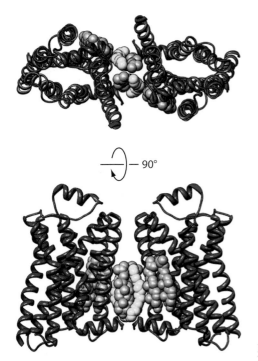

Figure 1

Structure of the human β$_2$-adrenergic receptor. The top (*top*) and side (*bottom*) views of the human β$_2$-adrenergic receptor (PDB entry 2RH1) (12) show that the contact between the two monomers (*red and blue ribbon models*) is mediated mainly by lipids, with two cholesterol molecules (*gold spheres*) and two palmitic acid molecules (*yellow spheres*) forming the majority of the interactions.

The refined, electron crystallographic structure of bacteriorhodopsin (bR) at 3.5 Å included 30 lipid chains (29) (**Figure 2a**). The lipids, however, could not be specified because the head groups were disordered in this structure. The bR structure subsequently obtained from lipidic cubic phase (LCP) crystallization at 1.55 Å revealed 18 tightly bound lipid chains, of which 13 were phytanyl lipids and 1 was a squalene (57) (**Figure 3a**). The 1.9 Å structure of the aquaporin-0 (AQP0)-mediated membrane junction obtained by electron crystallography revealed nine lipid molecules per monomer (26, 37) (**Figure 2b**). A layer of complete and partially resolved lipid molecules surrounds the voltage-dependent Kv1.2 channel (55, 56) (**Figure 3b**). In the case of the rotor (K ring) of

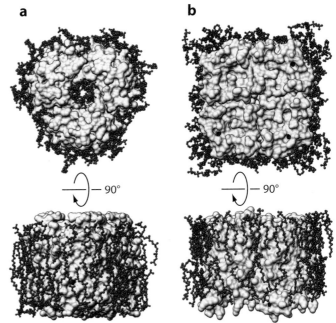

a   **b**

Figure 2

Top (*top panels*) and side (*bottom panels*) views of annular lipids in membrane protein structures determined by electron crystallography.
(*a*) Bacteriorhodopsin (PDB entry 2BRD) (29). (*b*) Aquaporin-0 (PDB entry 2B6O) (26). The proteins are shown in surface representation (*yellow*) and the lipid molecules are shown in ball-and-stick representation (*red*).

the V-type Na$^+$-ATPase from *Enterococcus hirae*, all the lipids associated with the external surface of the K ring were lost during purification, whereas nearly all lipids inside the K ring were retained and could be resolved in the crystal structure (65) (**Figure 3c**).

All these structures showed that the lipid acyl chains tend to occupy grooves in the protein surface, where they are engaged in hydrophobic interactions with apolar residues and backbone atoms. All the structures revealing annular lipids have in common that either the crystal packing was very tight, so that almost the complete space between proteins was occupied by ordered lipids, or, in the case of the V-type Na$^+$-ATPase rotor, the lipids were entrapped inside a protein ring. The tight packing of lipids between proteins or their entrapment inside a protein ring was likely the reason that the lipids could be resolved, because the mobility of these lipids must be substantially lower

bR:
bacteriorhodopsin

Lipidic cubic phase (LCP): a matrix used for membrane protein crystallization consisting of curved lipid bilayers whose structures follow infinitely periodic minimal surfaces

AQP: aquaporin

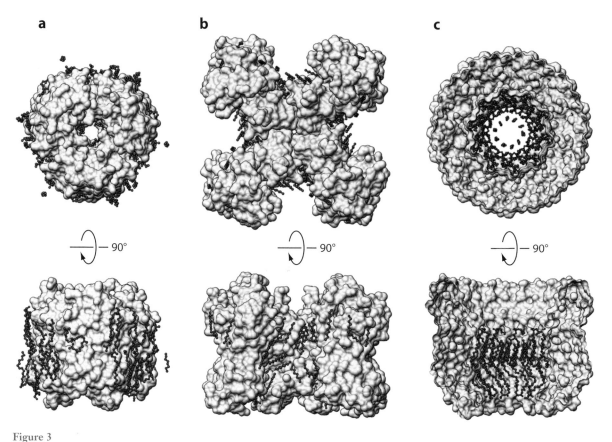

Figure 3

Top (*top panels*) and side (*bottom panels*) views of annular lipids in membrane protein structures determined by X-ray crystallography. (*a*) Bacteriorhodopsin (PDB entry 1C3W). (*b*) A chimeric voltage-dependent K$^+$ channel (PDB entry 2R9R) (56). (*c*) V-type Na$^+$-ATPase from *Enterococcus hirae* (PDB entry 2BL2) (65). The proteins are shown in surface representation (*yellow*) and the lipid molecules are shown in ball-and-stick representation (*red*).

DMPC: dimyristoyl phosphatidylcholine

Two-dimensional (2D) crystals: crystalline arrays of proteins typically just a single molecule thick, often produced by reconstituting membrane proteins into a lipid bilayer

than in an unrestrained biological membrane, in which the lipids are more mobile and can exchange with bulk lipids of the bilayer. Only two structures, those of AQP0 (26) and the K ring (65), also showed lipids outside the annulus of lipids that directly interact with the protein. In both structures, these representatives of bulk lipids were more mobile, as indicated by higher temperature factors, and thus less well defined than the annular lipids.

The structures of bR, the K ring, and Kv1.2 were obtained with protein preparations in which the proteins were not stripped of their associated lipids during purification. The lipids seen in the structures are therefore the native lipids from the donor mem-

brane. In contrast, AQP0 was fully delipidated during purification. The lipids in the structure thus represent dimyristoyl phosphatidylcholine (DMPC) molecules, the synthetic lipid used for two-dimensional (2D) crystallization. Because DMPC is not present in native lens membranes, from which AQP0 was purified, the lipid-protein interactions seen in the structure must be nonspecific in nature. Although the lipids seen in the X-ray and electron crystallography structures of bR were those of the native *Halobacterium salinarum* membrane, the structures also seem to show nonspecific lipid-protein interactions. When the two structures are overlaid, only few of the lipids overlap, suggesting that they randomly interact with the hydrophobic surface of bR. Aquaporins remain a special case, however, because they can form 2D crystals with different lipids. AQP1, for example, formed well-ordered 2D crystals with three different lipids, DMPC (41), *E. coli* polar lipids (64), and dioleyl phosphatidylcholine (DOPC) (76). Similarly, AQP0 forms well-ordered, crystalline arrays not only in DMPC, but also in a number of other lipids (R. Hite & T. Walz, unpublished results). By visualizing the structure of AQP0 in 2D crystals obtained with lipids with either different acyl chains or different head groups, it may thus become possible to understand how the bilayer structure adapts to accommodate the hydrophobic belt of membrane proteins, which lipid characteristics affect nonspecific interactions with membrane proteins, and whether the protein structure changes to adapt to lipid bilayers of different thicknesses.

BRIEF OVERVIEW OF METHODS USED TO STUDY THE STRUCTURE OF MEMBRANE PROTEINS IN A LIPIDIC ENVIRONMENT

Freeze-Fracture Electron Microscopy and Atomic Force Microscopy

Freeze-fracture electron microscopy (EM) was the first technique that made it possible to obtain structural information on membrane proteins in a lipid bilayer. When cells were fractured through the membrane and the exposed surfaces metal shadowed, membrane proteins became visible as particles protruding from the membrane (62). Although the methodology has been improved over the years, it remains a rather low-resolution technique. Still, freeze-fracture EM is routinely used to visualize membrane proteins reconstituted into liposomes (74), to study the assembly of membrane proteins into functional domains (86) (**Figure 4a,b**), and even to obtain limited information on the structure of membrane proteins (95). While freeze-fracture EM continues to play a role in structural studies of

membrane proteins in lipid bilayers, another technique, atomic force microscopy (AFM) (6), has gained in importance for such studies. Like freeze-fracture EM, AFM visualizes the surface of membrane proteins protruding from the membrane but at higher resolution and with a much higher signal-to-noise ratio. AFM can thus be used to visualize and characterize reconstituted membrane proteins (reviewed in Reference 63) and to elucidate the supramolecular organization of membrane proteins in native membranes, such as photosynthetic complexes in bacterial membranes (reviewed in Reference 81) and AQP0 and connexins in junctional domains of lens membranes (9) (**Figure 4c,d**).

Electron Crystallography

Freeze-fracture EM and AFM do not provide information on the transmembrane domains of integral membrane proteins. Henderson & Unwin pioneered electron crystallography to determine protein structures from 2D crystals (91) and produced a density map of bR at 7 Å resolution (34). This density map showed for the first time that transmembrane domains adopt an α-helical secondary structure. Henderson and coworkers continued to work on bR and produced the first atomic model by electron crystallography (33), which was later refined to higher resolution (44). Initially, electron crystallography was only used to study the structure of naturally occurring 2D crystals, and it only matured into a generally applicable structure determination technique when atomic structures were determined using artificially produced 2D crystals, the first of which was obtained with the plant light-harvesting complex II (49). While electron crystallography has been developed with bR, it may now be most intimately linked to structural studies of members of the AQP family of water channels (reviewed in References 3 and 28). The electron crystallographic structure of AQP1 (64) provided the first atomic model of an aquaporin, and the AQP4 structure (36) was the first atomic model of a mammalian multispan membrane

Freeze-fracture electron microscopy: electron microscopic imaging of surface structures by quick-freezing a specimen, fracturing it, and shadowing the fracture surfaces with heavy metal

EM: electron microscopy

Atomic force microscopy (AFM): imaging of surface structures by scanning a fine tip over the specimen and measuring the interaction forces

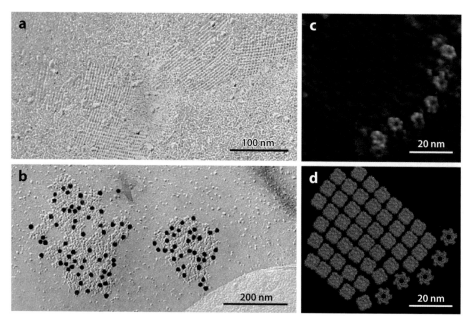

Figure 4

Freeze-fracture electron microscopy and atomic force microscopy (AFM). (*a, b*) Freeze-fracture electron microscopy of the AQP4 isoform AQP4M23, which assembles into square arrays. (*a*) Image of the exoplasmic fracture face of Chinese hamster ovary cells transfected with AQP4M23, showing the typical pit-like imprints of square arrays. (*b*) Immunogold labeling of the protoplasmic fracture face, showing the protein as protrusions from the membrane and labeled with gold particles. Adapted with permission from Reference 86, copyright 2008. Reprinted with permission from Elsevier. (*c, d*) AFM of a junctional domain in a native lens fiber cell membrane. (*c*) High-resolution AFM topograph of a junctional microdomain consisting of an AQP0 array terminated by a row of connexons. (*d*) Space-filling representation of the pseudoatomic model of the junctional microdomain deduced from the AFM topograph. Adapted from Reference 82.

protein obtained with recombinant protein. Finally, with a resolution of 1.9 Å the structure of AQP0 is the highest-resolution structure determined to date by electron crystallography, which revealed not only the structure of the protein but also water and lipid molecules (26).

X-Ray Crystallography

Electron crystallography was once the only technique used to study the structure of membrane proteins in a lipid environment at high resolution. The situation has changed with the introduction of LCP as a new matrix for 3D crystallization of membrane proteins (50). In the continuous bilayer formed by lipids in the LCP, membrane proteins can freely diffuse for

nucleation and crystal growth. This method was key for the successful crystallization of bacterial rhodopsins (5, 46, 57, 69). LCP in combination with a recombinant insertion of lysozyme was particularly powerful in determining the high-resolution structure of the human β_2-adrenergic G protein-coupled receptor (12). The method has some technical disadvantages and its application is thus limited. It does not work reliably at low temperatures because the hosting lipid turns solid. Its handling is difficult and time-consuming, thus making it arduous to test many different crystallization conditions. Crystals also tend to be small, hard to see, and difficult to extract from the LCP. However, many improvements have been made in the past few years to overcome these difficulties and limitations; in particular, new lipids have been designed to allow crystallization at low temperatures (60).

Lipid-detergent bicelles (18) and lipidic sponge phases (93) have been introduced as new matrices for protein crystallization; they are liquid and can therefore be used in standard vapor-diffusion, hanging-drop experiments. Bicelles are small bilayer disks into which membrane proteins can be incorporated to generate a more bilayer-like environment. Most recently the structure of the human β_2-adrenergic receptor was solved using lipid-detergent bicelles (73). A different approach to creating a membrane-like environment for 3D crystallization of membrane proteins is to add detergent-solubilized lipids to the protein, as has recently been done for a chimeric voltage-dependent K^+ channel (56).

Nuclear Magnetic Resonance Spectroscopy

The size of proteins that can be studied by NMR is limited to $\sim$40 kDa. Many membrane proteins are thus too big for NMR studies, especially because the detergent micelle needed to solubilize membrane proteins contributes to the size limitation. Nevertheless, NMR produced the structures of several small membrane proteins, such as the solution structure of the transmembrane domain of glycophorin A (58). Small β-barrel proteins, such as outer membrane protein (Omp) X and OmpA (4, 20), PagP (40), and OmpG (53), were the first intact membrane proteins whose structures were elucidated by NMR. With the exception of PagP, all the NMR structures were solved after X-ray crystal structures were already available. Recently, NMR produced the first de novo structures of helical membrane proteins, namely those of the full-length phospholamban homopentamer (67) and the transmembrane region of the tetrameric M2 proton channel of influenza A virus (83). Owing to increasingly stronger magnets, progress is now also being made in the structure determination of larger membrane proteins. For example, complete assignment of backbone reso-

nances and secondary structures of the 44-kDa trimeric diacylglycerol kinase (68) and the 68-kDa tetrameric KcsA potassium channel (13) has been accomplished, suggesting the feasibility of complete structure determination of these larger membrane proteins.

Methods are now also being developed that may make it possible to use NMR to determine membrane protein structure in a lipidic environment. One such method is to incorporate membrane proteins into fast tumbling bicelles (71, 80). One study compared the structure of a short peptide of the HIV-1 envelope protein gp41 solubilized either in detergent micelles or in small bicelles (14), but no high-resolution structure of a full-length membrane protein has so far been determined in bicelles [except for the small transmembrane domain of BNip3 (7)]. Solid-state NMR on oriented lipid bilayers is another NMR technique used to obtain structural information of membrane proteins in a lipidic environment. Lines in solid-state NMR spectra are less intense and significantly broader than lines in solution NMR spectra, causing severe problems in assigning the peaks. However, with NMR spectra obtained with polarization inversion spin exchange at the magic angle (72) it has been possible to determine the orientation of transmembrane α-helices with respect to the lipid bilayer [e.g., the helix orientation of the fd virus coat protein (97)]. Solid-state NMR measurements require much higher protein concentrations than solution NMR studies do. The best solid-state NMR spectra of membrane proteins have thus been obtained with crystalline samples, such as 2D crystals of OmpG (35). Nevertheless, many difficulties still need to be overcome before solid-state NMR can deliver the first high-resolution structure of a membrane protein embedded in a lipid bilayer.

INDIVIDUAL STEPS IN ELECTRON CRYSTALLOGRAPHY

At this point, electron crystallography may still be the most appropriate method to study the structure of membrane proteins in a lipidic

Bicelles: small lipid bilayer disks that form in certain lipid/amphiphile mixtures and can be used to incorporate membrane proteins for structural studies

LPR: lipid-to-protein ratio

environment (38). In the remainder of this review, we describe the individual steps involved in determining the structure of a membrane protein by electron crystallography and discuss challenges that need to be overcome to transform electron crystallography into a method that can rival X-ray crystallography.

Protein Preparation

Arguably the greatest problem in the structure determination of a membrane protein, by any technique, remains obtaining the large amounts of protein needed for structural studies. Expressing proteins, in particular mammalian membrane proteins, is a formidable challenge. Bacterial membrane proteins are somewhat easier to obtain in large quantities, and the first structures of members of many membrane protein families were thus obtained with proteins of bacterial origin. The structures of mammalian membrane proteins, however, are the most desirable ones, because they can serve as the basis for rational drug design. Many studies employing a variety of strategies have focused on the expression of mammalian membrane proteins in various systems (17, 87, 88), but no generally applicable method has yet been identified and the success ratio remains low.

Once expressed, membrane proteins have to be solubilized with detergent. The detergent has to efficiently extract the protein from the membrane and keep it stable in solution, often causing a problem for 3D crystallization owing to the high protein concentrations and long incubation times needed to grow 3D crystals. Furthermore, detergents with short aliphatic chains can destabilize membrane proteins, whereas the large micelles formed by long-chain detergents can interfere with 3D crystal contacts. Much lower protein concentrations suffice for 2D crystallization, and because lipids are added shortly after purification, the time the membrane protein is vulnerable is dramatically shorter. Hence, it is often possible to use more aggressive detergents to purify proteins for 2D than for 3D crystallization. Also, as the detergent is removed in the reconstitution

process, long-chain detergents have no adverse effect on 2D crystal formation.

2D Crystallization

2D crystallization is accomplished by mixing detergent-solubilized membrane proteins with detergent-solubilized lipids followed by removal of the detergent, using only little lipid to induce the formation of 2D crystals upon reconstitution. Removal of the detergent can be achieved by dialysis of the protein/lipid mixture against detergent-free buffer (42, 48). Detergent can also be removed by binding it to BioBeads (78) or by chelating it with methyl-β-cyclodextrin (84). Alternatively, the detergent concentration can simply be lowered below its critical micelle concentration (cmc) through dilution (75). The most critical time during the reconstitution process may be when the detergent reaches its cmc. At this time, the protein starts to be incorporated into a lipid bilayer, but the remaining detergent keeps the bilayer fluid, allowing the protein to easily diffuse and make contacts with neighboring proteins. It could thus be advantageous to cross the cmc slowly, extending the time during which the proteins can crystallize in the membrane. In dialysis, the speed of detergent removal is difficult to control, but dialysis can be slowed by lowering the temperature or by using dialysis buffers with decreasing detergent concentrations. Chelation and dilution allow for a finer control of detergent removal, but because these techniques have not yet been widely used, all the high-resolution structures to date were produced with 2D crystals obtained by dialysis.

A number of factors critical for the formation of 2D crystals include the lipid used for reconstitution, the lipid-to-protein ratio (LPR), and the composition and pH of the dialysis buffer. Less clear is the influence of other factors, such as the initial detergent concentration of the sample, a preincubation of the protein with the lipid prior to detergent removal, and temperature cycles during reconstitution. One of the main difficulties in defining the important factors for producing 2D crystals lies in the poor reproducibility of 2D crystallization experiments, which may be due to variations in the protein preparations, changes in the used lipid and detergent batches, or even pipetting inaccuracies resulting in slightly different LPRs. Unlike in the X-ray crystallography field, no screens or robots are commercially available for 2D crystallization. Several groups have been developing 2D crystallization screens and robots (11, 92). Such developments will certainly facilitate the production of 2D crystals, but experiments aimed at better understanding the 2D crystallization process and making it more reproducible may have the highest impact on overcoming the greatest hurdle in making electron crystallography as powerful as X-ray crystallography.

2D crystals can be vesicular, tubular, or sheet-like, and it is currently unclear what parameters determine the crystal morphology. Vesicular crystals can be useful for functional assays (94), but they are not ideal for structural studies because they tend to form folds upon adsorption to an EM grid, which reduces the useful area for data collection and image processing. In addition, diffraction patterns and Fourier transforms of images of vesicular 2D crystals usually show two sets of diffraction spots originating from the lower and the upper membrane of the collapsed vesicle. The spots of the two lattices can overlap, which causes difficulties for the subsequent data processing. Tubular crystals have the advantage that each image contains projections of the crystallized protein in many orientations, making it unnecessary to collect data from tilted specimens, but it is usually not possible to collect high-resolution electron diffraction patterns of tubular crystals. Thus, structures from tubular crystals have to be produced solely on the basis of image data, and so far only the tubular crystals formed by the nicotinic acetylcholine receptor have yielded an atomic model of a membrane protein (61, 90).

Crystalline sheets are ideal for high-resolution electron crystallographic studies. In addition to being well ordered, such crystals also need to have sufficiently large coherent areas (∼1 μm or larger) to be suitable for collecting high-resolution electron diffraction patterns. A problem that often occurs with crystalline sheets is stacking. If a defined number of crystalline sheets stack together with good alignment between the layers, such as in the case of the double-layered AQP0 crystals (26, 27) or the double-layered (89) and triple-layered (66) connexin 2D crystals, it is straightforward to reconstruct the structure of the protein forming the multilayered crystals. However, if the layers are not well aligned to each other or if the number of layers in the stack varies, structure determination becomes difficult, because only data of the same crystal type can be combined to produce a structure. Currently, there are no straightforward methods to determine the number of layers in a crystal, particularly in the case of electron diffraction patterns. In the case of the double-layered 2D crystals of AQP4, the two layers were not well aligned with respect to each other, and to overcome this problem, each crystal from which an electron diffraction pattern was recorded also had to be imaged. By attempting to combine the images, it was possible to identify images of a particular crystal subtype, which then made it possible to combine the corresponding electron diffraction patterns. Although this tour de force approach produced an atomic model of AQP4 from heterogeneous, double-layered 2D crystals (36), a straightforward method to determine a structure from 2D crystals with a variable number of layers remains to be developed.

Specimen Preparation

The two methods currently used to prepare 2D crystals for data collection are vitrification and sugar embedding. In vitrification (1), the crystals are applied to a carbon film and quickly plunged into liquid ethane, which is cooled to liquid nitrogen temperature. An advantage of vitrification is that the 2D crystals are only lightly attached to the carbon film, thus preventing the squashing of large protein domains protruding from the lipid bilayer. However, because the crystals are only loosely attached to the carbon film, the crystals often are not flat, which causes problems for data collection. Most atomic structures were thus determined using 2D crystals embedded in sugar. Because of the partial drying of the sample in this method, the crystals tend to lie flatter on the carbon film, thus increasing the chances of collecting high-resolution data from tilted specimens. Excessive drying, however, damages the crystals. Furthermore, because the crystals are more firmly attached to the carbon film, sugar embedding can cause protein domains protruding from the lipid bilayer to deform.

For high-resolution data collection from tilted specimens, the support to which the crystals are adsorbed has to be almost atomically flat. If the crystals are not flat, reflections in the direction perpendicular to the tilt axis become increasingly blurred with increasing resolution until they completely disappear. Factors that influence specimen flatness are the carbon film as well as the EM grid. Although several accounts describe how suitable carbon films can be produced (25, 31, 47), preparing an atomically flat carbon film remains an art form. Standard EM grids consist of copper, which has a thermal expansion coefficient that differs significantly from that of carbon. Upon cooling of a grid, the copper thus shrinks more than the carbon film, resulting in the formation of folds in the carbon film, a phenomenon known as cryo-crinkling (8). To overcome this problem, molybdenum grids have been introduced, as molybdenum has a thermal expansion coefficient close to that of carbon, thus preventing cryo-crinkling. In addition, new molybdenum grids now have a smooth surface, further preventing the occurrence of folds in the carbon film (21).

Considering the need for extremely flat specimens and perfect sugar embedding of the 2D crystals, preparing specimens for high-resolution electron crystallographic data collection is far from routine even for the most experienced practitioners. The difficulty in preparing the many good specimens needed to collect a full 3D data set is thus probably the second largest remaining hurdle in determin-

ing the structure of a membrane protein by electron crystallography.

Data Collection

Two types of data are collected in electron crystallography, images and electron diffraction patterns. Images contain both amplitude and phase information, and structures can be determined exclusively from image data (61), but electron diffraction patterns provide more accurate amplitude information. In addition, high-resolution diffraction patterns are much easier to collect than high-resolution images because diffraction is translation invariant and thus not sensitive to specimen movement, a major problem when collecting high-resolution images.

The greatest problem when collecting EM data of biological specimens is beam damage. To minimize beam damage, EM data is collected using low-dose procedures, in which the specimen is illuminated only with a low electron dose ($\sim$10 electrons/Å^2) during the actual recording of an image or diffraction pattern, and every 2D crystal is exposed only once. To further counteract the effects of beam damage, the specimen is cooled to low temperature. Although low temperature does not prevent the actual beam damage, diffusion slows down and fragments remain longer at the position where they are initially formed, thus slowing structural collapse of the specimen. Typically, specimens are cooled down to liquid nitrogen temperature, but a study by Fujiyoshi and coworkers measuring the decay of diffraction intensities with accumulating electron dose at different temperatures showed that liquid helium temperature provides even further cryo-protection, prompting the development of electron microscopes with specimen stages cooled with liquid helium (22). Although the benefit of liquid helium temperature is not generally accepted, virtually all atomic models of membrane proteins determined by electron crystallography to date are based on data collected at liquid helium temperature.

The electron microscope only provides projection images. To obtain information on the specimen in the direction parallel to the electron beam, images and electron diffraction patterns must be collected from tilted specimens. Collecting images of tilted specimens has proven particularly difficult because of the mechanical and thermal instabilities of side-entry specimen holders under tilted conditions and because of beam-induced movement. The stability of specimen stages has significantly improved with the introduction of top-entry stages and stage designs that thermally uncouple the specimen from the outside environment. Progress has also been made in overcoming specimen charging, the cause of beam-induced movement. The first solution to the problem was spot-scan imaging, in which the electron beam is reduced to a small diameter and scanned over the specimen, illuminating and imaging only one small specimen area at a time (16). An additional advantage of spot-scan imaging is that the spot can be scanned parallel to the tilt axis and the focus can be adjusted for each line, thus compensating for the focus gradient perpendicular to the tilt axis. Images of tilted specimens recorded with such dynamic focusing can then be processed in the same way as those of untilted specimens, simplifying the image processing procedure. An alternative approach to overcoming specimen charging is the carbon sandwich technique, in which the specimen is sandwiched between two carbon films (30). In the case of bR, this preparation technique increased the yield of useful images of highly tilted specimens from 2% to 90% (30). Recently, the use of conductive titanium-silicon (TiSi) metal glass films has been introduced as another solution to the charging problem (77). The use of TiSi with bR nearly eliminated the problem of beam-induced movement, resulting in a success rate of close to 100%. However, whether TiSi films are similarly successful for imaging other specimens remains to be tested.

Data Processing

Data processing with the Medical Research Council (MRC) software (15) has remained almost unchanged since electron crystallography was first established (2). Images are usually recorded on photographic film, which are then digitized. The digitized images are computationally unbent to correct for distortions in the lattice of the imaged 2D crystal, corrected for the contrast transfer function (CTF), and the phases of the reflections are extracted from calculated Fourier transforms of the unbent, CTF-corrected images. Intensities are obtained from electron diffraction patterns, which are usually recorded on a CCD camera, by masking out the beam stop, subtracting a radial background, and then integrating the pixel values of the diffraction spots from which the local background is subtracted. Phases and amplitudes from untilted and tilted images are combined to reconstruct the reciprocal lattice lines, which are then sampled, assuming a certain thickness for the 2D crystal. The resulting 3D phase and amplitude data set is converted into a density map by inverse Fourier transformation, and a model is built into the density map and refined with the standard software packages used in X-ray crystallography.

The algorithms used in electron crystallographic data processing proved to be robust and yielded a number of atomic structures, but data processing used to be time-consuming and tedious, and the programs were not particularly user friendly. The situation has changed with increasingly powerful computers and the *2dx* software package, which introduces a user-friendly interface for the MRC programs and allows for automation of the image processing (23, 24). New software is also being developed to facilitate and automate the processing of electron diffraction patterns (70). One trend now is to adapt methods used in single-particle EM to address problems in electron crystallography, such as the maximum-likelihood approach to improve the processing of poorly ordered 2D crystals (96) and single-particle refinement to address the lack of specimen flatness (45).

One strength of electron crystallography is that the phases can be directly obtained from the images, but collecting high-resolution images is much more challenging than recording

CTF: contrast transfer function

high-resolution diffraction patterns. Therefore, adapting indirect phasing methods from X-ray crystallography may provide a faster way to structure determination by electron crystallography. Early attempts to obtain phases by heavy metal labeling of bR were unsuccessful because the phasing power was too small to provide sufficiently accurate phases (10). In contrast, molecular replacement proved powerful and was used successfully to determine the structures of AQP0 (27) and AQP4 (36). Not yet fully explored are phase extension approaches, although early attempts seemed promising (85).

SUMMARY POINTS

1. With only four structures of membrane proteins with annular lipids, our structural knowledge of nonspecific lipid-protein interactions is limited.

2. Several methods have been used in X-ray crystallography to study membrane proteins in a lipidic environment: 3D crystallization in the presence of lipids, in LCP, and in lipid bicelles.

3. Methods are being developed in NMR spectroscopy to study membrane proteins in a lipidic environment: solution NMR of membrane proteins in bicelles and solid-state NMR of membrane proteins reconstituted in lipid bilayers.

4. Electron crystallography may currently still be the best-suited technique to study membrane proteins in a lipidic environment.

5. Electron crystallography is mainly limited by difficulties in expressing sufficiently large quantities of membrane proteins, in producing (and reproducing) well-ordered 2D crystals, and in preparing suitable EM specimens, less so by data collection and data processing.

6. Electron crystallography is not an all-or-nothing method, and even poorly ordered 2D crystals can provide structural information.

7. 2D crystals of some membrane proteins can be produced with different lipids, which may make it possible to systematically study nonspecific lipid-protein interactions.

FUTURE ISSUES

1. All techniques currently used to determine membrane protein structures, including electron crystallography, require a substantial amount of protein. Better expression systems are thus required, especially for mammalian membrane proteins.

2. Concerted efforts aimed at better understanding the 2D crystallization process and at improving the reproducibility of 2D crystallization are needed.

3. More work needs to be invested in improving the preparation of 2D crystals for high-resolution data collection, with the goal to reproducibly obtain very flat specimens and avoid beam-induced movement.

4. Considering the difficulties in high-resolution imaging of 2D crystals, alternative ways of obtaining high-resolution phases, such as phase extension approaches, should be explored.

5. Treating the unit cells of a 2D crystal as single particles may become an alternative method to process images of poorly ordered 2D crystals.

6. The future will certainly bring more automation of every step involved in electron crystallography, from screening 2D crystallization conditions to computational data processing.

DISCLOSURE STATEMENT

The authors are not aware of any biases that might be perceived as affecting the objectivity of this review.

ACKNOWLEDGMENTS

Electron crystallographic work on membrane proteins in the Walz laboratory is supported by National Institutes of Health grants R01 EY015107 (to T.W.), R01 GM082927 (to T.W.), and P01 GM062580 (to S.C. Harrison). T.W. is an Investigator of the Howard Hughes Medical Institute. S.R. was a Fellow of the German Academy of Sciences Leopoldina (BMBF-LPD 9901/8-163) and is now supported by the Emmy Noether Program of the German Research Foundation (RA 1781/1-1). The molecular EM facility at Harvard Medical School was established with a generous donation from the Giovanni Armenise Harvard Center for Structural Biology.

LITERATURE CITED

1. Adrian M, Dubochet J, Lepault J, McDowall AW. 1984. Cryo-electron microscopy of viruses. *Nature* 308:32–36

2. Amos LA, Henderson R, Unwin PN. 1982. Three-dimensional structure determination by electron microscopy of two-dimensional crystals. *Prog. Biophys. Mol. Biol.* 39:183–231

3. Andrews SA, Reichow LS, Gonen T. 2008. Electron crystallography of aquaporins. *IUBMB Life.* 60:430–36

4. Arora A, Abildgaard F, Bushweller JH, Tamm LK. 2001. Structure of outer membrane protein A transmembrane domain by NMR spectroscopy. *Nat. Struct. Biol.* 8:334–38

5. Belrhali H, Nollert P, Royant A, Menzel C, Rosenbusch JP, et al. 1999. Protein, lipid and water organization in bacteriorhodopsin crystals: a molecular view of the purple membrane at 1.9 Å resolution. *Structure* 7:909–17

6. Binnig G, Quate CF, Gerber C. 1986. Atomic force microscope. *Phys. Rev. Lett.* 56:930–33

7. Bocharov EV, Pustovalova YE, Pavlov KV, Volynsky PE, Goncharuk MV, et al. 2007. Unique dimeric structure of BNip3 transmembrane domain suggests membrane permeabilization as a cell death trigger. *J. Biol. Chem.* 282:16256–66

8. Booy FP, Pawley JB. 1993. Cryo-crinkling: what happens to carbon films on copper grids at low temperature. *Ultramicroscopy* 48:273–80

9. Buzhynskyy N, Hite RK, Walz T, Scheuring S. 2007. The supramolecular architecture of junctional microdomains in native lens membranes. *EMBO Rep.* 8:51–55

10. Ceska TA, Henderson R. 1990. Analysis of high-resolution electron diffraction patterns from purple membrane labelled with heavy-atoms. *J. Mol. Biol.* 213:539–60

11. Cheng A, Leung A, Fellmann D, Quispe J, Suloway C, et al. 2007. Towards automated screening of two-dimensional crystals. *J. Struct. Biol.* 160:324–31

12. Cherezov V, Rosenbaum DM, Hanson MA, Rasmussen SG, Thian FS, et al. 2007. High-resolution crystal structure of an engineered human β_2-adrenergic G protein-coupled receptor. *Science* 318:1258–65

13. Chill JH, Louis JM, Miller C, Bax A. 2006. NMR study of the tetrameric KcsA potassium channel in detergent micelles. *Protein Sci.* 15:684–98

14. Chou JJ, Kaufman JD, Stahl SJ, Wingfield PT, Bax A. 2002. Micelle-induced curvature in a water-insoluble HIV-1 Env peptide revealed by NMR dipolar coupling measurement in stretched polyacrylamide gel. *J. Am. Chem. Soc.* 124:2450–51

15. Crowther RA, Henderson R, Smith JM. 1996. MRC image processing programs. *J. Struct. Biol.* 116:9–16

16. Downing KH. 1991. Spot-scan imaging in transmission electron microscopy. *Science* 251:53–59

17. Eifler N, Duckely M, Sumanovski LT, Egan TM, Oksche A, et al. 2007. Functional expression of mammalian receptors and membrane channels in different cells. *J. Struct. Biol.* 159:179–93

18. **Faham S, Bowie JU. 2002. Bicelle crystallization: A new method for crystallizing membrane proteins yields a monomeric bacteriorhodopsin structure. *J. Mol. Biol.* 316:1–6**

19. Ferguson AD, Hofmann E, Coulton JW, Diederichs K, Welte W. 1998. Siderophore-mediated iron transport: crystal structure of FhuA with bound lipopolysaccharide. *Science* 282:2215–20

20. Fernández C, Hilty C, Bonjour S, Adeishvili K, Pervushin K, Wüthrich K. 2001. Solution NMR studies of the integral membrane proteins OmpX and OmpA from *Escherichia coli*. *FEBS Lett.* 504:173–78

21. Fujiyoshi Y. 1998. The structural study of membrane proteins by electron crystallography. *Adv. Biophys.* 35:25–80

22. Fujiyoshi Y, Mizusaki T, Morikawa K, Yamagishi H, Aoki Y, et al. 1991. Development of a superfluid helium stage for high-resolution electron microscopy. *Ultramicroscopy* 38:241–51

23. Gipson B, Zeng X, Stahlberg H. 2007. 2dx_merge: data management and merging for 2D crystal images. *J. Struct. Biol.* 160:375–84

24. Gipson B, Zeng X, Zhang ZY, Stahlberg H. 2007. *2dx*: user-friendly image processing for 2D crystals. *J. Struct. Biol.* 157:64–72

25. Glaeser RM. 1992. Specimen flatness of thin crystalline arrays: influence of the substrate. *Ultramicroscopy* 46:33–43

26. **Gonen T, Cheng Y, Sliz P, Hiroaki Y, Fujiyoshi Y, et al. 2005. Lipid-protein interactions in double-layered two-dimensional AQP0 crystals. *Nature* 438:633–38**

27. Gonen T, Sliz P, Kistler J, Cheng Y, Walz T. 2004. Aquaporin-0 membrane junctions reveal the structure of a closed water pore. *Nature* 429:193–97

28. Gonen T, Walz T. 2006. The structure of aquaporins. *Q. Rev. Biophys.* 39:361–96

29. **Grigorieff N, Ceska TA, Downing KH, Baldwin JM, Henderson R. 1996. Electron-crystallographic refinement of the structure of bacteriorhodopsin. *J. Mol. Biol.* 259:393–421**

30. Gyobu N, Tani K, Hiroaki Y, Kamegawa A, Mitsuoka K, Fujiyoshi Y. 2004. Improved specimen preparation for cryo-electron microscopy using a symmetric carbon sandwich technique. *J. Struct. Biol.* 146:325–33

31. Han BG, Wolf SG, Vonck J, Glaeser RM. 1994. Specimen flatness of glucose-embedded biological materials for electron crystallography is affected significantly by the choice of carbon evaporation stock. *Ultramicroscopy* 55:1–5

32. Harrenga A, Michel H. 1999. The cytochrome c oxidase from *Paracoccus denitrificans* does not change the metal center ligation upon reduction. *J. Biol. Chem.* 274:33296–99

33. **Henderson R, Baldwin JM, Ceska TA, Zemlin F, Beckmann E, Downing KH. 1990. Model for the structure of bacteriorhodopsin based on high-resolution electron cryo-microscopy. *J. Mol. Biol.* 213:899–929**

34. **Henderson R, Unwin PN. 1975. Three-dimensional model of purple membrane obtained by electron microscopy. *Nature* 257:28–32**

35. Hiller M, Krabben L, Vinothkumar KR, Castellani F, van Rossum BJ, et al. 2005. Solid-state magic-angle spinning NMR of outer-membrane protein G from *Escherichia coli*. *Chembiochem* 6:1679–84

36. Hiroaki Y, Tani K, Kamegawa A, Gyobu N, Nishikawa K, et al. 2006. Implications of the aquaporin-4 structure on array formation and cell adhesion. *J. Mol. Biol.* 355:628–39

37. Hite RK, Gonen T, Harrison SC, Walz T. 2008. Interactions of lipids with aquaporin-0 and other membrane proteins. *Pflugers Arch.* 456:651–61

38. Hite RK, Raunser S, Walz T. 2007. Revival of electron crystallography. *Curr. Opin. Struct. Biol.* 17:389–95

39. Hunte C, Richers S. 2008. Lipids and membrane protein structures. *Curr. Opin. Struct. Biol.* 18:406–11

18. This paper introduces lipid/detergent bicelles as a new matrix for 3D crystallization of membrane proteins.

26. The 1.9 Å resolution structure of AQP0 determined by electron crystallography revealed water molecules in the channel as well as nine lipid molecules surrounding each monomer.

29. This paper introduced refinement to electron crystallography, and the structure of bR at 3.5 Å resolution included 30 lipid chains.

33. The 4 Å resolution structure of bR represents the first atomic model obtained by electron crystallography.

34. The density map of bR at 7 Å resolution is the first one obtained by electron crystallography and showed for the first time that transmembrane domains adopt an α-helical secondary structure.

40. Hwang PM, Choy WY, Lo EI, Chen L, Forman-Kay JD, et al. 2002. Solution structure and dynamics of the outer membrane enzyme PagP by NMR. *Proc. Natl. Acad. Sci. USA* 99:13560–65

41. Jap BK, Li H. 1995. Structure of the osmo-regulated H_2O-channel, AQP-CHIP, in projection at 3.5 Å resolution. *J. Mol. Biol.* 251:413–20

42. Jap BK, Zulauf M, Scheybani T, Hefti A, Baumeister W, et al. 1992. 2D crystallization: from art to science. *Ultramicroscopy* 46:45–84

43. Jormakka M, Tornroth S, Byrne B, Iwata S. 2002. Molecular basis of proton motive force generation: structure of formate dehydrogenase-N. *Science* 295:1863–68

44. Kimura Y, Vassylyev DG, Miyazawa A, Kidera A, Matsushima M, et al. 1997. Surface of bacteriorhodopsin revealed by high-resolution electron crystallography. *Nature* 389:206–11

45. Koeck PJ, Purhonen P, Alvang R, Grundberg B, Hebert H. 2007. Single particle refinement in electron crystallography: a pilot study. *J. Struct. Biol.* 160:344–52

46. Kolbe M, Besir H, Essen LO, Oesterhelt D. 2000. Structure of the light-driven chloride pump halorhodopsin at 1.8 Å resolution. *Science* 288:1390–96

47. Koning RI, Oostergetel GT, Brisson A. 2003. Preparation of flat carbon support films. *Ultramicroscopy* 94:183–91

48. Kühlbrandt W. 1992. Two-dimensional crystallization of membrane proteins. *Q. Rev. Biophys.* 25:1–49

49. Kühlbrandt W, Wang DN, Fujiyoshi Y. 1994. Atomic model of plant light-harvesting complex by electron crystallography. *Nature* 367:614–21

50. **Landau EM, Rosenbusch JP. 1996. Lipidic cubic phases: a novel concept for the crystallization of membrane proteins. *Proc. Natl. Acad. Sci. USA* 93:14532–35**

51. Lee AG. 2003. Lipid-protein interactions in biological membranes: a structural perspective. *Biochim. Biophys. Acta* 1612:1–40

52. Lee AG. 2004. How lipids affect the activities of integral membrane proteins. *Biochim. Biophys. Acta* 1666:62–87

53. Liang B, Tamm LK. 2007. Structure of outer membrane protein G by solution NMR spectroscopy. *Proc. Natl. Acad. Sci. USA* 104:16140–45

54. Liu Z, Yan H, Wang K, Kuang T, Zhang J, et al. 2004. Crystal structure of spinach major light-harvesting complex at 2.72 Å resolution. *Nature* 428:287–92

55. Long SB, Campbell EB, Mackinnon R. 2005. Crystal structure of a mammalian voltage-dependent Shaker family K^+ channel. *Science* 309:897–903

56. **Long SB, Tao X, Campbell EB, MacKinnon R. 2007. Atomic structure of a voltage-dependent K^+ channel in a lipid membrane-like environment. *Nature* 450:376–82**

57. **Luecke H, Schobert B, Richter HT, Cartailler JP, Lanyi JK. 1999. Structure of bacteriorhodopsin at 1.55 Å resolution. *J. Mol. Biol.* 291:899–911**

58. MacKenzie KR, Prestegard JH, Engelman DM. 1997. A transmembrane helix dimer: structure and implications. *Science* 276:131–33

59. Marsh D. 2008. Protein modulation of lipids, and vice-versa, in membranes. *Biochim. Biophys. Acta* 1778:1545–75

60. Misquitta Y, Cherezov V, Havas F, Patterson S, Mohan JM, et al. 2004. Rational design of lipid for membrane protein crystallization. *J. Struct. Biol.* 148:169–75

61. **Miyazawa A, Fujiyoshi Y, Unwin N. 2003. Structure and gating mechanism of the acetylcholine receptor pore. *Nature* 423:949–55**

62. Moor H. 1969. Freeze-etching. *Int. Rev. Cytol.* 25:391–412

63. Müller DJ, Sapra KT, Scheuring S, Kedrov A, Frederix PL, et al. 2006. Single-molecule studies of membrane proteins. *Curr. Opin. Struct. Biol.* 16:489–95

64. Murata K, Mitsuoka K, Hirai T, Walz T, Agre P, et al. 2000. Structural determinants of water permeation through aquaporin-1. *Nature* 407:599–605

65. Murata T, Yamato I, Kakinuma Y, Leslie AG, Walker JE. 2005. Structure of the rotor of the V-Type Na^+-ATPase from *Enterococcus hirae*. *Science* 308:654–59

66. Oshima A, Tani K, Hiroaki Y, Fujiyoshi Y, Sosinsky GE. 2007. Three-dimensional structure of a human connexin26 gap junction channel reveals a plug in the vestibule. *Proc. Natl. Acad. Sci. USA* 104:10034–39

50. This paper introduces LCP as a new matrix for the crystallization of membrane proteins.

56. The structure of a chimeric voltage-dependent K^+ channel reveals the pore and voltage sensors embedded in a membrane-like arrangement of lipid molecules.

57. The bR structure obtained from LCP crystallization at 1.55 Å revealed 18 tightly bound lipid chains.

61. This paper presents an atomic model of the closed nicotinic acetylcholine receptor, obtained by EM of crystalline postsynaptic membranes.

67. Oxenoid K, Chou JJ. 2005. The structure of phospholamban pentamer reveals a channel-like architecture in membranes. *Proc. Natl. Acad. Sci. USA* 102:10870–75

68. Oxenoid K, Kim HJ, Jacob J, Sönnichsen FD, Sanders CR. 2004. NMR assignments for a helical 40 kDa membrane protein. *J. Am. Chem. Soc.* 126:5048–49

69. Pebay-Peyroula E, Rummel G, Rosenbusch JP, Landau EM. 1997. X-ray structure of bacteriorhodopsin at 2.5 Å from microcrystals grown in lipidic cubic phases. *Science* 277:1676–81

70. Philippsen A, Schenk AD, Signorell GA, Mariani V, Berneche S, Engel A. 2007. Collaborative EM image processing with the IPLT image processing library and toolbox. *J. Struct. Biol.* 157:28–37

71. Prosser RS, Evanics F, Kitevski JL, Al-Abdul-Wahid MS. 2006. Current applications of bicelles in NMR studies of membrane-associated amphiphiles and proteins. *Biochemistry* 45:8453–65

72. Ramamoorthy A, Opella SJ. 1995. Two-dimensional chemical shift/heteronuclear dipolar coupling spectra obtained with polarization inversion spin exchange at the magic angle and magic-angle sample spinning (PISEMAMAS). *Solid State Nuclear Magn. Reson.* 4:387–92

73. Rasmussen SG, Choi HJ, Rosenbaum DM, Kobilka TS, Thian FS, et al. 2007. Crystal structure of the human β_2 adrenergic G-protein-coupled receptor. *Nature* 450:383–87

74. Raunser S, Haase W, Bostina M, Parcej DN, Kühlbrandt W. 2005. High-yield expression, reconstitution and structure of the recombinant, fully functional glutamate transporter GLT-1 from *Rattus norvegicus*. *J. Mol. Biol.* 351:598–613

75. Rémigy HW, Caujolle-Bert D, Suda K, Schenk A, Chami M, Engel A. 2003. Membrane protein reconstitution and crystallization by controlled dilution. *FEBS Lett.* 555:160–69

76. Ren G, Reddy VS, Cheng A, Melnyk P, Mitra AK. 2001. Visualization of a water-selective pore by electron crystallography in vitreous ice. *Proc. Natl. Acad. Sci. USA* 98:1398–403

77. Rhinow D, Kühlbrandt W. 2008. Electron cryo-microscopy of biological specimens on conductive titanium-silicon metal glass films. *Ultramicroscopy* 108:698–705

78. Rigaud JL, Mosser G, Lacapere JJ, Olofsson A, Levy D, Ranck JL. 1997. Bio-Beads: an efficient strategy for two-dimensional crystallization of membrane proteins. *J. Struct. Biol.* 118:226–35

79. Robinson NC, Zborowski J, Talbert LH. 1990. Cardiolipin-depleted bovine heart cytochrome c oxidase: binding stoichiometry and affinity for cardiolipin derivatives. *Biochemistry* 29:8962–69

80. Sanders CR, Prosser RS. 1998. Bicelles: a model membrane system for all seasons? *Structure* 6:1227–34

81. Scheuring S. 2006. AFM studies of the supramolecular assembly of bacterial photosynthetic core-complexes. *Curr. Opin. Chem. Biol.* 10:387–93

82. Scheuring S, Buzhynskyy N, Jaroslawski S, Gonçalves RP, Hite RK, Walz T. 2007. Structural models of the supramolecular organization of AQP0 and connexons in junctional microdomains. *J. Struct. Biol.* 160:385–94

83. Schnell JR, Chou JJ. 2008. Structure and mechanism of the M2 proton channel of influenza A virus. *Nature* 451:591–95

84. Signorell GA, Kaufmann TC, Kukulski W, Engel A, Rémigy HW. 2007. Controlled 2D crystallization of membrane proteins using methyl-β-cyclodextrin. *J. Struct. Biol.* 157:321–28

85. Stroud RM, Agard DA. 1979. Structure determination of asymmetric membrane profiles using an iterative Fourier method. *Biophys. J.* 25:495–512

86. Suzuki H, Nishikawa K, Hiroaki Y, Fujiyoshi Y. 2008. Formation of aquaporin-4 arrays is inhibited by palmitoylation of N-terminal cysteine residues. *Biochim. Biophys. Acta* 1778:1181–89

87. Tate CG. 2001. Overexpression of mammalian integral membrane proteins for structural studies. *FEBS Lett.* 504:94–98

88. Tate CG, Haase J, Baker C, Boorsma M, Magnani F, et al. 2003. Comparison of seven different heterologous protein expression systems for the production of the serotonin transporter. *Biochim. Biophys. Acta* 1610:141–53

89. Unger VM, Kumar NM, Gilula NB, Yeager M. 1999. Three-dimensional structure of a recombinant gap junction membrane channel. *Science* 283:1176–80

90. Unwin N. 2005. Refined structure of the nicotinic acetylcholine receptor at 4Å resolution. *J. Mol. Biol.* 346:967–89

91. Unwin PN, Henderson R. 1975. Molecular structure determination by electron microscopy of unstained crystalline specimens. *J. Mol. Biol.* 94:425–40

92. Vink M, Derr K, Love J, Stokes DL, Ubarretxena-Belandia I. 2007. A high-throughput strategy to screen 2D crystallization trials of membrane proteins. *J. Struct. Biol.* 160:295–304

93. Wadsten P, Wöhri AB, Snijder A, Katona G, Gardiner AT, et al. 2006. Lipidic sponge phase crystallization of membrane proteins. *J. Mol. Biol.* 364:44–53

94. Walz T, Smith BL, Zeidel ML, Engel A, Agre P. 1994. Biologically active two-dimensional crystals of aquaporin CHIP. *J. Biol. Chem.* 269:1583–86

95. Zampighi GA, Kreman M, Lanzavecchia S, Turk E, Eskandari S, et al. 2003. Structure of functional single AQP0 channels in phospholipid membranes. *J. Mol. Biol.* 325:201–10

96. Zeng X, Stahlberg H, Grigorieff N. 2007. A maximum likelihood approach to two-dimensional crystals. *J. Struct. Biol.* 160:362–74

97. Zeri AC, Mesleh MF, Nevzorov AA, Opella SJ. 2003. Structure of the coat protein in fd filamentous bacteriophage particles determined by solid-state NMR spectroscopy. *Proc. Natl. Acad. Sci. USA* 100:6458–63

RELATED RESOURCES

Chou HT, Evans JE, Stahlberg H. 2007. Electron crystallography of membrane proteins. *Methods Mol. Biol.* 369:331–43

Glaeser RM, Downing KH, DeRosier D, Chiu W, Frank J. 2007. *Electron Crystallography of Biological Macromolecules.* NY: Oxford Univ. Press

Palsdottir H, Hunte C. 2004. Lipids in membrane protein structures. *Biochim. Biophys. Acta* 1666:2–18

Nuclear Envelope Formation: Mind the Gaps

Banafshé Larijani[1] and Dominic L. Poccia[2]

[1]Cell Biophysics Laboratory, Lincoln's Inn Fields Laboratories, Cancer Research UK, London WC2A 3PX, United Kingdom; email: banafshe.larijani@cancer.org.uk

[2]Department of Biology, Amherst College, Amherst, Massachusetts 01002; email: dlpoccia@amherst.edu

Annu. Rev. Biophys. 2009. 38:107–24

The *Annual Review of Biophysics* is online at biophys.annualreviews.org

This article's doi:
10.1146/annurev.biophys.050708.133625

Key Words

membrane fusion, lipid mass spectrometry, phosphoinositides, polybasic clusters

Abstract

During mitosis in metazoans, the nuclear envelope (NE) breaks down at prophase and reassembles at telophase. The regulation of NE assembly is essential to correct cell functioning. The complex issue of the regulation of NE formation remains to be solved. It is still uncertain that a single mechanism depicts NE formation during mitosis. The aim of this review is to address some of the cytological, biophysical, and molecular aspects of models of NE formation. Our emphasis is on the role of lipids and their modifying enzymes in envelope assembly. We consider how the NE can be used as a model in characterizing membrane dynamics during membrane fusion. Fusion mechanisms that give insight into the formation of the double membrane of the envelope are summarized. We speculate on the possible roles of phosphoinositides in membrane fusion and NE formation.

Contents

INTRODUCTION

The nuclear envelope (NE) of eukaryotic cells segregates the chromosomes and other nuclear materials such as nascent transcripts, transcription factors, and histones from the cytoplasm during interphase. During open mitosis in metazoans, the NE typically breaks down at prophase and reassembles at telophase. The formation of the NE is central to correct cell functioning. Understanding how its complex architecture is attained is of importance since correct NE formation is required of all cells that break down their NEs at mitosis, and the envelope is used as a prognostic tool for various forms of cancer such as prostate and small-cell lung carcinomas. In these cancers, NE lobulations and increased malleability are observed, respectively (17, 41, 112). Moreover, defects in the nuclear lamina and associated proteins of the inner nuclear membrane have been linked to disruption of basic cellular mechanisms, such as replication, required for normal cell cycle progression (35, 42, 59, 92), and to various human diseases including Emery-

NE: nuclear envelope

ER: endoplasmic reticulum

Dreifuss muscular dystrophy, Dunnigan-type familial partial lipodystrophy, and congenital cardiomyopathies (45, 49, 74, 107) in which normal gene expression is disrupted.

The NE consists of an inner and an outer nuclear membrane. The inner nuclear membrane is usually associated with a lamina composed of lamin proteins, and the outer nuclear membrane is generally continuous with the endoplasmic reticulum (ER) during interphase. In the past few years there has been much debate on the mechanisms involved in the assembly of the NE. Although some recent publications (4–6) suggest that the problem of the mechanism and dynamics of NE formation is solved, we believe that this complex issue has not been fully addressed (4–6, 39, 69). It is not certain if there is a single mechanism to explain NE formation during open mitosis, or if any one mechanism yet described is completely accurate. And the literature often emphasizes differences rather than aspects of NE formation common to various models.

In this review we attempt to address certain cytological, molecular, and biophysical aspects of models that describe NE assembly. In contrast to the typical focus on the involvement of structural proteins, we emphasize the role of lipids and their modifying enzymes in NE formation. We take into account how NE assembly can be used as a model for characterizing the molecular dynamics of membrane fusion. We discuss the pros and cons of the current models and suggest how these may be consolidated to explain NE formation. We summarize some mechanisms, likely operative in most membrane fusions, derived from studies of both synthetic and natural membranes that may give insight into how the NE formation may take place, and speculate on possible roles for phosphoinositides in membrane fusion suggested by recent work on the NE.

MODELS OF NUCLEAR ENVELOPE FORMATION

At first glance it would seem rather simple through light or electron microscopic analysis to describe how membranes re-enclose the chromosomes at the end of mitosis. However, traditional tissue-sectioning techniques often fail to distinguish membrane vesicles from cross sections of tubules, so although elongated or flattened ER cisternae can be identified, vesicles and cross sections of tubules are not easily delineated without three-dimensional reconstruction.

Current models of NE formation are variations of two principal models, each of which addresses the role of membrane fusion differently. The first model suggests that reorganization of the ER at mitosis encloses the chromosomes to form the NE without contribution of membrane vesicle fusion. This model proposes that binding of ER tubules to chromatin, perhaps tethered by proteins that will come to reside in the inner nuclear membrane, is followed by flattening of these membranes against the surface of the chromatin and ultimately almost fully enclosing it. Remaining "holes" or gaps are finally sealed by an unspecified mechanism, which might involve the incorporation of nuclear pore complexes. This model has been generated mainly from in vitro and a few in vivo experiments and has been suggested to apply to all metazoa. The role of membrane fusion in this model is to form or repair the ER, but fusion is not required for formation of the NE itself. Recent work on this model emphasizes proteins of the ER responsible for binding to chromatin (4).

The second model assumes that fusion of membrane vesicles and ER membranes results in envelopment of the chromatin and closure to give the double membrane of the NE, with or without nuclear pores. It potentially involves chromatin-vesicle, vesicle-vesicle, or vesicle-ER binding followed by fusion of the membranes. The topology of this membrane fusion would be different from that of typical membrane vesicle fusions such as those occurring during exocytosis or endocytosis, because it takes place on a chromatin substratum, so the end product is a double membrane separated by a cisternum corresponding to the vesicle contents or ER lumen instead of simply a larger single membrane containing the contents of the fusing vesicles. This model also derives from both in vivo and in vitro experiments, the latter primarily based on cellular homogenates in which the continuous ER is of necessity converted to vesicles. However, the model does not require ER-ER membrane fusion in vivo, only in vitro. Recent work on this model emphasizes a role of lipids and lipid-modifying enzymes in NE formation and postulates that the fusion machinery does not lie in the ER but in membrane vesicles distinct from it (57).

CYTOLOGY OF NUCLEAR ENVELOPE ASSEMBLY

The details of NE formation vary by organism, cell type, and developmental stage, making it difficult to generalize about its morphology. Some lower eukaryotes such as yeast do not undergo NE breakdown/re-formation at mitosis but instead form an intranuclear spindle in a so-called closed mitosis. A modified closed mitosis is seen in *Drosophila* early embryos in which NE breakdown occurs only at the poles at prometaphase, and a full second NE, lacking pores but containing inner nuclear membrane marker proteins (44), forms outside the chromosomes by metaphase to create a spindle envelope that breaks down at interphase (93). More typically in animal cells, full NE breakdown and formation occur in open mitosis.

In open mitosis, although it is agreed that most of the NE is derived from the ER, it remains contentious whether re-formation of the NE during telophase results from vesicle fusion, endoplasmic reticular envelopment, or some combination of the two (5, 39, 57, 77). Whether the ER typically remains intact or breaks down into smaller units at mitosis (or during male pronuclear envelope formation following fertilization) is not certain for all cell types, but there are clear examples of each in the literature. For example, in cultured Chinese hamster ovary cells the ER appears to remain continuous throughout mitosis (78). In cultured 3T3 cells the ER appears to break down into smaller fragments at mitosis and re-form later (53).

LBR: lamin B
receptor

Electron microscopic images of serial sections of HeLa cells (82) show that during prophase smooth membranes concentrate at the cell periphery defined by the poles at metaphase (**Figure 1***a*). It was suggested that "polar elements fuse," proceeding toward the central regions of the chromosomes. In any event, ER tubules appear to assemble around the chromosomes in these cells.

One example in which membrane fusion seems clearly involved in NE formation is in karyomeres, individual chromosomes in early embryos surrounded by intact nuclear membranes (108). Resolution of individual karyomeres into nuclei entails fusion of intact NEs with one another. One of the most detailed electron microscope studies of karyomeres is by Longo (63). During the first cell cycle of the sea urchin, karyomeres form between late anaphase and telophase, whereupon individual chromosomes become surrounded by NE double membranes with pores. These NEs then progressively fuse, forming single nuclei of the daughter cells. The strongest evidence of fusion of pre-existent membranes is the successive appearance of three structures: chromosomes with points of contact characterized by distinct inner membranes but continuous outer membranes, small regions of fusion of inner membranes at these points, and finally the merging of chromatin of two karyomeres surrounded by shared continuous inner and outer membranes (**Figure 1***b*).

Karyomere fusion of preformed NEs is reminiscent of fusion of male and female pronuclei to create the zygote nucleus (62, 65). In both pronuclei, membranes variously referred to as vesicles or tubules accumulate at the chromatin surface and appear to fuse. The electron microscope images suggest progressive fusion, which is further supported by the observation that application of the reducing agent dithiothreitol interferes with karyomere fusion, syngamy of the male pronucleus and the female pronucleus, and completion of cleavage furrow. The last two processes involve membrane fusion. Other processes of cell cycle progression, pronuclear migration, chromosome separation at mitosis, and chromosome condensation cycles are not affected by the treatment (86).

Although it seems clear that complete NEs are capable of fusing to one another, this leaves open the question of how the karyomere NEs are initially formed. In the frog *Xenopus laevis*, incomplete NEs with pores initially form around the apices of anaphase chromosomes (**Figure 1***c*), and by telophase all chromosomes are surrounded, indicating that envelopment, if it occurs, does not happen concomitantly throughout (73).

EXPERIMENTAL EVIDENCE OF THE MODELS

The models of how NEs form in vivo differ in two important respects: whether the ER remains intact or vesiculates prior to NE formation and whether fusion of any membranes contributes to or completes NE formation. Here we evaluate data supporting the models and point out where they might be reconciled.

ER Envelopment Models

Several integral nuclear membrane proteins come to reside in the ER during mitosis in cultured mammalian cells (109), giving rise to the idea that the NE does not retain identity during mitosis. A paper critical of the vesiculation model of NE formation is that of Ellenberg et al. (39), in which NE formation was followed in live cells by confocal microscopy. A series of fluorescence recovery after photobleaching (FRAP) experiments using GFP–lamin B receptor (LBR) as an inner nuclear membrane marker in COS-7 cells led to the important conclusion that in these cells, LBR, a protein of the inner nuclear membrane, resides in the ER during mitosis and later becomes immobilized when associated with chromatin at telophase, thus associating almost exclusively with the NE at interphase. Because LBR is an integral membrane protein that binds to heterochromatin, it could serve as an ideal anchor of that portion of the ER that contributes to the NE. A role for LBR in binding of ER to chromatin

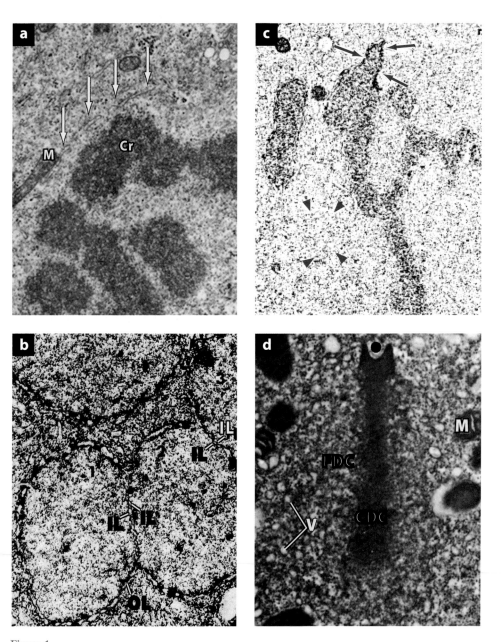

Figure 1

(*a*) Elongated cisternae (*yellow arrows*) in the vicinity of chromosomes (Cr) at anaphase in HeLa cells. Reproduced with permission from Reference 82. (*b*) Fusion of karyomere nuclear envelopes (NEs) at telophase of sea urchin embryos. Inner and outer lamina (membranes) are indicated (IL, OL). Reproduced with permission from Reference 63. (*c*) Formation of partial NEs with pores (*red arrows*) at tips of anaphase chromosomes in *Xenopus* early embryos. Arrowheads indicate numerous small membrane vesicles. Reproduced with permission from Reference 73. (*d*) Initial stages of NE formation in the male pronucleus of the sea urchin. V, putative vesicles; asterisk (*), region of sperm NE remnants; FDC, finely dispersed chromatin; CDC, coarsely dispersed chromatin. Reproduced with permission from Reference 65.

GTPγS:
nonhydrolyzable form
of GTP

was previously demonstrated by Collas et al. (28), who showed that an ER-derived vesicle fraction from sea urchin eggs required LBR to bind to chromatin during in vitro NE assembly reactions.

Ellenberg et al. (39) showed that from anaphase through cytokinesis tubules from the ER were "directly connected to immobilized GFP-LBR in membranes wrapping around the chromatin." Although their data establish that this inner nuclear membrane marker protein appears to reside in the ER during mitosis, the evidence does not exclude participation of membrane vesicles in NE re-formation, but only establishes that LBR [and other NE proteins redistributing with different kinetics (15)] does not accumulate in vesicles during mitosis. They conclude that "NE reassembly occurs by coalescence of ER elements as opposed to vesicles."

More recently, papers based primarily on in vitro observations state that "membrane fusion is not the principal mechanism of NE formation from an intact network but is required for ER reconstitution and maintenance" (4–6). Most of the experiments address binding of ER membranes to chromatin rather than completion of NE formation (4). It was proposed that the formation, expansion, and merging of NE sheets are driven by chromatin-mediated shaping of the ER network using the intrinsic capacity of the ER to form sheets. Specific mechanisms of gap closure in the ER lattice formed upon binding were not given, but nuclear pore formation was suggested.

The main evidence that fusion is not required for NE formation is the formation of a dextran impermeant barrier around sperm nuclei if the nuclei and GTPγS are added to extracts that have been allowed to preform ER in vitro, a process that does require GTP hydrolysis. However, because GTPγS and nuclei were apparently added to the reaction simultaneously and no further evidence was given of hydrolysis inhibition, it is not clear that hydrolysis was prevented during the critical period of NE formation. Fusion may occur early in the formation of NEs (22), before inhibition by GTPγS is complete. Furthermore, GTP hydrolysis acts as a timer that determines the frequency of fusion events (84). GTPγS may freeze monomeric GTP switches in an "on" or an "off" position, either promoting or inhibiting particular fusion events (84). Because GTPase activity is likely to be an upstream regulatory event of fusion (10), it is also not certain if the conditions employed during ER formation did not bypass the upstream regulation.

Many experiments did not use chromatin but artificial substrates such as double-stranded DNA on a glass surface in the presence of cytosol to study tubule spreading, or DNA-coated magnetic beads in the presence and absence of cytosol to evaluate NE formation by dextran exclusion. These substrates seem rather far removed from the physiologically relevant chromatin. Even though histone binding to the DNA occurs, there is no evidence of nucleosome or heterochromatin formation. None of these experiments employed GTPγS to inhibit fusion.

In addition to these technical issues, the enclosure models do not address several critical matters in NE assembly. First, formation of the NE is a regulated phenomenon. The phenomena shown in many of the in vitro experiments can occur through either charge effects or nonspecific binding of membranes to DNA. Furthermore, the ER is not a static structure in cells but participates in budding and fusion events. The entire ER network is also dynamic, losing and regaining continuity at different times and in different cell types (88, 101).

Proponents of the enclosure mechanisms minimize the importance of lipids in NE formation and suggest that the lipids or their modifying enzymes are not the rate-limiting step in the formation of the double membrane (4). We consider potential roles of lipids below.

Most important of all is the problem of closing the gaps that separate the tubules or sheets of ER adhering to the chromatin prior to full enclosure, a problem naturally solved by membrane fusion. To enclose an object (nucleus or chromosome) by a continuous membrane (the ER) such that the same surface (cytoplasmic face) of the membrane faces both the chromatin and the cytosol, at least one gap needs to be sealed. Evidence from in vitro data suggests that there are actually many such gaps present prior to enclosure (105, 106). That nuclear pores may seal the gaps of ER enveloping the chromosomes at telophase has sometimes been suggested by proponents of the enclosure model.

Details of nuclear pore structure and assembly have been recently reviewed (7). It is not clear how nuclear pores could be localized in the gaps formed by spreading ER or how they could capture the surrounding membranes to seal the gaps, although such a mechanism is theoretically possible (19). Likewise it is not clear how pores are inserted into expanding interphase NEs without disruption of the integrity of the envelope and whether such an insertion is similar to what occurs during mitosis. A possible mechanism of interphase pore insertion would involve fusion of the inner and outer nuclear membranes, forming a temporary gap to be filled somehow by a pore, but such a process has not been imaged.

What is the evidence that nuclear pores are required for NE formation? Depletion of the transmembrane nucleoporin POM121 from extracts inhibits NE formation. However, NEs can form in vitro without pore incorporation when additional nucleoporins are depleted from the extracts (8), which shows that pore formation is not an absolute requirement for closing gaps. Thus under some conditions assembly of nuclear pores is tightly integrated with NE assembly and possibly involved in its regulation. For example, in the presence of the Ca^{++} chelator BAPTA, nuclear pores do not assemble but NEs form in the Xenopus system (67). Immunodepletion of a component of the nuclear pore NUP107-160 complex results in NEs devoid of nuclear pores (43). Partial depletion of the same nuclear pore complex with RNAi in HeLa cells severely reduced pore density, and full immunodepletion in Xenopus extracts resulted in envelopes without pores (103). The ability to form NEs without pores suggests an alternative route to closing gaps must exist. For example, two membrane vesicle fractions can contribute to NE formation in Xenopus extracts. One yields poreless NEs; in combination they result in NEs with pores (85).

We believe that the enclosure model still lacks a definitive demonstration of how the gaps are closed. Although this problem can be solved by nuclear pore insertion, it is not known whether pore formation is typically responsible for closing the gaps or simply accompanies NE formation.

Membrane Fusion Models

The second set of models postulates that NE formation requires at least some membrane fusion. At one extreme is the idea that the NE forms entirely from vesicles, but it is also possible that vesicles are used only for filling the gaps in a network of ER attached to the chromatin. Models involving membrane fusion do not require vesiculation of the ER, because either non-ER-derived membranes may contribute to NE formation or vesicles from the ER may bud and fuse while the integrity of the rest of the ER remains uninterrupted.

Some support for the fusion idea comes from electron microscope images as outlined above. The second source of data depends on observations that vesicles largely derived from ER in cell-free systems (whose preparation artificially converts ER to vesicles through cell homogenization) fuse to create a NE. Two of the main cell-free systems used to support the fusion model are derived from oocytes of the frog X. laevis and sea urchin eggs of various species (29, 30, 106). It may be relevant that extensive ER rearrangement occurs following fertilization in these and other eggs in vivo (94).

In these cases, to support the vesicle model it is necessary to show that some vesicles that fuse and contribute to the NE are not ER derived, exist in vivo before homogenization, and are required for NE formation, because fusion of ER-ER vesicles may simply represent a reconstitution of the ER network. Recent data that satisfy these criteria are summarized below. They implicate a role for both lipids and

Nucleoporins: proteins of the nuclear pore

lipid-modifying enzymes, as well as regulatory mechanisms found in other cellular membrane fusion events.

Much of the initial information giving rise to the fusion models came from cell-free systems derived from *Xenopus* and other amphibians (76) and from mitotic mammalian tissue culture cell extracts (20). All these share similar properties. By far the most work has been done with amphibian extracts of meiotic oocytes to which sperm nuclei are exogenously added in a system pioneered by Lohka & Masui (61) that requires GTP hydrolysis (16). Ran GTPase and p97 ATPase are required at some steps (47, 48). Various membrane fractions that contribute to NE formation, including some lacking ER markers (38, 85, 99), have been isolated from this system.

Evidence is consistent with roles for both vesicles and protein fusion machinery in *Xenopus*. Cotter et al. (32) showed that breakdown of *Xenopus* NEs in metaphase extracts occurs both by formation of tubular structures with the ER and by vesiculation of the NE, offering the possibility that NE-derived vesicles are available for subsequent NE reassembly. A role for membrane fusion comes from a study by Baur et al. that implicates both SNAREs and the ATPase NSF in NE assembly (14). These proteins are involved in other membrane fusion events. SNAREs were inhibited by high concentrations of α-SNAP, which blocked NE formation but not ER formation. NSF was needed until full sealing was complete. Similarly, depletion of the t-SNARE syntaxin syn-4 in *Caenorhabditis elegans* blastomeres prevents NE formation (50).

Detailed biochemical information has been derived from the sea urchin cell-free systems that emphasize a role for lipids and lipid-modifying enzymes in NE formation. The sea urchin is an evolutionary outgroup to the chordates whose genome has recently been sequenced (90, 91). Sea urchins possess most gene families found in mammals, but these are simpler than those in vertebrate families. Sea urchin gene families are usually not redundant but retain the full complexity of the gene set.

Sea urchin eggs are haploid, arrested in G1 (G0), and activated by the fertilizing sperm. In vivo, the sperm nucleus enters the egg with a NE lacking pores (62, 66). Its envelope is rapidly disassembled, the membranes vesiculating as the sperm chromatin decondenses. During this process, remnants of the sperm NE at the tip and base of the conical nucleus, which line two cup-shaped cavities, are retained. Electron microscopy reveals apparent membrane vesicles, mostly derived from the egg ER, accumulating along the sides of the nucleus that then seem to fuse to form a NE with pores (**Figure 1*d***) (64). The polar sperm NE remnants are incorporated into the new male pronuclear envelope. During this time of disassembly and reassembly the ER itself disassembles and reassembles, losing overall continuity to become discontinuous 2 min following fertilization and re-establishing continuity by 8 min (96–98). Therefore it is conceivable that some ER-derived vesicles created at fertilization remain to contribute to the NE or that independent units of ER bind to the chromatin in vivo.

Much of this transition to a male pronuclear envelope is replicated in vitro. Incorporation of the remnants into the newly forming NE derived from egg membranes is regulated by GTP hydrolysis and requires PtdIns-3 kinase activity (57), also required for other membrane fusions such as in the endosomal membrane compartment.

In vitro evidence from this system indicates that non-ER-derived vesicles are required for NE formation. These fuse with the ER-derived vesicles (which are likely produced by homogenization) to contribute to the NE. Importantly, the non-ER vesicles exist separately from the ER in vivo and have a lipid and protein composition different from that of the ER (22). They do not contain the inner nuclear membrane protein LBR, which resides in the ER fraction (28).

The non-ER-derived membrane fraction called MV1 was distinguished initially as a low-density vesicle fraction lacking an ER enzyme marker protein found in the major membrane fraction MV2 (29). Subsequent demonstration that MV1 was highly enriched in phosphoinositides and PLCγ permitted its localization by

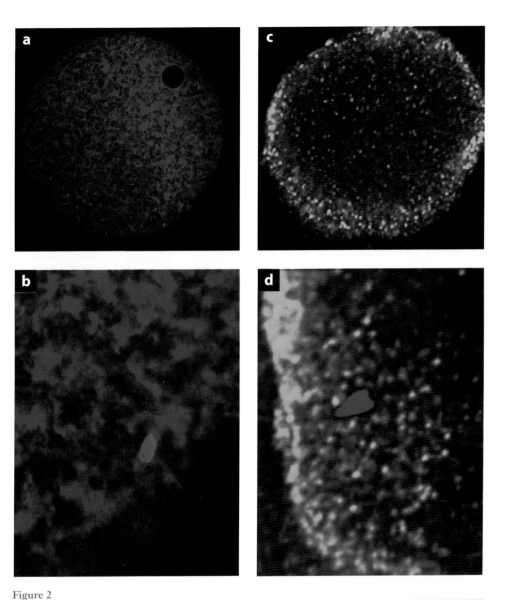

Figure 2

(*a, b*) Endoplasmic reticulum labeled with diIC$_{18}$ in unfertilized and 2-min post-fertilized sea urchin eggs (*red*). (Reproduced with permission from B. Fong & D. Poccia, unpublished data.) (*c, d*) PLCγ-positive vesicles in unfertilized and 2-min post-fertilized eggs. (Reproduced with permission from K. Han & D. Poccia, unpublished data.) For methods, see Reference 22. Male pronuclei are labeled blue with Hoechst 33342 in panels *b* and *d*.

immunofluorescence microscopy to ∼0.5-μm vesicles predominantly in the cortical region of eggs distinct from the ER (**Figure 2**) (22). The MV2 fraction is enriched in an ER marker protein, contains LBR, and has a lipid composition similar to that of mammalian ER (28, 29, 55, 58). LBR is required for binding of the ER fraction to nuclei, which occurs over the entire nuclear surface. In contrast, MV1 vesicles bind via peripheral proteins exclusively to the two poles of the nucleus, where the sperm NE remnants reside (28, 29).

PtdInsP:
phosphatidylinositol
phosphate

PtdInsP$_2$:
phosphatidylinositol
bisphosphate

DAG: diacylglycerol

The most remarkable property of MV1 is its phospholipid composition. It is highly enriched in total PtdIns. Initially two-dimensional NMR spectroscopy showed that the majority of phospholipids in MV1 were PtdIns species (58). Subsequently, by use of HPLC electrospray ionization tandem mass spectrometry, the species of phosphoinositides present in MV1 were quantified. MV1 phosphoinositide content is the highest reported for natural membranes: 20 mole% PtdInsP, 12% PtdInsP$_2$, and 10% PtdInsP$_3$, in addition to 20% PtdIns. The remainder is almost entirely phosphatidylcholine (**Figure 3**) (22).

High PtdInsP$_2$ levels in MV1 are accompanied by a >125-fold enrichment of PtdInsP$_2$-specific PLCγ compared with MV2. Early in NE formation, PLCγ is activated by phosphorylation. Because levels of the polyphosphoinositides in MV1 are much higher than typically employed in signaling mechanisms, it was postulated that the PIP$_2$ hydrolysis product diacylglycerol (DAG) might have a structural effect on the membrane, leading to localized fusion (25, 56, 79, 81). Consistent with this proposal, protein-free liposomes of phospholipid composition mimicking MV1 can substitute for MV1 in binding specificity and functionality in the NE assembly assay, indicating that binding is mediated by recruitment of peripheral proteins from cytosol (10). Substitution of DAG for PtdIns in the liposomes leads to fusion without GTP hydrolysis.

Vesicular models derived from in vitro systems have yielded a great deal of biochemical mechanistic detail. However, many aspects of the models also remain to be confirmed in vivo.

ROLES FOR PROTEINS AND PHOSPHOLIPIDS IN MEMBRANE FUSION

To interpret the potential consequences of phosphoinositide composition and the PI-PLC content of MV1, we review next some biophysical aspects of the role of lipids and proteins in membrane fusion. Then we consider the spe-

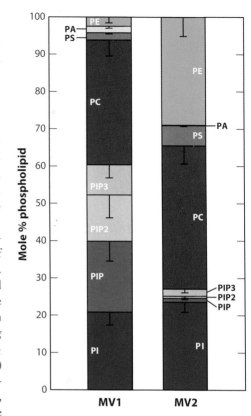

Figure 3

Phospholipid composition of sea urchin membrane vesicle fractions MV1 and MV2 by electrospray ionization tandem mass spectrometry (22). Abbreviations: PI, phosphatidylinositol; PIP, phosphatidylinositol phosphate; PIP2, phosphatidylinositol bisphosphate; PIP3, phosphatidylinositol trisphosphate; PC, phosphatidylcholine; PS, phosphatidylserine; PA, phosphatidic acid; PE, phosphatidylethanolamine.

cific implications of the phosphoinositides for membrane structure and potential fusion properties of membrane vesicles containing them. The aim is not a detailed synopsis of membrane fusion, as this subject has been discussed exhaustively in many other publications (23–26, 60, 79, 89, 95), but instead a summary of some important points involved in membrane fusion that may relate to NE formation.

The actual membrane fusion event is a local one that involves interactions in an area <20 nm in diameter, reorganizing lipids to form a transient highly curved intermediate referred to as a stalk (68). Experimental results show that the early stages of fusion forming the stalk are energetically less demanding than the opening of fusion pores and their subsequent expansion (24, 25, 54). The lipid-dependent fusion stage occurs prior to the lipid mixing that is necessary for completion of membrane fusion. However, to sustain nonlocal fusion (the final stages of fusion after lipid mixing), proteins are required to overcome energy barriers (24, 25, 54).

Proteins such as SNAREs have been suggested to provide forces through the formation of α-helical protein core interactions, which can overcome energy barriers that prevent close apposition of pre-fusion membranes (26). Whether SNAREs represent a minimal machinery for fusion in natural membranes as suggested by experiments incorporating them into model membranes in vitro (104) is still controversial, and some evidence indicates that SNAREs may have a role in docking only (70). Not only are SNAREs and Rab GTPases compartment specific, but so are phosphoinositide kinases and phosphatases (34), all of which may contribute to recognition and completion of correct fusions. It is certain that correct vesicle docking requires proteins. However, whether the core of the membrane intermediate leading to fusion is lipidic, proteinaceous, or proteolipidic is an open question (25, 110).

Membrane lipids can also alter the energetics of fusion. Lipids of negative curvature such as DAG, cholesterol, phosphadtidylethanolamine, and *cis*-unsaturated fatty acids are suggested to disrupt bilayers and induce biological fusion (18, 27, 31, 33). Studies of synthetic membranes indicate a potential role in fusion for DAG derived from PtdIns hydrolysis. Fusion can be induced in synthetic membranes containing phosphatidylcholine with bacterial phosphatidylcholine-specific phospholipase C (13, 75) or PtdIns with bacterial PI-specific PLC (100), each thought to result from the production of the membrane-destabilizing lipid DAG.

DAG has been implicated in membrane fusion and curvature during yeast vacuole formation, mast cell exocytosis, myoblast fusion, and echinocyte formation in erythrocytes (2, 3, 51, 52, 102). Additionally, there is evidence in yeast that DAG may interact with SNAREs and Rab proteins (51), potentially linking this lipid with proteins involved in fusion reactions. Unsaturated fatty acids have also been linked to formation of hemifused intermediates in neurosecretory granules (111).

IMPLICATIONS OF HIGH LEVELS OF POLYPHOSPHOINOSITIDES FOR FUSION AND ITS REGULATION

The atypical lipid composition of MV1 raises issues concerning its function in NE formation and potentially other membrane fusions. This has led to studying the physical properties of model membranes with elevated amounts of PtdIns and its phosphorylated forms. Generally, polyphosphoinositides are thought to be mostly transient signaling molecules, present in low concentrations with relatively short lifetimes (36). We suggest that phosphoinositides or their products may also affect membrane dynamics, possessing a dual role in both signaling and membrane structure (55, 57).

In addition to having high levels of phosphoinositides, MV1 is enriched in long-chain, polyunsaturated phospholipids. Although both PtdIns and phosphatidylcholine species are largely polyunsaturated, PtdIns species are predominantly diacyl, whereas phosphatidylcholine species are almost exclusively alkylacyl. In addition, $PtdInsP_2$ in MV1 is much more enriched in diacyl species than in alkylacyl species compared with the ER fraction (22).

These compositional differences have potentially important implications for phospholipase-mediated signaling mechanisms. DAG and alklyacylglycerol are expected to act differently as signaling molecules. Moreover, diacyl $PtdInsP_2$ is preferred by PI-PLC. The alkylacyl content may affect detergent solubility, but effects on fluidity are probably minimal (1).

On the other hand, polyunsaturated species increase membrane fluidity, which in turn may facilitate localized membrane fusion (56). The predominant 18:0/20:4 species is the preferred substrate for PI-PLC (10, 22). Under conditions promoting fusion, 18:0/20:4 DAG is produced (10) and 18:0/20:4 PtdInsP$_2$ is depleted (21).

Solid-state NMR spectroscopy studies of model membranes that have high levels of 18:0/20:4 DAG and polyunsaturated PtdIns indicate they are fluid even at low temperatures (56). PtdIns may provide a greater chain disorder of the membrane, which results in the reduction of membrane thickness (37, 40). This effect may be induced both by the headgroup and the fatty acid chains. It is well documented that augmentation of the unsaturated bonds induces a greater disorder in lipid bilayers (83, 95). This type of effect could be adapted for local fusion events that require a dynamic membrane interior (24, 25). ^{2}H solid-state phase transition spectra indicate that unsaturated DAG promotes nonlamellar structures (56), which may also be prerequisites for localized fusion (11, 12, 80).

From these studies a model has been proposed in which both polyunsaturated PtdIns and DAG are implicated in localized membrane fusion events. The 18:0/20:4 PtdIns in MV1 enhances membrane fluidity, priming the membranes, and subsequent formation of 18:0/20:4 DAG leads to a highly curved structure and the initiation of fusion (25, 56).

How can a membrane compartment with elevated concentrations of highly charged phosphoinositides be stable? What is the molecular mechanism by which vesicles such as MV1 could promote localized fusion?

Concerning stability, elevated concentrations of phosphoinositides could be neutralized by proteins with polybasic clusters such as GTPases (46) and MARCKS (71, 72). Because these levels of the phosphoinositides are much more elevated in MV1 than in the cytoplasmic leaflet of the plasma membrane (71, 72), it is

highly likely that proteins with polybasic clusters such as GTPases (Arf, Rac, Rho, and Ras) would interact strongly with the polyphosphoinositides (46) of MV1, resulting in membrane vesicles with net positive charge. Phosphoinositide levels could remain high in the vesicles even though they contain PLCγ because the enzyme is inactive until regulated upstream by GTP, PtdIns-3 kinase, or tyrosine kinase activation.

The attraction of proteins with polybasic clusters to MV1 may have a dual purpose. The first purpose would be to sequester the phosphoinositides at equilibrium, resulting in a net positive charge, and the second purpose might be to involve the GTPases in bending or deforming the membrane to prompt the initial stages of fusion (9, 25). Hydrophobic amino acids near the polybasic clusters can insert into the hydrophobic acyl chain region of the membrane (71). This would promote an effect known as the bilayer-couple mechanism (87). The penetration of these hydrophobic residues induces an increase in the area of the cytosolic leaflet of the membrane. Because the area of the opposite monolayer remains unchanged, the bilayer bends. The bilayer-couple mechanism will only be seen with proteins at high concentrations and/or high molar partition coefficients. Bending of the membrane toward a preferred curvature produces tension and deformation, which in turn could promote fusion (9).

Therefore, MV1 might be regarded as a membrane compartment that provides the destabilizing membrane proteins and lipids that would be necessary for deforming the membrane and inducing curvature that would facilitate localized membrane fusion. Moreover, the small vesicles of MV1 can be primed to unbending more easily than larger membranes can. This modification of the membrane may induce fusion by the depletion of lipids or proteins from one monolayer (24). So, MV1 could initiate localized membrane fusion by deforming the monolayer through the bilayer-coupling mechanism or by unbending it readily through the loss of proteins and lipids.

PERSPECTIVES AND SUMMARY

Careful comparison of the limited results so far available on the morphology and biochemistry of NE formation makes it possible to reconcile most of the data while allowing for variations due to cell type or stage. All models agree that ER contributes the majority of the membrane of the NE. Work supporting each model provides evidence of inner nuclear membrane proteins such as LBR residing in the ER or vesicles derived from it, conferring chromatin binding properties to these membranes, and eventually localizing them to the inner nuclear membrane. Neither requires nor rules out vesiculation of the ER in NE formation. Each requires sealing of membranes bound to chromatin.

The models differ concerning how gaps in the chromatin-bound membranes are sealed. The enclosure model is neutral or it postulates that gaps are closed by nuclear pores. The fusion model naturally accommodates enclosure by fusion, which, when occurring on the surface of chromatin, leads to closure of both inner and outer nuclear membranes. Given the range of cell types found in metazoans, and the limited number of systems studied so far, it would be premature to generalize a mechanism, and one must additionally allow for the possibility that both mechanisms could be utilized, either redundantly or cooperatively.

The atypical phosphoinositide composition of non-ER membrane fractions and their involvement in membrane fusion should lead to further investigations of the structures that these charged lipids may induce locally in membrane bilayers. Predictions from solid-state NMR studies indicate that elevated quantities of monophosphoinositides such as PtdIns(4)P will form hexagonal structures. Such membrane deformations may result in localized membrane fusion. A more detailed understanding of phosphoinositides as modulators of membrane structure will be important to pursue.

SUMMARY POINTS

1. Two main models are reported for NE assembly. The ER enclosure model postulates that gaps are closed by nuclear pores. The fusion model accommodates enclosure by membrane fusion.

2. From the vast range of cell types found in metazoans, and with the limited number of systems studied so far, it would be premature to generalize one mechanism and one should allow for the possibility of both mechanisms to be utilized, either redundantly or cooperatively.

3. Involvement of lipids in the regulation of NE assembly should not be dismissed. A non-ER membrane fraction that is PLCγ-enriched both in vivo and in vitro has elevated levels of phosphoinositides, is involved in NE formation, and possibly functions in other localized membrane fusion events.

4. Phosphoinositides are thought to be transient signaling molecules. However, with the finding of atypically elevated amounts of phosphoinositides (~60 mol%) in these non-ER membrane vesicles, it may be that phosphoinositides or their products affect membrane dynamics, giving them roles in both signaling and membrane structure.

DISCLOSURE STATEMENT

The authors are not aware of any biases that might be perceived as affecting the objectivity of this review.

ACKNOWLEDGMENTS

This work was supported by CR-UK core funding and an Amherst College Faculty Research Award from the H. Axel Schupf '57 Fund for Intellectual Life (DLP). We would like to thank Stuart McLaughlin for scientific discussions.

LITERATURE CITED

1. Ahyayauch H, Larijani B, Alonso A, Goni FM. 2006. Detergent solubilization of phosphatidylcholine bilayers in the fluid state: influence of the acyl chain structure. *Biochim. Biophys. Acta* 1758:190–96

2. Allan D, Michell RH. 1975. Accumulation of 1,2-diacylglycerol in the plasma membrane may lead to echinocyte transformation of erythrocytes. *Nature* 258:348–49

3. Allan D, Thomas P, Michell RH. 1978. Rapid transbilayer diffusion of 1,2-diacylglycerol and its relevance to control of membrane curvature. *Nature* 276:289–90

4. **Anderson DJ, Hetzer MW. 2007. Nuclear envelope formation by chromatin-mediated reorganization of the endoplasmic reticulum. *Nat. Cell Biol.* 9:1160–66**

5. Anderson DJ, Hetzer MW. 2008. Shaping the endoplasmic reticulum into the nuclear envelope. *J. Cell Sci.* 121:137–42

6. Anderson DJ, Hetzer MW. 2008. The life cycle of the metazoan nuclear envelope. *Curr. Opin. Cell Biol.* 20:386–92

7. Antonin W, Ellenberg J, Dultz E. 2008. Nuclear pore complex assembly through the cell cycle: regulation and membrane organization. *FEBS Lett.* 582:2004–16

8. Antonin W, Franz C, Haselmann U, Antony C, Mattaj IW. 2005. The integral membrane nucleoporin pom121 functionally links nuclear pore complex assembly and nuclear envelope formation. *Mol. Cell* 17:83–92

9. Antonny B. 2006. Membrane deformation by protein coats. *Curr. Opin. Cell Biol.* 18:386–94

10. Barona T, Byrne RD, Pettitt TR, Wakelam MJ, Larijani B, Poccia DL. 2005. Diacylglycerol induces fusion of nuclear envelope membrane precursor vesicles. *J. Biol. Chem.* 280:41171–77

11. Basanez G, Nieva JL, Goni FM, Alonso A. 1996. Origin of the lag period in the phospholipase C cleavage of phospholipids in membranes. Concomitant vesicle aggregation and enzyme activation. *Biochemistry* 35:15183–87

12. Basanez G, Nieva JL, Rivas E, Alonso A, Goni FM. 1996. Diacylglycerol and the promotion of lamellar-hexagonal and lamellar-isotropic phase transitions in lipids: implications for membrane fusion. *Biophys. J.* 70:2299–306

13. Basanez G, Ruiz-Arguello MB, Alonso A, Goni F, Karlsson G, Edwards K. 1997. Morphological changes induced by phospholipase C and by sphingomyelinase on large unilamellar vesicles: a cryo-transmission electron microscopy study of liposome fusion. *Biophys. J.* 72:2630–37

14. **Baur T, Ramadan K, Schlundt A, Kartenbeck J, Meyer HH. 2007. NSF- and SNARE-mediated membrane fusion is required for nuclear envelope formation and completion of nuclear pore complex assembly in *Xenopus laevis* egg extracts. *J. Cell Sci.* 120:2895–903**

15. Beaudouin J, Gerlich D, Daigle N, Eils R, Ellenberg J. 2002. Nuclear envelope breakdown proceeds by microtubule-induced tearing of the lamina. *Cell* 108:83–96

16. Boman AL, Delannoy MR, Wilson KL. 1992. GTP hydrolysis is required for vesicle fusion during nuclear envelope assembly in vitro. *J. Cell Biol.* 116:281–94

17. Broers JL, Ramaekers FC, Bonne G, Yaou RB, Hutchison CJ. 2006. Nuclear lamins: laminopathies and their role in premature ageing. *Physiol. Rev.* 86:967–1008

18. Burger KN. 2000. Greasing membrane fusion and fission machineries. *Traffic* 1:605–13

19. Burke B. 2001. The nuclear envelope: filling in gaps. *Nat. Cell Biol.* 3:E273–74

20. Burke B, Gerace L. 1986. A cell free system to study reassembly of the nuclear envelope at the end of mitosis. *Cell* 44:639–52

21. Byrne RD, Barona TM, Garnier M, Koster G, Katan M, et al. 2005. Nuclear envelope assembly is promoted by phosphoinositide-specific phospholipase C with selective recruitment of phosphatidylinositol-enriched membranes. *Biochem. J.* 387:393–400

4. States that membrane fusion is not the principal mechanism of NE formation but is only required for ER reconstitution and maintenance.

14. A role for typical membrane fusion proteins comes from this study, which implicates both SNAREs and the ATPase NSF in NE assembly in *Xenopus*.

22. Byrne RD, Garnier-Lhomme M, Han K, Dowicki M, Totty N, et al. 2007. PLCγ is enriched on poly-phosphoinositide-rich vesicles to control nuclear envelope assembly. *Cell. Signal.* 19:913–22

23. Chernomordik L, Kozlov MM, Zimmerberg J. 1995. Lipids in biological membrane fusion. *J. Membr. Biol.* 146:1–14

24. Chernomordik LV, Kozlov MM. 2003. Protein-lipid interplay in fusion and fission of biological membranes. *Annu. Rev. Biochem.* 72:175–207

25. Chernomordik LV, Kozlov MM. 2008. Mechanics of membrane fusion. *Nat. Struct. Mol. Biol.* 15:675–83

26. Chernomordik LV, Zimmerberg J, Kozlov MM. 2006. Membranes of the world unite! *J. Cell Biol.* 175:201–7

27. Churchward MA, Rogasevskaia T, Brandman DM, Khosravani H, Nava P, et al. 2008. Specific lipids supply critical intrinsic negative curvature—an essential component of native Ca^{2+} triggered membrane fusion. *Biophys. J.* 94:3976–86

28. Collas P, Courvalin JC, Poccia D. 1996. Targeting of membranes to sea urchin sperm chromatin is mediated by a lamin B receptor-like integral membrane protein. *J. Cell Biol.* 135:1715–25

29. Collas P, Poccia D. 1996. Distinct egg membrane vesicles differing in binding and fusion properties contribute to sea urchin male pronuclear envelopes formed in vitro. *J. Cell Sci.* 109(Pt. 6):1275–83

30. Collas P, Poccia DL. 1995. Formation of the sea urchin male pronucleus in vitro: membrane-independent chromatin decondensation and nuclear envelope-dependent nuclear swelling. *Mol. Reprod. Dev.* 42:106–13

31. Coorssen JR, Rand RP. 1990. Effects of cholesterol on the structural transitions induced by diacylglycerol in phosphatidylcholine and phosphatidylethanolamine bilayer systems. *Biochem. Cell Biol.* 68:65–69

32. Cotter L, Allen TD, Kiseleva E, Goldberg MW. 2007. Nuclear membrane disassembly and rupture. *J. Mol. Biol.* 369:683–95

33. Das S, Rand RP. 1986. Modification by diacylglycerol of the structure and interaction of various phospholipid bilayer membranes. *Biochemistry* 25:2882–89

34. De Matteis MA, Godi A. 2004. Protein-lipid interactions in membrane trafficking at the Golgi complex. *Biochim. Biophys. Acta* 1666:264–74

35. Dechat T, Vlcek S, Foisner R. 2000. Review: lamina-associated polypeptide 2 isoforms and related proteins in cell cycle-dependent nuclear structure dynamics. *J. Struct. Biol.* 129:335–45

36. Di Paolo G, De Camilli P. 2006. Phosphoinositides in cell regulation and membrane dynamics. *Nature* 443:651–57

37. Douliez JP, Leonard A, Dufourc EJ. 1995. Restatement of order parameters in biomembranes: calculation of C-C bond order parameters from C-D quadrupolar splittings. *Biophys. J.* 68:1727–39

38. Drummond S, Ferrigno P, Lyon C, Murphy J, Goldberg M, et al. 1999. Temporal differences in the appearance of NEP-B78 and an LBR-like protein during *Xenopus* nuclear envelope reassembly reflect the ordered recruitment of functionally discrete vesicle types. *J. Cell Biol.* 144:225–40

39. Ellenberg J, Siggia ED, Moreira JE, Smith CL, Presley JF, et al. 1997. Nuclear membrane dynamics and reassembly in living cells: targeting of an inner nuclear membrane protein in interphase and mitosis. *J. Cell Biol.* 138:1193–206

40. Faure C, Tranchant JF, Dufourc EJ. 1996. Comparative effects of cholesterol and cholesterol sulfate on hydration and ordering of dimyristoylphosphatidylcholine membranes. *Biophys. J.* 70:1380–90

41. Fischer AH, Taysavang P, Weber CJ, Wilson KL. 2001. Nuclear envelope organization in papillary thyroid carcinoma. *Histol. Histopathol.* 16:1–14

42. Gant TM, Harris CA, Wilson KL. 1999. Roles of LAP2 proteins in nuclear assembly and DNA replication: truncated LAP2beta proteins alter lamina assembly, envelope formation, nuclear size, and DNA replication efficiency in *Xenopus laevis* extracts. *J. Cell Biol.* 144:1083–96

43. Harel A, Orjalo AV, Vincent T, Lachish-Zalait A, Vasu S, et al. 2003. Removal of a single pore subcomplex results in vertebrate nuclei devoid of nuclear pores. *Mol. Cell* 11:853–64

44. Harel A, Zlotkin E, Nainudel-Epszteyn S, Feinstein N, Fisher PA, Gruenbaum Y. 1989. Persistence of major nuclear envelope antigens in an envelope-like structure during mitosis in *Drosophila melanogaster* embryos. *J. Cell Sci.* 94(Pt. 3):463–70

45. Hegele RA. 2000. The envelope, please: nuclear lamins and disease. *Nat. Med.* 6:136–70

22. This paper characterizes a non-ER membrane vesicle fraction with atypically elevated levels of phosphoinositides that is required for nuclear envelope formation in sea urchins.

39. An alternative to the vesiculation model of nuclear envelope breakdown and re-formation.

46. Heo WD, Inoue T, Park WS, Kim ML, Park BO, et al. 2006. PI(3,4,5)P3 and PI(4,5)P2 lipids target proteins with polybasic clusters to the plasma membrane. *Science* 314:1458–61

47. Hetzer M, Bilbao-Cortes D, Walther TC, Gruss OJ, Mattaj IW. 2000. GTP hydrolysis by Ran is required for nuclear envelope assembly. *Mol. Cell* 5:1013–24

48. Hetzer M, Meyer HH, Walther TC, Bilbao-Cortes D, Warren G, Mattaj IW. 2001. Distinct AAA-ATPase p97 complexes function in discrete steps of nuclear assembly. *Nat. Cell Biol.* 3:1086–91

49. Hutchison CJ, Alvarez-Reyes M, Vaughan OA. 2001. Lamins in disease: Why do ubiquitously expressed nuclear envelope proteins give rise to tissue-specific disease phenotypes? *J. Cell Sci.* 114:9–19

50. Jantsch-Plunger V, Glotzer M. 1999. Depletion of syntaxins in the early *Caenorhabditis elegans* embryo reveals a role for membrane fusion events in cytokinesis. *Curr. Biol.* 9:738–45

51. Jun Y, Fratti RA, Wickner W. 2004. Diacylglycerol and its formation by phospholipase C regulate Rab- and SNARE-dependent yeast vacuole fusion. *J. Biol. Chem.* 279:53186–95

52. Kennerly DA, Sullivan TJ, Sylwester P, Parker CW. 1979. Diacylglycerol metabolism in mast cells: a potential role in membrane fusion and arachidonic acid release. *J. Exp. Med.* 150:1039–44

53. Koch GL, Booth C, Wooding FB. 1988. Dissociation and reassembly of the endoplasmic reticulum in live cells. *J. Cell Sci.* 91(Pt. 4):511–22

54. Kozlovsky Y, Chernomordik LV, Kozlov MM. 2002. Lipid intermediates in membrane fusion: formation, structure, and decay of hemifusion diaphragm. *Biophys. J.* 83:2634–51

55. Larijani B, Barona TM, Poccia DL. 2001. Role for phosphatidylinositol in nuclear envelope formation. *Biochem. J.* 356:495–501

56. Larijani B, Dufourc E. 2006. Polyunsaturated phosphatidylinositol and diacylglycerol substantially modify the fluidity and polymorphism of bio-membranes: a solid state deuterium NMR study. *Lipids* 41:925–32

57. Larijani B, Poccia D. 2007. Protein and lipid signaling in membrane fusion: nuclear envelope assembly. *Signal. Transduct.* 7:142–53

58. Larijani B, Poccia DL, Dickinson LC. 2000. Phospholipid identification and quantification of membrane vesicle subfractions by two-dimensional ^{31}P-^{1}H nuclear magnetic resonance. *Lipids* 35:1289–97

59. Laskey RA, Gorlich D, Madine MA, Makkerh JP, Romanowski P. 1996. Regulatory roles of the nuclear envelope. *Exp. Cell Res.* 229:204–11

60. Leikin SL, Kozlov MM, Chernomordik LV, Markin VS, Chizmadzhev YA. 1987. Membrane fusion: overcoming of the hydration barrier and local restructuring. *J. Theor. Biol.* 129:411–25

61. Lohka MJ, Masui Y. 1983. Formation in vitro of sperm pronuclei and mitotic chromosomes induced by amphibian ooplasmic components. *Science* 220:719–21

62. Longo F. 1981. Regulation of pronuclear development. In *Bioregulators of Reproduction*, ed. G Jagiello, C Vogel, pp. 529–57. New York: Acad. Press

63. Longo FJ. 1972. An ultrastructural analysis of mitosis and cytokinesis in the zygote of the sea urchin, *Arbacia punctulata*. *J. Morphol.* 138:207–38

64. Longo FJ. 1976. Derivation of the membrane comprising the male pronuclear envelope in inseminated sea urchin eggs. *Dev. Biol.* 49:347–68

65. Longo FJ, Anderson E. 1968. The fine structure of pronuclear development and fusion in the sea urchin, *Arbacia punctulata*. *J. Cell Biol.* 39:339–68

66. Longo FJ, Anderson E. 1969. Sperm differentiation in the sea urchins *Arbacia punctulata* and *Strongylocentrotus purpuratus*. *J. Ultrastruct. Res.* 27:486–99

67. Macaulay C, Forbes DJ. 1996. Assembly of the nuclear pore: biochemically distinct steps revealed with NEM, GTP gamma S, and BAPTA. *J. Cell Biol.* 132:5–20

68. Markin VS, Kozlov MM, Borovjagin VL. 1984. On the theory of membrane fusion. The stalk mechanism. *Gen. Physiol. Biophys.* 3:361–77

69. Mattaj IW. 2004. Sorting out the nuclear envelope from the endoplasmic reticulum. *Nat. Rev. Mol. Cell Biol.* 5:65–69

70. Mayer A. 1999. Intracellular membrane fusion: SNAREs only? *Curr. Opin. Cell Biol.* 11:447–52

71. McLaughlin S, Murray D. 2005. Plasma membrane phosphoinositide organization by protein electrostatics. *Nature* 438:605–11

72. McLaughlin S, Wang J, Gambhir A, Murray D. 2002. PIP(2) and proteins: interactions, organization, and information flow. *Annu. Rev. Biophys. Biomol. Struct.* 31:151–75

73. Montag M, Spring H, Trendelenburg MF. 1988. Structural analysis of the mitotic cycle in pregastrula *Xenopus* embryos. *Chromosoma* 96:187–96

74. Nagano A, Arahata K. 2000. Nuclear envelope proteins and associated diseases. *Curr. Opin. Neurol.* 13:533–39

75. Nieva JL, Goni FM, Alonso A. 1989. Liposome fusion catalytically induced by phospholipase C. *Biochemistry* 28:7364–67

76. Poccia D, Collas P. 1996. Transforming sperm nuclei into male pronuclei in vivo and in vitro. *Curr. Top. Dev. Biol.* 34:25–88

77. Prunuske AJ, Ullman KS. 2006. The nuclear envelope: form and reformation. *Curr. Opin. Cell Biol.* 18:108–16

78. Puhka M, Vihinen H, Joensuu M, Jokitalo E. 2007. Endoplasmic reticulum remains continuous and undergoes sheet-to-tubule transformation during cell division in mammalian cells. *J. Cell Biol.* 179:895–909

79. Rand RP, Fang Y. 1993. Structure and energetics of phospholipid and diacylglycerol assemblies relative to membrane fusion. *Biochem. Soc. Trans.* 21:266–70

80. Rand RP, Parsegian VA. 1984. Physical force considerations in model and biological membranes. *Can. J. Biochem. Cell Biol.* 62:752–59

81. Rand RP, Reese TS, Miller RG. 1981. Phospholipid bilayer deformations associated with interbilayer contact and fusion. *Nature* 293:237–38

82. Robbins E, Gonatas NK. 1964. The ultrastructure of a mammalian cell during the mitotic cycle. *J. Cell Biol.* 21:429–63

83. Roos DS, Choppin PW. 1985. Biochemical studies on cell fusion. I. Lipid composition of fusion-resistant cells. *J. Cell Biol.* 101:1578–90

84. Rybin V, Ullrich O, Rubino M, Alexandrov K, Simon I, et al. 1996. GTPase activity of Rab5 acts as a timer for endocytic membrane fusion. *Nature* 383:266–69

85. Sasagawa S, Yamamoto A, Ichimura T, Omata S, Horigome T. 1999. In vitro nuclear assembly with affinity-purified nuclear envelope precursor vesicle fractions, PV1 and PV2. *Eur. J. Cell Biol.* 78:593–600

86. Schatten H. 1994. Dithiothreitol prevents membrane fusion but not centrosome or microtubule organization during the first cell cycles in sea urchins. *Cell Motil. Cytoskelet.* 27:59–68

87. Sheetz MP, Singer SJ. 1974. Biological membranes as bilayer couples. A molecular mechanism of drug-erythrocyte interactions. *Proc. Natl. Acad. Sci. USA* 71:4457–61

88. Shibata Y, Voeltz GK, Rapoport TA. 2006. Rough sheets and smooth tubules. *Cell* 126:435–39

89. Siegel DP, Epand RM. 1997. The mechanism of lamellar-to-inverted hexagonal phase transitions in phosphatidylethanolamine: implications for membrane fusion mechanisms. *Biophys. J.* 73:3089–111

90. Sodergren E, Shen Y, Song X, Zhang L, Gibbs RA, Weinstock GM. 2006. Shedding genomic light on Aristotle's lantern. *Dev. Biol.* 300:2–8

91. Sodergren E, Weinstock GM, Davidson EH, Cameron RA, Gibbs RA, et al. 2006. The genome of the sea urchin *Strongylocentrotus purpuratus. Science* 314:941–52

92. Spann TP, Moir RD, Goldman AE, Stick R, Goldman RD. 1997. Disruption of nuclear lamin organization alters the distribution of replication factors and inhibits DNA synthesis. *J. Cell Biol.* 136:1201–12

93. Stafstrom JP, Staehelin LA. 1984. Dynamics of the nuclear envelope and of nuclear pore complexes during mitosis in the *Drosophila* embryo. *Eur. J. Cell Biol.* 34:179–89

94. Stricker SA. 2006. Structural reorganizations of the endoplasmic reticulum during egg maturation and fertilization. *Semin. Cell Dev. Biol.* 17:303–13

95. Teague WE, Fuller NL, Rand RP, Gawrisch K. 2002. Polyunsaturated lipids in membrane fusion events. *Cell Mol. Biol. Lett.* 7:262–64

96. Terasaki M, Jaffe LA. 1991. Organization of the sea urchin egg endoplasmic reticulum and its reorganization at fertilization. *J. Cell Biol.* 114:929–40

97. Terasaki M, Jaffe LA. 1993. Imaging endoplasmic reticulum in living sea urchin eggs. *Methods Cell Biol.* 38:211–20

98. Terasaki M, Jaffe LA, Hunnicutt GR, Hammer JA. 1996. Structural change of the endoplasmic reticulum during fertilization: evidence for loss of membrane continuity using the green fluorescent protein. *Dev. Biol.* 179:320–28

99. Vigers GPA, Lohka MJ. 1991. A distinct vesicle population targets membranes and pore complexes to the nuclear envelope in *Xenopus* eggs. *J. Cell. Biol.* 112:545–56

100. Villar AV, Alonso A, Goni FM. 2000. Leaky vesicle fusion induced by phosphatidylinositol-specific phospholipase C: observation of mixing of vesicular inner monolayers. *Biochemistry* 39:14012–18

101. Voeltz GK, Prinz WA, Shibata Y, Rist JM, Rapoport TA. 2006. A class of membrane proteins shaping the tubular endoplasmic reticulum. *Cell* 124:573–86

102. Wakelam MJ. 1988. Myoblast fusion—a mechanistic analysis. *Curr. Top. Membr. Transp.* 32:87–112

103. Walther TC, Alves A, Pickersgill H, Loiodice I, Hetzer M, et al. 2003. The conserved Nup107-160 complex is critical for nuclear pore complex assembly. *Cell* 113:195–206

104. Weber T, Zemelman BV, McNew JA, Westermann B, Gmachl M, et al. 1998. SNAREpins: minimal machinery for membrane fusion. *Cell* 92:759–72

105. Wiese C, Goldberg MW, Allen TD, Wilson KL. 1997. Nuclear envelope assembly in *Xenopus* extracts visualized by scanning EM reveals a transport-dependent 'envelope smoothing' event. *J. Cell Sci.* 110(Pt. 13):1489–502

106. Wiese C, Wilson KL. 1993. Nuclear membrane dynamics. *Curr. Opin. Cell Biol.* 5:387–94

107. Wilkie GS, Schirmer EC. 2006. Guilt by association: the nuclear envelope proteome and disease. *Mol. Cell Proteomics* 5:1865–75

108. Wilson EB. 1925. *The Cell in Development and Heredity*. New York: Macmillan

109. Yang L, Guan T, Gerace L. 1997. Integral membrane proteins of the nuclear envelope are dispersed throughout the endoplasmic reticulum during mitosis. *J. Cell Biol.* 137:1199–210

110. Zimmerberg J. 2001. How can proteolipids be central players in membrane fusion? *Trends Cell Biol.* 11:233–35

111. Zimmerberg J, Chernomordik LV. 2005. Neuroscience. Synaptic membranes bend to the will of a neurotoxin. *Science* 310:1626–27

112. Zink D, Fischer AH, Nickerson JA. 2004. Nuclear structure in cancer cells. *Nat. Rev. Cancer* 4:677–87

The Interplay of Catalysis and Toxicity by Amyloid Intermediates on Lipid Bilayers: Insights from Type II Diabetes

James A. Hebda and Andrew D. Miranker

Department of Molecular Biophysics and Biochemistry, Yale University, New Haven, Connecticut 06520-8114; email: Andrew.Miranker@yale.edu

Annu. Rev. Biophys. 2009. 38:125–52

First published online as a Review in Advance on February 2, 2009

The *Annual Review of Biophysics* is online at biophys.annualreviews.org

This article's doi: 10.1146/annurev.biophys.050708.133622

Key Words

IAPP, amylin, protein folding, α-synuclein, Aβ, mechanism

Abstract

The dynamics, energies, and structures governing protein folding are critical to biological function. Amyloidoses are a class of disease defined, in part, by the misfolding and aggregation of functional protein precursors into fibrillar states. Amyloid fibers contribute to the pathology of many diseases, including type II diabetes, Alzheimer's, and Parkinson's. In these disorders, amyloid fibers are present in affected tissues. However, it has become clear that intermediate states, rather than mature fibers, represent the cytotoxic species. In this review, we focus particularly on lipid bilayer–bound intermediates. Remarkably, the precursors of these fibers are intrinsically disordered, and yet catalysis of β-sheet formation appears to be mediated by the stabilization of α-helical states. On the lipid bilayer, these intermediate species have been implicated as cytotoxic through elimination of ionic homeostasis. Recent advances are enabling insights at a molecular level that promise to provide meaningful targets for the development of therapeutics.

Contents

PROTEIN FOLDING DEFINED

Why is protein folding so captivating? This problem has drawn the attention of physical chemists for nearly a century and shows no sign of abating. The familiar refrain of the Levinthal Paradox is sufficient to draw the initial interest of a student, or an investigator from an unrelated discipline, but hardly accounts for the commitment of time and resources that have been invested and continue to be committed to the field. First, it is an intrinsically beautiful problem. In all of the sciences, singular solutions in the face of astronomical possibilities are appealing. Second, an understanding of protein folding has implications spanning engineering, medical, and basic biological sciences. An understanding of protein folding enables structure prediction that can affect the design of enzymes or materials (54). It can also allow rational determination of a functional protein structure in situations where experimental methods are inadequate (131).

The formations of compact globular states and multisubunit complexes have long been intimately related subdisciplines of protein folding. In 1953, Prof. Fred Richards used subtilisin to cleave ribonuclease into two polypeptides (124). Alone, neither polypeptide was folded to a compact state, and neither had significant nuclease activity. The breakthrough was the observation that a stoichiometric mix of these two peptides resulted in spontaneous self-assembly of a folded and active protein. Such an observation captivated the imagination because it offered the first molecular basis for connecting intermolecular interactions with the generation of biological activities, such as occurs in signal transduction. A modern manifestation of this is the folding of the phosphorylated kinase-inducible activation domain (pKID) of the transcription factor CREB. This domain is unstructured in the absence of the CREB binding protein. Upon exposure to the CREB binding protein, pKID associates and undergoes a series of conformational steps that result in a well-defined complex (143). Thus, folding paradigms dictate the structure and function of small globular proteins, as well as the establishment of both long-lived and transient molecular complexes. Protein folding may therefore be more appropriately and broadly defined as any molecular transition in which a change of state is accompanied by a significant change in the breadth of the conformational ensemble.

The evolution of the scientific investigation of folding has revealed an ever-increasing richness in polypeptide behavior. Initial biophysical investigations of folding centered on model systems. First, proteins that were abundant, compact, and reversibly folded, such as lysozyme and RNase A, were studied (20). Then investigation expanded to include proteins that were folded and stable only in membranes, such as glycophorin (128). Many of the most important insights into these fields stemmed from investigations that seemed to break these newly established rules. A dramatic example of this is α-lytic protease, which appears to fold in a manner contradictory to the 1972 Nobel Prize–winning insights of Anfinsen. Haber & Anfinsen found that the native state of a protein represented the minimum free energy conformation (60). In contrast, α-lytic protease is regarded as kinetically rather than thermodynamically isolated (137) (i.e., it adopts a fold that is not at the free energy minimum). Assembly mechanisms of these systems are such that once the native conformation is achieved, the transition state barrier becomes too great for unfolding to occur to any significant extent. Other challenges to folding paradigms have also broadened our views. It now appears that many proteins are natively unfolded in their apo form (151). Such intrinsically disordered proteins demonstrate that obtaining a functional, defined structure may be dependent upon information outside the primary amino acid sequence. Our understanding of the interplay of sequence and structure must accommodate oligomerization if we are to obtain a biologically relevant view of protein folding.

As novel findings develop into independent subdisciplines of protein folding, it is important to appreciate that many of the observables represent limiting behavior. Proteins can but need not fold in the absence of a binding partner. Proteins, such as actin, can exist as a monomer or filament. Other systems, such as bacterial toxins, can be soluble in aqueous solution or on bilayers. The fold of a protein is dependent on primary structure, yet great sequence variability can be found within nearly identical folds. The study of systems that can adopt various folding pathways is required to observe the full spectrum of protein conformational possibilities.

PROTEIN MISFOLDING AS A BASIS FOR DISEASE

Why is protein misfolding so captivating? As one of the subdisciplines of protein folding, misfolding represents a recognition by scientists that the complexity of functional conformational change is not always satisfied spontaneously. An early example of this is the folding of luciferase from bacteria (5). This two-subunit protein folds to a correct and active form only if the subunits are refolded together. Subunits folded independently of one another and then mixed do not give rise to an active species. The origins of this observation were driven because the domains are structurally homologous. Refolding of one domain independently of the other produces homodimers that do not readily dissociate. More recently, it has been suggested that homomeric complexes experience selective pressure to diversify to avoid nonnative oligomeric states (162). This hypothesis was drawn from the observation that titan domains are homologous but not identical. Incorrect self-association was shown to be far more probable if the sequences of adjacent domains are identical. The myriad of selective pressures on proteins is not readily satisfied. Fold, turnover, function, and solubility all make competing demands on sequence. As a result, protein misfolding is a far more prevalent phenomenon than the study of globular model systems has suggested.

Misfolding events can result in a toxic gain of function. Toxicity is often associated with the irreversible formation of a particular class of aggregates known as amyloids. Amyloid fibers are homomeric, linear aggregates of protein that have been implicated in many disease states, including Alzheimer's, Parkinson's, and type II diabetes. In each case, a normally soluble protein, beta amyloid peptide (Aβ), α-synuclein, and islet amyloid polypeptide (IAPP), respectively, self-assembles into β-sheet-rich fibers in a process that is associated with cell death

Aβ: beta amyloid peptide

IAPP: islet amyloid polypeptide

(45, 59, 148). In the case of Aβ and α-synuclein, this results in neurodegeneration. In IAPP, this results in reduction in the mass of the insulin-producing, pancreatic islet β-cells. Collectively, type II diabetes, Alzheimer's, and Parkinson's affect tens of millions of people in the United States alone. A remarkable feature of amyloid toxicity is that it appears to be a generic property of polypeptides (118, 141). It also appears to be a generic property of polypeptides that they can be induced to form amyloid structures (145). Given these observations, it seems likely that an increasing number of diseases will be reported to have amyloid formation as a component of pathology.

Amyloid fiber formation captures many of the paradigms of folding discussed above, in which conformational behavior straddles the limiting regimes of folding. Many amyloid precursors are soluble in aqueous and membrane environments. They can be intrinsically disordered, only becoming structured upon forming fibrillar states. Last, the resulting fibers are kinetically isolated. Amyloid formation is the subject of frequent reviews from many perspectives. Here, we focus on the intermediates of fiber formation that interact with lipid membranes. Such interactions have the capacity to induce structure in the protein, affect preamyloid and amyloid assembly, and disrupt membrane integrity. Most recently these intermediates have been implicated as causal agents of toxicity in living cells. To date, over 10 amyloid systems have been demonstrated to act on biological membranes (99). Our focus is primarily on IAPP from type II diabetes; however, we draw parallels from the Aβ-peptide in Alzheimer's and from α-synuclein in Parkinson's. In all three diseases, the precursor is unfolded or, at most, partially structured in solution. Upon exposure to the membrane, these proteins bind the membrane and become structured. α-Helical membrane-bound states, correlated to amyloid formation, have been observed for each of these amyloidogenic proteins. These similarities suggest a commonality of mechanism, which when understood will help elucidate the relationship between amyloid formation and toxicity.

ISLET AMYLOID POLYPEPTIDE IN TYPE II DIABETES

IAPP, which is also called amylin, is a peptide hormone normally cosecreted with insulin by the β-cells of the pancreas (25, 59, 70, 108). In type II diabetes, IAPP deposits as amyloid fibers evident in the extracellular spaces of the pancreas. Loss of β-cell mass is a central feature of diabetes pathology and is coincident with amyloid plaque formation. In its soluble form, IAPP is an unstructured, C-terminally amidated 37-residue peptide with an intrinsic disulfide that exists between residues 2 and 7 (**Figure 1a**). Its function is unknown but is suggested to include regulation of gastric emptying into the duodenum (169) and/or paracrine feedback for the regulation of insulin secretion (110). Many factors suggest an intimate relationship between IAPP and insulin. Expression is coregulated; the peptides are processed by a common set of convertases, copackaged, and subjected to degradation by the insulin-degrading enzyme (IDE) (136). Type II diabetes is preceded by systemic insulin resistance, which can be overcome by increased secretion of insulin. Increased insulin secretion is necessarily accompanied by increased secretion of IAPP. The development of toxicity by IAPP results in a progressive reduction in the population of β-cells available to meet this increased demand. This accounts for the later stage pathology of diabetes in which patients require insulin supplementation. It is also plausible that IAPP has a causal role in diabetes pathology. A loss of pulsatile release of insulin from β-cells is noted in human IAPP transgenic rats and has been suggested to result in the manifestation of systemic insulin resistance (110).

Wild-type, unmodified human IAPP readily forms fibers at 10 μM concentrations on the minutes to hours timescale upon dilution from denaturant into physiological buffer solution (**Figure 2a**). As IAPP is a secreted protein, its normal concentrations are orders of

a

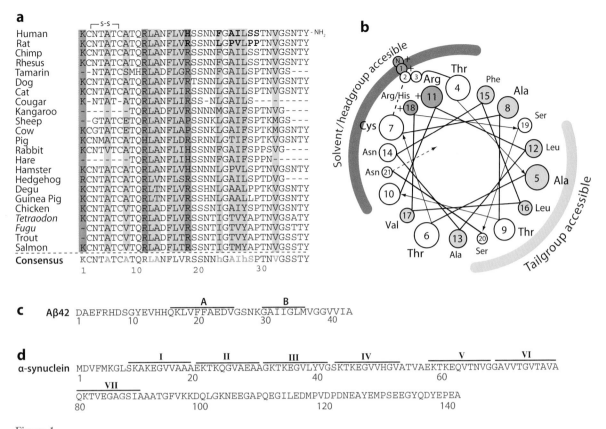

Human	KCNTATCATQRLANFLVHSSNNFGAILSSTNVGSNTY-NH₂
Rat	KCNTATCATQRLANFLVRSSNNLGPVLPPTNVGSNTY
Chimp	KCNTATCATQRLANFLVRSSNNFGAILSSTNVGSNTY
Rhesus	KCNTATCATQRLANFLVRSSNNFGAILSSTNVGSNTY
Tamarin	--NTATCSMHRLADFLGRSSNNFGAILSPTNVGS---
Dog	KCNTATCATQRLANFLVRTSNNLGAILSPTNVGSNTY
Cat	KCNTATCATQRLANFLIRSSNNLGAILSPTNVGSNTY
Cougar	K-NTAT-ATQRLANFLIRSS-NLGAILS--------
Kangaroo	--------TQRLADFLVRSNNNMGAIFSPTNVG----
Sheep	--GTATCETQRLANFLAPSSNKLGAIFSPTKMGS---
Cow	KCGTATCETQRLANFLAPSSNKLGAIFSPTKMGSNTY
Pig	KCNMATCATQHLANFLDRSRNNLGTIFSPTKVGSNTY
Rabbit	KCNTVTCATQRLANFLIHSSNNFGAIFSPPSVG----
Hare	KCNTVTCATQRLANFLIHSSNNFGAIFSPPN------
Hamster	KCNTATCATQRLANFLVHSNNNLGPVLSPTNVGSNTY
Hedgehog	RCNTATCATQRLVNFLSRSSNNLGAILSPTDVG----
Degu	KCNTATCATQRLTNFLVRSSHNLGAALPPTKVGSNTY
Guinea Pig	KCNTATCATQRLTNFLVRSSHNLGAALPTDVGSNTY
Chicken	KCNTATCVTQRLADFLVRSSSNIGAIYSPTNVGSNTY
Tetraodon	KCNTATCVTQRLADFLVRSSNTIGTVYAPTNVGSATY
Fugu	-CNTATCVTQRLADFLVRSSNTIGTVYAPTNVGSTTY
Trout	-CNTATCVTQRLADFLTRSSNTIGTVYAPTNVGSSTY
Salmon	KCNTATCVTQRLADFLTRSSNTIGTMYAPTNVGSSTY
Consensus	KCNTATCATQRLANFLVRSSNNhGAIhSPTNVGSSTY

1   10   20   30

c

Aβ42 DAEFRHDSGYEVHHQKLVFFAEDVGSNKGAIIGLMVGGVVIA

A (over residues ~17-22), B (over residues ~30-36)

1   10   20   30   40

d

α-synuclein MDVFMKGLSKAKEGVVAAAEKTKQGVAEAAGKTKEGVLYVGSKTKEGVVHGVATVAEKTKEQVTNVGGAVVTGVTAVA

I   II   III   IV   V   VI

1   20   40   60

VII

QKTVEGAGSIAAATGFVKKDQLGKNEEGAPQEGILEDMPVDPDNEAYEMPSEEGYQDYEPEA

80   100   120   140

Figure 1

Primary and secondary structure of amyloid proteins. (*a*) The multiple sequence alignment for species variants of islet amyloid polypeptide (IAPP). The six differences between human and rat IAPP are indicated with bold lettering. Human IAPP readily forms amyloid under physiological conditions, whereas the variant from rat does not. Positively charged residues are highlighted in blue, and residues that are generally hydrophobic are highlighted green. Apparent consensus is indicated. Lowercase *b* indicates hydrophobic residue. (*b*) Helical wheel projection of human and rat IAPP through residue 21. Residues exposed to either water-soluble (*orange*) or lipid-soluble (*green*) relaxation agents (2) are indicated as arcs. Residues in blue are positively charged, green are hydrophobic, and white are polar. (*c, d*) Primary sequence of human Aβ42 and α-synuclein. Regions that adopt α-helical repeats in Aβ (27, 77) and the α-11/3 helical repeats in α-synuclein are underlined.

magnitude higher. The fact that amyloid is not ubiquitously observed in healthy individuals rests in part because IAPP fiber formation is inhibited by its cosecretant, insulin (92, 96). Insulin is expressed in 50–100 molar excess over IAPP. Remarkably, substoichiometric levels of insulin are sufficient to suppress the amyloid process in solution (**Figure 2*a***) (98). Perturbations in normal insulin-IAPP physiology by environmental effects may therefore trigger IAPP-related pathology.

Aβ-PEPTIDE IN ALZHEIMER'S DISEASE

The Aβ-peptide is initially expressed as part of a transmembrane anchor of the 770-residue secreted protein, amyloid precursor protein (APP) (45, 146). Familial forms of Alzheimer's disease have been linked to APP, yet the majority of Alzheimer's disease cases are sporadic. The link between Alzheimer's disease and APP can be traced to the generation and accumulation of an amyloidogenic proteolytic fragment

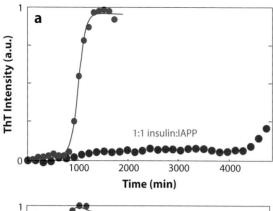

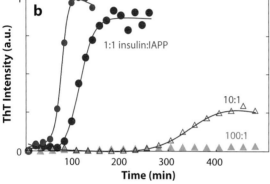

Figure 2

Effects of anionic liposomes and insulin on islet amyloid polypeptide (IAPP) fiber formation kinetics. (*a*) In the absence of liposomes (*blue*), 10 uM IAPP in aqueous buffer, pH 7.4, at 25°C converts to fibers in ~15 h. Here, fiber formation is detected in real time by the inclusion and monitoring of an amyloid-specific fluorescent dye, thioflavin T (ThT). Under these conditions, stoichiometric inclusion of insulin effectively inhibits fiber formation (*red*). (*b*) In the presence of 1.3 mM extruded DOPG liposomes, 10 μM IAPP fiber formation is accelerated to a ~1 h (*blue*) timescale. The capacity of insulin to inhibit fiber formation is markedly reduced. Figure adapted from Reference 92.

α-SYNUCLEIN IN PARKINSON'S DISEASE

Parkinson's disease patients suffer debilitating motor and cognitive neurological degeneration as a result of cell death of dopamanergic neurons in the substantia nigra (24, 42). The pathology of this disease is characterized, in part, by the presence of characteristic intracellular deposits known as Lewy bodies (42). α-synuclein is the most prominent protein in these deposits and is associated with the neurodegeneration that gives rise to clinical symptoms (37). That α-synuclein can be causal in disease has been suggested in a variety of ways. Notably, in some familial forms of the disease, duplication of the α-synuclein gene has been observed (19). Although familial forms tie protein to disease, the majority of cases are sporadic (61). The precise function of this 140-residue protein is not yet clear (**Figure 1*d***). The protein is predominantly unstructured in solution but adopts α-helical structure upon binding membranes (7, 55). It is at the membrane where function is generally asserted, e.g., as a mediator of vesicular trafficking (28).

THE AMYLOID HYPOTHESIS

Investigations looking to explain the observed correlation of amyloid plaques and disease have led to what is known as the amyloid hypothesis. The amyloid hypothesis was well supported by two classes of observation. First, pathology in any given amyloidosis is associated with the deposition of a specific protein precursor into particular tissues. For example, plaque deposits in Alzheimer's disease patients are predominantly formed from Aβ-peptide and deposited into the interstitial space around brain neurons (10, 45). Second, familial forms of the disease often map either directly to the protein precursor or to machinery specific to the expression and proteostasis of that precursor (4). For example, the numerous inherited mutations that map to the Aβ-peptide or to the secretases that process its precursor, APP, have been associated with the early onset of

of APP. Proteolytic processing releases the soluble domain of APP, called A4, but also produces peptides that are variable in length (26). The most amyloidogenic forms of these peptides are 40 and 42 residues in length, termed Aβ40 and Aβ42, respectively (**Figure 1*c***). Aβ40 is far less amyloidogenic, and it has been suggested that an increase in Aβ42 relative to Aβ40 is a key trigger in Alzheimer's disease pathology (73). Aggregation of Aβ42 peptides into amyloid aggregates is associated with neuronal degeneration, manifesting into the dementia that characterizes Alzheimer's disease patients (45).

dementia (15, 43). In IAPP, spontaneous forms of diabetes arise only in species in which IAPP is amyloidogenic (human, cat, and monkey) (84). The assignment of amyloid to disease states has also been strongly supported by animal model systems. For example, differentially amyloid-prone mutants of Aβ42 in transgenic *Drosophila melanogaster* showed a positive correlation between pathogenic phenotype and in vitro aggregation propensity (104). Such studies leave little doubt that the potential for amyloid fiber formation is directly coupled to specific disease states.

AMYLOID FIBER STRUCTURE

Amyloid fibers are long, linear, unbranched structures varying in width from 10 to 20 nm (**Figure 3**) (106). Electron microscopy (EM) and atomic force microscopy (AFM) studies reveal these to be composed of bundles of two or more protofilaments typically of the order of 5–10 nm in width. Each of these filaments is composed of a core of β-sheets. Depending on the precursor, the β-strands of these sheets can be connected by short loops, or significant portions of the precursor protein may reside outside the fiber core. Solid-state NMR studies of IAPP have indicated a strand-loop-strand morphology for monomers within the fiber state (**Figure 4a,b**). IAPP showed β-sheet structure between residues 8–17 and 28–37 (103). This conformation is analogous to that shown for Aβ40, in which structural constraints are consistent with β-sheet structure from 10–23 and 28–40 (119). Similarity in structure between IAPP and Aβ has also been implied, as a nonamyloidogenic variant of IAPP inhibits Aβ aggregation and cell toxicity (165). Elements of this similarity are perhaps functionally evident in the atomic structure of the IDE (136). IDE was crystallized with IAPP or Aβ bound as substrate, with structure apparent for residues 1–6 and 11–23 for bound IAPP. Correspondingly, in Aβ-IDE, structure was observed for residues 1–3 and 16–23 (**Figure 4e**).

The β-strands of an amyloid fiber are arranged (by definition) in a cross β pattern,

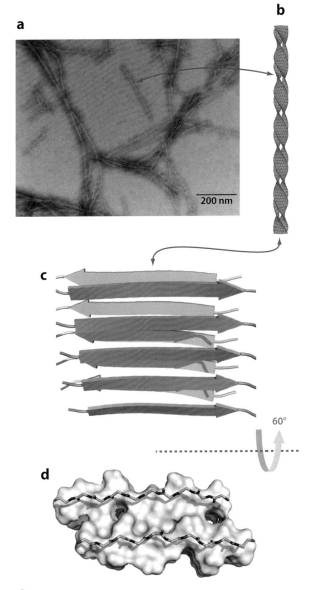

Figure 3

Schematic of the typical ultrastructural features of amyloid fibers. (*a*) Representative negative stain transmission electron micrograph of islet amyloid polypeptide (IAPP) fibers produced by dilution of dimethyl sulfoxide stock solution of IAPP into aqueous buffer, pH 7.4, containing 150 mM KCl at 25°C. Amyloid fibers are typically composed of two or more protofilaments wound in a helical repeat (80). (*b*) Schematic of an isolated protofilament. Protofilaments may be composed of two or more β-sheets. (*c*) The β-strands of protofilaments are arrayed orthogonal to the long axis of the fiber, with backbone hydrogen bonding running parallel to this axis. (*d*) Viewed end on, typical amyloid β-sheets form interfaces with ~10 Å spacing. Here, the main chain of the endmost strands is shown with sticks, and the remainder of the fiber is represented as a surface.

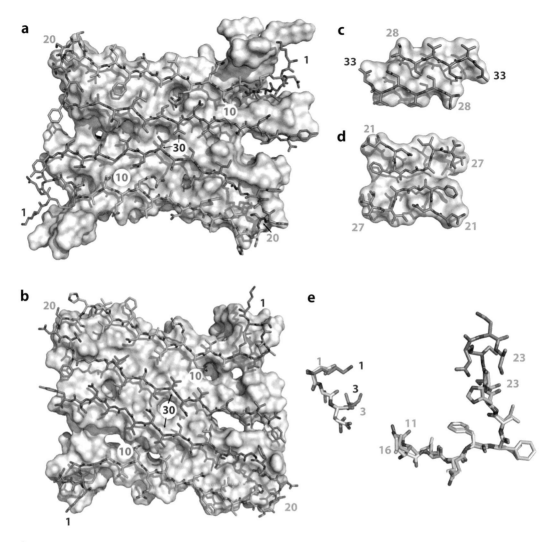

Figure 4

Molecular models of islet amyloid polypeptide (IAPP) peptides. (*a, b*) Two alternative models of IAPP fiber structure consistent with solid-state NMR-derived constraints (103). The β-sheets appear to be composed of two regions separated by a loop. This is analogous to the structure models asserted for Aβ fiber structure (105, 150) (*c, d*) Fiber structures suggested by atomic structures of IAPP subpeptides using crystals of IAPP$_{28-33}$ and IAPP$_{21-27}$ (160). (*e*) Overlay of IAPP and Aβ in the binding site of the insulin-degrading enzyme. Density was evident and enabled modeling of residues 1–3 and 11–23 of IAPP and 1–3 and 16–23 of Aβ (136). For all panels, IAPP is color coded by sequence, with 1–7 in blue, 8–19 in green, 20–29 in orange, and 30–37 in purple. Fiber models in panels *a–d* are shown end on, as per **Figure 3d**.

AFM: atomic force microscopy

in which the strands run orthogonal to the long axis of the fiber while backbone hydrogen bonding runs parallel (**Figure 3c**) (150). The intrasheet interactions include backbone-backbone hydrogen bonds between adjacent β-strands, as well as amino acid side chain interactions that include aromatic stacking, van der Waals contacts, and hydrogen bonding networks. Intrasheet backbone-backbone hydrogen bonding both defines β-sheet structure and provides a reason why nearly any sequence can adopt amyloid structure. Additional driving forces for amyloid formation that affect stability and kinetics are provided by side chain interactions. The identity and arrangement of side chains within amyloid strands account for much of the variation in amyloid morphology, stability, and kinetics between amyloid systems (156). These variations are limited by the arrangement of strands within the β-sheet core of the amyloid fiber.

Intersheet interactions are predominantly formed from side chain–side chain and side chain–main chain interactions, both of which play an important structural role. Stability may be derived from the close interdigitation of the side chains of opposing sheets, as has been observed in structural analyses (129, 150). Indeed, it has been suggested that stability and structure of amyloid fibers are dominated by zippers of strands composed of a short sequence of amino acids ($\sim$10 residues) (113). Longer peptides that adopt strand-turn-strand structure result in fibers that are significantly more stable than fibers formed by short peptides. For example, $IAPP_{20-29}$ requires 10-fold more peptide than full-length IAPP does to efficiently form fibers (127). The main cause for this may be a synergistic contribution of stability when two strands are bound within the same amyloid fiber. If one strand were to dissociate, the other strand would tether that sequence to the fiber, resulting in rapid reassociation due to extraordinarily high, effective local concentration. In addition, once mature fibers have formed, the close interactions of β-sheets exclude water, making the barrier to dissociation very high. These driving forces result in a highly stable basic structure that many sequences can easily adopt.

AMYLOID INTERMEDIATES ARE THE TOXIC SPECIES

It is challenging to understand how mature amyloid fibers themselves could be toxic. Amyloid fibers are stable, proteinaceous structures. One could imagine that such structures should be more likely to be used by nature for structural scaffolds, rather than give rise to toxicity. Functional amyloids have been discovered in bacteria and mammals and include Pmel17,

an amyloid-forming protein in mammals that is used to template melanin deposition (50, 51). In *Escherichia coli*, the protein CsgA is secreted and nucleated into amyloid on the surface of the cell by a second factor, CsgB. These structures serve to facilitate colonization of the host organism (6, 18). Amyloids also provide a basis for the development of nanomaterials (21). For example, protein gels formed from amyloid peptides are being developed for use as three-dimensional scaffolds for tissue culture (48). It has been asserted that cytotoxicity is a general property of peptide amyloid precursors (118). It therefore seems unlikely that such three-dimensional scaffolds should be effective cell culture substrates unless the fibers themselves are nontoxic.

In recent years it has become clear that the amyloid hypothesis must be amended. This adjustment stems from indications that fibers themselves are not causal to disease pathology. For example, the correlation of the extent of Aβ deposits with dementia is poor. Indeed, postmortem investigation in non-Alzheimer's disease adults showed that Aβ plaques are nevertheless accumulated (33). Similarly, amyloid deposition in type II diabetes is evident in >90% of patients (64, 83). However, this fails to explain the fact that the remaining 10% of patients do not present significant amyloid deposition (59). Such observations do not refute the hypothesis. Rather, they suggest that intermediate structures of amyloid formation are more relevant to pathology. Thus, whereas these diseases can be ascribed to particular proteins, the fibrillar form of the protein may not and likely is not the toxic state. This positive but poor correlation of amyloid deposition with pathology has understandably led to attention being given to intermediate states as the potentially toxic species (99).

Laboratories have succeeded, in recent years, in developing stable preparations of protein states that may represent stabilized intermediates of fiber formation. In Aβ, for example, solutions of stabilized oligomeric species are prepared by incubating monomer, taking fractions before fiber formation occurs, and then subjecting the sample to centrifugation and/or size exclusion (97). These species reduce cell viability in standardized assays by ~50% at concentrations of 0.5–10 μM. Other studies have enabled the generation of antibodies with apparent specificity for intermediate states (86). Such antibodies have a protective effect against toxicity by oligomeric amyloid precursors in systems that include IAPP, α-synuclein, and Aβ (87). The study of amyloid formation kinetics and requisite intermediates is critical. This involves developing an understanding of the structures and stabilities of intermediate states, as well as the molecular mechanisms of interactions that give rise to their toxicity.

A ROLE FOR AMYLOID INTERMEDIATES IN IAPP CYTOTOXICITY

The implication of intermediate states of IAPP in toxicity by in vitro studies has been supported by a number of recently developed animal models. Five separate mouse and rodent lines transgenic for human IAPP have been developed independently by three groups. These animal lines, which have recently been reviewed (111), differ in their genetic background, expression levels of IAPP, and homozygosity. Interestingly, while only some of these animals develop amyloids (161), all more readily develop diabetes-like symptoms accompanied by loss of β-cell mass. For example, the transgenic human IAPP (HIP) rat develops diabetes spontaneously between 5 and 10 months of age. This is accompanied by a 60% loss of β-cell mass that is correlated with IAPP deposition and hyperglycemia (110). This is remarkable because a systemic pathology has resulted from the introduction of a single amyloidogenic peptide. In nontransgenic rodent models of diabetes (Zucker diabetic fatty rats) an increase in β-cell death has been noted even though amyloid deposition is not observed (46). This suggests that rodent IAPP, under proper environmental conditions, may be able to form toxic species. As with Aβ in Alzheimer's disease, plaques of IAPP are not necessarily the origin of β-cell toxicity.

A number of studies have localized oligomeric intermediates to biological membranes. For example, oligomer-specific antibodies have been used to suggest that amyloid oligomers are localized to cell membranes (57, 86, 97). In other studies, exogenous addition of IAPP has been shown to induce abnormal morphology in the plasma membrane (75). Such addition of prefibrilar IAPP oligomers to cultured cells has also been observed to trigger disruption of cell-cell adhesion, perturbation of insulin secretion, and apoptosis (126). Addition of oligomer-specific antibodies (which reduce toxicity of exogenously added amyloid oligomers) does not prevent toxicity in cells expressing human IAPP, suggesting toxic oligomers can form inside cells and do not need to be secreted and form on the exterior plasma membrane (100). These studies clearly implicate membranes as the target for toxicity in animal models and cell culture.

The suggested origins of amyloid induced toxicity are varied. Possibilities include the formation of reactive oxygen species (68) or the induction of apoptosis via endoplasmic reticulum (ER) stress (63, 116). ER stress has also been noted in transgenic hIAPP rodent, as evidenced by upregulation of stress markers such as X-box binding protein and active caspase-12 (65, 66). Furthermore, it has been shown that Aβ can induce apoptosis through a Toll-like receptor pathway in cultured neuronal cells (144). Recent work examining the generic toxicity of amyloid precursors has shown that amyloid peptides formed from L-amino acids are equally as toxic to cells as D-amino acid polypeptides (118). This suggests that a direct amyloid-receptor interaction is unlikely to mediate cell death. Finally, it remains a viable hypothesis that fibers themselves are toxic. Indeed, recent efforts in Aβ have suggested that toxicity is both mediated by and dependent on the substructure of the fiber surface (107, 168). The most likely cause of toxicity, however, is still the disruption of membrane integrity. For IAPP, toxicity could be the result of membrane disruption, at any point along the secretory pathway, from ER to Golgi, granule, or plasma membrane.

Disruption of cellular compartmentalization and loss of chemical potential across a membrane are potent explanations for the origins of cellular stress that can lead to apoptosis.

Model membrane studies strongly support the hypothesis that amyloid intermediates are the cause of cellular toxicity. Recent in vitro studies have shown a shared ability of amyloid intermediates to render bilayers permeable to species as small as calcium ions and as large as dyes such as calcein (622 Da). For example, $A\beta$ has been suggested to form Ca^{2+} channels in single-channel measurement studies (3, 101, 122). Ca^{2+} homeostasis is believed to play a significant role in Alzheimer's disease pathology (8). In contrast to oligomers, neither monomeric precursors nor mature $A\beta$ fibers appear to permeabilize synthetic vesicles. This accurately reflects corresponding observations of cell toxicity by the same states (62). IAPP toxicity was linked to β-cell apoptosis through its ability to permeabilize membranes, as evidenced by membrane bilayer capacitance changes using single-channel conductivity measurements (75, 112). Thus, the localization of oligomeric species to biological membranes in vivo, the toxicity of intermediate states to cell culture, and the capacity of the same intermediate states to render biological membranes leaky combine to generate a compelling case for the loss of membrane integrity as the origin of cytotoxicity.

MECHANISM OF MEMBRANE DISRUPTION

The mechanism and structures by which intermediates render membranes leaky is not yet clear. Amyloid intermediates are generally amphipathic structures. As such, they may have the capacity to insert into membranes, lay on top of membranes, or even mimic the behavior of cell-penetrating peptides (99). Oligomerization in the context of a membrane has the potential to generate protein-stabilized pores. Still others have identified the capacity of intermediates to remove lipid components from the bilayer by a detergent-like mechanism. There is as yet no consensus on which perturbations are relevant to disease.

For the purpose of this review, we delineate between three structurally divergent modes of toxicity: carpeting, detergent effects, and pore formation. We simplify this complex and broad subject matter (67, 81, 88, 157) as follows. We define carpeting as the binding of prefibrilar states to one leaf of a bilayer. This causes asymmetric pressure to be present between the two leaves. Relaxation of this pressure, proximal or distal to the protein, is accompanied by leakage of small molecules. Detergent effects are caused by a hole forming where lipid has been removed from the bilayer by the protein acting as a surfactant. Removal of lipid may occur at one leaf, resulting in asymmetric pressure between the two leaves, as with carpeting. Removal of the outer leaf may result in transient membrane thinning, facilitating ions or small molecules to traverse the barrier. Alternatively, both leaves may be extracted by the protein, resulting in transient formation of a hole. Pore formation is caused by protein spanning the bilayer, resulting in a hole that is surrounded and stabilized by protein. Carpeting and detergent-based disruptions are large enough to let ions or small molecules pass through the bilayer. Pores, such as the antimicrobial peptides melittin and alamethicin (67), can facilitate the transport of small molecules. Thus, pore structures formed by amyloid intermediates can also be expected to be able to do this.

SURFACE CATALYSIS OF AMYLOID FORMATION

The kinetics of amyloid formation are characterized by nucleation-dependent assembly (52, 121, 127). Briefly, a precursor placed into amyloidogenic conditions is initially quiescent, remaining dispersed and soluble for a period of time called the lag phase. Conversion to the fibrillar form occurs in an apparently cooperative manner, with material converting to amyloid often on a timescale comparable to or even shorter than the lag phase itself (**Figure 2a**). Sigmoidal kinetic behavior alone is not sufficient to indicate a nucleation-dependent process (52). As most methods of monitoring fiber formation observe the product (fiber), any sequential assembly mechanism has the potential to yield a sigmoidal reaction profile. Nucleation represents the presence of conformational and/or oligomeric species that are on-pathway but energetically unfavorable. These states become progressively more stable as a result of conversion of the nucleus to a fiber. Subsequent extension of the fiber with additional precursor is an energetically favorable process. The presence of nucleation dependence is typically assayed by assessing the capacity of preformed amyloid material to act as seed for accelerating the amyloid process.

As the nucleus of fiber formation is a high-energy state, reaction conditions that can stabilize the nucleus can also serve as catalysts of the process. One of the most important conditions is the presence of a surface. In all in vitro experiments, surfaces are present at the vessel wall and at the air-buffer interface. The latter is important given the particular sensitivity of fibrillogenesis kinetics to agitation. Furthermore, amyloid peptides have an intrinsic capacity to adopt a β-strand conformation at air-water interfaces (130). In recent work, we assessed the surface catalysis of amyloid formation by $IAPP_{20-29}$ by

using a buffer-dichloromethane interface (127). Insights resulted, in part, from our observation that both the reaction order and enthalpy of fiber nucleation were the same in the presence and absence of this interface. This finding suggested that the walls of fibers themselves could act as catalytic surfaces upon which nuclei are stabilized.

Biological membranes represent an abundant surface with the potential to serve as a catalyst for amyloid formation. Indeed, lipid bilayers dramatically accelerate the conversion of IAPP to β-sheet-rich amyloid fibers (**Figures 2** and **5a,c**) (91). The capacity of phospholipid bilayers to act as catalysts in kinetic experiments has been widely reported by other laboratories for IAPP (13, 39, 78, 102), α-synuclein (47, 170), and Aβ (12, 23). Supported bilayers have also been used extensively to characterize fiber formation in these systems through the use of imaging techniques such as atomic force microscopy (166, 167) and fluorescence microscopy (34, 163). These studies uniformly show that amyloid fibers become observable much earlier in bilayer-containing samples than in control incubations.

Surface catalysis needs to consider three elements. First, adsorption to a surface greatly increases local concentration. As amyloid

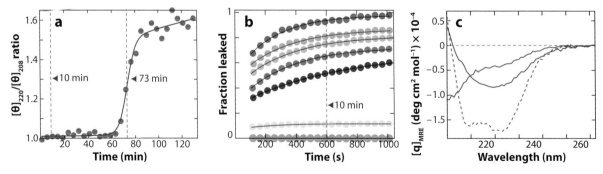

Figure 5

Human islet amyloid polypeptide (hIAPP). (*a, b*) Kinetics of fiber formation in a liposome-catalyzed fiber formation reaction (*a*) under conditions similar to the monitoring of loss of bilayer integrity (*b*). (*a*) Kinetic trace of a 20 μM IAPP reaction in the presence of 340 μM DOPG liposomes. Fiber formation was monitored by the ratio of changes to far-UV circular dichroism (CD) at 220 and 208 nm. β-sheet formation is not detectable until >60 min. (*b*) Several representative assays of leakage of the dye calcein from 420 μM DOPG liposomes induced by 2- to 8-μM hIAPP. Solution conditions are matched to those shown in (*a*) and show leakage occurring on a 10-min timescale. (*c*) Far-UV CD spectra of 20 μM IAPP in aqueous buffer before (*blue solid line*) exposure to liposomes, after bilayer binding (*blue dashed line*), and after lipid-catalyzed fiber formation (*black line*). Figure adapted from Reference 90.

formation is a polymerization, reaction orders are invariably >0, resulting in high sensitivity to protein concentration. Second, adsorption to a surface imparts order to the peptide structure and order to interpeptide interactions. Such order represents a significant loss of entropy. The extent to which this ordering is consistent with amyloid structures lends additional importance to the physicochemical properties of the surface. In our study examining catalysis of IAPP$_{20-29}$, for example, catalysis was specific for dichloromethane-water interfaces and was wholly absent from hexane-water interfaces. Aβ is also greatly affected by the interface of water with chloroform, hexafluoroisopropanol, or sodium dodecyl sulfate micelles (114, 123). Third, to qualify as a catalyst, the surface must be regenerated. Mechanistically, we suggested that this arises because the intrinsic twist of a β-sheet (∼15° per strand) is incompatible with adsorption to a planar two-dimensional surface (91). The free energy of an untwisted β-sheet adsorbed to a surface becomes higher than that of the same sheet twisted and in solution. Evidence for such regeneration of the lipid bilayer catalyst can be seen in imaging of IAPP from the lipid surfaces that catalyzed its formation.

THE ROLE OF MEMBRANE COMPOSITION

Amyloid-forming peptides readily bind to anionic lipid bilayers prior to their conversion to amyloid. In our own work, we showed that both human and rodent IAPP can readily bind to liposomes (90, 91). Importantly, binding occurred well within the lag phase of subsequent amyloid assembly. Aβ and α-synuclein also bind to bilayers on timescales that are much faster than their rate of conversion (11, 12). Thus, it is important to consider the capacity of different membrane constituents to affect protein-lipid affinity, lateral diffusion, and the effect of membrane components on the peptide conformation and orientation relative to the bilayer plane. Alterations in any of these properties can greatly affect the capacity of the membrane to act as a catalyst.

A critical factor affecting affinity is the net surface charge of the lipid bilayer. Using mixtures of anionic and zwitterionic headgroups, we determined that IAPP possessed a transition from unbound to bound states between 10% and 30% anionicity (91). Physiological membrane anionicities are estimated to be in this range (32, 149). This suggested that alterations in lipid metabolism, the extent of surface sialic acid incorporation into glycolipids and glycoproteins, and in vivo ionic strength are important factors in affecting amyloid pathology. For solution biophysical studies, consideration of the techniques and conditions used is particularly important. For example, IAPP can permeabilize wholly zwitterionic membranes (120). This implies leakage under conditions in which we report no apparent binding. Our assays, however, rely on slow off-rates, which may not affect leakage assays. Consider the bacterial toxin magainin. This protein, like IAPP, preferentially binds anionic but not zwitterionic bilayers (109). However, the atomic structure of the lipid-bound form was determined by using NMR in the presence of zwitterionic membranes (153). Plainly, the results of different laboratories cannot be compared without careful consideration of techniques employed, concentrations of peptides and lipids used, and buffer conditions (i.e., salt, pH).

Surface charge represents only one element of a much broader need for examination of membrane composition. A wide examination of this issue is evident in the published record for α-synuclein. The proposed functions of α-synuclein imply a direct association with synaptic vesicle pools. As a result, the effect of differential lipid composition has been extensively studied (7). In particular, α-synuclein has demonstrated specificity in its affinity for different bilayer headgroups. For example, α-synuclein binds PIP$_2$ and phosphatidic acid (PA) headgroups significantly better than it binds phosphatidylglycerol (PG) headgroups, even though all are negatively charged. Lipid structure, in the form of microdomains (49, 95) and bilayer curvature (115), can change the binding affinity of α-synuclein.

CD: circular
dichroism

Sphingomyelin-cholesterol mixtures have also been successfully employed to suggest specific binding of α-synuclein to lipid microdomains (53). This suggests that there are specific, well-defined targets of α-synuclein binding in vivo, and that these events can only be studied under well-controlled model membrane systems.

Membrane composition effect studies are limited for IAPP. Most of these studies, including our own, have focused only on charge density using anionic PA and PG mixed with zwitterionic phosphatidylcholine (PC) headgroup-containing lipids. This paucity of data results in part because IAPP is not perceived to have physiologically relevant membrane-bound activities. As IAPP is a secreted protein, it is expressed, processed through the ER and Golgi, and packaged proximal to the secretory granule membrane at concentrations in the approximate millimolar range. Given that in vitro binding studies of IAPP are generally performed at approximate micromolar concentrations, we have argued that IAPP should be regarded as having a biologically relevant membrane-bound function (92). It is intriguing, for example, that in vitro IAPP can specifically enable the recruitment of anionic liposomes to insulin microcrystals (92). Performing membrane composition studies on IAPP may therefore provide important insights into IAPP native function as well as its misfolding.

A second factor affecting bilayer affinity is the concentrations of species used. The apparent membrane binding affinity of IAPP increases with protein concentration (90). Global analysis of this cooperativity suggested a model in which there is a nucleated transition from a discretely sized (assumed to be monomeric) monodisperse state to a heterogeneously sized membrane-bound aggregated state (90). Such transitions are likely to be general among lipid-catalyzed amyloid-forming proteins. Similarly, the bilayer binding properties of Aβ are consistent with the presence of cooperativity and/or conformational change (152). As oligomer populations in solution or on the membrane will be affected by concentration, consideration of pro-

tein concentration is critical to understanding the mechanisms of both amyloid assembly and membrane-mediated cytotoxicity.

α-HELICAL STRUCTURE CHARACTERIZES MEMBRANE-BOUND STATES

IAPP adopts helical conformational states upon lipid binding. Initial evidence for this came from circular dichroism (CD) measurements using SDS micelles (69) and phospholipid bilayers (78). The transition from random coil to α-helix is readily observed in far-UV CD spectra (**Figure 5c**) and IR spectroscopy (102). In general, the induction of helical structure in an amphipathic peptide by interaction with surfactants or phospholipids should not be surprising. Such observations particularly stand out in the amyloid field because amyloid is a β-sheet structure. Stabilization of an α-helical state seems an unlikely mechanism for catalysis of fibrillogenesis.

The capacity of IAPP to form α-helices on membranes is a property shared with Aβ and α-synuclein. Aβ is a random coil in solution but may transiently adopt structured populations (30, 125, 134). Aβ adopts a helical structure in the presence of trifluoroethanol (TFE) (29, 138, 142) or SDS (27, 77, 135), or when bound to membranes (12, 79). The membrane-bound α-helical state of Aβ was initially considered off-pathway to amyloid formation. For example, the mutant V18A of Aβ40 adopts a more pronounced helical state and yet fiber formation is greatly inhibited (138). However, there are equally reasonable assertions that helicity is on-pathway. In solution-phase fiber formation studies, Aβ was suggested to proceed through an α-helical intermediate (89) before populating a series of oligomeric intermediate states, followed by adoption of a β-sheet structure (12). α-synuclein is similarly unstructured in solution and adopts an α-helical state, as determined by NMR secondary shifts, when bound to micelles and liposomes (9, 16, 38). Recent elegant work using EPR indicates that α-synuclein forms an extended helical structure up to at least residue 90 (76). These studies further support the suggestion (38) that α-synuclein forms α11/3 helices (in which three turns span 11 residues), incorporating its seven 11-mer repeats (**Figure 1d**). Whether α-synuclein forms one extended helix or two helices separated by a kink is debated (9, 17). Recently, EPR findings have suggested that the native membrane-bound conformation of α-synuclein may be a bent structure, in which the first 100 residues form two antiparallel helices (35). In the case of α-synuclein, such structural insights carry broad implications. This helical form may allow the appropriate alignment of hydrophobic and charged residues to ensure a functional ability to bind convex (i.e., curved) vesicles. The helical states of IAPP, Aβ, and α-synuclein represent important areas of structural investigation, as there are both functional and pathological consequences to their formation.

The primary sequence of IAPP can be divided into several regions (**Figures 1a,c and 4a**). IAPP is predominantly random coil in solution, α-helical when bound to lipid, and β-sheet when in the amyloid state. Considerable information is known about the secondary structures adopted by IAPP in each of these states. The first seven residues adopt a structured subdomain that, due to the presence of a disulfide bond between residues 2 and 7, is incapable of adopting an extended β-strand conformation. Elimination of this subdomain has a modest but important effect of apparently reducing fiber-dependent nucleation processes (94). If full-length IAPP supports lateral nucleation off the fiber walls, as we have reported for IAPP$_{20-29}$, then elimination of residues 1–7 would strongly affect surface properties of the IAPP fiber. In addition, this subdomain includes two positive charges relevant to membrane binding. The mutant K1E, for example, strongly affects the capacity of lipid bilayers to catalyzed IAPP fiber formation (91). These observations implicate the N terminus as a participant in membrane binding.

The helical region of IAPP has been mapped to residues 5–20. Helical prediction algorithms and comparisons with structures known for the homologue calcitonin gene-related peptide (CGRP) have suggested that this region forms a helix (154, 155). Recently, we showed using solution NMR on soluble states of rat IAPP that the region 5–22 strongly, albeit transiently, samples α-helical states (159). Mutational analysis dramatically supports the importance of this region to helix formation as well as the importance of helix formation to amyloid assembly in solution. Residue N14 resides in the middle of the putative helical domain (**Figure 1a,b**). Swapping the side chains of residue N14 with L12 or with L16 results in a peptide with unchanged amino acid content, but with greatly diminished intrinsic helicity (93). Importantly, amyloid formation is all but eliminated by these mutations. In contrast, mutation at eight other sites, including three alternative Asn residues, resulted in only modest alteration of solution-phase fiber formation kinetics (93). Residues 5–20 also represent the domain that adopts helical structure upon binding a lipid bilayer (2). In addition, the subpeptide IAPP$_{1-19}$ appears to adopt a helical structure upon binding a lipid bilayer. This suggests that helix formation is independent of the remainder of the sequence. Importantly, no subsequent transition to β sheet was observed in this subpeptide (13). These observations strongly implicate a role for helicity in the solution-phase and lipid-phase fiber formation of IAPP. Furthermore, it suggests that solution-phase and lipid-phase fiber formation are mechanistically similar.

The remaining portions of IAPP, residues 22–37, appear to be wholly unstructured both in solution and on membranes. Only upon conversion to the amyloid state is regular structure observed in this region (103) (**Figure 4**). Isolated subpeptides support the observations made on full-length IAPP. Peptides representing residues 8–20 and 30–37 are capable of converting to an amyloid state, albeit at higher concentrations or on longer timescales than the full-length protein (71, 72). The region 20–29 similarly converts to a β-strand conformation for human, but not rat, sequence variants.

The most detailed structural analysis of this has emerged recently with the publication of the atomic structures for $IAPP_{21-27}$, NNFGAIL, and $IAPP_{28-33}$, SSTNVG (160). Both sequences are capable of adopting β-sheet-rich conformations with tight packing. In the case of NNFGAIL, the structure contains a non-canonical bent conformation (**Figure 4d**). This is perhaps consistent with this region serving as a turn in models of the full-length protein fiber based on solid-state NMR (103) (**Figure 4a,b**).

The sequence of IAPP may be meaningfully divided into four subdomains. $IAPP_{1-7}$ is noncanonical. $IAPP_{5-20}$ alternatively adopts α-helical or β-strand states. $IAPP_{20-29}$ is critical to the capacity of IAPP to form amyloid; however, its conversion likely includes changes from random coil to distorted or noncanonical β-strand conformation. Finally, residues $IAPP_{30-37}$ undergo a random coil to β-strand conformational change that is accompanied by the burial of the C terminus (117).

The structure, density, and organization of prefibrillar IAPP on membranes are relevant to its subsequent amyloid assembly. Initial efforts on lipid-catalyzed amyloid formation suggested that catalysis is maximal at an IAPP/lipid ratio of 1:12 (91). This density suggests that there is insufficient surface area for the entire IAPP sequence to lie on the liposome. Recent work using quenching agents on nitroxide-labeled IAPP enabled the determination that residues 1–20 interact with lipid bilayers (2). Quenching showed periodicity consistent with the α-helix residing at the surface of the lipid bilayer (**Figure 1b**). This structural model is analogous to a study from the same group on α-synuclein that showed that α-synuclein forms a helical structure that lays parallel to the membrane (76). It is important to consider the conditions under which such studies were performed. In particular, the protein/lipid ratio was 1:1000. This indicates that the study was conducted under conditions in which IAPP is not forming helical aggregates. There is potential for additional structures and/or conformations to be adopted either at higher protein/lipid ratios or in the presence of alternative lipid compositions. For example, NMR and neutron reflectivity studies suggest that Aβ alternatively adopts surface-bound and partially inserted conformations dependent on lipid composition. The greater the anionic/zwitterionic ratio, the more Aβ is inserted (12, 22). For IAPP, it remains to be seen if, and to what extent, lateral-to-transmembrane transitions occur and to what extent regions outside residues 1–20 are involved.

Intermediate, membrane-bound states of amyloid-forming peptides have been visualized by AFM and EM. Such observations are challenging to make because the states are small and heterogeneous. Sample preparation is typically achieved by applying preformed oligomers to lipid bilayers. Given the sensitivity of amyloid peptides to membrane environment, and the alternative capacity of amyloid peptides to adopt lateral versus transmembrane states, variation in sample handling is likely to be important. In general, AFM studies have yielded the most evocative images. Irregular, annular oligomers have been noted and reproduced by several laboratories for several amyloid-forming proteins including IAPP, α-synuclein, and Aβ (122). These structures are heterogeneously populated and vary in size depending on the particular system being observed. The relationship of such species to underlying secondary structure and oligomeric states detected by indirect methods has yet to be established. Such structural observations are suggestive of a pore-like or channel-like mechanism of leakage.

Surface-bound oligomerization of IAPP involves residues outside the helical domain. The helical region of hIAPP and rat IAPP differ at a single amino acid, H18R. Given the importance of electrostatic interactions to bilayer binding, it is surprising that hIAPP binds membranes containing 30% DOPG, whereas rat IAPP does not. Indeed, at mildly acidic pH, where the histidine is protonated, hIAPP shows dramatically greater affinity than rat IAPP does (90). Global analysis enabled the quantitative (albeit model-dependent) assessment of binding parameters. Despite the differences in apparent affinity of these two proteins, the partition coefficients for hIAPP and rat IAPP to the membrane surface were effectively identical. Given the close similarity of the protein sequence helical domain, this is a gratifying result. Instead, apparent differences in affinity were localized to the free energy of nucleation to a helical aggregated state. For human, this was ~-1 kcal mol^{-1} and for rat, it was $\sim+0.5$ kcal mol^{-1} (90). Other studies have shown that hIAPP$_{1-19}$ binds as tight or tighter than full-length hIAPP (40). Thus, it appears that the N-terminal region of IAPP drives binding, whereas residues C-terminal to the α-helix are disruptive to membrane binding for rat IAPP. Oligomerization potential is strongly attenuated by residues outside the helical domain of IAPP.

Lipid binding accelerates fiber formation, but the structural basis for this effect remains unclear. Acceleration of amyloid formation is correlated with the amount of helical aggregated IAPP, rather than the total amount of IAPP in the system (90). Such a correlation strongly suggests that helical oligomers represent an on-pathway intermediate species. Furthermore, oligomeric assemblies have been directly detected on lipid bilayers by cross-linking (90). Similar results have been reported for Aβ, which shows helical sampling prior to fiber formation in solution and acceleration of fiber formation by 30% anionic DPPG liposomes (22). Seemingly in contrast to IAPP and Aβ, α-synuclein binds lipid bilayers and adopts helical structure, but fiber formation is not necessarily accelerated. At a lipid/protein ratio greater than 95:1, α-synuclein fiber formation is inhibited. Under matched solution conditions, but with a lipid/protein ratio of 25:1 or less, fiber formation is accelerated (170). For all three systems, conditions that result in strongly stabilized helices can be expected to disfavor fiber formation. Similarly, conditions that result in dilution of precursor across a large surface area of bilayer can be expected to disfavor fiber formation. The phase diagram for all three systems can reasonably be expected to include a common set of regimes. The conditions under which a peptide samples different elements of phase space will be system specific.

DISRUPTION OF LIPID BILAYERS BY AMYLOID INTERMEDIATES

The putative relationship between membrane-bound amyloid intermediates and membrane disruption can be categorized on the basis of size and specificity of the perturbation and timescale. The relevant timescales are usefully divided into three qualitative regions. Membrane leakage may occur upon precursor binding to the membrane, after binding but within the lag phase of fiber formation, or coincident with fiber formation. Within each of these time frames, there are many considerations. Membrane binding by itself is a complex process with the potential for multiple phases. These phases can include initial adsorption, adoption of substructures, and formation of higher-order interactions (41, 132). Within the lag phase, preamyloid intermediates may form. Alternatively, small populations of membrane-bound amyloid fibers may form that escape direct detection by the tools of ensemble measurement. Leakage that is coincident with fiber formation may represent direct disruption of the membrane by amyloid fibers, or disruption associated with the assembly process. Finally, it may also represent a point in time where preamyloid intermediates are maximally populated. Here, we consider a range of reported experimental observations and speculate how these reflect on alternative mechanisms.

IAPP can cause large perturbations to bilayers. When IAPP binds 100% DOPG liposomes, membrane integrity is disrupted such that calcein, a 622-Da molecule, can escape encapsulation. This is detected as an increase of fluorescence, as dye is encapsulated at concentrations that result in autoquenching. The profile of this leakage is complex (**Figure 5b**). A fraction of the leakage occurs in the dead time. This is followed by single exponential kinetic profiles that can plateau at less than 100%. Importantly, the timescales of these events are faster than the amyloid transition (**Figure 5a**). Leakage during the dead time of such an experiment might be a result of carpeting or bilayer thinning from initial surface adsorption.

Subsequent, exponential leakage could be the result of detergent-like removal of lipid, a continuation of carpeting effects, or pore formation by early oligomers of IAPP fiber formation.

Leakage occurs when α-helical states are significantly populated (90). Under similar conditions, lipid damage caused by rat IAPP was inhibited at salt concentrations, whereas hIAPP was not affected (58). This likely reflects the higher membrane affinity of the human protein. The importance of the helical region has also been corroborated by studies showing that subpeptides of the helical region could independently cause leakage (13). Binding studies suggest that IAPP binds to lipids initially in a monodisperse fashion (40) and then forms α-helical aggregates (90). Leakage that is coincident with monomeric binding is inconsistent with a detergent-like model playing a role in early leakage events. The fact that membrane perturbation properties can be maintained without the C-terminal portion (40) suggests a simple mode of toxicity, such as carpeting. Importantly, global analysis of surface-bound oligomerization indicates that a $\sim$10-fold-higher protein/lipid ratio was required for monomeric states to cause equivalent leakage induced by helical aggregates (90). This suggests a mode of leakage that reflects a pore or detergent mechanism of membrane disruption. Given that β-cell depletion has been noted in nontransgenic, diabetic rodent models (46), modes of membrane disruption mediated by helical states seem likely.

Membrane perturbation can also occur coincident with amyloid formation (39). In such cases, leakage need not be apparent during the lag phase of fiber formation. We surmise that in these cases conditions may not favor dense α-helical oligomers of IAPP to form. Intriguingly, membrane perturbation could be induced by accelerating fibrillogenesis via seeding (39). Equivalent addition of preformed fibers did not cause leakage. Thus, leakage may be the result of growing fibers exerting force on membranes. The conformational changes of a peptide bound to a bilayer can reasonably be expected to induce transient perturbation as the individual phospholipids relax to accommodate alternative protein structures. In a recent and elegant study, IAPP fibers have been observed to incorporate lipid molecules from supported bilayers as they mature and leave the bilayer (34). This supports earlier work on IAPP that showed that membrane disruption may be caused when lipid is extracted from the bilayer by growing fibers (139). These data suggest that amyloid fiber formation may contribute to leakage by a detergent-like mechanism.

Small oligomers may play a role in membrane damage induced by fibers. Immature fibers are less stable than mature amyloid fibers and have been observed by AFM to fragment upon exposure to lipid monolayers (158). Uptake of lipid by such species could explain the detergent-like disruptions that have been reported. Alternatively, as amyloid fibers themselves are surfaces, their walls could act as sinks from which oligomeric species could dissociate and interact with the bilayer. Indeed, variations in the surface structure of Aβ fibers have been correlated to in vitro toxicity (133, 168). Fibers may therefore retain toxicity as either a property of their formation or in their capacity to generate or stabilize oligomeric species states.

Preformed oligomeric species can produce bilayer perturbations via pore formation. This has been noted for IAPP, Aβ, and α-synuclein, as well as other systems (56). Annular structures have been observed by several groups (101, 122, 140, 147). They show irregular morphologies loosely resembling transmembrane channels. Modeling, for example in Aβ, has been used to propose how amyloid proteins could be arranged to form such structures (74). Pores that stabilize large holes can be expected to allow small molecules as well as ions to pass through the membrane. Smaller holes may only allow species such as ions to pass. In small pores, the arrangement of side chains and backbone carbonyls may give rise to bias. Aβ, for example, shows some bias for divalent species such as Ca^{2+} (101). Others have suggested, however, that Aβ does not generate a stable water-filled hole. Rather, localized displacement of lipid by protein could lower the dielectric barrier sufficiently to lose homeostasis in cells (152). Additional structural investigations, ideally in parallel with leakage measurements, are needed to fully describe the contribution of pore formation to amyloid-related toxicity.

The structures sampled by preamyloid states on membranes dictate their mode of toxicity. Numerous studies have indicated the potential of multiple mechanisms to disrupt cell viability. We have also stated that membrane composition and protein concentration play crucial roles in determining the energetics, populations, and conformations of lipid-bound states. It may therefore be inappropriate to seek a singular characterization of amyloid-induced membrane perturbation. Rather, condition dependence affects an interplay between alternative helical bound states, helix-sheet transition, and β-sheet intermediates that could result in several modes of membrane disruption.

Conformations that induce leakage can also serve to catalyze the fiber formation process. One possibility is that the α-helical domain of IAPP forms aggregates on the lipid surface that align peptides and promote β-sheet formation within C-terminal residues. Oligomerization of helical states could be accompanied by constant sampling of transmembrane configurations. Such transitions have been noted for Aβ-lipid interactions (12). High-density binding of IAPP could result in a carpet-like mode of leakage while promoting nucleation through high local concentration and parallel alignment of monomers (**Figure 6**). High-density binding with partially transmembrane peptides is also a favorable condition for the development of an amyloid pore, similar to the mechanism seen with some antimicrobial peptides (14). Pores themselves could be helical or sampling β-sheet structure similar to conformations found in amyloid fibers (74). The binding promiscuity of the open end of a β-sheet may link the driving forces of fiber formation, as well as account for some causes of toxicity (31). Toxic states of amyloid may share some of the same characteristic structures required for membrane-catalyzed fiber formation.

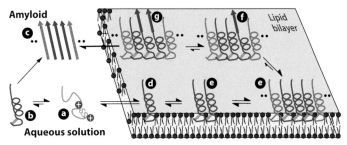

Figure 6

Model of lipid-catalyzed fiber formation. Aqueous-phase human islet amyloid polypeptide (hIAPP) is predominantly random coil in solution (*a*) but transiently samples α-helical states (*b*). Solution-phase IAPP can partition onto the membrane and form bilayer-bound monodisperse α-helical states (*d*). These freely diffusing lipid-bound states can nucleate to form heterogeneously sized helical aggregated states (*e*). These states transiently sample β-strand conformation at residues C-terminal to the α-helical domain (*f*). Local concentration and orientation effects catalyze nucleation of amyloid formation (*g*). Nucleation is followed by extension, which converts the helical domain to a β-strand structure in a process that includes release of mature amyloid to solution (*c*). Leakage by alternative mechanisms may accompany each of these transitions. Figure adapted from Reference 90.

THE IMPORTANCE OF MECHANISM IN THE DESIGN OF PHARMACOLOGICAL INTERVENTION

An understanding of mechanism is key to the effective development of therapeutics. The practical consequences of our current biophysical understanding of amyloid toxicity are twofold. First, therapeutics must prevent accumulation of toxic intermediate species. This has recently been addressed, for example, in the development of therapeutics of transthyretin (82) and lysozyme amyloidoses (36). In these systems, the precursor is a folded protein that must partially unfold to adopt an amyloidogenic state. Small molecules or antibodies that stabilize the folded conformation diminish the population of such states. In the case of IAPP, Aβ, and α-synuclein, the solution-phase precursor state is unstructured and is therefore a more challenging target for such an approach. Second, a mechanistically targeted therapeutic need only make modest alterations to the levels and localization of amyloidogenic species. The amyloid peptides discussed in this work are all expressed and turned over in abundance in healthy individuals. A small alteration in balance is likely to be sufficient to assist in vivo machinery involved in the clearance of misfolded aggregates (4). These considerations can be addressed more directly as the mechanism(s) of assembly becomes better defined.

The first mechanistically inspired amyloid inhibitors were based on the premise of crystal poisoning. Amyloid fibers can reasonably be thought of as one-dimensional crystals. Growth can therefore be disproportionally affected by the introduction of a small amount of a variant of the precursor in which small alterations have been made. For Aβ, numerous peptide inhibitors based on modification to the subsequence LVFFA have been analyzed and shown to serve as effective inhibitors of the formation of mature fibers in vitro (44). For IAPP, a number of studies have also taken such an approach. This has included the use of IAPP variants in which amide groups at residues 24 and 26 have been N-methylated. As a result, the amide nitrogen of the backbone can no longer serve as a hydrogen bond donor (85, 164). Similarly, the mutation I26P results in a variant of hIAPP that can abrogate fiber formation by the wild-type peptide (1). Such studies indicate that a mechanistic understanding can inform on therapeutic design. Targeting the fibers themselves may not, however, be an effective strategy in vivo, as this may have the adverse effect of enhancing the population of amyloid intermediates.

The observation of membrane-bound intermediates such as those reviewed here yields numerous novel opportunities. Suitable targets can include stability of the helical state, helix-helix interactions, protein-membrane adsorption, and the interconversion of adsorbed states from lateral to transmembrane orientations. To our knowledge, these states have not been directly targeted for screening; however, the promise is plain. For example, the published record for α-synuclein indicates an environmental dependence on the correlation of helical states with amyloidogenicity. Stabilization of a helical state or a reduction in helix-helix interaction energies can be expected to accelerate or alternatively inhibit subsequent amyloid conversion depending on the balance of energy. For IAPP, the capacity of hIAPP versus rat IAPP to assemble into a membrane-permeabilizing state rests in part on the energy of nucleating a membrane-bound helical aggregated state. The use of a crystal poison for this helical nucleation step may therefore serve as an effective inhibitor of the formation of toxic species. Finally, there may be protein-specific processes that can be capitalized upon. For example, the capacity of insulin to inhibit fiber formation by IAPP is greatly reduced in bilayer-catalyzed reactions (**Figure 2b**). This suggests that molecules that displace IAPP from the membrane surface might act synergistically with endogenous insulin. Mechanism can and should be used in the design of therapeutic strategy.

SUMMARY

The interaction of amyloid precursors with lipid bilayer surfaces has recently emerged as common origin for the catalysis of amyloid formation and for the formation of cytotoxic species. The range of potential structures and processes associated with these preamyloid states is remarkable. Adsorbed species can adopt helical conformations. These states can self-associate and can alternate between lateral and transmembrane states as well as α-helical and β-sheet conformations. These species can be monomeric or form self-associated oligomers that result in catalysis of amyloid formation through a combination of local concentration effects and a reduction of entropy due to the intrinsic orienting properties of surface binding. The origins of membrane integrity loss are similarly complex. Evidence in the published record suggests that several mechanisms may be active depending on conditions. How such observations reflect events in vivo is not yet clear. The potential of simplifying and explaining these myriad of possibilities rests in comparison of several systems. In this regard, IAPP has asserted itself not only as a system central to diabetes pathology, but also as a model system appropriate for the study of several neurodegenerative disorders such as Alzheimer's and Parkinson's. Structural and kinetic studies have greatly informed on the mechanism of fiber formation. Membrane-associated kinetic, structural, and leakage studies have elucidated the existence of intermediate states. Mutants and chemically modified variants of IAPP provide novel tools to observe and perturb intermediate forma-tion. Spectroscopic and imaging methods have also advanced, enabling structurally specific insights to be made. With the emergence of several transgenic model systems, bio-physical investigations can be paralleled with biological testing. Such investigations into membrane-bound states and toxicity will serve as a cornerstone for the broader understanding of amyloid formation, membrane interactions, and cytotoxicity.

DISCLOSURE STATEMENT

The authors are not aware of any biases that might be perceived as affecting the objectivity of this review.

ACKNOWLEDGMENTS

We thank Prof. L. Rhoades, N. Last, G. Manley, and M. Calabrese for critical reading of this manuscript. We thank R. Tycko for generously providing coordinates used to prepare **Figure 4a,b**. This work supported, in part, by NIH (DK54899, GM084391, DK079829, and AG031612).

LITERATURE CITED

1. Abedini A, Meng F, Raleigh DP. 2007. A single-point mutation converts the highly amyloidogenic human islet amyloid polypeptide into a potent fibrillization inhibitor. *J. Am. Chem. Soc.* 129:11300–1

2. Apostolidou M, Jayasinghe SA, Langen R. 2008. Structure of alpha-helical membrane-bound human islet amyloid polypeptide and its implications for membrane-mediated misfolding. *J. Biol. Chem.* 283:17205–10

3. Arispe N, Pollard HB, Rojas E. 1994. The ability of amyloid beta-protein [A beta P (1-40)] to form Ca2+ channels provides a mechanism for neuronal death in Alzheimer's disease. *Ann. N. Y. Acad. Sci.* 747:256–66

4. Balch WE, Morimoto RI, Dillin A, Kelly JW. 2008. Adapting proteostasis for disease intervention. *Science* 319:916–19

5. Baldwin TO, Ziegler MM, Chaffotte AF, Goldberg ME. 1993. Contribution of folding steps involving the individual subunits of bacterial luciferase to the assembly of the active heterodimeric enzyme. *J. Biol. Chem.* 268:10766–72

6. Barnhart MM, Chapman MR. 2006. Curli biogenesis and function. *Annu. Rev. Microbiol.* 60:131–47

7. Beyer K. 2007. Mechanistic aspects of Parkinson's disease: alpha-synuclein and the biomembrane. *Cell Biochem. Biophys.* 47:285–99

8. Bezprozvanny I, Mattson MP. 2008. Neuronal calcium mishandling and the pathogenesis of Alzheimer's disease. *Trends Neurosci.* 31:454–63

9. Bisaglia M, Tessari I, Pinato L, Bellanda M, Giraudo S, et al. 2005. A topological model of the interaction between alpha-synuclein and sodium dodecyl sulfate micelles. *Biochemistry* 44:329–39

10. Blennow K, de Leon MJ, Zetterberg H. 2006. Alzheimer's disease. *Lancet* 368:387–403

11. Bokvist M, Grobner G. 2007. Misfolding of amyloidogenic proteins at membrane surfaces: the impact of macromolecular crowding. *J. Am. Chem. Soc.* 129:14848–49

12. Bokvist M, Lindstrom F, Watts A, Grobner G. 2004. Two types of Alzheimer's beta-amyloid (1-40) peptide membrane interactions: aggregation preventing transmembrane anchoring versus accelerated surface fibril formation. *J. Mol. Biol.* 335:1039–49

13. Brender JR, Lee EL, Cavitt MA, Gafni A, Steel DG, Ramamoorthy A. 2008. Amyloid fiber formation and membrane disruption are separate processes localized in two distinct regions of IAPP, the type-2-diabetes-related peptide. *J. Am. Chem. Soc.* 130:6424–29

14. Brogden KA. 2005. Antimicrobial peptides: pore formers or metabolic inhibitors in bacteria? *Nat. Rev. Microbiol.* 3:238–50

15. Brouwers N, Sleegers K, Van Broeckhoven C. 2008. Molecular genetics of Alzheimer's disease: an update. *Ann. Med.* 18:1–22

16. Bussell R Jr, Eliezer D. 2003. A structural and functional role for 11-mer repeats in alpha-synuclein and other exchangeable lipid binding proteins. *J. Mol. Biol.* 329:763–78

17. Chandra S, Chen X, Rizo J, Jahn R, Sudhof TC. 2003. A broken alpha-helix in folded alpha-synuclein. *J. Biol. Chem.* 278:15313–18

18. Chapman MR, Robinson LS, Pinkner JS, Roth R, Heuser J, et al. 2002. Role of *Escherichia coli* curli operons in directing amyloid fiber formation. *Science* 295:851–55

19. Chartier-Harlin MC, Kachergus J, Roumier C, Mouroux V, Douay X, et al. 2004. Alpha-synuclein locus duplication as a cause of familial Parkinson's disease. *Lancet* 364:1167–69

20. Chen Y, Ding F, Nie H, Serohijos AW, Sharma S, et al. 2008. Protein folding: then and now. *Arch. Biochem. Biophys.* 469:4–19

21. Cherny I, Gazit E. 2008. Amyloids: not only pathological agents but also ordered nanomaterials. *Angew. Chem. Int. Ed. Engl.* 47:4062–69

22. Chi EY, Ege C, Winans A, Majewski J, Wu G, et al. 2008. Lipid membrane templates the ordering and induces the fibrillogenesis of Alzheimer's disease amyloid-beta peptide. *Proteins* 72:1–24

23. Choo-Smith LP, Garzon-Rodriguez W, Glabe CG, Surewicz WK. 1997. Acceleration of amyloid fibril formation by specific binding of Abeta-(1-40) peptide to ganglioside-containing membrane vesicles. *J. Biol. Chem.* 272:22987–90

24. Chua CE, Tang BL. 2006. Alpha-synuclein and Parkinson's disease: the first roadblock. *J. Cell Mol. Med.* 10:837–46

25. Clark A, Nilsson MR. 2004. Islet amyloid: a complication of islet dysfunction or an aetiological factor in type 2 diabetes? *Diabetologia* 47:157–69

26. Cole SL, Vassar R. 2008. The role of APP processing by BACE1, the beta-secretase, in Alzheimer's disease pathophysiology. *J. Biol. Chem.* 283:29621–25

27. Coles M, Bicknell W, Watson AA, Fairlie DP, Craik DJ. 1998. Solution structure of amyloid beta-peptide(1-40) in a water-micelle environment. Is the membrane-spanning domain where we think it is? *Biochemistry* 37:11064–77

28. Cooper AA, Gitler AD, Cashikar A, Haynes CM, Hill KJ, et al. 2006. Alpha-synuclein blocks ER-Golgi traffic and Rab1 rescues neuron loss in Parkinson's models. *Science* 313:324–28

29. Crescenzi O, Tomaselli S, Guerrini R, Salvadori S, D'Ursi AM, et al. 2002. Solution structure of the Alzheimer amyloid beta-peptide (1-42) in an apolar microenvironment. Similarity with a virus fusion domain. *Eur. J. Biochem.* 269:5642–48

30. Danielsson J, Andersson A, Jarvet J, Graslund A. 2006. 15N relaxation study of the amyloid beta-peptide: structural propensities and persistence length. *Magn. Reson. Chem.* 44(Spec. No.):S114–21

31. De Simone A, Esposito L, Pedone C, Vitagliano L. 2008. Insights into stability and toxicity of amyloid-like oligomers by replica exchange molecular dynamics analyses. *Biophys. J.* 95:1965–73

32. Diaz GB, Cortizo AM, Garcia ME, Gagliardino JJ. 1988. Lipid composition of normal male rat islets. *Lipids* 23:1125–28

33. Dickson DW. 1997. The pathogenesis of senile plaques. *J. Neuropathol. Exp. Neurol.* 56:321–39

34. Domanov YA, Kinnunen PK. 2008. Islet amyloid polypeptide forms rigid lipid-protein amyloid fibrils on supported phospholipid bilayers. *J. Mol. Biol.* 376:42–54

35. Drescher M, Veldhuis G, van Rooijen BD, Milikisyants S, Subramaniam V, Huber M. 2008. Antiparallel arrangement of the helices of vesicle-bound alpha-synuclein. *J. Am. Chem. Soc.* 130:7796–97

36. Dumoulin M, Last AM, Desmyter A, Decanniere K, Canet D, et al. 2003. A camelid antibody fragment inhibits the formation of amyloid fibrils by human lysozyme. *Nature* 424:783–88

37. El-Agnaf OM, Irvine GB. 2000. Review: formation and properties of amyloid-like fibrils derived from alpha-synuclein and related proteins. *J. Struct. Biol.* 130:300–9

38. Eliezer D, Kutluay E, Bussell R Jr, Browne G. 2001. Conformational properties of alpha-synuclein in its free and lipid-associated states. *J. Mol. Biol.* 307:1061–73

39. Engel MF, Khemtemourian L, Kleijer CC, Meeldijk HJ, Jacobs J, et al. 2008. Membrane damage by human islet amyloid polypeptide through fibril growth at the membrane. *Proc. Natl. Acad. Sci. USA* 105:6033–38

40. Engel MF, Yigittop H, Elgersma RC, Rijkers DT, Liskamp RM, et al. 2006. Islet amyloid polypeptide inserts into phospholipid monolayers as monomer. *J. Mol. Biol.* 356:783–89

41. Engelman DM, Chen Y, Chin CN, Curran AR, Dixon AM, et al. 2003. Membrane protein folding: beyond the two stage model. *FEBS Lett.* 555:122–25

42. Fahn S. 2003. Description of Parkinson's disease as a clinical syndrome. *Ann. N. Y. Acad. Sci.* 991:1–14

43. Fawzi NL, Kohlstedt KL, Okabe Y, Head-Gordon T. 2008. Protofibril assemblies of the Arctic, Dutch, and Flemish mutants of the Alzheimer's Abeta1-40 peptide. *Biophys. J.* 94:2007–16

44. Findeis MA, Musso GM, Arico-Muendel CC, Benjamin HW, Hundal AM, et al. 1999. Modified-peptide inhibitors of amyloid beta-peptide polymerization. *Biochemistry* 38:6791–800

45. Finder VH, Glockshuber R. 2007. Amyloid-beta aggregation. *Neurodegener. Dis.* 4:13–27

46. Finegood DT, McArthur MD, Kojwang D, Thomas MJ, Topp BG, et al. 2001. Beta-cell mass dynamics in Zucker diabetic fatty rats. Rosiglitazone prevents the rise in net cell death. *Diabetes* 50:1021–29

47. Fink AL. 2006. The aggregation and fibrillation of alpha-synuclein. *Acc. Chem. Res.* 39:628–34

48. Flamia R, Salvi AM, D'Alessio L, Castle JE, Tamburro AM. 2007. Transformation of amyloid-like fibers, formed from an elastin-based biopolymer, into a hydrogel: an X-ray photoelectron spectroscopy and atomic force microscopy study. *Biomacromolecules* 8:128–38

49. Fortin DL, Troyer MD, Nakamura K, Kubo S, Anthony MD, Edwards RH. 2004. Lipid rafts mediate the synaptic localization of alpha-synuclein. *J. Neurosci.* 24:6715–23

50. Fowler DM, Koulov AV, Alory-Jost C, Marks MS, Balch WE, Kelly JW. 2006. Functional amyloid formation within mammalian tissue. *PLoS Biol.* 4:e6

51. Fowler DM, Koulov AV, Balch WE, Kelly JW. 2007. Functional amyloid—from bacteria to humans. *Trends Biochem. Sci.* 32:217–24

52. Frieden C. 2007. Protein aggregation processes: in search of the mechanism. *Protein Sci.* 16:2334–44

53. Gai WP, Yuan HX, Li XQ, Power JT, Blumbergs PC, Jensen PH. 2000. In situ and in vitro study of colocalization and segregation of alpha-synuclein, ubiquitin, and lipids in Lewy bodies. *Exp. Neurol.* 166:324–33

54. Gazit E. 2007. Self-assembled peptide nanostructures: the design of molecular building blocks and their technological utilization. *Chem. Soc. Rev.* 36:1263–69

55. George JM, Jin H, Woods WS, Clayton DF. 1995. Characterization of a novel protein regulated during the critical period for song learning in the zebra finch. *Neuron* 15:361–72

56. Glabe CG. 2006. Common mechanisms of amyloid oligomer pathogenesis in degenerative disease. *Neurobiol. Aging* 27:570–75

57. Gong Y, Chang L, Viola KL, Lacor PN, Lambert MP, et al. 2003. Alzheimer's disease-affected brain: Presence of oligomeric A beta ligands (ADDLs) suggests a molecular basis for reversible memory loss. *Proc. Natl. Acad. Sci. USA* 100:10417–22

58. Green JD, Kreplak L, Goldsbury C, Li Blatter X, Stolz M, et al. 2004. Atomic force microscopy reveals defects within mica supported lipid bilayers induced by the amyloidogenic human amylin peptide. *J. Mol. Biol.* 342:877–87

59. Haataja L, Gurlo T, Huang CJ, Butler PC. 2008. Islet amyloid in type 2 diabetes, and the toxic oligomer hypothesis. *Endocr. Rev.* 29:303–16

60. Haber E, Anfinsen CB. 1962. Side-chain interactions governing the pairing of half-cystine residues in ribonuclease. *J. Biol. Chem.* 237:1839–44

61. Halliday GM, McCann H. 2008. Human-based studies on alpha-synuclein deposition and relationship to Parkinson's disease symptoms. *Exp. Neurol.* 209:12–21

62. Heinitz K, Beck M, Schliebs R, Perez-Polo JR. 2006. Toxicity mediated by soluble oligomers of beta-amyloid(1-42) on cholinergic SN56.B5.G4 cells. *J. Neurochem.* 98:1930–45

63. Hiddinga HJ, Eberhardt NL. 1999. Intracellular amyloidogenesis by human islet amyloid polypeptide induces apoptosis in COS-1 cells. *Am. J. Pathol.* 154:1077–88

64. Hoppener JW, Ahren B, Lips CJ. 2000. Islet amyloid and type 2 diabetes mellitus. *N. Engl. J. Med.* 343:411–19

65. Huang CJ, Haataja L, Gurlo T, Butler AE, Wu X, et al. 2007. Induction of endoplasmic reticulum stress-induced beta-cell apoptosis and accumulation of polyubiquitinated proteins by human islet amyloid polypeptide. *Am. J. Physiol. Endocrinol. Metab.* 293:E1656–62

66. Huang CJ, Lin CY, Haataja L, Gurlo T, Butler AE, et al. 2007. High expression rates of human islet amyloid polypeptide induce endoplasmic reticulum stress mediated beta-cell apoptosis, a characteristic of humans with type 2 but not type 1 diabetes. *Diabetes* 56:2016–27

67. Huang HW, Chen FY, Lee MT. 2004. Molecular mechanism of peptide-induced pores in membranes. *Phys. Rev. Lett.* 92:198304

68. Huang X, Moir RD, Tanzi RE, Bush AI, Rogers JT. 2004. Redox-active metals, oxidative stress, and Alzheimer's disease pathology. *Ann. N. Y. Acad. Sci.* 1012:153–63

69. Hubbard JA, Martin SR, Chaplin LC, Bose C, Kelly SM, Price NC. 1991. Solution structures of calcitonin-gene-related-peptide analogues of calcitonin-gene-related peptide and amylin. *Biochem. J.* 275(Pt. 3):785–88

70. Hull RL, Westermark GT, Westermark P, Kahn SE. 2004. Islet amyloid: a critical entity in the pathogenesis of type 2 diabetes. *J. Clin. Endocrinol. Metab.* 89:3629–43

71. Jaikaran ET, Clark A. 2001. Islet amyloid and type 2 diabetes: from molecular misfolding to islet pathophysiology. *Biochim. Biophys. Acta* 1537:179–203

72. Jaikaran ET, Higham CE, Serpell LC, Zurdo J, Gross M, et al. 2001. Identification of a novel human islet amyloid polypeptide beta-sheet domain and factors influencing fibrillogenesis. *J. Mol. Biol.* 308:515–25

73. Jan A, Gokce O, Luthi-Carter R, Lashuel HA. 2008. The ratio of monomeric to aggregated forms of Abeta 40 and Abeta 42 is an important determinant of Abeta aggregation, fibrillogenesis, and toxicity. *J. Biol. Chem.* 283:28176–89

74. Jang H, Zheng J, Lal R, Nussinov R. 2008. New structures help the modeling of toxic amyloidbeta ion channels. *Trends Biochem. Sci.* 33:91–100

75. Janson J, Ashley RH, Harrison D, McIntyre S, Butler PC. 1999. The mechanism of islet amyloid polypeptide toxicity is membrane disruption by intermediate-sized toxic amyloid particles. *Diabetes* 48:491–98

76. Jao CC, Der-Sarkissian A, Chen J, Langen R. 2004. Structure of membrane-bound alpha-synuclein studied by site-directed spin labeling. *Proc. Natl. Acad. Sci. USA* 101:8331–36

77. Jarvet J, Danielsson J, Damberg P, Oleszczuk M, Graslund A. 2007. Positioning of the Alzheimer Abeta(1-40) peptide in SDS micelles using NMR and paramagnetic probes. *J. Biomol. NMR* 39:63–72

78. Jayasinghe SA, Langen R. 2005. Lipid membranes modulate the structure of islet amyloid polypeptide. *Biochemistry* 44:12113–19

79. Ji SR, Wu Y, Sui SF. 2002. Cholesterol is an important factor affecting the membrane insertion of beta-amyloid peptide (A beta 1-40), which may potentially inhibit the fibril formation. *J. Biol. Chem.* 277:6273–79

80. Jimenez JL, Nettleton EJ, Bouchard M, Robinson CV, Dobson CM, Saibil HR. 2002. The protofilament structure of insulin amyloid fibrils. *Proc. Natl. Acad. Sci. USA* 99:9196–201

81. Joh NH, Min A, Faham S, Whitelegge JP, Yang D, et al. 2008. Modest stabilization by most hydrogen-bonded side-chain interactions in membrane proteins. *Nature* 453:1266–70

82. Johnson SM, Wiseman RL, Sekijima Y, Green NS, Adamski-Werner SL, Kelly JW. 2005. Native state kinetic stabilization as a strategy to ameliorate protein misfolding diseases: a focus on the transthyretin amyloidoses. *Acc. Chem. Res.* 38:911–21

83. Kahn SE, Andrikopoulos S, Verchere CB. 1999. Islet amyloid: a long-recognized but underappreciated pathological feature of type 2 diabetes. *Diabetes* 48:241–53

84. Kapurniotu A. 2001. Amyloidogenicity and cytotoxicity of islet amyloid polypeptide. *Biopolymers* 60:438–59

85. Kapurniotu A, Schmauder A, Tenidis K. 2002. Structure-based design and study of non-amyloidogenic, double N-methylated IAPP amyloid core sequences as inhibitors of IAPP amyloid formation and cytotoxicity. *J. Mol. Biol.* 315:339–50

86. Kayed R, Head E, Thompson JL, McIntire TM, Milton SC, et al. 2003. Common structure of soluble amyloid oligomers implies common mechanism of pathogenesis. *Science* 300:486–89

87. Kayed R, Sokolov Y, Edmonds B, McIntire TM, Milton SC, et al. 2004. Permeabilization of lipid bilayers is a common conformation-dependent activity of soluble amyloid oligomers in protein misfolding diseases. *J. Biol. Chem.* 279:46363–66

88. Killian JA, Nyholm TK. 2006. Peptides in lipid bilayers: the power of simple models. *Curr. Opin. Struct. Biol.* 16:473–79

89. Kirkitadze MD, Condron MM, Teplow DB. 2001. Identification and characterization of key kinetic intermediates in amyloid beta-protein fibrillogenesis. *J. Mol. Biol.* 312:1103–19

90. Knight JD, Hebda JA, Miranker AD. 2006. Conserved and cooperative assembly of membrane-bound alpha-helical states of islet amyloid polypeptide. *Biochemistry* 45:9496–508

91. Knight JD, Miranker AD. 2004. Phospholipid catalysis of diabetic amyloid assembly. *J. Mol. Biol.* 341:1175–87

92. Knight JD, Williamson JA, Miranker AD. 2008. Interaction of membrane-bound islet amyloid polypeptide with soluble and crystalline insulin. *Protein Sci.* 17:1850–56

93. Koo BW, Hebda JA, Miranker AD. 2008. Amide inequivalence in the fibrillar assembly of islet amyloid polypeptide. *Protein Eng. Des. Sel.* 21:147–54

94. Koo BW, Miranker AD. 2005. Contribution of the intrinsic disulfide to the assembly mechanism of islet amyloid. *Protein Sci.* 14:231–39

95. Kubo S, Nemani VM, Chalkley RJ, Anthony MD, Hattori N, et al. 2005. A combinatorial code for the interaction of alpha-synuclein with membranes. *J. Biol. Chem.* 280:31664–72

96. Kudva YC, Mueske C, Butler PC, Eberhardt NL. 1998. A novel assay in vitro of human islet amyloid polypeptide amyloidogenesis and effects of insulin secretory vesicle peptides on amyloid formation. *Biochem. J.* 331(Pt. 3):809–13

97. Lambert MP, Viola KL, Chromy BA, Chang L, Morgan TE, et al. 2001. Vaccination with soluble Abeta oligomers generates toxicity-neutralizing antibodies. *J. Neurochem.* 79:595–605

98. Larson JL, Miranker AD. 2004. The mechanism of insulin action on islet amyloid polypeptide fiber formation. *J. Mol. Biol.* 335:221–31

99. Lashuel HA, Lansbury PT Jr. 2006. Are amyloid diseases caused by protein aggregates that mimic bacterial pore-forming toxins? *Q. Rev. Biophys.* 39:167–201

100. Lin CY, Gurlo T, Kayed R, Butler AE, Haataja L, et al. 2007. Toxic human islet amyloid polypeptide (h-IAPP) oligomers are intracellular, and vaccination to induce anti-toxic oligomer antibodies does not prevent h-IAPP-induced beta-cell apoptosis in h-IAPP transgenic mice. *Diabetes* 56:1324–32

101. Lin H, Bhatia R, Lal R. 2001. Amyloid beta protein forms ion channels: implications for Alzheimer's disease pathophysiology. *FASEB J.* 15:2433–44

102. Lopes DH, Meister A, Gohlke A, Hauser A, Blume A, Winter R. 2007. Mechanism of islet amyloid polypeptide fibrillation at lipid interfaces studied by infrared reflection absorption spectroscopy. *Biophys. J.* 93:3132–41

103. Luca S, Yau WM, Leapman R, Tycko R. 2007. Peptide conformation and supramolecular organization in amylin fibrils: constraints from solid-state NMR. *Biochemistry* 46:13505–22

104. Luheshi LM, Tartaglia GG, Brorsson AC, Pawar AP, Watson IE, et al. 2007. Systematic in vivo analysis of the intrinsic determinants of amyloid beta pathogenicity. *PLoS Biol.* 5:e290

105. Luhrs T, Ritter C, Adrian M, Riek-Loher D, Bohrmann B, et al. 2005. 3D structure of Alzheimer's amyloid-beta(1-42) fibrils. *Proc. Natl. Acad. Sci. USA* 102:17342–47

106. Makin OS, Serpell LC. 2005. Structures for amyloid fibrils. *FEBS J.* 272:5950–61

107. Martins IC, Kuperstein I, Wilkinson H, Maes E, Vanbrabant M, et al. 2008. Lipids revert inert Abeta amyloid fibrils to neurotoxic protofibrils that affect learning in mice. *EMBO J.* 27:224–33

108. Marzban L, Park K, Verchere CB. 2003. Islet amyloid polypeptide and type 2 diabetes. *Exp. Gerontol.* 38:347–51

109. Matsuzaki K. 1999. Why and how are peptide-lipid interactions utilized for self-defense? Magainins and tachyplesins as archetypes. *Biochim. Biophys. Acta* 1462:1–10

110. Matveyenko AV, Butler PC. 2006. Beta-cell deficit due to increased apoptosis in the human islet amyloid polypeptide transgenic (HIP) rat recapitulates the metabolic defects present in type 2 diabetes. *Diabetes* 55:2106–14

111. Matveyenko AV, Butler PC. 2006. Islet amyloid polypeptide (IAPP) transgenic rodents as models for type 2 diabetes. *Ilar J.* 47:225–33
112. Mirzabekov TA, Lin MC, Kagan BL. 1996. Pore formation by the cytotoxic islet amyloid peptide amylin. *J. Biol. Chem.* 271:1988–92
113. Nelson R, Sawaya MR, Balbirnie M, Madsen AO, Riekel C, et al. 2005. Structure of the cross-beta spine of amyloid-like fibrils. *Nature* 435:773–78
114. Nichols MR, Moss MA, Reed DK, Hoh JH, Rosenberry TL. 2005. Rapid assembly of amyloid-beta peptide at a liquid/liquid interface produces unstable beta-sheet fibers. *Biochemistry* 44:165–73
115. Nuscher B, Kamp F, Mehnert T, Odoy S, Haass C, et al. 2004. Alpha-synuclein has a high affinity for packing defects in a bilayer membrane: a thermodynamics study. *J. Biol. Chem.* 279:21966–75
116. O'Brien TD, Butler PC, Kreutter DK, Kane LA, Eberhardt NL. 1995. Human islet amyloid polypeptide expression in COS-1 cells. A model of intracellular amyloidogenesis. *Am. J. Pathol.* 147:609–16
117. Padrick SB, Miranker AD. 2001. Islet amyloid polypeptide: identification of long-range contacts and local order on the fibrillogenesis pathway. *J. Mol. Biol.* 308:783–94
118. Pastor MT, Kummerer N, Schubert V, Esteras-Chopo A, Dotti CG, et al. 2008. Amyloid toxicity is independent of polypeptide sequence, length and chirality. *J. Mol. Biol.* 375:695–707
119. Petkova AT, Buntkowsky G, Dyda F, Leapman RD, Yau WM, Tycko R. 2004. Solid state NMR reveals a pH-dependent antiparallel beta-sheet registry in fibrils formed by a beta-amyloid peptide. *J. Mol. Biol.* 335:247–60
120. Porat Y, Kolusheva S, Jelinek R, Gazit E. 2003. The human islet amyloid polypeptide forms transient membrane-active prefibrillar assemblies. *Biochemistry* 42:10971–77
121. Powers ET, Powers DL. 2008. Mechanisms of protein fibril formation: nucleated polymerization with competing off-pathway aggregation. *Biophys. J.* 94:379–91
122. Quist A, Doudevski I, Lin H, Azimova R, Ng D, et al. 2005. Amyloid ion channels: a common structural link for protein-misfolding disease. *Proc. Natl. Acad. Sci. USA* 102:10427–32
123. Rangachari V, Moore BD, Reed DK, Sonoda LK, Bridges AW, et al. 2007. Amyloid-beta(1-42) rapidly forms protofibrils and oligomers by distinct pathways in low concentrations of sodium dodecylsulfate. *Biochemistry* 46:12451–62
124. Richards FM. 1958. On the enzymic activity of subtilisin-modified ribonuclease. *Proc. Natl. Acad. Sci. USA* 44:162–66
125. Riek R, Guntert P, Dobeli H, Wipf B, Wuthrich K. 2001. NMR studies in aqueous solution fail to identify significant conformational differences between the monomeric forms of two Alzheimer peptides with widely different plaque competence, A beta(1-40)(ox) and A beta(1-42)(ox). *Eur. J. Biochem.* 268:5930–36
126. Ritzel RA, Meier JJ, Lin CY, Veldhuis JD, Butler PC. 2007. Human islet amyloid polypeptide oligomers disrupt cell coupling, induce apoptosis, and impair insulin secretion in isolated human islets. *Diabetes* 56:65–71
127. Ruschak AM, Miranker AD. 2007. Fiber-dependent amyloid formation as catalysis of an existing reaction pathway. *Proc. Natl. Acad. Sci. USA* 104:12341–46
128. Sachs JN, Engelman DM. 2006. Introduction to the membrane protein reviews: the interplay of structure, dynamics, and environment in membrane protein function. *Annu. Rev. Biochem.* 75:707–12
129. Sawaya MR, Sambashivan S, Nelson R, Ivanova MI, Sievers SA, et al. 2007. Atomic structures of amyloid cross-beta spines reveal varied steric zippers. *Nature* 447:453–57
130. Schladitz C, Vieira EP, Hermel H, Mohwald H. 1999. Amyloid-beta-sheet formation at the air-water interface. *Biophys. J.* 77:3305–10
131. Schueler-Furman O, Wang C, Bradley P, Misura K, Baker D. 2005. Progress in modeling of protein structures and interactions. *Science* 310:638–42
132. Seelig J. 2004. Thermodynamics of lipid-peptide interactions. *Biochim. Biophys. Acta* 1666:40–50
133. Seilheimer B, Bohrmann B, Bondolfi L, Muller F, Stuber D, Dobeli H. 1997. The toxicity of the Alzheimer's beta-amyloid peptide correlates with a distinct fiber morphology. *J. Struct. Biol.* 119:59–71
134. Sgourakis NG, Yan Y, McCallum SA, Wang C, Garcia AE. 2007. The Alzheimer's peptides Abeta40 and 42 adopt distinct conformations in water: a combined MD/NMR study. *J. Mol. Biol.* 368:1448–57
135. Shao H, Jao S, Ma K, Zagorski MG. 1999. Solution structures of micelle-bound amyloid beta-(1-40) and beta-(1-42) peptides of Alzheimer's disease. *J. Mol. Biol.* 285:755–73

136. Shen Y, Joachimiak A, Rosner MR, Tang WJ. 2006. Structures of human insulin-degrading enzyme reveal a new substrate recognition mechanism. *Nature* 443:870–74

137. Sohl JL, Jaswal SS, Agard DA. 1998. Unfolded conformations of alpha-lytic protease are more stable than its native state. *Nature* 395:817–19

138. Soto C, Castano EM, Frangione B, Inestrosa NC. 1995. The alpha-helical to beta-strand transition in the amino-terminal fragment of the amyloid beta-peptide modulates amyloid formation. *J. Biol. Chem.* 270:3063–67

139. Sparr E, Engel MF, Sakharov DV, Sprong M, Jacobs J, et al. 2004. Islet amyloid polypeptide-induced membrane leakage involves uptake of lipids by forming amyloid fibers. *FEBS Lett.* 577:117–20

140. Srinivasan R, Marchant RE, Zagorski MG. 2004. ABri peptide associated with familial British dementia forms annular and ring-like protofibrillar structures. *Amyloid* 11:10–13

141. Stefani M, Dobson CM. 2003. Protein aggregation and aggregate toxicity: new insights into protein folding, misfolding diseases and biological evolution. *J. Mol. Med.* 81:678–99

142. Sticht H, Bayer P, Willbold D, Dames S, Hilbich C, et al. 1995. Structure of amyloid A4-(1-40)-peptide of Alzheimer's disease. *Eur. J. Biochem.* 233:293–98

143. Sugase K, Dyson HJ, Wright PE. 2007. Mechanism of coupled folding and binding of an intrinsically disordered protein. *Nature* 447:1021–25

144. Tang SC, Lathia JD, Selvaraj PK, Jo DG, Mughal MR, et al. 2008. Toll-like receptor-4 mediates neuronal apoptosis induced by amyloid beta-peptide and the membrane lipid peroxidation product 4-hydroxynonenal. *Exp. Neurol.* 213:114–21

145. Tartaglia GG, Pawar AP, Campioni S, Dobson CM, Chiti F, Vendruscolo M. 2008. Prediction of aggregation-prone regions in structured proteins. *J. Mol. Biol.* 380:425–36

146. Thinakaran G, Koo EH. 2008. Amyloid precursor protein trafficking, processing, and function. *J. Biol. Chem.* 283:29615–19

147. Thundimadathil J, Roeske RW, Jiang HY, Guo L. 2005. Aggregation and porin-like channel activity of a beta sheet peptide. *Biochemistry* 44:10259–70

148. Tofaris GK, Spillantini MG. 2007. Physiological and pathological properties of alpha-synuclein. *Cell Mol. Life Sci.* 64:2194–201

149. Turk J, Wolf BA, Lefkowith JB, Stump WT, McDaniel ML. 1986. Glucose-induced phospholipid hydrolysis in isolated pancreatic islets: quantitative effects on the phospholipid content of arachidonate and other fatty acids. *Biochim. Biophys. Acta* 879:399–409

150. Tycko R. 2006. Molecular structure of amyloid fibrils: insights from solid-state NMR. *Q. Rev. Biophys.* 39:1–55

151. Uversky VN, Oldfield CJ, Dunker AK. 2008. Intrinsically disordered proteins in human diseases: introducing the D2 concept. *Annu. Rev. Biophys.* 37:215–46

152. Valincius G, Heinrich F, Budvytyte R, Vanderah DJ, McGillivray DJ, et al. 2008. Soluble amyloid {beta} oligomers affect dielectric membrane properties by bilayer insertion and domain formation: Implications for cell toxicity. *Biophys. J.* 95:4845–61

153. Wakamatsu K, Takeda A, Tachi T, Matsuzaki K. 2002. Dimer structure of magainin 2 bound to phospholipid vesicles. *Biopolymers* 64:314–27

154. Westermark P, Engstrom U, Johnson KH, Westermark GT, Betsholtz C. 1990. Islet amyloid polypeptide: pinpointing amino acid residues linked to amyloid fibril formation. *Proc. Natl. Acad. Sci. USA* 87:5036–40

155. Westermark P, Wernstedt C, O'Brien TD, Hayden DW, Johnson KH. 1987. Islet amyloid in type 2 human diabetes mellitus and adult diabetic cats contains a novel putative polypeptide hormone. *Am. J. Pathol.* 127:414–17

156. Wetzel R, Shivaprasad S, Williams AD. 2007. Plasticity of amyloid fibrils. *Biochemistry* 46:1–10

157. White SH, von Heijne G. 2005. Transmembrane helices before, during, and after insertion. *Curr. Opin. Struct. Biol.* 15:378–86

158. Widenbrant MJ, Rajadas J, Sutardja C, Fuller GG. 2006. Lipid-induced beta-amyloid peptide assemblage fragmentation. *Biophys. J.* 91:4071–80

159. Williamson JA, Miranker AD. 2007. Direct detection of transient alpha-helical states in islet amyloid polypeptide. *Protein Sci.* 16:110–17

160. Wiltzius JJ, Sievers SA, Sawaya MR, Cascio D, Popov D, et al. 2008. Atomic structure of the cross-beta spine of islet amyloid polypeptide (amylin). *Protein Sci.* 17:1467–74

161. Wong WP, Scott DW, Chuang CL, Zhang S, Liu H, et al. 2008. Spontaneous diabetes in hemizygous human amylin transgenic mice that developed neither islet amyloid nor peripheral insulin resistance. *Diabetes* 57:2737–44

162. Wright CF, Teichmann SA, Clarke J, Dobson CM. 2005. The importance of sequence diversity in the aggregation and evolution of proteins. *Nature* 438:878–81

163. Yagi H, Ban T, Morigaki K, Naiki H, Goto Y. 2007. Visualization and classification of amyloid beta supramolecular assemblies. *Biochemistry* 46:15009–17

164. Yan LM, Tatarek-Nossol M, Velkova A, Kazantzis A, Kapurniotu A. 2006. Design of a mimic of non-amyloidogenic and bioactive human islet amyloid polypeptide (IAPP) as nanomolar affinity inhibitor of IAPP cytotoxic fibrillogenesis. *Proc. Natl. Acad. Sci. USA* 103:2046–51

165. Yan LM, Velkova A, Tatarek-Nossol M, Andreetto E, Kapurniotu A. 2007. IAPP mimic blocks Abeta cytotoxic self-assembly: Cross-suppression of amyloid toxicity of Abeta and IAPP suggests a molecular link between Alzheimer's disease and type II diabetes. *Angew. Chem. Int. Ed. Engl.* 46:1246–52

166. Yip CM, Darabie AA, McLaurin J. 2002. Abeta42-peptide assembly on lipid bilayers. *J. Mol. Biol.* 318:97–107

167. Yip CM, McLaurin J. 2001. Amyloid-beta peptide assembly: a critical step in fibrillogenesis and membrane disruption. *Biophys. J.* 80:1359–71

168. Yoshiike Y, Akagi T, Takashima A. 2007. Surface structure of amyloid-beta fibrils contributes to cytotoxicity. *Biochemistry* 46:9805–12

169. Young A, Pittner R, Gedulin B, Vine W, Rink T. 1995. Amylin regulation of carbohydrate metabolism. *Biochem. Soc. Trans.* 23:325–31

170. Zhu M, Li J, Fink AL. 2003. The association of alpha-synuclein with membranes affects bilayer structure, stability, and fibril formation. *J. Biol. Chem.* 278:40186–97

Advances in High-Pressure Biophysics: Status and Prospects of Macromolecular Crystallography

Roger Fourme,[1] Eric Girard,[1,2] Richard Kahn,[2]
Anne-Claire Dhaussy,[3] and Isabella Ascone[1]

[1] Synchrotron-SOLEIL, BP48 Saint Aubin, 91192 Gif sur Yvette, France;
email: roger.fourme@synchrotron-soleil.fr

[2] Institut de Biologie Structurale, UMR 5075 CEA-CNRS-UJF-PSB, 38027 Grenoble,
France

[3] CRISMAT, Ensicaen, 14000 Caen, France

Annu. Rev. Biophys. 2009. 38:153–71

First published online as a Review in Advance on
February 9, 2009

The *Annual Review of Biophysics* is online at
biophys.annualreviews.org

This article's doi:
10.1146/annurev.biophys.050708.133700

Key Words

high-pressure macromolecular crystallography, HPMX, protein,
virus, nucleic acid, energy landscape, diamond anvil cell

Abstract

A survey of the main interests of high pressure for molecular biophysics
highlights the possibility of exploring the whole conformational space
using pressure perturbation. A better understanding of fundamental
mechanisms responsible for the effects of high pressure on biomolecules
requires high-resolution molecular information. Thanks to recent
instrumental and methodological progress taking advantage of the remarkable adaptation of the crystalline state to hydrostatic compression, pressure-perturbed macromolecular crystallography is now a full-
fledged technique applicable to a variety of systems, including large
assemblies. This versatility is illustrated by selected applications, including DNA fragments, a tetrameric protein, and a viral capsid. Binding of compressed noble gases to proteins is commonly used to solve
the phase problem, but standard macromolecular crystallography would
also benefit from the transfer of experimental procedures developed for
high-pressure studies. Dedicated short-wavelength synchrotron radiation beamlines are unarguably required to fully exploit the various facets
of high-pressure macromolecular crystallography.

Contents

INTRODUCTION

High pressure has been used for about a century in physics and chemistry to study the physical properties of matter in various states and chemical reactions. The field is extremely active, for example in geophysics, and this is reflected by the chronic oversubscription of multipurpose high-pressure (HP) beamlines at synchrotron radiation facilities. The situation is different in the biosciences. The interest of pressure was confined for a long time on marine biology and deep-sea diving. After all, more than half (≈62%) of the total volume of biosphere is at a depth in excess of 1000 m (i.e., submitted to a pressure higher than 10 MPa) (28). A new aspect of deep-sea biology was added with the discovery that some organisms not only survive but thrive under extreme conditions (30, 31),

and the occurrence of obligate barophile organisms (i.e., not able to grow at ambient pressure) was firmly established (71). The range of pressure and extent of biosphere are further increased when including Earth's crust and the bacterial organisms within. The broader theme of life adaptation under extreme conditions encompasses basic questions such as the onset of life on Earth and its possible occurrence on other stellar objects. HP bioscience is a field in which basic research, applied research, and industrial applications are in close interaction. Hydrostatic pressure has been broadly used by industry and biotechnology, for instance, for food extraction and/or preservation since the late 1980s (6, 29, 50). Medical and pharmaceutical applications of high pressure directed at viruses (53), bacteria (61), and prions (10) are

Synchrotron radiation: electromagnetic radiation emitted by relativistic electrons submitted to an acceleration in the magnetic field of, for example, the periodic magnetic structure of an undulator

expanding. Fundamental studies on the effects of pressure on biosystems were pioneered by the physicist Percy W. Bridgman (9), then continued during several decades with scarce publications of a small community, and has emerged fully since the 1960s and 1970s (28). For a long period, sample environment was an overwhelming problem. Nowadays, most biophysical methods have been adapted to high pressure, so that accurate and reliable data are more routinely available. Overall, the scientific gathering has been rich on pressure unfolding, dissociation of multimeric proteins, and changes in conformational states, ligand binding, DNA transcription, and enzyme kinetics. These studies, although diverse, have in common the exploitation of some unique features of HP perturbation, in particular the fact that it changes the Gibbs free energy (G) through merely pressure work—mechanical energy—in relation to volume changes of the system. This crucial point deserves further exploration.

KEYS FOR TRAVEL THROUGH THE ENERGY LANDSCAPE

Proteins, as well as other biological macromolecules, are multiconformational (1). They have been designed by natural selection not just for the ground or native state but also for higher-energy, nonnative states involved in function, folding, and unfolding. These conformers are distributed into a wide conformational space, from the folded state to the unfolded state(s). Their free energies are different, but differences are at most a few kilocalories per mole. They also have characteristic partial molar volumes (the effective volumes of protein conformers in an aqueous environment including the contribution from hydration). A volume theorem of protein (38), stating that the partial molar volume of a protein decreases in parallel with the decrease of its conformational order, has been proposed. This empirical rule is based on a variety of observations, with no exceptions at least at normal temperature, and may be used as a guiding principle for experimentalists exploring the multiple

conformational nature of globular proteins with variable-pressure perturbation (2). The partial volume difference ΔV between the unfolded and folded states is negative in the vast majority of cases and is in the range 20–100 ml mole^{-1} (i.e., 34–170 Å^3 per molecule) (12, 56), which gives the upper limit of ΔV for all conformers of a protein. In order to appreciate the energetic significance of a particular ΔV, a volume change of 20 ml mole^{-1} at 293 K and 122 MPa corresponds to RT (0.582 kcal mole^{-1}). If a chemical system at equilibrium experiences a change in concentration, temperature, volume, or total pressure, then the equilibrium shifts to partially counteract the imposed change (Le Chatelier-Braun principle): Accordingly, pressure displaces equilibrium toward more compact phases. This displacement can be done either by elastic compression of the conformers present in the solution, or by shift of equilibria among conformers in favor of lower partial-volume conformers. Increasing pressure ultimately leads to destruction of the tertiary structure by denaturation, preceded in the case of oligomeric proteins by dissociation of the quaternary structure in smaller units (67, 68).

Consider a hypothetical protein solution with two subensembles of conformers: the native subensemble (N) and another subensemble (I) attached to a partially unfolded conformer.

Gibbs free energy (G): thermodynamical potential measuring the useful work obtainable from an isothermal isobaric work. A general rule of thumb is under conditions of constant P and T, the equilibrium condition corresponds to the state of the system with the lowest G

N and I conformers differ in topology, free energy ($\Delta G = G_I - G_N$), and partial molar volume ($\Delta V = V_I - V_N$). Under an isothermal compression at p

$$\Delta G = G_I - G_N = \Delta G(p_0) + \Delta V_0(p - p_0)$$
$$- \frac{1}{2}(\Delta\beta)V_0(p - p_0)^2, \qquad 1.$$

where the subscript zero is related to values at ambient pressure and temperature, β is the isothermal compressibility ($\beta = -(1/V)(\partial V/\partial P)_T$), and $\Delta\beta$ is the variation of β between p and p_0. The term proportional to $\Delta\beta$ can often be neglected in a relatively narrow pressure range. Then,

$$\Delta G \approx \Delta G(p_0) + \Delta V_0(p - p_0). \qquad 2.$$

Higher-energy state: state with higher (i.e., less negative) G than the fundamental (native) state

In an isothermal variable-pressure experiment, the protein gains only a few kilocalories per mole through the mechanical energy supplied by pressure. Application of pressure affects internal interactions exclusively by changes in the distances (volumes) of the components, whereas the total energy of the system remains almost constant (51).

In an isobaric variable-temperature experiment, the situation is different. Free-energy difference ΔG between the two conformers results from the balance of the enthalpic and entropic terms, ΔH and $-T\Delta S$, which differ only in second-order (or higher) term. In a thermal transition, each term increases substantially with temperature ($\Delta H > 100$ kcal mole^{-1}), so that the system gains a substantial amount of internal energy and entropy (54). Further, any subtle conformational variations existing at physiological temperature are likely to be smeared out before they are detected. The first conclusion is obvious: Pressure is a mild and efficient way to "scan" the conformational space of macromolecular systems.

The equilibrium constant K between the two conformers is related to the free-energy difference by the relation $K = [I]/[N] = \exp(-\Delta G/RT)$. In Equation 2, $\Delta G(p_0)$ is >0 (as conformer N is more stable than conformer I at ambient pressure). As p increases, the second term, which is <0 ($\Delta V_0 < 0$), will

counterbalance $\Delta G(p_0)$, so that K increases. Let us take a representative example at T = 293 K, with $\Delta G(p_0) = 2$ kcal mole^{-1}, $\Delta V_0 = -50$ ml mole^{-1} (-5.10^{-5} m^3 mole^{-1}), and p = 4 kbar (4.10^8 Pa). With these values, the (N) state is overwhelming (97%) at ambient pressure, whereas the more compact (I) state is dominant (83%) at high pressure. So, the second important conclusion about pressure is that it amplifies the population of higher-energy states that are normally hidden by the dominant population of the native state (4) up to the point at which their study becomes possible. Such selective population enhancement was demonstrated first by NMR on several proteins (reviews in References 2, 3, and 39). Accordingly, pressure perturbation is a unique tool to explore the conformational space of proteins and other biomolecules, from the native substate to the higher-energy substates (including those that are functionally relevant) and finally to the unfolded states. This exploration is selective in the sense that, at least in principle, various substates can be promoted and investigated in a well-defined thermodynamic state.

THE QUEST FOR HIGH-RESOLUTION STRUCTURAL INFORMATION

Interpretation of experimental results in the broader context of protein structure and dynamics is still far from being accomplished. On the one hand, the domain is still challenging when compared with other domains of materials science. Proteins are extremely complex objects. Their stability results from the delicate balance between a variety of interactions in which relations of water molecules with macromolecules play a central role. On the other hand, there is a lack of high-resolution structural information on pressure-perturbed macromolecules. Pressure modifies volume (i.e., distances between components), such that an accurate description of these changes is crucial to understand these mechanisms. Most results on the effect of pressure on biomolecules have been obtained with spectroscopic methods (in particular fluorescence and infrared spectroscopy) and, less commonly, with low-spatial-resolution structural techniques based on solution X-ray or neutron scattering. The adaptation of higher-resolution structural biology methods, NMR, and macromolecular crystallography (MX) has been a major step forward. The online cell high-pressure (HP) NMR technique with heteronuclear 2D or multidimensional NMR capability has been recently introduced to protein studies (5, 70). When pressure-induced chemical shifts are used for structural determination (55), structures in solution can be determined with atomic details along with their thermodynamic stability. HPNMR is presently limited by the pressure range (≈300 MPa) and molecular weight of macromolecules. MX is unique in that it can reveal high-resolution 3D structures of macromolecules of almost arbitrary complexity. The structure of hen egg-white lysozyme at 100 MPa, published in 1987 (41),was the first achievement of MX under high hydrostatic pressure (HPMX), completed by analysis of pressure effects on solvent in lysozyme crystals (42). Then, the structure of sperm whale myoglobin at 14.5 MPa was reported in 1988 (63, 64). Until 2001, only compressibility measurements up to 1 GPa of the unit cell of two forms of hen egg-white lysozyme (34) and a report on unit cell changes in hexameric cobalt insulin pressurized up to 3 MPa (52) were published. This nearly complete absence of new data during more than a decade came about presumably because the expected scientific outcome of HPMX and its possible impact on standard structural work were not, at the time, strong or clear enough to encourage biocrystallographers to add the deterrent complexity of HP environment to standard crystallographic experiments. The unique application of HPMX, which was popular during this period, was the fabrication of heavy-atom derivatives using noble gas binding at moderate pressure (<6 MPa) to solve the phase problem in protein crystallography (57). Since 2001, progress in both instrumentation and methodology relieved most technical problems of HPMX (20), except for the constraint

of specialized equipment, which can be found only at few HP synchrotron radiation beamlines. Furthermore, HP biophysics in general, and pressure-perturbed MX in particular, open new avenues for research.

MACROMOLECULAR CRYSTALS UNDER HIGH HYDROSTATIC PRESSURE

The crystal is a prerequisite for crystallography and is the origin of both power and constraints of this method. The crystalline state is a homogeneous and reproducible medium (for given crystallization conditions) that contains between 20% and 80% of liquid phase. The semiliquid/semisolid state characteristic of macromolecular crystals has several features of interest for HP studies (21). As crystallization medium (mother liquor) and crystal solvent communicate through intracrystal channels, the biopolymer is submitted to hydrostatic compression. The crystalline state is both sufficiently rigid for long-range order needed by diffraction and sufficiently plastic to accommodate fairly large pressure-induced volume variations of the unit cell and its components and to relax to a state of minimum energy after perturbation. Accordingly, these crystals, which are often seen as fragile, can withstand relatively large changes in cell parameters and conformational changes while keeping an excellent 3D order, as evidenced first by compressibility measurements of lysozyme crystals up to 1 GPa in a diamond anvil cell (DAC) (34). Starting from crystals grown by standard techniques (e.g., vapor diffusion method), we, as other authors, have compressed a variety of crystals from A-DNA (27) and B-DNA (T. Prangé, E. Girard, R. Kahn & R. Fourme, unpublished results) models and monomeric (21) and oligomeric (15) proteins in pressures ranging from ambient to 2 GPa. The mother liquor in crystallization drops was used as compression medium with minor changes in composition (generally increasing the percentage of the precipitating agent to better stabilize the crystal). Pressure was ramped smoothly at a typical rate of

(HP)NMR: (high-pressure) nuclear magnetic resonance

(HP)MX: (high-pressure) macromolecular crystallography

Diamond anvil cell (DAC): device consisting of two opposing diamonds with a sample compressed between the culets

1–2 GPa h^{-1}. In all cases in which the sample loaded in the pressure cell diffracted well at ambient pressure, the crystal quality remained excellent upon compression as probed by mosaicity and resolution. Increasing pressure invariably finally led to loss of diffraction. For oligomeric proteins and assemblies, alteration of 3D order was related to dissociation in subunits and was observed at relatively low pressure (15, 22), but crystals of bovine superoxide dismutase, a dimeric protein, withstand pressures of at least 1 GPa (20). For nucleic acid crystals, diffraction was observed up to about 2 GPa (27). In one case, degradation of 3D order was related to wedge effects in the tight crystal packing (27) and, in another case (study of B-DNA dodecamer; T. Prangé, E. Girard, R. Kahn & R. Fourme, unpublished results), to the crystal packing with molecules arranged in parallel linear arrays, which was unfavorable to compression. The mother liquor used as compression medium must remain in a liquid state to ensure hydrostaticity. The solidification pressure of pure water at room temperature is about 1 GPa, but solutions containing organic compounds such as methylpentane diol can be used at much higher pressure, at least 2 GPa. So, this condition is not a severe limitation. In conclusion, instrumentation for HPMX should be designed for the pressure range from ambient to about 2–2.5 GPa.

During our investigations of the capsid of *Cowpea mosaic virus* (CpMV), we found that pressure induces a transition that can be described as a reduction of degrees of freedom in the packing of quasi-spherical viral capsids leading to a spectacular improvement in 3D order (22, 26). Beyond this particular transition, the possibility to induce transitions between different subensembles within the crystalline matrix opens a wealth of possibilities to HPMX (24). Several questions are raised. The image derived from a crystallographic study is an electron density averaged over trillions of molecules. To interpret this image, crystallographers assume implicitly that their sample is monodisperse and contains exclusively the native conformer, with only a few local variations that can be detected

and analyzed in high-resolution studies. The derivation of structural information on a nonnative conformer requires that (*a*) the population of this conformer is a large fraction of the total and (*b*) the crystal accommodates the transition and still diffracts to high resolution.

Crystallization is a classical way to remove impurities (minority species) from a sample. This led us to formulate the hypothesis that a crystal might act as a conformation filter (24). At higher pressure, the conformational equilibrium may be shifted beyond the native ensemble, so that higher-energy substates may become dominant. The periodic arrangement of units within the lattice is a driving force to reach a free-energy well, which could enhance conformational homogeneity within the sample by increasing the percentage of the dominant species. If this is true, the promotion of higher-energy substates might be more efficient in the crystalline state than in solution. Temperature annealing at ambient pressure, which exploits the intrinsic plasticity of macromolecular crystals, is commonly used to improve 3D order. We plan to combine temperature and pressure annealing to improve both monodispersity and quality of higher-energy conformer crystals by using a purposely designed instrumentation that is under construction.

MATERIALS AND METHODS OF HIGH-PRESSURE MACROMOLECULAR CRYSTALLOGRAPHY

From the Beryllium Cell to DAC

Pioneering HPMX work was performed with a beryllium cell filled with a liquid pressurized by a hand-driven device (40), and variants of this device are used by the CHESS group (13, 65). The beryllium cell has a large aperture, but the pressure limit is about 200 MPa, the sample is not visible, and the polycrystalline beryllium wall produces diffuse scattering and powder diffraction rings at high resolution. The use of another device, the DAC, was a crucial step for the progress of HPMX. DAC is a simple

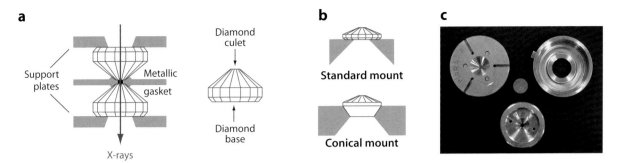

Figure 1

(*a*) Principle of a diamond anvil cell (DAC). Pressure is generated between the culets of two diamond anvils and can reach high values (69). The diamond is transparent in the visible range and has a good transmission for wavelengths most appropriate for high-pressure macromolecular crystallography (HPMX) (0.7–0.3 Å). A metallic gasket inserted between the two diamonds is used to confine the sample (66). The thrust applied to anvils may be generated by screws (47), screw and levers (69), or a built-in pneumatic device (19, 43). (*b*) Diamond anvils with a conical mount provide a useful aperture larger than the conventional mount, together without loss of pressure range (8). (*c*) DAC optimized for HPMX (25) using a metallic membrane inflated with helium (43) and conical diamond mount (useful opening angle about 82°). Adapted with permission from Reference 25.

and versatile device that consists of two opposed single crystals of diamond (69) that squeeze a metal gasket with a hole (66) containing a pressure-transmitting medium, a pressure sensor, and the sample (**Figure 1**). The hydrostatic pressure on the sample depends mainly on the squeezing thrust applied on the area of diamond culets. The first application of DAC to HPMX was the determination of the unit cell compressibility of two crystalline forms of hen egg-white lysozyme, without subsequent data collection (34). Since 2001, using ultrashort wavelength X-rays, diffraction data from a variety of compressed macromolecular crystals have been routinely collected. Because of this instrumentation and associated methodology, HPMX is now a full-fledged technique (24, 25). This situation, together with a better appraisal of the interest of pressure perturbation for molecular biophysics, has created a favorable context for the application of HPMX to a variety of biologically relevant questions.

Instrumentation at ESRF and SOLEIL

ID27, at the European Synchrotron Radiation Facility (ESRF), is a HP microdiffraction beamline (48) that has been adapted to HPMX (**Figure 2**). Two in-vacuum undulators with a magnetic period of 23 mm are bright sources of

X-rays in the short and ultrashort wavelength domains. In the HPMX mode, the wavelength is selected by a monolithic Si two-reflection monochromator; focusing mirrors are not used,

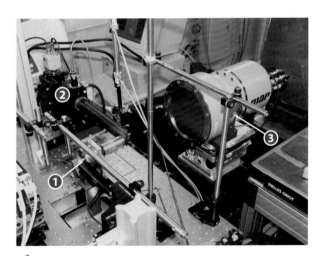

Figure 2

Partial view of the high-pressure microdiffraction beamline ID27 at the ESRF, showing the high-pressure macromolecular crystallography (HPMX) equipment. Monochromatized synchrotron radiation comes from the left. The diamond anvil cell (DAC) (part 1) is installed on the goniometer in a normal-beam geometry (here the cell axis and the X-ray beam direction coincide). The DAC oscillates about the vertical axis of the goniometer during data collection. The optical system (part 2) is shown in the position for sample viewing and collection of ruby fluorescence; it is translated sideways during data collection. Diffraction images are currently collected with a standard single-element charge-coupled device detector (part 3).

Short (ultrashort)
wavelength X-rays:
X-rays with a
wavelength
≈ 0.5 (0.3) Å

ESRF: European
Synchrotron Radiation
Facility

so that the low divergence of undulator radiation is preserved. Beam dimensions are adjusted by a pair of slits. The DAC is settled on a three-axis goniometer mounted onto a stack of three orthogonal translations. A normal-beam geometry is used. Pressure measurement is based on the pressure-dependent wavelength shift of the R-fluorescence lines of ruby (18). An optical system (Betsa, France) is used both for viewing the crystal and for collecting the laser-excited fluorescence from the ruby chip. The area detector was initially a MAR345 imaging plate. In this case, the X-ray wavelength was adjusted at the K-absorption edge of barium (0.331 Å) to maximize signal-to-noise ratio on the detector (20). We currently use a MAR165 charge-coupled device detector, which has a much faster readout than the imaging plate, with the wavelength usually adjusted at the K-absorption edge of iodine (0.374 Å) for calibration purposes. This detector is not optimal because the detective quantum efficiency at short and ultrashort wavelengths is low. Our recent instrumental developments include (*a*) a DAC in which thrust is generated by a pneumatic system with a piston (19) instead of a toroidal membrane (43), ensuring higher sensitivity and improved reversibility (this instrument is being adapted to the PROXIMA I MX beamline at the synchrotron radiation facility SOLEIL), and (*b*) DAC handling by the robotized sample-loading system on the French MX beamline BM30-FIP at ESRF.

Sample Loading

Crystals are grown by usual methods such as the vapor diffusion technique. The first step of sample loading in the DAC is the preparation of a gasket. A piece of copper or Inconel® (typical thickness of 200 µm) is inserted between the diamonds, the cap is screwed, and the gasket is indented by applying an appropriate thrust. Then the cap is removed and a hole is drilled by electroerosion. The final compression chamber is a cylindrical cavity with a thickness of about 180 µm and a diameter of 350 µm. The gasket is then placed on the diamond in the

initial position, using the indentation as a guide. A drop of mother liquor is deposited on the hole and a crystal fished out of one of the crystallization drops is gently pushed into the cavity together with a tiny ruby sphere. The cap is screwed and a modest thrust seals the cavity. Finally, the DAC is placed on the goniometer. The X-ray beam cross-section is adjusted at typically 50×50 µm^2. Pressure is ramped by increasing thrust. Once the desired pressure has been obtained, the DAC position is adjusted so that the center of the cavity is exactly at the intersection of the X-ray axis and the vertical rotation axis of the goniometer. The complexity of diffraction patterns from macromolecular crystals is an advantage. In effect, a single still picture of the crystal collected with the area detector contains so many Bragg reflections that it is generally sufficient to determine the unit cell parameters, select possible space groups, and determine the orientation matrix. The setup is then ready for data collection.

Data Collection

Data are collected by the rotation method. Electronic pictures (frames) are recorded using adjacent rotations of 0.5° to 1.0° about the vertical axis of the goniometer. DAC is translated by small amounts (typically 50 µm) after a few frames to irradiate successively fresh crystal zones. With this procedure it is generally possible to extract large amounts of data from each sample in spite of collection at room temperature (**Figure 3*a***). If only small crystals are available, several samples may be loaded within the compression cavity (**Figure 3*b***).

In general, crystals from the same batch have the same habit and take a limited number of orientations with respect to diamond culets. The worse case is plate-shaped crystals. This may be a serious problem for data completeness because some regions of reciprocal space are thus not accessible. In such case, we use diamond splinters introduced into the cavity (**Figure 3*c***) to set different orientations (25).

X-ray diffraction images are usually quite good. The tight collimation of the X-ray beam

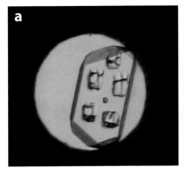 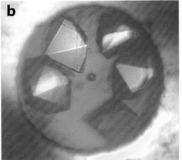 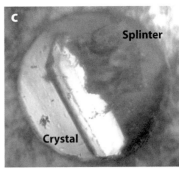

Figure 3

(*a*) Multiple impacts of the X-ray beam on a crystal mounted in the cylindrical compression chamber (diameter 300 μm) drilled in the gasket. The diamond anvil cell was translated by small amounts (typically 50 μm) every 10° rotation to irradiate successively fresh zones. With this procedure, it is generally possible to extract large amounts of data from each sample in spite of collection at room temperature. (*b*) Multiple small crystals mounted in the compression chamber, an alternative to multiple translations of a larger single crystal. (*c*) Diamond splinter introduced into the compression cavity to modify the sample orientation with respect to diamond culets. This procedure is crucial to get high-completeness data from anisotropic (e.g., plate-shaped) crystals. Adapted with permission from Reference 25.

avoids diffraction rings from the polycrystalline gasket (**Figure 4**). Measuring integrated intensities of Bragg reflections proceeds in parallel with the acquisition of frames using the program XDS (32). The sparse diffraction spots from diamonds are automatically identified and discarded at this stage. A routine that corrects experimental structure factor amplitudes for the effects of absorption of X-rays by diamonds can be applied prior to scaling. The quality of data, evaluated as usual by data completeness, symmetry-averaged reliability (Rsym) factors, resolution, and results of refinement (e.g., crystallographic R- and Rfree-factors, errors on bond lengths, and bond angles), usually meets standards of conventional crystallography.

SELECTED APPLICATIONS OF HIGH-PRESSURE MACROMOLECULAR CRYSTALLOGRAPHY

The Double-Helix Base-Paired Architecture is a Molecular Spring

Until recently, the behavior of nucleic acids under high pressure had been investigated by spectroscopic techniques in solution only. The d(GGTATACC) octanucleotide was selected

because of the high stability of its crystalline A-DNA form, its strong base stacking, and its particularly well-defined hydrogen-bonding network. The crystal structure at atmospheric pressure was known (60). Molecules pack in infinite superhelices of duplexes down the unique axis of the hexagonal P6₁ space group. The central channel of the superhelix has the appropriate diameter to host oriented molecules of B-DNA that are reminiscent of the A-B equilibrium in solution. The fiber diffraction pattern of B-DNA is superimposed on the Bragg reflections of the crystallized A-DNA (16). Accordingly, we had at hand a system in which both A- and B-forms of DNA can be simultaneously monitored against external hydrostatic pressure.

Single crystals of the nucleotide were gradually compressed from ambient pressure to 2 GPa. The compressibility curve was derived from measurements of unit cell parameters. This compressibility is positive up to about 1.5 GPa, with the largest value (0.215 GPa⁻¹) at ambient pressure. Beyond 1.6 GPa, the compressibility is negative and the crystal quality deteriorates, with a complete loss of diffraction at about 2 GPa. Four complete diffraction data sets at 1.60–1.65 Å resolution were recorded at ambient pressure, 0.55, 1.04, and

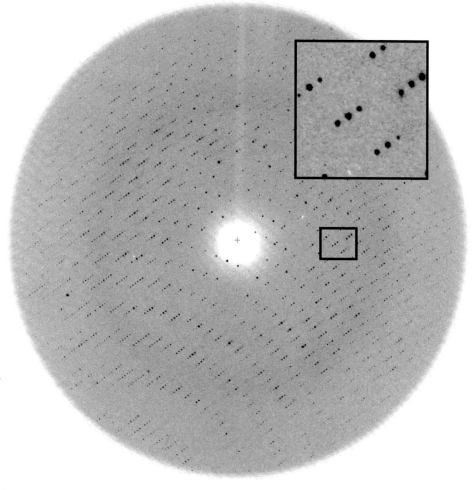

Figure 4

Typical X-ray diffraction image of a crystal of protein (cellulase from *Pseudoalteromonas haloplanktis*, a piezo-psychrophilic bacteria) compressed at 175 MPa; storage ring parameters: energy 6 GeV; intensity 160 mA; X-ray wavelength 0.331 Å; slits 50 μm × 50 μm; oscillation range 0.5°; exposure time 60 s; MAR345 imaging plate detector; resolution 1.8 Å. Compton scattering by diamond contributes to a continuous and relatively weak background. Overall, signal-to-noise ratio of diffraction spots is high (see insert).

1.39 GPa, and the structures were fully refined. The double helix features a large axial compressibility of the molecule. The base-stacking shrinkage is 2.6 Å for the full octamer length (a relative contraction of 11%). The average base-pair step varies from 2.92 to 2.73 Å. The transversal compressibility is negligible. The molecule reacts under high pressure as a molecular spring, and during compression the geometry of Watson-Crick base-pairings (which carry genetic information) is preserved (**Figure 5**). A wedge effect due to packing is probably at the origin of crystal degradation beyond 1.6 GPa. The strong meridional reflections of the fibrous B-DNA are visible up to at least 2 GPa, which shows that the double helix is preserved in this pressure range. They monitor the shortening of the average base-pair step, with a decrease from 3.34 Å at ambient pressure to 3.07 Å at 2 GPa. We concluded that the double-helix topology is remarkably adapted to high

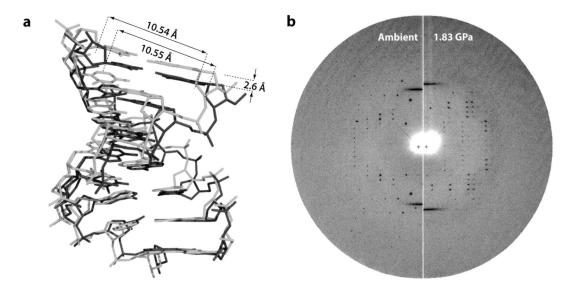

Figure 5

(a) d(GGTATACC) structure. Superimposition of the A-DNA duplex at ambient pressure (*green*) and 1.39 GPa (*red*) showing the high axial compressibility of this molecule and the quasi-invariance of diameter. (b) Fiber diffraction pattern of B-DNA superimposed to the diffraction pattern of A-DNA crystal at ambient pressure (*left*) and 1.83 GPa (*right*). At 1.83 GPa, the two B-DNA meridian reflections still persist over the degrading A-DNA diffraction pattern. Adapted with permission from Reference 27.

pressure. We suggested that the adaptation of this architecture to harsh conditions may have played an important role at the prebiotic stage and in the first steps of the emergence of life (27).

Crystals of the Drew-Dickerson dodecamer d(CGCGTTAACGCG) (17) have been selected to study the behavior of B-DNA in the crystalline state. We found that diffraction is lost at moderate pressure, around 400 MPa. This loss is not due to denaturation of the double helix, but to disruption of the packing. In the orthorhombic $P2_12_12_1$ space group, molecules are arranged in parallel files, a packing that is easily perturbed by pressure (T. Prangé, E. Girard, R. Kahn & R. Fourme, unpublished results). Both octamer and dodecamer studies show the influence of packing in stability of compressed crystals.

Conformational Substates in a Small Monomeric Protein

HPMX up to 150 MPa with a beryllium cell was used to probe effects of pressure on sperm whale myoglobin structure (65). A comparison of pressure effects with those seen at low pH suggests that structural changes under pressure are interpretable as a shift in the populations of conformational substates. Pressure modifies populations of those conformations and crystallographic results most likely reflect the ensemble average of the change in substate populations. This paper also introduced HP cooling as an alternative to room-temperature beryllium cell-based techniques, and it evaluated whether low temperatures successfully lock-in relevant pressure-induced structural changes when cooled under pressure.

The Pressure-Induced Structure-Function Modifications and Dissociation of an Oligomeric Enzyme

Another important effect of pressure is the destabilization of the quaternary structure of oligomeric proteins. A simple explanation of the mechanism by which pressure leads to dissociation of such systems has been proposed

(67): Compression of the weaker protein-protein interactions results in preferential destabilization of the subunit interactions and shifts the equilibrium toward the formation of the shorter and stronger subunit-water bonds. It is also possible to invoke the formal Le Chatelier-Braun's principle. There is a net volume decrease at dissociation that stems primarily from filling with solvent the structural voids that exist in the oligomer and on the electrostriction of solvent when charged groups at the monomer surface are exposed after dissociation; hydrophobic interactions may play some role in the volume change, but it is considered a minor effect for proteins (68). This destabilization occurs in most cases at fairly low pressure (100–300 MPa), before denaturation of the tertiary structure of subunits.

A multifaceted study of the effect of pressure on an oligomeric protein, urate oxidase (UOx) from *Aspergillus flavus*, has been performed. UOx belongs to the purine degradation pathway. It catalyzes the oxidation of uric acid to a primary reaction intermediate, 5-hydroxyisourate, with release of hydrogen peroxide at physiological or basic pH. It is functionally active as a homotetramer. This enzyme, which is not produced in humans, is used to resolve hyperuricemic disorders that can occur in chemotherapy treatments in which DNA is massively degraded. The structure of the 135-kDa UOx tetramer complexed with the competitive inhibitor 8-azaxanthin was first determined at atmospheric pressure, using orthorhombic crystals (space group I222) diffracting at high resolution (15). The asymmetric unit contains one monomer, and subunits (A, B, C, D) are related by the crystallographic twofold axis. The A-B (or C-D) dimer consists of a 16-stranded beta-barrel structure with eight alpha-helices outside the barrel. The association of two such dimers forms the active homotetramer crossed by a tunnel 50 Å long and 12 Å in diameter along the twofold axis. The four active sites in each tetramer are located at the interfaces of A-B and C-D dimers, which have by far the largest contact zones (**Figure 6**). These crystals were selected for HPMX study.

Although this enzyme is stable at ambient conditions, it is sensitive to high pressure. Diffraction is lost at about 200 MPa. A complete data set at 2.3 Å resolution was collected at 140 MPa, and the structure was solved and refined using these data (HP structure). A reference structure was also obtained at atmospheric pressure (AP structure). Final R-factors are good: 14.1% and 16.1% for the AP and HP structures, respectively. A careful comparison of structures showed many classical effects of elastic compression that we have also observed in other studies such as CpMV. They include the reduction of cavity volumes, the overall reduction of buried areas, and a tighter packing of building blocks (here tetramers) at high pressure. In contrast, abnormal features were observed, suggesting that the onset of the dissociation of tetramers had been captured (15).

After these incentive results, pressure perturbation on UOx, either native or liganded by uric acid substrate or a substrate-like inhibitor, was investigated thoroughly by X-ray

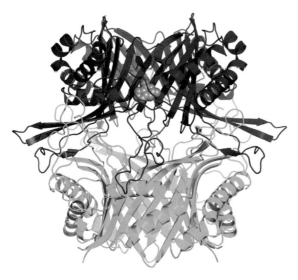

Figure 6

Structure at 140 MPa of the urate oxidase homotetramer complexed with the competitive inhibitor 8-azaxanthin (15). Subunits A and B are colored red and blue to highlight their interface, which hosts the active site indicated by the presence of the inhibitor molecule (*spheres*). The C-D dimer is colored green.

crystallography at higher pressure and higher resolution, HP X-ray small angle scattering, HP fluorescence spectroscopy, and measurements of catalytic activity both under pressure and after decompression. A global model for pressure-induced dissociation, coherent with all measurements, was proposed. The tetramer dissociates irreversibly into monomers without dimeric intermediates; ligand binding has a stabilizing, concentration-dependent effect on the stability of the tetramer. Subtle structural changes under pressure reveal not only premises of tetramer dissociation and ligand debinding, but also higher-energy conformational substates related to an enzymatic mechanism (E. Girard, S. Marchal, J. Perez, S. Finet, R. Kahn, et al., manuscript submitted). These results illustrate how pressure perturbation, by promoting higher-energy states, can reveal conformations of high biological relevance. They also demonstrate how the combination of various biophysical techniques with HPMX may be fruitful.

Elastic Compression of a Viral Capsid

CpMV is a plant virus. The 3D structure of the capsid had been determined at ambient pressure in the cubic body-centered space group I23 (a = 317 Å) and refined at 2.8 Å resolution (44). In crystallization drops, only a small fraction of crystals conform to characteristics of the I23 space group, in which Bragg reflections with h + k + l odd are strictly forbidden. These crystals have a low mosaicity and diffract at 2.6–2.8 Å resolution, which reveals a good long-range order. For most crystals in the drops, diffraction patterns show weak reflections with h + k + l odd; the mosaicity is large and the resolution is modest (4–5 Å). These features have been interpreted by small random rotations of quasi-spherical capsids about their center of gravity (45), which perturb the periodicity of the I23 structure. This disorder has plagued previous structural studies at ambient pressure.

We found that compressing disordered crystals beyond about 240 MPa systematically removed odd reflections, dramatically reduced mosaicity, and improved resolution (22, 26). In this way, I23 crystals are systematically obtained, which diffract up to about 400 MPa. We collected a highly complete data set at 330 MPa. The structure was fully refined at 2.8 Å resolution (HP structure). Using original diffraction data from Lin et al. (44), the I23 structure was refined at ambient pressure by using the same programs (AP structure) and served as a reference. The HP and AP structures were compared. The most interesting effects of pressure on the capsid include the compression of the capsid, the large reduction of the cavity volumes, the increase of buried surfaces, and the shortening of H-bond lengths by about 0.1 Å. GPa^{-1} on average. This latter value is significant because it is based on a histogram of H-bond length variations between AP and HP incorporating about 600 values, and it is in agreement with results from HPNMR (4) and simulations (11). All these features are characteristic of elastic compression within the same conformation subset. Further, the origin of disorder was confirmed and described in terms of intercapsid interactions. High pressure produces shorter interactions and new connections mediated by water molecules, thus restoring the body-centered lattice, the long-range order in the crystal, and high-quality diffraction. This work was the first HPMX study of a virus and showed that high-quality HP data can be collected on a large assembly.

IMPACT OF HIGH PRESSURE ON CONVENTIONAL MACROMOLECULAR CRYSTALLOGRAPHY

Low-Pressure Gas Binding to Proteins

Gas ligands such as oxygen, carbon monoxide, nitric oxide, xenon, and krypton interact weakly with the host protein owing to their small sizes and hydrophobicity. Pressurizing native crystals with gas at low pressure (<10 MPa) may increase binding and reveal binding sites and the pathway of gases (49).

Xenon or krypton reversibly bound in hydrophobic cavities of pressurized proteins can be exploited to solve the crystallographic phase problem. Indeed, they can be used to make good heavy-atom derivatives with significant anomalous scattering. The use of xenon was introduced in 1965 (59). The method was revisited, made user-friendly, and is now commonly used (58).

Pressure as a Tool for Improving Order in Macromolecules and Crystals

The effects of pressure and temperature on equilibria or kinetics are antagonistic in molecular terms: As follows from the principle of microscopic ordering, isothermal compression leads to an ordering of molecules or a decrease in the entropy of a system. The resolution that can be obtained in a crystallographic experiment depends on the quality of crystal order. The quality of pre-existing crystals may be improved by compression in the DAC, with subsequent data collection at high pressure. A clear-cut, albeit particular example of the ordering effect in the packing of CpMV capsids (26) has been reported above. Application of pressure and pressure annealing might become a standard way of improving the long-range order of crystals. There are also protein structures in which a portion of electron density is not visible because of disorder. Such problems might be cured by compression, as steric constraints are increased. Systematic investigation of ordering effects of high pressure is needed, but the rarity of synchrotron radiation beamtime for such studies is a problem.

Growing crystals is still the major bottleneck of MX. Pressure may be one important variable in protein crystallization. In hydrostatic conditions, any pressure variation affects the whole system uniformly and rapidly. Comparatively to huge investments in high-throughput crystallization facilities and low-gravity experiments, efforts to study ab initio nucleation and crystallization in pressurized solutions have been limited (33, 46, 62). A systematic program for crystal growth using pressure perturbation would be of interest.

HP Cooling

Another application of high pressure in which diffraction data are collected at ambient pressure consists of cooling a crystal while it is compressed in helium gas up to about 500 MPa and then releasing pressure. This method does not need cryoprotectants and commonly improves crystal diffraction with respect to standard cryocooling (37). There is evidence that the formation of high-density amorphous ice is responsible for this improvement (35). The method is compatible with the use of noble gases for phasing (36).

The goal is to freeze-in pressure-induced collective movements of the polypeptide chain. Subsequent data collection is performed on cooled sample at atmospheric pressure. Radiation damage is strongly reduced with respect to room temperature data collection.

Responses of sperm whale myoglobin crystals to active compression and after HP cooling were accurately compared (65). Similar responses were found for several major collective displacements, but not for others. Presumably, modifications due to high pressure were only partially preserved by HP cooling and superimpose with those induced by cooling. HP cooling has been used to correlate alterations of the structure of a fluorescent protein, citrine, with the accompanying spectral shift (7).

Transfer of Experimental Procedures

The constraints of HPMX, including data collection at room temperature, led us to carefully examine how the signal-to-noise ratio in data collection can be improved and more generally how data collection efficiency, defined as the information content collected per unit volume of crystal (23), can be increased. Our results suggest that some experimental conditions used during HPMX data collection (particularly the use of a quasi-parallel beam of ultrashort wavelength with narrow collimation, irradiation of successive crystal zones, a large detector placed at a long crystal-to-detector distance) should also be considered for standard data collection with crystals either cryocooled or at room temperature.

CONCLUSION

HP perturbation is only marginally used by the community of macromolecular crystallography. This situation could evolve in the coming years. On the one hand, the unique advantages of HP—in particular for exploring conformational states of macromolecules—are now better appraised. On the other hand, HPMX has reached the status of a full-fledged technique that is relatively simple to use on specialized synchrotron radiation beamlines. Narrow gap undulators on intermediate energy storage rings such as SOLEIL are bright sources of short wavelength X-rays usable for HPMX data collection as an alternative to ultrashort wavelengths available only at high-energy facilities (ESRF, APS, SPring 8, PETRA III). The association of HPMX with other techniques of structural biophysics, in particular small-angle X-ray scattering and NMR, opens new routes to study problems of high biological relevance.

SUMMARY POINTS

1. Macromolecular crystals are remarkably suitable for HP studies.

2. HPMX is now a full-fledged method applicable to a broad range of systems, from small biomolecules to large assemblies. The quality and accuracy of data often meet standards of conventional crystallography.

3. In addition to the use of noble gases to solve the phase problem, standard MX would also benefit from methodology developed for HP studies.

FUTURE ISSUES

1. Understanding the fundamental mechanisms of the effect of pressure on macromolecular structures requires high-resolution structural information, which is still scarce. HPMX studies, coupled with HPNMR, need to be developed.

2. The study of biologically relevant higher-energy conformers is probably the most important prospect of HPMX.

3. Understanding the molecular basis of adaptation to extreme conditions is important for both fundamental science and applications.

4. The development of HPMX for (*a*) HP studies, (*b*) the improvement of crystal quality in otherwise standard MX, and (*c*) the transfer of HPMX methodology to conventional MX requires dedicated synchrotron radiation beamlines designed to operate with short and ultrashort wavelengths.

5. Systematic programs on crystallization in compressed solution should be undertaken.

6. The development of simulations guided by HPMX and HPNMR structural data, using massively parallel computing, is needed.

DISCLOSURE STATEMENT

The authors are not aware of any biases that might be perceived as affecting the objectivity of this review.

ACKNOWLEDGMENTS

ESRF (Grenoble, France) was a firm basis for the development of HPMX, in particular in the context of a long-term project (MX421, from 2005 to 2008). We thank M. Mezouar, principal beamline scientist of the ESRF ID27 beamline, whose support and collaboration since 2000 have been invaluable. We thank the scientific and technical staff of the beamline, in particular W. Crichton, and the ESRF BLISS group. Some results reported in this review have been obtained in the course of collaborative work with other scientists. We are especially indebted to N. Colloc'h (urate oxidase project) and T. Prangé (nucleic acids project). Finally, we thank J.-C. Chervin and B. Couzinet at the IMPMC (Paris) for their contribution to instrumental developments.

LITERATURE CITED

1. Anfinsen CB. 1973. Principles that govern the folding of protein chains. *Science* 181:223–30
2. Akasaka K. 2003. Exploring the entire conformational space of proteins by high-pressure NMR. *Pure Appl. Chem.* 75(7):927–36

3. A fundamental article to grasp the interest of high-pressure studies on biomolecules.

3. **Akasaka K. 2006. Probing conformational fluctuations of proteins by pressure perturbation. *Chem. Rev.* 106:1814–35**
4. Akasaka K, Li H, Yamada H, Li R, Thoresen T, Woodward CK. 1999. Pressure response of protein backbone structure. Pressure-induced amide 15N chemical shifts in BPTI. *Protein Sci.* 8:1946–53
5. **Akasaka K, Yamada H. 2001. On-line cell high-pressure nuclear magnetic resonance technique: application to protein studies. *Methods Enzymol.* 338:134–58**

5. A review on instrumentation and applications of high-pressure NMR to biophysics.

6. Balny C, Hayashi R, Heremans K, Masson P, eds. 1992. *High Pressure and Biotechnology*. London: John Libbey
7. Barstow B, Ando N, Kim CU, Gruner SM. 2008. Alteration of citrine structure by hydrostatic pressure explains the accompanying spectral shift. *Proc. Natl. Acad. Sci. USA* 105:13362–66
8. Boehler R, de Hantsetters K. 2004. New anvil designs in diamond-cells. *High Press. Res.* 24:1–6
9. **Bridgman PW. 1914. The coagulation of albumin by pressure. *J. Biol. Chem.* 19:511–12**

9. The pioneering work on the effect of high-pressure on biosystems.

10. Brown P, Meyer R, Cardone F, Pocchiari M. 2003. Ultra high pressure inactivation of prion infectivity in processed meat: a practical method to prevent human infection. *Proc. Natl. Acad. Sci. USA* 100:6093–97
11. Canalia M, Malliavin TE, Kremer W, Kalbitzer HR. 2004. Molecular dynamics simulations of HPr under hydrostatic pressure. *Biopolymers* 74:377–88
12. Chalikian TV, Breslauer KJ. 1996. On the volume changes accompanying conformational transitions of biopolymers. *Biopolymers* 39:619–25
13. Collins MD, Hummer G, Quillin ML, Matthews BW, Gruner SM. 2005. Cooperative water filling of a nonpolar protein cavity observed by high-pressure crystallography and simulation. *Proc. Natl. Acad. Sci. USA* 102(46):16668–71
14. Colloc'h N, El Hajji M, Bachet B, L'Hermite G, Schiltz M, et al. 1997. Crystal structure of the protein drug urate oxidase-inhibitor complex at 2.05 Å resolution. *Nat. Struct. Biol.* 4:947–52
15. Colloc'h N, Girard E, Dhaussy A-C, Kahn R, Ascone I, et al. 2006. High-pressure macromolecular crystallography: the 140-MPa crystal structure at 2.3 Å resolution of urate oxidase, a 135-kDa tetrameric assembly. *Biochim. Biophys. Acta* 1764:391–97
16. Doucet J, Benoit J-P, Cruse WBT, Prangé T, Kennard O. 1989. Coexistence of A- and B-form DNA in a single crystal lattice. *Nature* 337:190–93
17. Drew HR, Wing RM, Takano T, Broka C, Tanaka, et al. 1981. Structure of a B-DNA dodecamer: conformation and dynamics. *Proc. Natl. Acad. Sci. USA* 78:2179–83

18. Forman RA, Piermarini GJ, Barnett JD, Block S. 1972. Pressure measurement made by the utilization of ruby sharp-line luminescence. *Science* 176:284–85

19. Fourme R. 1968. Appareillage pour études radiocristallographiques sous pression et à température variable. *J. Appl. Crystallogr.* 1:23–29

20. **Fourme R, Ascone I, Kahn R, Girard E, Hoerentrup C, et al. 2001. High pressure protein crystallography: instrumentation, methodology and results of data collection on lysozyme crystals. *J. Synchrotron Radiat.* 8:1149–56**

21. Fourme R, Ascone I, Kahn R, Girard E, Mezouar M, et al. 2003. New trends in macromolecular crystallography at high hydrostatic pressure. In *Advances in High Pressure Bioscience and Biotechnology II*, ed. R Winter, pp. 161–70. Berlin: Springer

22. Fourme R, Ascone I, Kahn R, Mezouar M, Bouvier P, et al. 2002. Opening the high-pressure domain beyond 2 kbar to protein and virus crystallography. *Structure* 10:1409–14

23. Fourme R, Girard E, Kahn R, Ascone I, Mezouar M, et al. 2003. Using a quasi-parallel X-ray beam of ultrashort wavelength for high-pressure virus crystallography: implications for standard macromolecular crystallography. *Acta Crystallogr. D* 59:1767–72

24. Fourme R, Girard E, Kahn R, Dhaussy A-C, Mezouar M, et al. 2006. High-pressure macromolecular crystallography (HPMX): status and prospects. *Biochim. Biophys. Acta* 1764:384–90

25. Girard E, Dhaussy A-C, Couzinet B, Chervin J-C, Mezouar M, et al. 2007. Toward full-fledged high-pressure macromolecular crystallography (HPMX). *J. Appl. Crystallogr.* 40:912–18

26. **Girard E, Kahn R, Mezouar M, Dhaussy A-C, Lin T, et al. 2005. The first crystal structure of a macromolecular assembly under high pressure: CpMV at 330 MPa. *Biophys. J.* 88:3562–71**

27. **Girard E, Prangé T, Dhaussy A-C, Migianu E, Lecouvey M, et al. 2007. Adaptation of base-paired double-helix to extreme hydrostatic pressure. *Nucleic Acids Res.* 35(14):4800–8**

28. Gross M, Jaenicke R. 1994. The influence of high hydrostatic pressure on structure, function and assembly of proteins and protein complexes. *Eur. J. Biochem.* 221:617–30

29. Hayashi R. 1989. Application of high pressure to food processing and preservation: philosophy and development. In *Engineering and Food 2*, ed. WEL Spiess, H Schubert, pp. 815–26. Amsterdam: Elsevier

30. Horikoshi K, Grant WD. 1998. *Extremophiles. Microbial Life in Extreme Environments.* New York: Wiley-Liss

31. Jannasch HW, Taylor CD. 1984. Deep sea microbiology. *Annu. Rev. Microbiol.* 38:487–514

32. Kabsch W. 1993. Automatic processing of rotation diffraction data from crystals of initially unknown symmetry and cell constants. *J. Appl. Crystallogr.* 26:795–800

33. Kadri A, Lorber B, Charron C, Robert MC, Capelle B, et al. 2005. Crystal quality and differential crystal-growth behaviour of three proteins crystallized in gel at high hydrostatic pressure. *Acta Crystallogr. D* 61:784–88

34. Katrusiak A, Dauter Z. 1996. Compressibility of lysozyme crystals by X-ray diffraction. *Acta Crystallogr. D* 52:607–8

35. Kim CU, Chen YF, Tate MW, Gruner SM. 2008. Pressure-induced high-density ice in protein crystals. *J. Appl. Crystallogr.* 41:1–7

36. Kim CU, Hao Q, Gruner SM. 2005. Solution of protein crystallographic structures by high-pressure cryocooling and noble-gas phasing. *Acta Crystallogr. D* 62:687–94

37. Kim CU, Kapfer R, Gruner SM. 2005. High pressure cooling of crystals without cryoprotectants. *Acta Crystallogr. D* 61:881–90

38. Kitahara R, Yamada H, Akasaka K, Wright PE. 2002. High pressure NMR reveals that apomyoglobin is an equilibrium mixture from the native to the unfolded. *J. Mol. Biol.* 320:311–19

39. Kitahara R, Yokoyama S, Akasaka K. 2005. NMR snapshots of a fluctuating protein structure: ubiquitin at 30 bar-3 kbar. *J. Mol. Biol.* 347:277–85

40. Kundrot CE, Richards FM. 1986. Collection and processing of X-ray diffraction data from protein crystals at high pressure. *J. Appl. Crystallogr.* 19:208–13

41. **Kundrot CE, Richards FM. 1987. Crystal structure of hen egg-white lysozyme at a hydrostatic pressure of 1000 atmospheres. *J. Mol. Biol.* 193:157–70**

42. Kundrot CE, Richards FM. 1988. Effect of hydrostatic pressure on the solvent in crystals of hen egg-white lysozyme. *J. Mol. Biol.* 200(2):401–10

20. Materials and methods of HPMX using DAC and ultrashort wavelength synchrotron radiation.

26. The first HPMX structure of a complex molecular assembly.

27. The first HPMX structure of an A-DNA model.

41. The first HPMX structure of a protein.

43. Le Toullec R, Pinceaux J-P, Loubeyre P. 1988. The membrane diamond anvil cell: a new device for generating continuous pressure and temperature variations. *High Press. Res.* 1:77–90

44. Lin T, Chen Z, Usha R, Stauffacher CV, Dai JB, et al. 1999. The refined crystal structure of *Cowpea mosaic virus*. *Virology* 265:20–34

45. Lin T, Schildkamp W, Bister K, Doerschuk PC, Somayazulu M, et al. 2005. The mechanism of high-pressure-induced ordering in a macromolecular crystal. *Acta Crystallogr. D* 61(6):737–43

46. Logtenberg EHP, Meersman F, Rubens P, Heremans K, Frank J. 2000. Influence of pressure on the crystallisation and dissolution of protein crystals. *High Press. Res.* 19(Pt. 2):675–80

47. Merrill L, Bassett WA. 1974. Miniature diamond anvil pressure cell for single crystal X-ray diffraction studies. *Rev. Sci. Instrum.* 45:290–94

48. Mezouar M, Crichton WA, Bauchau S, Thurel F, Witsch H, et al. 2005. Development of a new state-of-the-art beamline optimized for monochromatic single-crystal and powder X-ray diffraction under extreme conditions at the ESRF. *J. Synchrotron Rad.* 12:659–64

49. Montet Y, Amara P, Volbeda A, Vernede X, Hatchikian EC. 1997. Gas access to the active site of Ni-Fe hydrogenases probed by X-ray crystallography and molecular dynamics. *Nat. Struct. Biol.* 4:523–26

50. Mozhaev VV, Heremans K, Franks J, Masson P, Balny C. 1994. Exploiting the effect of high hydrostatic pressure in biotechnological applications. *Trends Biotechnol.* 12:493–501

51. Mozhaev VV, Heremans K, Frank J, Masson P, Balny C. 1996. High pressure effects on protein structure and function. *Proteins* 24:81–91

52. Nicholson JM, Körber FC, Lambert SJ. 1996. Application of moderate hydrostatic pressure induces unit-cell changes in rhomboedral insulin. *Acta Crystallogr. D* 52:1012–15

53. Oliveira AC, Ishimaru D, Goncalves RB, Smith TJ, Mason P, et al. 1999. Low temperature and pressure stability of picornaviruses: implications for virus uncoating. *Biophys J.* 76:1270–79

54. Privalov PL, Gill SJ. 1988. Stability of protein structure and hydrophobic interaction. *Adv. Protein Chem.* 39:191–234

55. Refaee M, Tezuka T, Akasaka K, Williamson MP. 2003. Pressure-dependent change in the solution structure of hen egg-white lysozyme. *J. Mol. Biol.* 327:857–65

56. Royer C. 2002. Revisiting volume changes in pressure-induced protein folding. *Biochim. Biophys. Acta* 1595:201–9

57. Schiltz M, Fourme R, Prangé T. 2003. The use of noble gases xenon and krypton as heavy atoms in protein structure determination for phasing. *Methods Enzymol.* 374:83–119

58. Schiltz M, Prangé T, Fourme R. 1994. On the preparation and X-ray data collection of isomorphous xenon derivatives. *J. Appl. Crystallogr.* 27:950–60

59. Schoenborn BP, Watson HC, Kendrew JC. 1965. Binding of xenon to sperm whale myoglobin. *Nature* 207:28–30

60. Shakked Z, Rabinovich D, Kennard O, Cruse WBT, Salisbury SA, et al. 1983. Sequence-dependent conformation of an A-DNA double helix. The crystal structure of the octamer d(G-G-T-A-T-A-C-C). *J. Mol. Biol.* 166:183–201

61. Silva CMS, Giongo V, Simpson AJG, da Silvas Camargo ER, Silva JL, et al. 2001. Effects of hydrostatic pressure on the *Leptospira interrogans*: high immunogenicity of the pressure-inactivated serovar *hardjo*. *Vaccine* 19(11–12):1511–14

62. Suzuki Y, Sazaki G, Miyashita S, Sawada T, Tamura K, et al. 2002. Protein crystallization under high pressure. *Biochim. Biophys. Acta* 1595(1–2):345–56

63. Tilton RF. 1988. A fixture for X-ray crystallographic studies of biomolecules under high gas-pressure. *J. Appl. Crystallogr.* 21:4–9

64. Tilton RF, Petsko GA. 1988. A structure of sperm whale myoglobin at a nitrogen-gas pressure of 145 atmospheres. *Biochemistry* 27(17):6574–82

65. Urayama P, Phillips GN, Gruner SM. 2002. Probing substates in sperm whale myoglobin using high-pressure crystallography. *Structure* 10:51–60

66. Van Valkenburg A. 1962. Visual observations of high pressure transitions. *Rev. Sci. Instrum.* 33:1462

67. Weber G. 1992. Thermodynamics of the association and the pressure dissociation of oligomeric proteins. *J. Phys. Chem.* 97:7108–15

68. Weber G. 1992. *Protein Interactions*, pp. 211–14. New York: Chapman & Hall

69. Weir CE, Lippincott ER, Van Valkenburg A, Bunting EN. 1959. Infrared studies in the 1-micron to 15-micron region to 30000 atmospheres. *J. Res. Natl. Bur. Stand. Phys. Chem.* 63A:55–62

70. **Yamada H, Nishikawa K, Honda M, Shimura T, Akasaka K, et al. 2001. Pressure-resisting cell for high-pressure, high-resolution nuclear magnetic resonance measurements at very high magnetic fields. *Rev. Sci. Instrum.* 72:1463–71**

71. Yayanos AA, Delong EF. 1987. Deep-sea bacterial fitness to environmental temperatures and pressures. In *Current Perspectives in High Pressure Biology*, ed. HW Jannasch, RE Marquis, AM Zimmerman, pp. 17–32. London: Academic

70. Landmark paper on HPNMR instrumentation.

Imaging Transcription in Living Cells

Xavier Darzacq,[1,2] Jie Yao,[1,3] Daniel R. Larson,[1,4]
Sébastien Z. Causse,[1,2] Lana Bosanac,[1,2]
Valeria de Turris,[1,4] Vera M. Ruda,[1,2]
Timothee Lionnet,[1,4] Daniel Zenklusen,[1,4]
Benjamin Guglielmi,[1,3] Robert Tjian,[1,3]
and Robert H. Singer[1,4]

[1]Janelia Farm Research Consortium on Imaging Transcription, Janelia Farm Research Campus, Howard Hughes Medical Institute, Ashburn, Virginia 20147

[2]Imagerie Fonctionelle de la Transcription, Ecole Normale Supérieure CNRS UMR 8541, 75230 Paris cedex 05, France; email: darzacq@ens.fr

[3]Howard Hughes Medical Institute, Department of Molecular and Cell Biology, University of California, Berkeley, California 94720; email: tijcal@berkeley.edu

[4]Department of Anatomy and Structural Biology, Gruss-Lipper Biophotonics Center, Albert Einstein College of Medicine of Yeshiva University, Bronx, New York 10461; email: rhsinger@aecom.yu.edu

Annu. Rev. Biophys. 2009. 38:173–96

First published online as a Review in Advance on February 3, 2009

The *Annual Review of Biophysics* is online at biophys.annualreviews.org

This article's doi:
10.1146/annurev.biophys.050708.133728

Key Words

modeling, RNA synthesis in live cells, polymerase kinetics

Abstract

The advent of new technologies for the imaging of living cells has made it possible to determine the properties of transcription, the kinetics of polymerase movement, the association of transcription factors, and the progression of the polymerase on the gene. We report here the current state of the field and the progress necessary to achieve a more complete understanding of the various steps in transcription. Our Consortium is dedicated to developing and implementing the technology to further this understanding.

Contents

INTRODUCTION

With the uncovering of the ever-growing fraction of the animal genome that is transcribed, transcription is more than ever the centerpiece of cell metabolism. Through biochemical analysis and genetics, many if not most of the proteins implicated in transcription have been identified. Decades of in vitro studies determined that the transcription process could be separated into three steps: preinitiation complex formation, initiation, and elongation. Each one of these steps may be subjected to regulation, accounting for the fine-tuning of gene expression. But biochemistry tells us only what is possible, not what actually happens, in the very specific milieu of a living cell.

It has become feasible, using the new advances in microscopy, to interrogate the processes that make up transcription and break them down into their component parts. Accurate quantification is possible due to technology that has evolved over the years to detect and measure photons. The components of the transcription reaction can then be assigned rate constants describing their forward and reverse rates. As a result of these analyses, new models have arisen to fit these data. Among those are the observations that for some models the transcriptional complex can be transient, existing only for a few seconds, and that the entire process is inefficient, yet other factors can be stably associated for hours. Understanding the kinetic components that give rise to these disparate time constants will be an important function of the new technologies.

This review is dedicated to exploring the work that is contributing to the real-time analysis of transcription, an aspect that contributes data not addressed by chromatin immunoprecipitation, microarray studies, or other bulk assays that cannot resolve the events occurring in single cells. Many cell and tissue models are described including bacteria, polytene chromosomes, various reporter genes, and cell lines. Different approaches include fluorescence recovery after photobleaching (FRAP), fluorescence correlation spectroscopy (FCS), fluorescence resonance energy transfer (FRET), and multiphoton microscopy (MPM) (see sidebar, Methods Used to Analyze Transcription in Living Cells). This is a new field that is rapidly emerging, and these initial forays represent the beginning of a new territory in the area of gene expression research.

The review is organized by the model systems that have contributed to studies in live cell imaging: naturally occurring and artificial gene arrays, viral genes, steroid receptor responsive genes, and single copy endogenous genes. There are four sections: imaging gene arrays, imaging the nuclear organization of transcription, imaging single copy genes, and the analysis of imaging data.

IMAGING GENE ARRAYS

Some gene families or duplicated genes in vertebrate genomes have naturally regrouped into tandem arrayed genes. These genes therefore lie as neighbors on the same locus

FRAP: fluorescence recovery after photobleaching

FCS: fluorescence correlation spectroscopy

FRET: fluorescence resonance energy transfer

MPM: multiphoton microscopy

in the genome and are often under the control of the same transcriptional regulators. In addition, the fruit fly (*Drosophila melanogaster*) contains giant polytene chromosomes in certain larval cell types where the entire genome is multiplied. This spatial clustering of genes offers researchers many opportunities to study transcription, especially when using fluorescence microscopy, in which accumulation of fluorescence means a better signal-to-noise ratio. Another advantage of such arrays is that when studying a fluorescent molecule that interacts directly or indirectly with the arrays, the majority of the fluorescence will be due to interacting molecules (in contrast to freely diffusing molecules, which might be preponderant at a nonamplified locus). This increases the resolution on binding events that reflect catalytic activities. So far three cases of amplified genes have been used to study transcription as illustrated in **Figure 1**: polytene salivary gland chromosomes in *Drosophila* (95), the ribosomal DNA nucleolar clusters (36), and artificially developed mammalian gene arrays (68). These examples are covered more fully below.

Drosophila melanogaster Polytene Chromosomes

Drosophila polytene chromosomes are found in many larval cell types formed by endoreduplication during development, i.e., these cells undergo DNA replication without cell division. Polytene chromosomes from salivary gland cells contain approximately 1000 copies of DNA. Condensed and decondensed chromatin form unique band and interband structures that can be distinguished with a light microscope. The chromosome banding patterns were categorized and named by Bridges (13) and have been used as a marker for cytogenetic localization of individual genes along the polytene chromosomes (**Figure 2a**). Examining the unique cytogenetic pattern has allowed early genetic mapping such as gene deletion, gene duplication, chromosome translocation, and inversion. Furthermore, localizing protein factors on polytene chromosomes with antibody-

METHODS USED TO ANALYZE TRANSCRIPTION IN LIVING CELLS

Image Sampling

Cells imaged under wide-field fluorescence microscopy using long exposure times can yield images of transcription factors specifically bound to the relatively stationary DNA in contrast to the unresolved undersampled free transcription factors. Tracking of individual transcription factors was possible using stroboscopic laser excitation.

FRAP

FRAP involves the irreversible photobleaching of a specified area in the nucleus by a focused laser of biologically nonabsorptive wavelength. The diffusion of surrounding fluorescent molecules into the bleached spot portrays a recovery characteristic of the movement of tagged fluorophores.

FLIP

By the method of FLIP, a specified cell area is repeatedly bleached and the loss of fluorescence of surrounding areas are monitored. This gives a more exact way to analyze the mobility of a protein which is likewise involved in binding. Dissociation kinetics of proteins from compartments can thus be determined more precisely.

FCS

FCS is a tool for binding measurements. A laser beam is focused in the cell within a femtoliter range volume. Fluctuations in fluorescence signals are measured over short periods of time, the output reflecting movement of labeled proteins through the volume. This gives a direct measurement of concentration, diffusion constants and binding constants. Generally, FCS is well suited to fast processes on a scale of milliseconds, whereas FRAP is better suited to slower processes on the scale of seconds. Cross-correlation FCS can likewise be used to determine physical interaction between two species by simultaneous comparison of their fluctuation traces.

based immunostaining techniques has provided a means to study protein-DNA interaction in vivo.

The naturally amplified chromatin template in *Drosophila* polytene chromosomes provides an opportunity to overcome the sensitivity

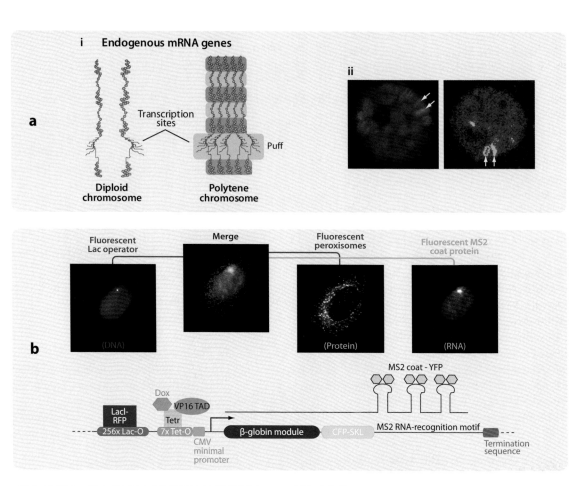

i   **Endogenous mRNA genes**

Transcription sites

Diploid chromosome

Polytene chromosome

Puff

ii

a

b

Fluorescent Lac operator

Merge

Fluorescent peroxisomes

Fluorescent MS2 coat protein

(DNA)

(Protein)

(RNA)

MS2 coat - YFP

Dox

VP16 TAD

LacI-RFP

TetR

256x Lac-O

7x Tet-O

CMV minimal promoter

β-globin module

CFP-SKL

MS2 RNA-recognition motif

Termination sequence

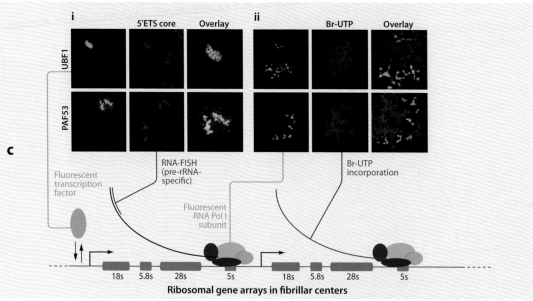

c

i   5'ETS core   Overlay   **ii**   Br-UTP   Overlay

UBF1

PAF53

Fluorescent transcription factor

RNA-FISH (pre-rRNA-specific)

Br-UTP incorporation

Fluorescent RNA Pol I subunit

18s   5.8s   28s   5s   18s   5.8s   28s   5s

Ribosomal gene arrays in fibrillar centers

limitation in visualizing transcription factors associated with endogenous gene loci in living cells. However, imaging transcription factors in living salivary gland tissues had been challenged by the thickness and optical properties of the tissue samples. Recently, Webb, Lis, and colleagues (94–96) have reported that MPM provides the experimental capability of resolving individual genetic loci (**Figure 2*b–e***) and studying dynamic interactions of a green fluorescent protein (GFP)-fused heat shock factor (HSF) or RNA polymerase II (Pol II) with active *hsp*70 gene loci in living tissues and in real time.

This is an elegant system for studying the dynamics of transcription regulation in vivo. The power of this approach lies in the combination of naturally amplified templates, *Drosophila* transgenic techniques, and MPM imaging, which provides optical sectioning deep within living tissues (99). The rapid and robust heat shock gene activation has allowed unambiguous localization of endogenous *hsp*70 loci and further assisted the visualization of the associated factors (**Figure 2*f,g***). Furthermore, multicolor fluorescent proteins and mutant gene alleles can be introduced by simple genetic crosses.

Some novel insights of transcription dynamics have arisen from the application of this method. For instance, the transient association (a few seconds) of a transcription activator with a gene promoter, the so called hit-and-run model (54), has been thought to be universal for all activators. However, in *Drosophila* polytene chromosomes, HSF is transiently associated with *hsp*83 gene loci (half life about 10 s)

before heat shock but becomes stably associated with *hsp*70 gene loci (half-life > 5 min) during heat shock (**Figure 2*h,i***) (95). Therefore, some transcription factors, such as HSF, are stably associated with their target genes and can provide a stable platform that supports multiple rounds of transcription. This may be a relatively common property of strong promoters, as suggested by other reports. (*a*) A stable association with DNA in vivo has been reported for yeast Gal4 activator (60); (*b*) higher transcription levels are correlated with a more stable association of Pol I subunits with rRNA genes (36). In addition, Pol II exhibits efficient recruitment to the locus and enters into elongation during early heat shock, and it is locally recycled during late heat shock (94). Furthermore, this MPM-based FRAP assay has been coupled with chromatin immunoprecipitation assays to study the dynamics of distinct populations of Pol II molecules: During early heat shock, FRAP was performed on *hsp*70 genes when the productively elongating forms of Pol II were eliminated by inhibiting P-TEFb kinase. This study shows that the remaining transcriptionally engaged/paused Pol II molecules near the transcription start site of the genes are stably associated with *hsp*70 genes (61). Although this experimental system will continue to provide unique insights into the dynamic mechanisms of gene regulation in *Drosophila*, it would be of interest to determine to what extent the transcription kinetics at micron-scale gene structures in polytene cells resembles the kinetics at nanoscale gene structures in diploid cells.

GFP: green fluorescent protein

HSF: heat shock factor

Pol I (II): RNA polymerase I (II)

Figure 1

Three models for visualizing transcription in vivo. (*a*) *Drosophila* polytene chromosome. (*i*) Diagram of diploid chromosome and polytene chromosome. Active transcription site of heat shock gene *hsp*70 loci at 87A and 87C exhibits "chromosome puffs" (*yellow arrows*). (*ii*) (*Left*) *hsp*70 loci are enriched with Pol II (*red*: Hoechst33342 stain; *green*: EGFP-Rpb3). (Reproduced with permission from Reference 95.) (*Right*) Upon heat shock, EGFP-Rpb3 exhibits a strong doublet at 87A and 87C loci that can be recognized in polytene nucleus (*arrows*). Image in pseudocolor. (*b*) An artificial reporter gene was stably inserted as multiple copies into a single locus in the genome of U2OS cells. The locus of integration was tracked with the use of a Lac-repressor-fused RFP. The MS2 stem loop structures on the RNA, which bind stably to MS2 coat proteins, fused with YFP, allowing the investigator to localize single molecules of mRNA in vivo (78). (*c*) Ribosomal RNA transcription occurring inside fibrillar centers in CMT3 cells. (*i*) PAF53, a Pol I subunit, and UBF1, a rDNA transcription factor, colocalize with nascent rRNA in fibrillar centers (*green*: PAF53-GFP or UBF1-GFP; *red*: fluorescent in situ hybridization probe directed to rRNA 5'ETS core). (*ii*) GFP-tagged PAF53 and UBF1 accumulate in active rRNA transcription sites (*green*: PAF53-GFP or UBF1-GFP; *red*: Bromo-UTP stain) (Adapted with permission from Reference 24). Abbreviations: EGFP, enhanced green fluorescent protein; ETS, external transcribed spacer; Pol I (II), polymerase I (II).

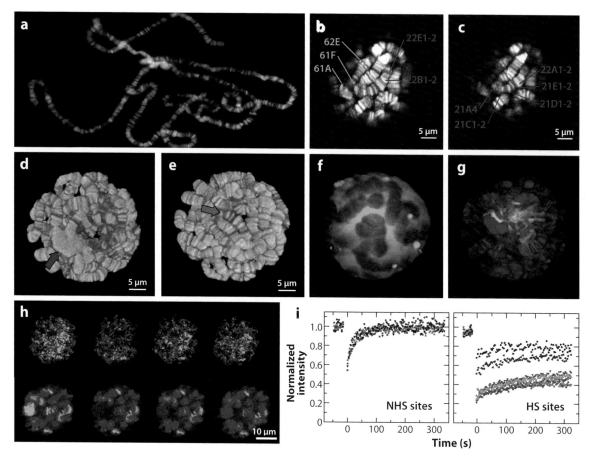

Figure 2

Imaging transcription kinetics at *Drosophila* polytene chromosomes. (*a*) An image of *Drosophila* polytene chromosomes spread onto a glass slide and stained by Hoechst33342. (*b, c*) Optical sections of polytene chromosomes in a live salivary gland stained with Hoechst33342. The z-distance is 0.5 µm. Labels identify specific bands on chromosome arms 2L (*red*) and 3L (*blue*). (*d, e*) Three-dimensional reconstructions of a polytene nucleus. Red and blue arrows indicate the centromeric region and telomere, respectively. Panels *b–e* are adapted with permission from Reference 95. (*f, g*) HSF expression and localization in salivary glands at NHS (*f*) and after HS (*g*). Shown are the three-dimensional reconstructions of the two-photon optical sections of polytene nuclei expressing HSF-EGFP (*green*) stained with Hoechst33342 (*red*). (*h, i*) FRAP of HSF-EGFP at endogenous gene loci in salivary glands. (*h*) Intensity images of a nucleus of a heat-shocked salivary gland in pseudocolor during FRAP. (*i*) FRAP curves at NHS and HS sites. (Adapted with permission from References 95 and 96). Abbreviations: EGFP, enhanced green fluorescent protein; FRAP, fluorescence recovery after photobleaching; HS, heat shock; HSF, heat shock factor; NHS, non-heat shock.

Natural Tandem Repeated Genes

The Misteli laboratory used the ribosomal genes' organization into such arrays to study Pol I transcription. Their work offered the first attempt to measure kinetic rates for a RNA polymerase in living cells. They proposed an elongation rate of 5.7 kb per minute. They also observed that different subunits of the Pol I complex were loaded onto the genes with

different kinetic rates, which can be interpreted as a pioneering subunit or subcomplex interacting with the DNA, serving as a docking platform for the other subunits (24). More recently, the same group analyzed the order of assembly of Pol I components and the regulation of this assembly during G_1/G_0 and S phases. According to their findings, Pol I is assembled in an on-the-spot stepwise process that reflects transcriptional efficiency (36). This is in contrast to in vitro studies that show that a pre-assembled RNA polymerase holoenzyme can be recruited to a promoter site and efficiently transcribe an RNA molecule (22), or that there is no exchange between Pol I subunits in yeast (Pol I remains intact without subunit exchange through multiple rounds of transcription in *Saccharomyces cerevisiae*) (22, 75). Similar studies have yet to be done with Pol II, whose assembly might also be influenced by the promoter sequence of the studied gene and the various specific transcriptional regulators involved.

Another example of natural array is the *CUP1* locus in the baker's yeast *S. cerevisiae* (47). *CUP1* exists as a small natural tandem array of 10 copies, transcribed by the Pol II. The transcriptional activator Ace1 binds to *CUP1* promoter in the presence of copper and activates transcription. Fusion of three copies of GFP to Ace1 enabled visualization of the *CUP1* array in live cells. The behavior of Ace1-GFP on the *CUP1* promoter has been monitored by FRAP. A complete fluorescence recovery occurred within 2 min after photobleaching. It is similar but somewhat slower than that observed for other transcriptional activators. Furthermore, at longer timescales a slow cycling of Ace1-GFP binding to *CUP1* can be detected. After computational analysis, Ace1 behavior is compatible with a model in which the slow cycle reflects the number of accessible binding sites at promoters and each accessible site can be bound by fast cycling molecules. It is suggested that the oscillation of histone occupancy at the locus accounts for cycling accessibility. At this promoter the fast cycle is responsible for transcription initiation and the slow cycle for adjusting the amount of mRNA synthesis. This simple natural model can be combined with powerful yeast genetics to explore further the implication of transcription factors and chromatin remodeling in the kinetics of transcription.

Gene arrays therefore offer a huge number of possibilities when it comes to studying transcription and related processes. A recent estimate suggested that such tandem gene arrays represent 14% of all genes in vertebrate genomes, although most are made of only two genes (62). That means that these gene arrays are available for investigators who wish to study transcription in a natural genomic context. However, these natural gene arrays offer low control over the fundamental mechanisms involved in transcription. A number of teams have therefore developed artificial gene arrays in which the reporter genes were tinkered with to study specific core mechanisms. These modified arrays represent a good compromise between natural conditions and control by the investigator.

Artificial Gene Arrays

The first purpose of an artificial array created by Tsukamoto et al. (89) was to study chromatin remodeling during transcriptional activation. By inserting this Tet-inducible reporter gene containing Lac operator sites into large arrays, they observed changes in the chromatin upon activation and verified that the fluorescent protein encoded was correctly expressed. This system was then improved by Janicki et al. (43), who inserted 24 repeats of the MS2 bacteriophage replicase translational operator, which allowed them to visualize the mRNA transcribed from the genes. This work has permitted a real-time parallel analysis of transcription and modifications in the chromatin, notably by following histone 3.3 depositions and HP1α depletion (43). The same system was then used to estimate various kinetic steps of transcription (21). Other studies on chromatin remodeling made use of a gene array first described in 2000 by McNally (54), made up of mouse mammary tumor virus long terminal repeats (MMTV-LTRs), which can be activated by glucocorticoid receptors (GRs) (**Figure 3d**) (45, 54, 58, 80). These studies have given great insight into the importance of various chromatin remodelers such as BRM and BRG1 during transcription. Finally, whether natural or artificial, gene arrays could be of great use in the study of many other nuclear processes, and the use of this powerful tool has only begun to reveal new and interesting observations.

MMTV: mouse mammary tumor virus

LTR: long terminal repeat

GR: glucocorticoid receptor

One of the most surprising results concerning transcription obtained with gene arrays was that transcription in vivo is an inefficient process. The studies concerning Pol I by Dundr and colleagues (24) show that only about 10%

of polymerases in the fibrillar centers are actually engaged in elongation. Similarly, Darzacq and colleagues (21) have shown that transcription in a Pol II array of genes was also inefficient, with only 1% of binding events resulting in the production of a complete mRNA. In the same study, they have determined the kinetics of different steps of transcription using FRAP experiments on an artificial gene array. They estimated that the elongation rate of Pol II is around 75 nucleotides per second, slightly slower than the 90 nucleotides per second published for Pol I (21, 24). These estimates are in accordance with previous findings of 50 and 100 nucleotides per second, respectively, using an independent method (76).

Viruses

Among viral species there exists a great variety of genomic structures and, consequently,

a

pExo-MS2X24

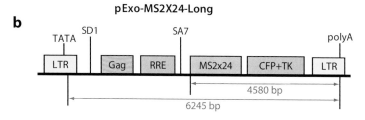

b

pExo-MS2X24-Long

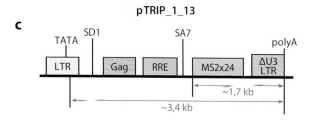

c

pTRIP_1_13

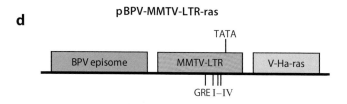

d

pBPV-MMTV-LTR-ras

e

pHIV-LTR-CFP-SKL

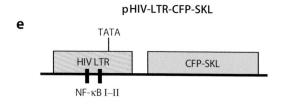

Figure 3

Diagrams of vectors containing viral sequences used to produce tandem arrays for real-time live cell imaging. (*a*) Vector used by Molle and colleagues to determine the residency times of Tat and Cdk9 on the HIV-1-LTR-driven transgene (57). (*b, c*) Modifications of the vector presented in panel *a* used in Reference 8 to measure transcription elongation rate. (*d*) MMTV-LTR containing glucocorticoid receptor (GR) binding sites developed by McNally and colleagues was used to determine various kinetic parameters of gene activation by hormones (45, 54, 58, 80). Similarly, the vector presented in panel *e* has been used to study NF-κB interaction with its binding sites (9). The original vector designations and references in which they were first described are given. The kinetic parameters of transcription were obtained on tandem arrays in each of these vectors. Systems are discussed in the text. Diagrams are not drawn to scale. (For a compilation of other existing gene arrays, see Reference 68.) Abbreviations: BPV: bovine papilloma virus; Gag: retroviral gene coding for internal structural proteins; GRE: glucocorticoid-responsive element; NF-κB: nuclear factor κB; RRE: Rev-responsive element; SA7: splice acceptor 7; SD1: splice donor 1; SKL: Ser-Lys-Leu peroxisome-targeting peptide; v-Ha-ras: Harvey viral ras.

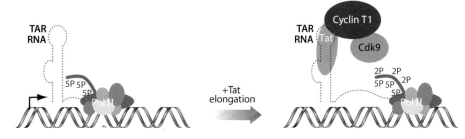

Figure 4

The role of the TAR:Tat:P-TEFb complex in HIV transcription. RNA polymerase II (Pol II) is stalled soon after initiation of transcription of the HIV provirus. The viral protein Tat recruits host cellular P-TEFb to the nascent stem-bulge-loop leader RNA, TAR (*trans*-activation responsive). The elongation factor P-TEFb composed of the cyclin T and the kinase Cdk9 phosphorylates the C-terminal domain repeats of the large subunit of Pol II at the Ser2 positions to stimulate processive elongation.

mechanisms of replication and gene expression. Only a small part of all known viruses depends on host cell polymerases in both replication and transcription.

For retroviruses, such as HIV-1 and MMTV, replication is the process whereby genome-sized RNA, which also functions as mRNA, is produced by host cell Pol II from the provirus integrated in the host genome. Thus, transcription is a means of their replication, as well as gene expression, and its tight regulation is important for the viral life cycle.

Some viruses use their own proteins to modify and redirect the activity of host cell's transcriptional machinery. In the case of HIV-1, transcription is activated by the viral protein Tat, which recruits the elongation factor P-TEFb [consisting of cyclin T and Cdk9, which phosphorylates the C-terminal domain of the large subunit of Pol II] to the nascent stem-bulge-loop leader RNA, TAR (*trans*-activation responsive) (**Figure 4**) (44). Recently, cell lines have been created harboring tandem arrays of a reporter that carries the elements required for HIV-1 RNA production (**Figure 3a–c**) (8, 57). In these cells the dynamics of the TAR:Tat:P-TEFb complex components has been analyzed by FRAP at the transcription sites visualized by expressing a nuclear MS2 phage coat protein (MS2cp) fusion with a fluorescent protein. Comparison of Cdk9-GFP dynamics at sites activated by Tat or phorbol 12-myristate 13-acetate/ionomycin showed that Cdk9 residency

time at the HIV-1 transcription site was several times longer in the presence of Tat than in the absence of Tat (71 s and 11 s, respectively) and that it was similar to the residency time measured for Tat-GFP itself (55 s), suggesting that significant fractions of Tat and Cdk9 are present at the site as parts of the same complex, likely interacting with elongating Pol II (57). The transcription elongation rate measured by FRAP on MS2-GFP on the same HIV-1 tandem array and its variants with a longer transcribed region or without the U3 region in the 3′ LTR (which is required for efficient transcript 3′-end formation) (**Figure 3a–c**) was estimated to be approximately 1.9 kb min^{-1} (8). The use of this HIV-MS2 tandem array also allowed the estimation of the Pol II residency time at the transcription site and its comparison to RNA production rates. The authors calculated a total polymerase residency time of 333 s, of which 114 s were attributed to elongation, 63 s to 3′-end processing and/or transcript release, and 156 s to polymerase remaining on the gene after RNA release (8).

Unlike P-TEFb, which stays at the transcription site induced by Tat for approximately 1 min (**Figure 4**), other transcription factors interacting with viral promoters interact transiently with the arrays. For example, mRFP-tagged NF-κB proteins interact with a multicopy array of transgenes containing the HIV 5′ LTR for only a few seconds (**Figure 3e**) (9). FRAP of GFP-tagged GRs at the tandem array

MS2cp: coat protein of MS2 phage

containing MMTV-LTR promoters showed even shorter residency times (**Figure 3d**) (54, 80).

NUCLEAR ORGANIZATION OF TRANSCRIPTION

Transcription Factories

Pol II is a multisubunit enzyme responsible for the transcription of most eukaryotic genes. The composite holoenzyme generated by the association of Pol II with other large complexes involved in related functions such as capping, splicing, and polyadenylation ensures the efficient production of mature transcripts. The key element necessary for coupling transcription with all the maturation steps is the large subunit of Pol II and in particular its C-terminal domain, essential for tethering the different machineries and regulating them temporally. The discovery of this intricate network gave rise to the idea that a specialized molecular machine is assembled at the site of transcription, nucleating from the promoter of an active gene. Because several of these factories may cluster together to ensure high local concentrations and therefore efficient interactions with all the partners involved, it is important to understand how different transcription units are transcribed and how their identity, nuclear surroundings, and positions could affect their expression.

Over the past fifteen years Cook and collaborators have put forward the concept of a superstructure called transcription factories, an assemblage of transcription and RNA-processing enzymes containing multiple genes. Before the advent of new visualizing methodologies, the transcription sites in mammalian cells were marked by elongation of nascent RNA in the presence of [^{3}H]uridine, [^{32}P]uridine, or Br-UTP, and subsequent observation at the fluorescent or electron microscope (39, 41, 42, 66). With these techniques they were able to see multiple nuclear foci sensitive to α-amanitin and containing splicing components (41). Those foci remained visible also after nucleolytic removal of most of the chro-

matin, highlighting the presence of an underlining structure responsible for the clustering of transcription units in which transcripts are both synthesized and processed (39). Moreover, these results are consistent with polymerases confined by the nucleoskeleton into factories and transcription occurring as templates slide past attached polymerases (40). Quantitative analysis (42) also showed that a typical factory contains approximately 30 engaged polymerases. Because two-thirds or more transcription units are associated with one polymerase at any time, each factory could contain at least 20 different transcription units.

A recent work on transcription factories (27) increased the resolution obtained with the electron microscope by coupling this technique with electron spectroscopic imaging. Electron spectroscopic imaging is a high-resolution and potent ultrastructural method that can be used to map atomic distribution in unstained preparations. Combining immunolabeling of the newly synthesized BrU-RNA with the distribution maps of nitrogen (N) and phosphorus (P) enabled specific atomic signature marking, allowing these nucleoplasmic sites to be identified. Template and nascent RNAs were attached to the surface of enormous protein-rich structures 87 nm in diameter and with a mass of 10 MDa. These structures appear porous, large enough to contain all the different protein complexes required for the complete maturation of the transcript. This finding suggested the idea that the polymerase was anchored, probably at the surface of the core, and that the DNA diffuses or loops to come in contact with a specific factory. Eskiw et al. (27) suggest that only a minority of all the machinery in the site is active, but that the high local concentrations will guarantee robust and efficient processivity.

Other questions concern how many transcription factories exist in a cell and how they should be classified. The first level of organization is the division of the three polymerases into different factories (67, 93). A more complicated issue is determining the influence of the genes' characteristics (promoter or presence of introns) on their arrangement within the nucleus. Using replicating minichromosomes from Cos7 cells analyzed by FISH (fluorescent in situ hybridization) and 3C (chromosome conformation capture), Xu & Cook (93) examined whether the factories were specialized and the importance of the genes' distinctive characteristics. Their results confirmed that plasmids were concentrated in transcribing foci and that those being copied by different polymerases were not transcribed by the same foci. Moreover, units transcribed by Pol II, with different promoters (CMV and U2) or with the same promoter but with or without an intron in the coding sequence, are seen in nonoverlapping foci.

Even more intriguing are the results from live cells where the fluorescent tagging system of Pol II large subunit has been exploited (87). The GFP-tagged version was stably expressed in a Chinese hamster ovary cell line bearing a temperature-sensitive Pol II mutant, tsTM4, and it was observed that the fluorescent version was functional and normally assembled in the complex and rescued the phenotype. Because each factory contains only a few polymerases, it would be difficult to image those foci in live cells; however, significant results could be obtained by the study of polymerase kinetics in the nucleus of living cells. FRAP and FLIP (fluorescence loss in photobleaching) experiments in the nucleus gave important information on fractions of the enzyme in different states (48). They revealed two kinetic components in the Pol II population: A fast mobile component showed that ~75% of the molecules were diffusing freely and the immobile component showed that ~25% of the molecules were transiently immobile with a $t_{1/2}$ of ~20 min. This latter fraction was likely the active one, since incubation with DRB (5,6-dichloro-1-β-D-ribofuranosyl-benzimidazole), a potent inhibitor of elongation, eliminated it. Their model of the transcription cycle supports the idea that the enzyme spends most of the time diffusing and exchanging between the nucleoplasm and a promoter or a transcription factory. Once bound, a third of the time is mainly dedicated to elongation. With the improved use of this photobleaching technique,

a more detailed analysis also resolved a third component resistant to DRB but sensitive to heat shock, representing the bound but not yet engaged fraction (37).

Whether these factories exist in most cells is a question that needs to be addressed with more sensitive technologies. Given that resolution problems pervade the experiments concerning testing of this concept, it will fall to the more quantitative, high-resolution methods to determine whether there simply exist gene-rich regional concentrations of transcription, or whether the factories are truly higher-order structures.

FISH: fluorescent in situ hybridization

Gene Positioning

The influence of the position of a gene with respect to the nuclear periphery on transcriptional competence has been extensively studied in recent years. Historically, the nuclear periphery has been seen as a nuclear substructure enriched in heterochromatin and thereby an area of transcriptional repression. However, data from yeast showing that active genes are often found in the nuclear periphery and in association with the nuclear pore complex led to a series of studies investigating the influence of nuclear positioning on transcriptional competence (3, 17).

In an early study, Cabal et al. (14) showed that the yeast *GAL1* gene changed its position from a mostly internal position to a preferential location at the nuclear periphery when the gene was activated, supporting the idea that the nuclear periphery harbors active genes. To do this, they used a fluorescently labeled *GAL1* locus in living cells by inserting an array of 112 TetO operators downstream of the *GAL1* gene, which upon coexpression of GFP-tagged TetR turns fluorescent (14). Nuclear positioning and movement of a locus were then followed using 4D live cell microscopy. Importantly, they found that the movement of the locus was not fully constrained but restricted to a 2D sliding movement at the nuclear envelope and was suggested to act as a gating mechanism to allow efficient mRNA processing and export.

Genes in yeast have been analyzed using this technique and were shown to move to the nuclear periphery upon activation (2, 12, 14, 23, 88). The requirements for the translocations, however, were often gene specific. In addition to components of the nuclear pore complex, promoters or elements in the 3′UTR, the SAGA complex of transcription factors, and components of the mRNA export machinery are involved (2, 12, 14, 23, 74, 88). Similarly, gene movement to the periphery has been suggested to occur before transcription starts for some genes, but it has also been suggested to occur as a result of transcription for other genes (12, 14, 23, 74, 88). It still remains to be shown if general principles exist that mediate the perinuclear localization of active genes in yeast and what fraction of genes use this mechanism to regulate their expression. Peripheral localization has also been suggested to mediate epigenetic memory over many generations (11).

These data from yeast led to the question whether such a mechanism may exist in higher eukaryotes. In yeast and in higher eukaryotes, chromatin loci in general are not statically positioned within the nucleus. In yeast as well as in higher eukaryotes, chromatin during interphase is mobile but mostly constrained within a radius of approximately 0.5–1 μm. That is less than 1% of the volume of a typical 10-μm spherical mammalian nucleus but half of the diameter of a yeast nucleus (52). In yeast, if a locus is located at the nuclear periphery, diffusion and the accessibility of binding sites at the nuclear periphery might be sufficient to allow tethering, as most genes likely encounter the nuclear periphery at least occasionally. In higher eukaryotes, however, if such events existed, it might require a more active movement, as a locus would have to move several microns to attach to the nuclear periphery or to nuclear pores. Peripheral heterochromatin is often interrupted at nuclear pores, indicating the presence of euchromatin in the vicinity of nuclear pores and making it possible that, like in yeast, active genes might get tethered to nuclear pore complexes to stimulate expression.

Recent live cell studies suggested that repositioning of genes from or to the nuclear periphery might have some influence on gene expression in higher eukaryotes, but that it might not be a major factor mediating gene expression. Imaging the naturally amplified *Drosophila* polytene nuclei in living salivary gland tissues by MPM did not reveal a preferred localization of the loci upon transcription induction. The genes could be found in the nuclear interior as well as at the nuclear periphery (94). Consistently, a GFP-tagged locus tethered to the nuclear periphery by a lamin B1 fusion maintained its transcriptional competence, indicating that sole peripheral or internal/central nuclear positioning does not influence transcription (51). However, another study suggested that expression of a subset of genes can reversibly be suppressed when tethered to the periphery, whereas many genes are not affected (30). Using DNA FISH, Reddy et al. (70) showed that genes can be silenced when targeted to the inner nuclear membrane. Together these results suggest that the nuclear periphery is not incompatible with active transcription but that it is not a primary determinant of whether genes are active. Different *cis*- and *trans*-acting factors are likely to determine whether peripherally localized genes in higher eukaryotes can be transcribed. However, chromatin movements in higher eukaryotes seem to actively play a role in regulating gene expression. Chromatin can frequently exhibit long-range movements of >2 μm during the cell cycle (90). Migration of an interphase chromosome site from the nuclear periphery to the interior has been observed 1–2 h after targeting a transcriptional activator to this site, showing a contrary localization to that in yeast (19). More surprisingly this movement was perturbed in specific actin and myosin I mutants, suggesting some kind of motor-driven movement. Similarly, actin-dependent intranuclear repositioning occurs with the U2 snRNA gene locus (25). If and how motor proteins mediate such long-distance chromatin movements still remain to be determined.

IMAGING ENDOGENOUS GENES

Steroid Receptors

Perhaps the most well-studied transcription factors of endogenous genes in living cells are nuclear receptor (NR) regulated. These ligand-activated transcription factors constitute the nuclear hormone receptor superfamily and are involved in regulating a vast array of eukaryotic genes. NR transcription is initiated by agonist binding to the receptor, forming either a homodimer or heterodimer complex. The corepressors (histone deacetylases, NR-specific corepressors) associated with the dimer are then replaced by coactivators such as histone acetylases (SRC/p160 family or CBP/p300) and histone methylases (CARM-1, PRMT-1). In addition, ATP-coupled chromatin remodeling complexes (SWI/SNF) are recruited. Eventually, the basal transcription machinery is assembled, followed by the initiation of Pol II. After initiation, transcription can be influenced by NR factors such as vitamin D receptor interacting protein and thyroid-associated protein (38). Thus, NR transcription is an excellent model system for observing the cooperative interactions among enhancers, repressors, transcription factors, and basal transcription components (63). The view that has emerged from live cell studies utilizing fluorescence techniques such as FRAP, FRET, and FCS is that these NR complexes are highly dynamic: Individual species have dwell times on the order of seconds to minutes. However, these same complexes can result in cycles of transcriptional progression that can last hours or days (56). NR-regulated transcription is therefore dynamically responsive to changes in agonist concentration and also capable of long-term changes of gene expression.

Live cell studies of NR-regulated transcription can be divided into those that study nuclear dynamics in general and those that focus on a particular locus. The first approach provides information about multiple possible transcription sites within the nucleus in addition to nonspecific interactions. The second approach has the benefit of providing specific information about interactions and dynamics at an active transcription site but usually requires modification of the locus—either multimerization of an endogenous gene (54) or creation of an artificial locus (85). The first example of this approach, which has been used by a number of investigators since its inception, was a large tandem array of a mouse mammary tumor virus/Harvey viral ras (MMTV/v-Ha-ras) reporter, which contains about 200 copies of the LTR and thus includes 800 to 1200 binding sites for the GR (54). This same array has been used for FRAP studies of the GR (6, 45, 55, 83), the androgen receptor (AR) (50), and the progesterone receptor (PR) (69). For each of those receptors, an agonist-dependent decrease in receptor mobility (increase in $t_{1/2}$) was observed [GR, $t_{1/2}$: 1–1.6 s (55); AR, $t_{1/2}$: 0.2–3.6 s (50); PR, $t_{1/2}$: 0.6–3.7 s; (69)]. A similar agonist-dependent decrease in mobility was also observed for general nuclear bleaching of the estrogen receptor [ER, $t_{1/2}$: 0.8–5.9 s (85)]. These observations demonstrate that the recovery time reflects the interaction of the NR with the locus in a specific fashion. In fact, Schaaf & Cidlowski (73) demonstrated that higher-affinity ligands result in slower recovery times, and Kino et al. (49) directly showed a positive correlation between FRAP $t_{1/2}$ times and transcriptional activity, with higher transcriptional activity corresponding to longer effective recovery times.

In contrast, other receptors do not show an agonist-dependent increase in $t_{1/2}$ for general nuclear recovery. The retinoic acid receptor (RAR), the thyroid hormone receptor (TR), the peroxisome proliferator-activated receptor (PPAR), and the retinoid X receptor (RXR) all have the same recovery time with or without ligand [RAR, $t_{1/2}$: 1.9–2.3 s; TR, $t_{1/2}$: 1.8–1.8 s (53); PPAR, $t_{1/2}$: 0.13–0.15 s; RXR, $t_{1/2}$: 0.2–0.25 s (29)]. In the case of PPAR, this lack of measurable difference may reflect some constitutive activity of the receptor (29).

In all FRAP experiments, the recovery dynamics will reflect both specific and nonspecific interactions. In the case of transient

NR: nuclear receptor

PR: progesterone receptor

transfections, in which an excess of receptor may be present, nonspecific interactions are likely to be a significant contribution to the dynamics for both locus-specific recovery and general nuclear recovery. The recovery curve is likely a convolution of more than one kinetic process. In computational models of AR dynamics, the recovery was separated into two distinct kinetic components: a fast component (due to diffusion or transient binding) of 1–5 s and a slow component of ~60 s (28, 50). This slow component presumably represents a longer-lived interaction in the vicinity of the gene such as with chromatin or nuclear matrix (28, 49, 55, 69, 73, 83), although the nature of this interaction is not clear and may vary between receptors.

In addition to receptor dynamics, several studies have addressed the kinetic behavior of coactivators involved in NR-regulated transcription. Becker et al. (6) observed the receptor coactivator GRIP1 (glucocorticoid receptor interacting protein 1) at the active MMTV array and measured a recovery time that was indiscernible from the GR $t_{1/2}$ (5 s), suggesting that the binding and release of these proteins may be coupled. CBP and SRC-1 (ER coactivators) have $t_{1/2}$ times of 4 s and 8 s, respectively (85); BRM and BRG1, subunits of the SWI/SNF chromatin remodeling complex, have $t_{1/2}$ times of 2 s and 4 s, respectively (45).

Taken together, the remarkable aspect of these data is that these recovery times are all less than or equal to 11 s (**Table 1**). Consider, for example, a typical NR transcription complex: NR $t_{1/2} = 5$ s, SRC1 $t_{1/2} = 8$ s, CBP $t_{1/2} = 4$ s (85), BRM $t_{1/2} = 2$ s, BRG1 $t_{1/2} = 4$ s (45), and GRIP1 $t_{1/2} = 5$ s (6). The only molecular species that has a dwell time on the order of minutes is the elongating polymerase ($t_{1/2} \sim 5$ min) (6). How might these transient interactions lead to transcriptional cycles that are observed in the timescale of hours? One idea that has been proposed is that of a transcriptional ratchet, in which permanent changes—methylation, acetylation, phosphorylation—accumulate at a transcription site as a result of the transient interactions

described above (56). There are several suggestive directions about how such long-lived interactions might occur. SRC1 recovery becomes progressively slower at longer times after stimulation of ER with estradiol ($t_{1/2} = 8.0$–30.2 s) (85); chromatin decondensation seems to depend on polymerase elongation (59). Live cell experiments that follow the change in dynamics over an induction period are likely to be informative as well.

Imaging a Single Gene

Imaging the transcription of a single gene is potentially a powerful approach because it obviates the averaging inherent in gene array studies. This way, the behavior of individual transcription units can be quantified and their variability assessed. However, this has been difficult to achieve because of technical challenges: specifically detecting the desired locus and then observing the small numbers of factors involved in transcribing a single gene.

When a major challenge must be overcome, the tool of choice in vivo is fluorescence microscopy. Although a single fluorescent protein molecule can be detected when immobilized on a surface, it is difficult to resolve in the context of a living cell, where it undergoes fast diffusion or transport and where the fluorescent background can be high. So far, only a few experiments have managed to provide direct observation of gene expression at the single gene level.

A series of recent experiments demonstrated that it is possible to detect single protein products resulting from the expression of a single gene in live bacteria (15, 18, 97). From the distribution of proteins synthesized over time, it is then possible to test different models of transcription. In the first experiment (15), the reporter was a β-galactosidase protein, which produces a fluorescent product upon hydrolysis of a synthetic substrate. Hydrolysis of a large number of substrate molecules by a single enzyme provides the signal amplification necessary to observe a single protein. By observing discrete values in the rate of hydrolysis, the

Table 1 FRAP experiments summarized

Receptor	Cofactor/mutation	Nuclear/array	Ligand	Agonist = 1; partial antagonist = 2; antagonist = 3	$t_{1/2}$ (s)	Immobile fraction[b]	Reference
AR		MMTV array	None		0.20		(50)
AR		MMTV array	R1881	1	3.60		
AR		MMTV array	DHT	1	5.30		
AR		MMTV array	TST	1	5.00		
AR		MMTV array	Bicalutamide	3	0.50		
AR		MMTV array	OHF	3	0.50		
AR		MMTV array	CPA	2	1.10		
AR		MMTV array	RU486	2	4.30		
AR		Nuclear	R1881	1	<5.90[a]		(28)
ER		Nuclear	None		0.80		(86)
ER		Nuclear	Estradiol	1	5.90		
ER		Nuclear	4-HT	2	5.30		
ER		Nuclear	ICI	3	–		
ER	SRC	Nuclear	Estradiol	1	9.80		
ER		Artificial array—LacO	Estradiol	1	–		
ER	SRC	Artificial array—LacO	None		2.10		(85)
ER	SRC	Artificial array—LacO	Estradiol	1	8.00		
ER	CBP	Artificial array—LacO	Estradiol	1	4.20		
ER		Nuclear	None		1.60	0.14	
ER		Nuclear	Estradiol	1	5.80	0.44	(53)
GR		MMTV array	Dexamethasone	1	5.00		(6)
GR	GRIP (p160 family)	MMTV array	Dexamethasone	1	5.00		
GR	RNAPII	MMTV array	Dexamethasone	1	300.00		
GR		MMTV array	RU486	3	<5		(84)
GR		MMTV array	Corticosterone	3	<5		
GR		Nuclear	None		1.00		(73)
GR		Nuclear	Dexamethasone	1	1.90		
GR		Nuclear	Triamcinolone	1	1.87		
GR		Nuclear	Corticosterone	1	1.13		
GR		Nuclear	RU486	3	1.43		
GR		Nuclear	ZK98299	3	1.07		
GR		Nuclear	Dexamethasone	1	0.43–>1.11		(49)
GR		MMTV array	Dexamethasone	1	1.60		(55)
GR	407C-GR mutant	MMTV array	Dexamethasone	1	1.34		

(*Continued*)

Table 1 (*Continued*)

Receptor	Cofactor/ mutation	Nuclear/array	Ligand	Agonist = 1; partial antagonist = 2; antagonist = 3	$t_{1/2}$ (s)	Immobile fraction[b]	Reference
GR	N525-GR mutant	MMTV array	None	1	0.69		
GR		MMTV array	RU486	3	0.82		
GR		MMTV array	Dex-Mes	2	1.40		
GR	BRG1	MMTV array	Dexamethasone	1	3.90		(45)
GR	BRM	MMTV array	Dexamethasone	1	1.95		
RAR		Nuclear	None		1.90	0.18	(53)
RAR		Nuclear	Retinoic acid	1	2.30	0.18	
TR		Nuclear	None		1.80	0.12	(53)
TR		Nuclear	Triiodithyronine	1	1.80	0.14	
PR		MMTV array	None		0.60		(69)
PR		MMTV array	R5020	1	3.70		
PR		MMTV array	RU486	3	11.00		
PR		MMTV array	ZK98299	3	1.80		
PPAR α		Nuclear	None		0.13	0.017	(29)
PPAR α		Nuclear	wy14643	1	0.15	0.024	
PPAR α		Nuclear	None		0.10	0.014	
PPAR α		Nuclear	L-165041	1	0.16	0.03	
PPAR α		Nuclear	None		0.10	0.05	
PPAR α		Nuclear	Rosiglitazone	1	0.12	0.033	
RXR α		Nuclear	None		0.20	0.019	
RXR α		Nuclear	9-*cis* retinoic acid	1	0.25	0.064	

[a]Two-component fit.

[b]Immobile fraction is only reported in a subset of studies.

Abbreviations: AR, androgen receptor; Dex-Mes, dexamethasone mesylate; ER, estrogen receptor; GR, glucocorticoid receptor; GRIP, glucocorticoid receptor interacting protein; MMTV, mouse mammary tumor virus; PPAR α, peroxisome proliferator-activated receptor α; PR, progesterone receptor; RAR, retinoic acid receptor; RNAPII, RNA polymerase II; RXR α, retinoid X receptor α; TR, thyroid hormone receptor.

authors could indeed resolve single protein numbers. In subsequent experiments, the reporter was a fluorescent protein fused to a membrane protein (18, 97). When bound to the membrane, the protein is slowly diffusing and it is therefore possible to accumulate enough fluorescence to resolve a single protein.

These experiments studied reporter genes under the control of the Lac repressor. In this classic system, two operator sequences on the DNA can be bound by a tetramer repressor. Upon full induction, the repressor unbinds the DNA and the cell fully expresses the *lac*

genes downstream. In the absence of inducer, protein is produced in infrequent bursts (0.5–1 per cell cycle) in which a few (2–4) proteins are produced. The distribution of the number of proteins produced per burst is consistent with a model in which a burst results from the transcription of a single mRNA molecule, finally yielding a few proteins. In the regime of moderate inducer concentration, both low-expressing cells (0–10 proteins) and high-expressing ones (hundreds of proteins) are observed. This bimodal distribution results from the presence of frequent, small bursts (similar to the noninduced state) and infrequent, large bursts of protein production. The authors proposed a model in which the small bursts consist of a partial dissociation of the repressor from one of the operator sequences; one RNA molecule is transcribed typically before the repressor binds back the operator sequence. In contrast, the large bursts correspond to full dissociation of the repressor from both operator sequences. In this case, many mRNA molecules are transcribed before another repressor binds the DNA, leading to the production of a large number of proteins.

A similar detection approach was used to study transcription factor dynamics in *Escherichia coli* at the single molecule level (27). Lac repressor (Lac I) molecules fused to a fluorescent protein could be detected when bound to their promoter sequence by imaging for long periods of time (~1 s) to average out the background of freely diffusing molecules. This made it possible to measure the kinetics of association of Lac I to its promoter in vivo. The authors also used short light excitation pulses to characterize the diffusion of the repressor as well as its nonspecific binding to DNA. From these results emerged a picture of Lac I dynamics: Searching for its target sequence, the protein spends 87% of its time in short events (<5 ms), where it is nonspecifically bound to DNA and undergoes 1D diffusion while it scans the DNA. These short events are separated by periods where the repressor diffuses in three dimensions between different DNA segments.

It is also possible to directly visualize mRNA molecules using a technique that exploits the high affinity between RNA stem loops and the bacteriophage MS2 coat protein (7). By introducing repeats of the stem loop coding sequence in the desired gene, and expressing the MS2 coat protein fused to a fluorescent protein, one can detect single mRNA molecules. This technique has been used in numerous studies to characterize single mRNA motion and localization in *E. coli* (34), yeast (5, 7), mammalian cells (33, 72, 77), and *Drosophila* oocytes (31, 91, 98).

Golding et al. (35) used this technique to study transcription in *E. coli* by utilizing an inducible reporter gene under the control of the P$_{lac/ara}$ promoter. By measuring the distribution of mRNA molecules per cell, the authors tested two models for transcription. The simplest model, in which transcription events were randomly initiated according to a Poisson process (with a constant probability per unit time), could not fit the data; a more elaborate model, in which the gene can switch between an "off" state (no transcription takes place) and an "on" state (transcription is randomly initiated) successfully described the data. The gene stays "on" for typically 6 min, during which it produces approximately two transcripts. In contrast, the "off" state lasts much longer (~37 min), which results in a burst-like transcription behavior, even in full-induction conditions.

Bursts of transcription have also been observed on *dscA*, an endogenous developmental gene in the social amoeba *Dictyostelium discoideum* (20). Although the authors could not detect single mRNA particles, they could resolve the sites of transcription because of the high fluorescence accumulated by the multiple nascent mRNAs. As differentiation occurred, they could observe the *dscA* gene switch between the "on" and "off" states, which displayed similar lifetimes (5.2 and 5.8 min, respectively), in contrast with the *E. coli* result. Over the course of development, variation was only observed in the fraction of the population expressing the gene, but the lifetime of the "on" and "off" states remained constant. In addition, the authors observed transcriptional memory, as a gene was more likely to enter the "on" state if it had been transcribing before than when it was undergoing de novo transcription.

In spite of their quantitative differences, both studies could be modeled the same way, using a simple two-state system. The nature of the event(s) that dictates the transition between the "on" and "off" states has yet to be discovered, but it could consist of DNA conformational change and/or chromatin remodeling, binding (or release) of an activator (or repressor), or transcription pausing/ reinitiation.

These studies demonstrate the potential of single-molecule techniques in studying transcription in vivo and open avenues for future research. Upcoming directions could involve expanding these observations to different systems and/or trying to combine imaging of different factors at a given locus.

ANALYSIS OF KINETICS

Mobility of Transcription Factors

Transcription factor mobility represents the process of a genome-wide search for specific target sites. Since the discovery of the GFP, more precise observations of nuclear protein mobility have been enabled. In vivo techniques identify populations of transcription factors on a real-time timescale, breaking down many assumptions previously held about this topic.

Dynamics

The use of fluorescent proteins made it possible to conduct experiments on the transcription factors of yeast (47), bacteria (26), and mammalian cells (54), enlightening many aspects of the dynamic behavior of transcription factors, from Brownian motion to anomalous diffusion to cyclic binding (dynamic equilibria) at binding sites to dynamic complex formation. Transcription factors are generally impaired in their diffusion throughout the nucleus by unspecific interactions with other nuclear components (64). Furthermore, a tagged nuclear protein might exhibit more than one apparent diffusion constant due to complex formation. Most models of FRAP have been applied to homogeneously and globally distributed binding sites, easily approximated by diffusion.

The pioneering work on the construction of localized cluster binding sites in the genome (54, 89) made it possible to address specific binding of transcription factors in the nucleus. Sprague et al. (80) used such a system to prove that recoveries resulting from bleaching of the tandem array area could not fit a model accounting for only diffusion. The new model involves "on" and "off" rates of the transcription factor's binding to the promoter array and provides information on the binding dynamics of the system. A similar construct has been used by the Singer laboratory (21) to analyze transcriptional mechanisms by bleaching the Pol II and nascent RNAs on a tandem array. This particular system was demonstrated to be kinetically independent of the availability of the freely diffusing components, therefore making it possible to disregard the diffusing component (80) and to use first-order differential equations to model the reactions.

The work of Natoli's laboratory (9) on NF-κB promoter binding microdynamics showed that stable bindings were actually states of dynamic equilibrium between promoter-bound and nucleoplasmic dimers (**Figure 3e**). In a subsequent study, Karpova et al. (47) showed that the yeast transcription factor Ace1p fit an accessibility model in which the slow cycle of binding reflects the number of accessible binding sites at promoters and each accessible site can be bound by fast-cycling molecules.

Complex formation of transcription factors on their promoters is likewise of a highly dynamic nature: The factor and its partners do not associate with and are not released from target promoters as a single and stable complex (9, 36, 71). Bosisio et al. (9) specified that NF-κB residence time on specific sites defined a stochastic window during which general transcription factors and possibly additional activators must collide with the same regulatory region for transcription to occur. Furthermore, Gorski et al. (36) successfully proved the role of complex formation in regulating transcription. Remodeling factors also play a critical role in the regulation of gene expression and in governing the dynamics of transcription factors (46, 47).

Modeling

FRAP analysis quantifies and sets a kinetic model characterized by parameters of diffusion coefficients, chemical rates, and residence times translatable into differential equations. Taking into account only diffusion, a convenient way of displaying fluorescence recovery curves is the form defining the mobile fraction:

$$f_K(t) = [F_K(t) - F_K(0)]/[(F_K(\infty) - F_K(0)].$$

Under assumptions of a Gaussian intensity profile laser, this gives the closed form solution of the normalized fluorescence recovery:

$$F_K(t) = (q\, P_0 C_0/A)v K^{-v}\Gamma(v)P(2K|v),$$

where P_0 designates the total laser power, K the bleaching parameter, C_0 the initial fluorophore concentration, and A the attenuation factor of beam during the observation of the FRAP recovery. q is the product of all quantum efficiencies of light absorption, emission, and detection. Furthermore, parameter v and the characteristic diffusion time τ_D are given by $v = (1 + 2t/\tau_D)^{-1}$ and $\tau_D = w^2/4D$, respectively, where t is time and w is the radius of the laser beam at e^{-2} intensity/height. $\Gamma(v)$ is given by the gamma function (4). $P(2K|v)$ is the probability distribution tabulated in Reference 1.

However, this might be only an approximation when it comes to systems in which specific binding cannot be ignored. A complete solution to FRAP reaction-diffusion equations has been proposed (81) in which various special cases of FRAP with binding (diffusion/binding dominant) can be covered by a set of differential equations including the diffusion terms given above, as well as the chemical kinetics of binding.

Whether a dynamic system is diffusion limited depends on the magnitude of two parameters: diffusion time and association rate. The relative magnitude of these two parameters reflects potential interplay between diffusion and binding and thus determines whether a recovery is diffusion coupled or uncoupled (79). A simple method of testing whether a system is diffusion limiting is to vary the spot size of the bleach: If the recovery is dependent on the spot size, the system is diffusion limiting and must be included in the analysis (83).

Apart from diffusion modeling with differential equations, Rino et al. (71) successfully portrayed modeling of splicing protein kinetics in the nucleus with a method involving k_{on} and k_{off} rates in a Monte Carlo simulation. Other types of modeling might explain dynamic behaviors (for more details, see Reference 65).

CONCLUSION

Controlled manipulation of the biological system by using drugs has proved to be a useful tool to perturb biological mechanisms in order to obtain deeper understanding of the mechanisms involved (32, 21). Other biological manipulation stems from the construction of binding defective mutants of the transcription factors under analysis (46, 82). Photoactiveable and photoswitchable fluorophores have a particular advantage, however, in that they can be used with single-cell, single-molecule sensitivity, and they produce photons, which are easily measured, quantified, and converted to the dynamic behavior of transcription (92). Further, hyper-resolution techniques provide a tool to produce single-molecule dynamic measurements at the subdiffraction level. Using these approaches, it will be possible to answer crucial questions about how transcription factor dynamics regulate gene expression, how transcription factors sort the right genes, and how they search for their targets. Short residence times, stochastic formation of complexes, anomalous diffusion with continuous assembly, and disassembly of the transcription factors is only the beginning of a complex story about the dynamic behavior of transcription factors.

THE FUTURE

The current conclusions regarding transcription dynamics are based mainly on synthetic genes and cell lines that give us some insight into the processes involved in gene expression. However, the next important step is to apply the technology to minimally perturbed systems, endogenous genes, and primary cells or tissues. To achieve this we will need more sensitive systems capable of processing weak signals. Additionally, high-speed imaging will be required to separate transient and rapid events from the diffusional rates occurring in the background. Brighter fluorochromes with lower photobleaching or novel labeling systems capable of multiplexing will also be required. Finally, the digitization of the data will allow for the type of mining that is common with microarray databases, but required now are algorithms capable of extracting data from large image sets, particularly those that contain 4D information (a time series in three dimensions). The field therefore will assemble expertise from engineers, computer scientists, chemists, physicists, and biophysicists. As these explorations evolve, they will lead to leaps in understanding the biological basis of gene expression.

SUMMARY POINTS

1. MPM imaging of *Drosophila* polytene loci provides an experimental system of analyzing transcription factor dynamics and function at specific native gene loci in vivo and in real time.

2. NR-induced/regulated transcription is an excellent model system for observing the cooperative interactions among enhancers, repressors, transcription factors, and basal transcription components.

3. The view that has emerged from live cell studies, utilizing fluorescence techniques such as FRAP, FRET, and FCS, is that these NR complexes are highly dynamic: Individual species have dwell times on the order of seconds to minutes.

4. Single-molecule studies of a single gene, if technically challenging, offer the most detailed framework on which to test the models of transcription.

5. The cell may organize transcription machinery to efficiently produce mature transcripts.

DISCLOSURE STATEMENT

The authors are not aware of any biases that might be perceived as affecting the objectivity of this review.

ACKNOWLEDGMENTS

We would like to thank Gerry Rubin and Kevin Moses for the support of this Consortium and John Lis and Watt Webb for reading portions of this manuscript and giving permission for use of published figures. Support for personnel is from the HHMI, the NIH, and the CNRS. The authors thank Shailesh Shenoy for his help in preparing the manuscript.

LITERATURE CITED

1. Abramowitz M, Stegun IA. 1965. *Handbook of Mathematical Functions*. New York: Dover

2. Abruzzi KC, Belostotsky DA, Chekanova JA, Dower K, Rosbash M. 2006. 3′-end formation signals modulate the association of genes with the nuclear periphery as well as mRNP dot formation. *EMBO J.* 25:4253–62

3. Akhtar A, Gasser SM. 2007. The nuclear envelope and transcriptional control. *Nat. Rev. Genet.* 8:507–17

4. Axelrod D, Koppel DE, Schlessinger J, Elson E, Webb WW. 1976. Mobility measurement by analysis of fluorescence photobleaching recovery kinetics. *Biophys. J.* 16:1055–69

5. Beach DL, Salmon ED, Bloom K. 1999. Localization and anchoring of mRNA in budding yeast. *Curr. Biol.* 9:569–78

6. Becker M, Baumann C, John S, Walker DA, Vigneron M, et al. 2002. Dynamic behavior of transcription factors on a natural promoter in living cells. *EMBO Rep.* 3:1188–94

7. Bertrand E, Chartrand P, Schaefer M, Shenoy SM, Singer RH, Long RM. 1998. Localization of ASH1 mRNA particles in living yeast. *Mol. Cell* 2:437–45

8. Boireau S, Maiuri P, Basyuk E, de la Mata M, Knezevich A, et al. 2007. The transcriptional cycle of HIV-1 in real-time and live cells. *J. Cell Biol.* 179: 291–304

9. Bosisio D, Marazzi I, Agresti A, Shimizu N, Bianchi ME, Natoli G. 2006. A hyperdynamic equilibrium between promoter-bound and nucleoplasmic dimers controls NF-kappaB-dependent gene activity. *EMBO J.* 25:798–810

10. Bres V, Yoh SM, Jones KA. 2008. The multi-tasking P-TEFb complex. *Curr. Opin. Cell Biol.* 20:334–40

11. Brickner DG, Cajigas I, Fondufe-Mittendorf Y, Ahmed S, Lee P-C, et al. 2007. H2A.Z-mediated localization of genes at the nuclear periphery confers epigenetic memory of previous transcriptional state. *PLoS Biol.* 5:e81

12. Brickner JH, Walter P. 2004. Gene recruitment of the activated INO1 locus to the nuclear membrane. *PLoS Biol.* 2:e342

13. Bridges CB. 1935. Salivary chromosome maps. *J. Hered.* 26:60–64

14. Cabal GG, Genovesio A, Rodriguez-Navarro S, Zimmer C, Gadal O, et al. 2006. SAGA interacting factors confine subdiffusion of transcribed genes to the nuclear envelope. *Nature* 441:770–73

15. Cai L, Friedman N, Xie XS. 2006. Stochastic protein expression in individual cells at the single molecule level. *Nature* 440:358–62

16. Carter DR, Eskiw C, Cook PR. 2008. Transcription factories. *Biochem. Soc. Trans.* 36:585–89

17. Casolari JM, Brown CR, Komili S, West J, Hieronymus H, Silver PA. 2004. Genome-wide localization of the nuclear transport machinery couples transcriptional status and nuclear organization. *Cell* 117:427–39

18. **Choi PJ, Cai L, Frieda K, Xie XS. 2008. A stochastic single-molecule event triggers phenotype switching of a bacterial cell. *Science* 322:442–46**

19. Chuang C-H, Carpenter AE, Fuchsova B, Johnson T, de Lanerolle P, Belmont AS. 2006. Long-range directional movement of an interphase chromosome site. *Curr. Biol.* 16:825–31

20. **Chubb JR, Trcek T, Shenoy SM, Singer RH. 2006. Transcriptional pulsing of a developmental gene. *Curr. Biol.* 16:1018–25**

21. **Darzacq X, Shav-Tal Y, de Turris V, Brody Y, Shenoy SM, et al. 2007. In vivo dynamics of RNA polymerase II transcription. *Nat. Struct. Mol. Biol.* 14:796–806**

22. Darzacq X, Singer RH. 2008. The dynamic range of transcription. *Mol. Cell* 30:545–46

23. Dieppois G, Iglesias N, Stutz F. 2006. Cotranscriptional recruitment to the mRNA export receptor Mex67p contributes to nuclear pore anchoring of activated genes. *Mol. Cell. Biol.* 26:7858–70

24. **Dundr M, Hoffmann-Rohrer U, Hu Q, Grummt I, Rothblum LI, et al. 2002. A kinetic framework for a mammalian RNA polymerase in vivo. *Science* 298:1623–26**

25. Dundr M, Ospina JK, Sung M-H, John S, Upender M, et al. 2007. Actin-dependent intranuclear repositioning of an active gene locus in vivo. *J. Cell Biol.* 179:1095–103

26. **Elf J, Li GW, Xie XS. 2007. Probing transcription factor dynamics at the single-molecule level in a living cell. *Science* 316:1191–94**

27. Eskiw CH, Rapp A, Carter DR, Cook PR. 2008. RNA polymerase II activity is located on the surface of protein-rich transcription factories. *J. Cell Sci.* 121:1999–2007

28. Farla P, Hersmus R, Geverts B, Mari PO, Nigg AL, et al. 2004. The androgen receptor ligand-binding domain stabilizes DNA binding in living cells. *J. Struct. Biol.* 147:50–61

29. Feige JN, Gelman L, Tudor C, Engelborghs Y, Wahli W, Desvergne B. 2005. Fluorescence imaging reveals the nuclear behavior of peroxisome proliferator-activated receptor/retinoid X receptor heterodimers in the absence and presence of ligand. *J. Biol. Chem.* 280:17880–90

30. Finlan LE, Sproul D, Thomson I, Boyle S, Kerr E, et al. 2008. Recruitment to the nuclear periphery can alter expression of genes in human cells. *PLoS Genet.* 4:e1000039

31. Forrest KM, Gavis ER. 2003. Live imaging of endogenous RNA reveals a diffusion and entrapment mechanism for nanos mRNA localization in *Drosophila*. *Curr. Biol.* 13:1159–68

18. The number of proteins expressed from a gene under the control of LacO is linked to dissociation of the repressor.

20. This work demonstrates the stochastic behavior of an endogenous, developmentally regulated gene.

21. This work demonstrates the power of combining live cell imaging of transcription with mathematical modeling to determine rate constants for each of the various components of the process.

24. This is the first work to measure polymerase kinetics.

26. Specific binding, diffusion tracking, and colocalization of the transcription factor to its DNA binding domains are measured.

32. Fromaget M, Cook PR. 2007. Photobleaching reveals complex effects of inhibitors on transcribing RNA polymerase II in living cells. *Exp. Cell Res.* 313:3026–33

33. Fusco D, Accornero N, Lavoie B, Shenoy SM, Blanchard JM, et al. 2003. Single mRNA molecules demonstrate probabilistic movement in living mammalian cells. *Curr. Biol.* 13:161–67

34. Golding I, Cox EC. 2004. RNA dynamics in live *Escherichia coli* cells. *Proc. Natl. Acad. Sci. USA* 101:11310–15

34. A real-time quantitative study of the expression of a single gene in bacteria demonstrating burst-like transcription.

36. This work demonstrates the power of combining live cell imaging of transcription with mathematical modeling to determine rate constants for each of the various components of the process.

47. This work demonstrates the rapid turnover of transcription factors at the promoter.

54. This seminal paper was the first to show the dynamic behavior of transcription factors at an active gene locus in a living cell.

35. Golding I, Paulsson J, Zawilski SM, Cox EC. 2005. Real-time kinetics of gene activity in individual bacteria. *Cell* 123:1025–36

36. Gorski SA, Snyder SK, John S, Grummt I, Misteli T. 2008. Modulation of RNA polymerase assembly dynamics in transcriptional regulation. *Mol. Cell* 30:486–97

37. Hieda M, Winstanley H, Maini P, Iborra FJ, Cook PR. 2005. Different populations of RNA polymerase II in living mammalian cells. *Chromosome Res.* 13:135–44

38. Hinojos CA, Sharp ZD, Mancini MA. 2005. Molecular dynamics and nuclear receptor function. *Trends Endocrinol. Metab.* 16:12–18

39. Iborra FJ, Pombo A, Jackson DA, Cook PR. 1996. Active RNA polymerases are localized within discrete transcription "factories" in human nuclei. *J. Cell Sci.* 109(Pt. 6):1427–36

40. Iborra FJ, Pombo A, McManus J, Jackson DA, Cook PR. 1996. The topology of transcription by immobilized polymerases. *Exp. Cell Res.* 229:167–73

41. Jackson DA, Hassan AB, Errington RJ, Cook PR. 1993. Visualization of focal sites of transcription within human nuclei. *EMBO J.* 12:1059–65

42. Jackson DA, Iborra FJ, Manders EM, Cook PR. 1998. Numbers and organization of RNA polymerases, nascent transcripts, and transcription units in HeLa nuclei. *Mol. Biol. Cell* 9:1523–36

43. Janicki SM, Tsukamoto T, Salghetti SE, Tansey WP, Sachidanandam R, et al. 2004. From silencing to gene expression: real-time analysis in single cells. *Cell* 116:683–98

44. Jeang KT, Xiao H, Rich EA. 1999. Multifaceted activities of the HIV-1 transactivator of transcription, Tat. *J. Biol. Chem.* 274:28837–40

45. Johnson TA, Elbi C, Parekh BS, Hager GL, John S. 2008. Chromatin remodeling complexes interact dynamically with a glucocorticoid receptor-regulated promoter. *Mol. Biol. Cell* 19:3308–22

46. Karpova TS, Chen TY, Sprague BL, McNally JG. 2004. Dynamic interactions of a transcription factor with DNA are accelerated by a chromatin remodeller. *EMBO Rep.* 5:1064–70

47. Karpova TS, Kim MJ, Spriet C, Nalley K, Stasevich TJ, et al. 2008. Concurrent fast and slow cycling of a transcriptional activator at an endogenous promoter. *Science* 319:466–69

48. Kimura H, Sugaya K, Cook PR. 2002. The transcription cycle of RNA polymerase II in living cells. *J. Cell Biol.* 159:777–82

49. Kino T, Liou SH, Charmandari E, Chrousos GP. 2004. Glucocorticoid receptor mutants demonstrate increased motility inside the nucleus of living cells: time of fluorescence recovery after photobleaching (FRAP) is an integrated measure of receptor function. *Mol. Med.* 10:80–88

50. Klokk TI, Kurys P, Elbi C, Nagaich AK, Hendarwanto A, et al. 2007. Ligand-specific dynamics of the androgen receptor at its response element in living cells. *Mol. Cell Biol.* 27:1823–43

51. Kumaran RI, Spector DL. 2008. A genetic locus targeted to the nuclear periphery in living cells maintains its transcriptional competence. *J. Cell Biol.* 180:51–65

52. Lanctot C, Cheutin T, Cremer M, Cavalli G, Cremer T. 2007. Dynamic genome architecture in the nuclear space: regulation of gene expression in three dimensions. *Nat. Rev. Genet.* 8:104–15

53. Maruvada P, Baumann CT, Hager GL, Yen PM. 2003. Dynamic shuttling and intranuclear mobility of nuclear hormone receptors. *J. Biol. Chem.* 278:12425–32

54. McNally JG, Muller WG, Walker D, Wolford R, Hager GL. 2000. The glucocorticoid receptor: rapid exchange with regulatory sites in living cells. *Science* 287:1262–65

55. Meijsing SH, Elbi C, Luecke HF, Hager GL, Yamamoto KR. 2007. The ligand binding domain controls glucocorticoid receptor dynamics independent of ligand release. *Mol. Cell Biol.* 27:2442–51

56. Metivier R, Reid G, Gannon F. 2006. Transcription in four dimensions: nuclear receptor-directed initiation of gene expression. *EMBO Rep.* 7:161–67

57. Molle D, Maiuri P, Boireau S, Bertrand E, Knezevich A, et al. 2007. A real-time view of the TAR:Tat:P-TEFb complex at HIV-1 transcription sites. *Retrovirology* 4:36

58. Muller WG, Rieder D, Karpova TS, John S, Trajanoski Z, McNally JG. 2007. Organization of chromatin and histone modifications at a transcription site. *J. Cell Biol.* 177:957–67

59. Muller WG, Walker D, Hager GL, McNally JG. 2001. Large-scale chromatin decondensation and recondensation regulated by transcription from a natural promoter. *J. Cell Biol.* 154:33–48

60. Nalley K, Johnston SA, Kodadek T. 2006. Proteolytic turnover of the Gal4 transcription factor is not required for function in vivo. *Nature* 442:1054–57

61. Ni Z, Saunders A, Fuda NJ, Yao J, Suarez JR, et al. 2008. P-TEFb is critical for the maturation of RNA polymerase II into productive elongation in vivo. *Mol. Cell Biol.* 28:1161–70

62. Pan D, Zhang L. 2008. Tandemly arrayed genes in vertebrate genomes. *Comp. Funct. Genomics* 2008:545269

63. Perissi V, Rosenfeld MG. 2005. Controlling nuclear receptors: the circular logic of cofactor cycles. *Nat. Rev. Mol. Cell Biol.* 6:542–54

64. Phair RD, Misteli T. 2000. High mobility of proteins in the mammalian cell nucleus. *Nature* 404:604–9

65. Phair RD, Misteli T. 2001. Kinetic modelling approaches to in vivo imaging. *Nat. Rev. Mol. Cell Biol.* 2:898–907

66. Pombo A, Cook PR. 1996. The localization of sites containing nascent RNA and splicing factors. *Exp. Cell Res.* 229:201–3

67. Pombo A, Jackson DA, Hollinshead M, Wang Z, Roeder RG, Cook PR. 1999. Regional specialization in human nuclei: visualization of discrete sites of transcription by RNA polymerase III. *EMBO J.* 18:2241–53

68. Rafalska-Metcalf IU, Janicki SM. 2007. Show and tell: visualizing gene expression in living cells. *J. Cell Sci.* 120:2301–7

69. Rayasam GV, Elbi C, Walker DA, Wolford R, Fletcher TM, et al. 2005. Ligand-specific dynamics of the progesterone receptor in living cells and during chromatin remodeling in vitro. *Mol. Cell Biol.* 25:2406–18

70. Reddy KL, Zullo JM, Bertolino E, Singh H. 2008. Transcriptional repression mediated by repositioning of genes to the nuclear lamina. *Nature* 452:243–47

71. Rino J, Carvalho T, Braga J, Desterro JM, Luhrmann R, Carmo-Fonseca M. 2007. A stochastic view of spliceosome assembly and recycling in the nucleus. *PLoS Comput. Biol.* 3:2019–31

72. Rook MS, Lu M, Kosik KS. 2000. CaMKIIalpha 3′ untranslated region-directed mRNA translocation in living neurons: visualization by GFP linkage. *J. Neurosci.* 20:6385–93

73. Schaaf MJ, Cidlowski JA. 2003. Molecular determinants of glucocorticoid receptor mobility in living cells: the importance of ligand affinity. *Mol. Cell Biol.* 23:1922–34

74. Schmid M, Arib G, Laemmli C, Nishikawa J, Durussel T, Laemmli UK. 2006. Nup-PI: the nucleopore-promoter interaction of genes in yeast. *Mol. Cell* 21:379–91

75. Schneider DA, Nomura M. 2004. RNA polymerase I remains intact without subunit exchange through multiple rounds of transcription in *Saccharomyces cerevisiae*. *Proc. Natl. Acad. Sci. USA* 101:15112–17

76. Sehgal PB, Derman E, Molloy GR, Tamm I, Darnell JE. 1976. 5,6-Dichloro-1-Beta-D-ribofuranosylbenzimidazole inhibits initiation of nuclear heterogeneous RNA chains in HeLa cells. *Science* 194:431–33

77. Shav-Tal Y, Darzacq X, Shenoy SM, Fusco D, Janicki SM, et al. 2004. Dynamics of single mRNPs in nuclei of living cells. *Science* 304:1797–800

78. Shav-Tal Y, Singer RH, Darzacq X. 2004. Imaging gene expression in single living cells. *Nat. Rev. Mol. Cell Biol.* 5:855–61

79. Sprague BL, McNally JG. 2005. FRAP analysis of binding: proper and fitting. *Trends Cell Biol.* 15:84–91

80. Sprague BL, Muller F, Pego RL, Bungay PM, Stavreva DA, McNally JG. 2006. Analysis of binding at a single spatially localized cluster of binding sites by fluorescence recovery after photobleaching. *Biophys. J.* 91:1169–91

81. Sprague BL, Pego RL, Stavreva DA, McNally JG. 2004. Analysis of binding reactions by fluorescence recovery after photobleaching. *Biophys. J.* 86:3473–95

82. Sprouse RO, Shcherbakova I, Cheng H, Jamison E, Brenowitz M, Auble DT. 2008. Function and structural organization of Mot1 bound to a natural target promoter. *J. Biol. Chem.* 283:24935–48

83. Stavreva DA, McNally JG. 2004. Fluorescence recovery after photobleaching (FRAP) methods for visualizing protein dynamics in living mammalian cell nuclei. *Methods Enzymol.* 375:443–55

84. Stavreva DA, Muller WG, Hager GL, Smith CL, McNally JG. 2004. Rapid glucocorticoid receptor exchange at a promoter is coupled to transcription and regulated by chaperones and proteasomes. *Mol. Cell Biol.* 24:2682–97

85. Stenoien DL, Nye AC, Mancini MG, Patel K, Dutertre M, et al. 2001. Ligand-mediated assembly and real-time cellular dynamics of estrogen receptor alpha-coactivator complexes in living cells. *Mol. Cell Biol.* 21:4404–12

86. Stenoien DL, Patel K, Mancini MG, Dutertre M, Smith CL, et al. 2001. FRAP reveals that mobility of oestrogen receptor-alpha is ligand- and proteasome-dependent. *Nat. Cell Biol.* 3:15–23

87. Sugaya K, Vigneron M, Cook PR. 2000. Mammalian cell lines expressing functional RNA polymerase II tagged with the green fluorescent protein. *J. Cell Sci.* 113(Pt. 15):2679–83

88. Taddei A, Van Houwe G, Hediger F, Kalck V, Cubizolles F, et al. 2006. Nuclear pore association confers optimal expression levels for an inducible yeast gene. *Nature* 441:774–78

89. Tsukamoto T, Hashiguchi N, Janicki SM, Tumbar T, Belmont AS, Spector DL. 2000. Visualization of gene activity in living cells. *Nat. Cell Biol.* 2:871–78

90. Walter J, Schermelleh L, Cremer M, Tashiro S, Cremer T. 2003. Chromosome order in HeLa cells changes during mitosis and early G1, but is stably maintained during subsequent interphase stages. *J. Cell Biol.* 160:685–97

91. Weil TT, Forrest KM, Gavis ER. 2006. Localization of bicoid mRNA in late oocytes is maintained by continual active transport. *Dev. Cell* 11:251–62

92. Xie XS, Choi PJ, Li GW, Lee NK, Lia G. 2008. Single-molecule approach to molecular biology in living bacterial cells. *Annu. Rev. Biophys.* 37:417–44

93. Xu M, Cook PR. 2008. Similar active genes cluster in specialized transcription factories. *J. Cell Biol.* 181:615–23

94. Yao J, Ardehali MB, Fecko CJ, Webb WW, Lis JT. 2007. Intranuclear distribution and local dynamics of RNA polymerase II during transcription activation. *Mol. Cell* 28:978–90

95. This work first describes the method of MPM imaging *Drosophila* polytene chromosomes to study transcription factor kinetics at endogenous gene loci.

95. **Yao J, Munson KM, Webb WW, Lis JT. 2006. Dynamics of heat shock factor association with native gene loci in living cells. *Nature* 442:1050–53**

96. Yao J, Zobeck KL, Lis JT, Webb WW. 2008. Imaging transcription dynamics at endogenous genes in living *Drosophila* tissues. *Methods* 45:233–41

97. Yu J, Xiao J, Ren X, Lao K, Xie XS. 2006. Probing gene expression in live cells, one protein molecule at a time. *Science* 311:1600–3

98. Zimyanin VL, Belaya K, Pecreaux J, Gilchrist MJ, Clark A, et al. 2008. In vivo imaging of oskar mRNA transport reveals the mechanism of posterior localization. *Cell* 134:843–53

99. Zipfel WR, Williams RM, Webb WW. 2003. Nonlinear magic: multiphoton microscopy in the biosciences. *Nat. Biotechnol.* 21:1369–77

A Complex Assembly Landscape for the 30S Ribosomal Subunit

Michael T. Sykes and James R. Williamson

Departments of Molecular Biology and Chemistry and The Skaggs Institute for Chemical Biology, The Scripps Research Institute, La Jolla, CA 92037; email: sykes@scripps.edu, jrwill@scripps.edu

Annu. Rev. Biophys. 2009. 38:197–215

First published online as a Review in Advance on February 2, 2009

The *Annual Review of Biophysics* is online at biophys.annualreviews.org

This article's doi:
10.1146/annurev.biophys.050708.133615

Key Words

30S subunit, ribosome assembly, RNA folding, protein binding, cooperativity

Abstract

The ribosome is a complex macromolecular machine responsible for protein synthesis in the cell. It consists of two subunits, each of which contains both RNA and protein components. Ribosome assembly is subject to intricate regulatory control and is aided by a multitude of assembly factors in vivo, but can also be carried out in vitro. The details of the assembly process remain unknown even in the face of atomic structures of the entire ribosome and after more than three decades of research. Some of the earliest research on ribosome assembly produced the Nomura assembly map of the small subunit, revealing a hierarchy of protein binding dependencies for the 20 proteins involved and suggesting the possibility of a single intermediate. Recent work using a combination of RNA footprinting and pulse-chase quantitative mass spectrometry paints a picture of small subunit assembly as a dynamic and varied landscape, with sequential and hierarchical RNA folding and protein binding events finally converging on complete subunits. Proteins generally lock tightly into place in a 5′ to 3′ direction along the ribosomal RNA, stabilizing transient RNA conformations, while RNA folding and the early stages of protein binding are initiated from multiple locations along the length of the RNA.

Contents

INTRODUCTION

The ribosome is an intricate molecular machine that is responsible for the synthesis of new proteins in the cell. The bacterial 70S ribosome consists of two parts, a small (30S) subunit composed of a single strand of RNA and 21 proteins, and a large (50S) subunit composed of two strands of RNA and 34 proteins. Assembling these components into intact subunits is a carefully controlled and choreographed process that is both demanding on the cell and efficient by necessity. The high-resolution atomic structures of both the 30S (63) and 50S (6) subunits were solved in 2000, followed by atomic resolution structures of the 70S ribosome (40, 49, 50), offering a detailed look at the end product of ribosome assembly. While the structures are useful for interpreting the extensive biochemical data concerning ribosome assembly, little information about the assembly process itself can be gained from their inspection. Ribosome assembly has been the subject of scrutiny for

several decades by a variety of methods including RNA chemical footprinting, equilibrium binding assays, mass spectrometry, and simulation, primarily in vitro. Although the story of ribosome assembly is far from complete, great strides have been made in advancing the knowledge of how these individual protein and RNA molecules come together to form complete and functional ribosomes.

RIBOSOME BIOGENESIS IN BACTERIA

Synthesis and assembly of an intact ribosome in bacterial cells occurs in a series of steps, including rRNA transcription, ribosomal protein synthesis, RNA processing, RNA folding, protein binding, and protein and RNA modification. The nature and regulation of many of these steps have been extensively reviewed elsewhere and are presented only briefly here with appropriate references. Ribosomal RNA (rRNA) is transcribed as a single transcript including both the 16S rRNA for the small subunit and the 23S and 5S rRNAs for the large subunit in what is the rate-limiting step for ribosome synthesis in vivo (39). The transcription of the rRNA operon is tightly coupled to growth rate and subject to regulation by multiple factors (27, 39). At the same time, ribosomal proteins are synthesized, some of which regulate not only their own translation but that of proteins sharing the same operon (64). Both ribosomal proteins (5, 27, 41) and rRNA (11, 27, 62) are chemically modified, and although not all such modifications appear to be essential, several modifications of the 23S rRNA in particular are necessary to avoid assembly defects (27). The single rRNA primary transcript is processed into mature 5S, 16S, and 23S rRNAs by a series of RNAses (27, 52). Finally, the rRNAs and proteins fold and assemble in a hierarchical manner into complete subunits, in a process governed by a set of over 20 assembly factors (27, 62). These essential steps are not strictly sequential, and rRNA folding, protein binding, and rRNA processing occur both in parallel and cotranscriptionally (12, 30).

rRNA: ribosomal RNA

16S rRNA: the RNA component of the 30S subunit

RECONSTITUTION OF RIBOSOMAL SUBUNITS

With the array of assembly factors present in vivo it is incredible that an in vitro reconstitution of intact ribosomal subunits using only the rRNA and proteins is even possible. Nevertheless, in vitro reconstitution of 30S subunits occurs with both reasonable speed and efficiency in the total absence of assembly factors, albeit significantly more slowly than in vivo. Reconstitution of the larger 50S subunits is comparatively slow and inefficient (17, 33) and as a result has not been studied as extensively. This review primarily focuses on the extensive body of work on 30S ribosome assembly in vitro.

In the 1970s, seminal work by Nomura and colleagues on 30S subunit assembly resulted in a complete assembly map for the 30S subunit (21, 31). Reconstitution of 30S subunits using different subsets of the total complement of proteins allowed the hierarchy of binding dependencies to be determined, as represented in what is now known as the Nomura assembly map. Protein S1 binds relatively weakly to the 30S subunit and exchanges rapidly, so it is typically excluded (61). The 30S assembly map shown in **Figure 1a** is modified from the original map by Nomura to reflect the domain structure of the 30S subunit and to indicate the updated location of S13 in the S7 assembly branch, recently determined by RNA footprinting (18). From the assembly map the 30S ribosomal proteins can be classified according to their physical location and order in the binding hierarchy. Proteins bind principally to one of the 5′, central, or 3′ domains of the rRNA and are designated as primary (binding directly to the rRNA), secondary (binding is dependent on primary binding proteins), or tertiary, (binding is dependent on secondary binding proteins). To this day, the Nomura assembly map offers one of the most useful overviews of 30S subunit assembly and is widely reproduced. Herold & Nierhaus developed a similar assembly map for the 50S subunit (22),

30S subunit: the small subunit of the ribosome, consisting of a single RNA and 21 proteins

RNA footprinting: a method used to probe the folding of an RNA by determining the accessibility of a nucleotide to hydroxyl radical cleavage or chemical modification

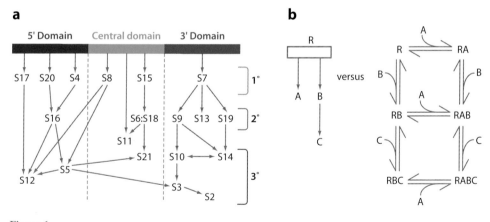

Figure 1

The Nomura assembly map for the 30S subunit. (*a*) A modified version of the traditional view of the Nomura assembly map that has been reorganized according to domains, with arrows indicating the facilitating effect of binding between proteins. Proteins can be categorized as 5′, central, or 3′ domain proteins, and either primary binding proteins (1°), secondary binding proteins (2°), or tertiary binding proteins (3°), the last two of which depend upon proteins from the previous category for binding to the 16S rRNA. S11 is categorized in a species-dependent manner, acting as a tertiary binding protein in *Escherichia coli*, but as a primary binding protein in *Aquifex aeolicus* (45). (*b*) A comparison between a hypothetical assembly map (*left*) and an equivalent assembly mechanism (*right*) of a quaternary RABC complex between RNA (R) and three proteins (A, B, and C). Each reaction step needs to be characterized by thermodynamic equilibrium constants and both forward and reverse rate constants.

but assembly of the large ribosomal subunit is not so clearly organized by structural domains and has many more proteins that interact in a more complex binding hierarchy.

While the Nomura map has constituted the overview of the 30S subunit assembly process for over 30 years, there are a number of shortcomings with this representation. First, the Nomura map is not a mechanism, but rather a diagram containing logical information about dependencies during assembly. An example of a true mechanism compared with an equivalent assembly map is shown in **Figure 1***b*. The assembly map does not contain information about the specific complexes that are present or the equilibria present. Second, all of the information in the Nomura map is thermodynamic, and it offers no information about the kinetics of protein binding or the specific energy changes that occur. Third there is no information about the folding pathway of the 16S rRNA that constitutes two-thirds of the mass of the 30S subunit. Subsequent work has attempted to shed light on these and other questions using a variety of methods to probe 30S assembly in vitro. The balance of this review focuses on these experiments.

A NEW REPRESENTATION FOR 16S rRNA

The 16S rRNA is usually represented as the phylogenetically derived secondary structure (10), or as a schematic ribbon and stick representation of the 3D structure, but neither is ideal. The former lacks information about the relative positions of the helices in space and the latter does not convey detailed nucleotide-level information well when rendered in 2D. The secondary structure can be annotated with interaction networks that show base-base connections through space (29), but the result is unwieldy for RNAs as large as the 16S. A hybrid representation has been developed, which displays the RNA helices in 2D but arranged according to their positions in the 3D structure. Axes for 51 helices in the 16S rRNA were defined from their 3D coordinates, all-versus-all distances were calculated between the ends of these axes, and the program NEATO (16) was used to generate an energy-minimized 2D projection. The resulting schematic is similar to the canonical view used to display the 30S subunit in 3D. A comparison of a phylogenetic secondary structure, the hybrid representation, and a cartoon rendering of the 30S subunit is provided in **Figure 2**. Each illustrates the domain structure of the 30S subunit. The hybrid representation contains less detailed information than the secondary structure but manages to capture the overall structure and shape of the 30S subunit in a much simpler way than the 3D image. Depth information is preserved by shading the helices in accordance with their depth in the 3D structure, with darker helices in back and lighter

---

Figure 2

A comparison of different representations of the 30S subunit. (*Left*) A traditional secondary structure diagram of the 16S rRNA. (*Bottom, center*) A 3D image of the 30S subunit including proteins. (*Right*) A 2D projection of the 16S rRNA helices. Interhelical distances are calculated from the coordinates of the helix axes in the 3D structure, and these distances are optimized for projection into 2D by the program NEATO (16), such that their layout is faithful to the 3D structure. Helices are shown as cylinders, capped by a semicircle when the two helical strands are contiguous. A black dot indicates the first nucleotide of the helix, and gray lines indicate connecting strands between helices. Helices are shaded in the hybrid representation according to their position along the axis normal to the page. Darker colors are farther away, while lighter colors are closer. These positions correspond to the actual positions in the 3D structure and are also used to order the rendering of helices in the cases in which overlap occurs. Proteins in the 3D image are shown in gray. Both the secondary structure and the hybrid representation have helices numbered according to Brimacombe numbers as well as their first nucleotide. The secondary structure is the simplest representation but fails to capture any information about the actual shape of the 30S subunit, and annotation with the tertiary RNA contacts results in a congested diagram. The hybrid representation blends the simplicity of the secondary structure with 3D information and captures the overall shape of the subunit. The secondary structure is based on one at the Comparative RNA Web Site and Project (10), and the 3D image was generated with PyMol (13).

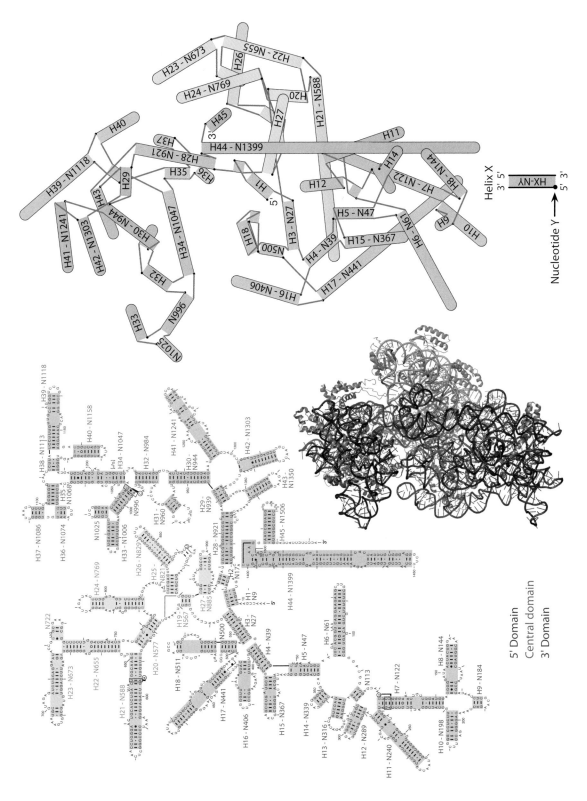

Helix X

$\frac{3' \quad 5'}{HX-NY}$ 5' 3'

Nucleotide Y →

5' Domain
Central domain
3' Domain

helices in front. Templates for this representation both with and without helix numbering are available as supplementary material (**Supplemental Figure 1**; follow the **Supplemental Material link** from the Annual Reviews home page at **http://www.annualreviews.org**).

BINDING OF PRIMARY PROTEINS

Six of the ribosomal proteins of the 30S subunit (S4, S7, S8, S15, S17, and S20) are primary binding proteins, capable of binding directly to the 16S rRNA. The RNA contacts made by these and the rest of the ribosomal proteins are shown in **Figure 3a**, overlaid on a hybrid schematic representation of the 16S rRNA. A new representation of the assembly map, where the proteins are overlaid on a hybrid schematic representation of the 16S rRNA according to their approximate positions in the 30S subunit, is shown in **Figure 3b**.

The low-temperature binary complex of the 5′ domain protein S4 and the 16S rRNA undergoes a temperature-dependent conformational change upon heating, as shown by chemical footprinting of the RNA (43). After the conformational change additional nucleotides are protected by S4, indicating a possible strengthening of the association by broadening the scope of the protein-RNA contacts. These additional contacts are implicated in the binding of tertiary protein S12 and secondary protein S16 to the RNA, both of which depend on S4 according to the Nomura map (**Figures 1a** and **3b**). A binary complex of S4 and the 16S rRNA in which the two components have been heated separately does not display the same pattern of protection, suggesting that the protein confers the conformational change directly to the RNA, allowing subsequent proteins to bind and ensuring the hierarchy of assembly. Subsequent studies of the remaining five primary binding proteins indicated a range of behavior (15). The other two 5′ domain primary proteins S17 and S20 bound to the 16S rRNA in a temperature-independent manner, while the central domain

primary proteins S8 and S15 conferred only small differences to the chemical footprint of the binary complex upon heating. The 3′ domain primary protein S7 behaved much in the same manner as S4, with significant changes to the chemical footprint of the binary complex with RNA upon heating. Taken together, these results suggest that one of the roles of the primary proteins is to promote certain conformations of the RNA and allow the binding of subsequent proteins.

Site-directed hydroxyl radical footprinting studies on the environment surrounding S15 indicated that although S8 may not mediate secondary protein binding, it does influence the organization of the RNA surrounding S15 (25), and that even proteins located a significant distance away from S15 were able to affect its environment (26). Similar studies on the environment of S20 revealed that its contacts with the 5′ domain RNA were formed early in the assembly process, whereas its contacts with the 3′ minor domain (helix H44–N1399) were not formed until later in the process, consistent with a 5′ to 3′ directionality for assembly.

INDEPENDENTLY FOLDING DOMAINS

In addition to reconstitutions of the entire 30S subunit, individual domains can be assembled independently using domain fragments of 16S rRNA. Three different studies have demonstrated the assembly of the 5′ domain rRNA with proteins S4, S16, S17, and S20 (60); the central domain rRNA of *Thermus thermophilus* with proteins S6, S8, S11, S15, and S18 (4); and the 3′ domain rRNA with proteins S2, S3, S7, S9, S10, S13, S14, and S19 (48). In each case, the assembled domains were similar in conformation to their counterparts in intact 30S subunits. However, the tertiary binding proteins in the 5′ and central domain reconstitutions predicted by the Nomura map (**Figure 1**) did not bind stably to the isolated domains, indicating dependencies on contacts to RNA or proteins bound in other domains. The dependence

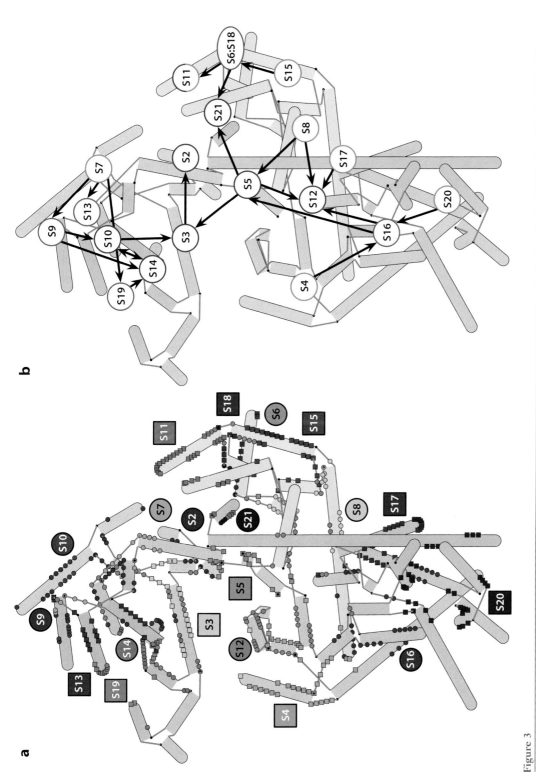

Figure 3

Ribosomal protein-RNA contacts. (*a*) These contacts are mapped onto a hybrid 2D representation of the 16S rRNA. A contact is annotated on a residue basis whenever any non-hydrogen atoms from a nucleotide and an amino acid residue are within 4 Å of each other. The cases in which a single nucleotide contacts multiple proteins are indicated in gray. Protein labels are placed near the primary sites of contact. (*b*) The Nomura map overlaid on a hybrid schematic representation of the 16S rRNA. Labels for the proteins are located according to their approximate position in the 3D structure of the 30S subunit. The primary binding proteins appear to bind in the periphery of the 30S subunit, and the tertiary proteins cluster around the cleft containing the decoding site.

of the 3' protein S3 on the 5' protein S5 in the Nomura map is not absolute, because S3 is incorporated into the 3' domain reconstitution. The 5' domain rRNA is capable of folding into a native-like conformation in the presence of magnesium ions, but in the absence of ribosomal proteins (2). Tertiary contacts virtually identical to those found in intact 30S subunits were observed, suggesting that RNA folding forms the basis for protein binding during 5' domain assembly.

CENTRAL DOMAIN ASSEMBLY

Perhaps the best understood region of the 30S subunit, in terms of assembly kinetics and thermodynamics, is the central domain. In the Nomura map proteins S15 and S8 are the primary binding proteins, and although S8 influences the RNA structure of the central domain (25), it is protein S15 that initiates a cascade of central domain protein binding by S6, S18, S11, and S21. An extensive series of biophysical studies has been carried out using fragments of the central domain RNA and individually purified proteins to better understand the molecular basis for the observed hierarchy.

The minimal binding site for the protein S15 was localized to a three-helix junction region in the central domain, helices H20, H21, and H22 (N577, N588, and N655 based on their starting nucleotide). The minimal site contains all the nucleotides implicated in S15 binding as determined by chemical probes and hydroxyl radical footprinting (44, 53). S15 binds to the minimal three-helix junction with a K_d value of ~35 nM using components from *Bacillus stearothermophilus* (7, 8). Binding of S15 induces a conformational change in the three-helix junction, suggested by an accelerated mobility of the RNA-protein complex in a nondenaturing polyacrylamide gel (7). Quantitative hydrodynamic experiments using transient electric birefringence demonstrated coaxial stacking of helices H21 and H22, while helices H20 and H22 formed an acute angle upon S15 binding (34). This conformational change implied that one of the functions of

S15 binding was to organize a critical region in the central domain to initiate assembly.

A series of deletion RNA constructs for the central domain was prepared to identify the minimal binding region for the remaining central domain proteins. Surprisingly, almost half of the central domain can be deleted while still retaining binding of proteins S15, S6, S18, and S11. A minimal fragment composed of two adjacent three-helix junctions binds to S15, S6, and S18. The X-ray crystal structure of the quaternary complex was solved, revealing a structure essentially identical to the corresponding region in the complete 30S subunit (3, 63).

In the structure of the S15:S6:S18 RNA complex, a number of important conclusions can be drawn that provide a molecular picture of specific features of the Nomura assembly map. Proteins S6 and S18 form an intimate heterodimer, which explains their completely cooperative association to 16S rRNA. There are no protein-protein interactions observed between S15 and either of the downstream binding proteins, indicating that the thermodynamic effect on binding observed in the Nomura map must be mediated by stabilization of the RNA tertiary structure by binding to protein S15. A surprising interhelix base pair was revealed in the structure between nucleotides A665 and G724. This tertiary base pair is apparently unstable or not formed in the absence of S15 binding, and it is required for stable binding of the S6:S18 heterodimer.

From these studies, a generalized model for ribosome assembly can be proposed. The RNA tertiary structure in 16S rRNA is not stable in the absence of supporting protein contacts. Transient RNA folding can create the binding site for a primary binding protein, and the local RNA tertiary structure is stabilized by protein binding, consolidating the RNA folding gains. Binding of the primary binding protein sets up the next RNA conformational change, creating the binding site for a secondary binding protein. Thus, the assembly proceeds by an alternating series of RNA conformational changes and protein binding events, sequentially stabilizing the final RNA tertiary structure.

CHEMICAL PROBING OF RIBOSOMAL INTERMEDIATES

When 30S particles are reconstituted at low temperatures ($\sim$0°C), a 21S reconstitution intermediate (RI) is obtained, rather than complete 30S subunits (56). RI lacks the tertiary binding proteins in the central and 5′ domains (S2, S3, S10, S14, and S21) and is not competent to form a complete small subunit. Incubation of RI at high temperatures ($\sim$40°C) produces a second intermediate, RI*, that is characterized by a different sedimentation coefficient (26S). Conversion of RI to RI* is believed to occur by a large conformational rearrangement of the rRNA, because of the observed difference in sedimentation between the two particles, which have the same rRNA and protein composition. Once RI* has been formed, it is competent to bind the remaining five ribosomal proteins to make complete 30S subunits. From these intermediates a simplified assembly framework has been suggested that is composed of three steps: (a) 16S rRNA and 15 ribosomal proteins form RI, (b) conformational change where RI becomes RI*, and (c) RI* and five ribosomal proteins form the 30S subunit.

RNA footprinting can be used to quantify the accessibility of nucleotides to hydroxyl radical cleavage or chemical modification. An early series of studies described the results of both untargeted chemical footprinting (32) and site-directed hydroxyl radical footprinting from modified ribosomal proteins (19, 20, 44). More recent work has focused on assembly intermediates, including two studies that explore the changes that occur in 16S rRNA over the course of the transitions to RI, RI*, and 30S (23, 24). Comparing the accessibility to chemical modification of individual nucleotides between consecutive species, the changes to the 16S that occur during an assembly step are revealed in the differences. The data for each of the three transitions are summarized in

Figure 4. Nucleotides exhibit either a protection from or an enhancement to chemical modification, resulting from either RNA folding or unfolding or protein binding. The primary differences observed for the 16S to RI transition are protections for nucleotides that are uniformly spread across the rRNA, suggestive of widespread RNA folding and protein binding and consistent with the addition of 15 ribosomal proteins (44). For the RI to RI* transition changes are primarily localized in the central and 3′ domains, and for the RI* to 30S transition most of the changes occur in the 3′ domain, suggesting a general 5′ to 3′ directionality for assembly. Of particular note is a series of enhancements in the core of the subunit that occurs in the RI to RI* transition. Because proteins do not dissociate during this step, it is clear that a major conformational rearrangement of RNA takes place at this time, resulting in the exposure of those nucleotides to chemical modification, consistent with the changes to sedimentation coefficient that have been observed.

The RI to RI* transition exhibits the hallmarks of a kinetic folding trap. RNA secondary structures are extremely stable thermodynamically, and consequently, misfolded structures with incorrect base pairing can be stable. Slow folding and kinetic traps are characteristic of the folding pathways of large RNAs, such as the *Tetrahymena* Group I intron and RNase P (38, 57, 58). Typically, lower magnesium ion concentrations, addition of denaturants, and higher temperatures accelerate the refolding reactions that must occur to escape from stable misfolded kinetic traps (14, 28, 36, 47, 51, 54). The RI to RI* transition is consistent with the observed folding of these other large RNAs, and heating is required to complete the transition. Even though kinetic traps are commonly observed in RNA folding reactions, they do not appear to be essential features of the folding mechanism, as rapidly folding sequences or mutants can usually be identified (35, 37, 46, 59). The possible significance of the RI to RI* transition is discussed in the context of these and other data below.

Reconstitution intermediate (RI): an intermediate observed during in vitro 30S subunit assembly at low temperature ($\sim$0°C)

RI*: a conformational variant of RI obtained by heating RI to high temperature ($\sim$40°C)

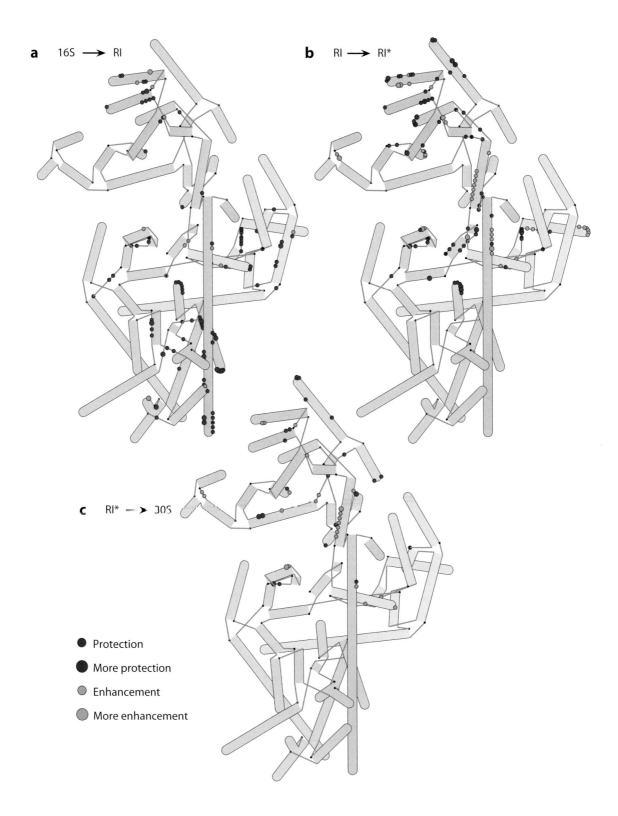

a   16S ⟶ RI

b   RI ⟶ RI*

c   RI* ⟶ 30S

● Protection
● More protection
○ Enhancement
○ More enhancement

KINETIC STUDIES OF PROTEIN BINDING

An isotope pulse-chase experiment was developed to monitor the assembly kinetics of all 30S proteins simultaneously. The 16S rRNA is first incubated with a pulse of isotopically labeled ribosomal proteins for varying lengths of time and then chased with an excess of unlabeled proteins, allowing the reaction to reach completion. The relative amount of unlabeled and labeled protein in the resulting 30S ribosomal subunits can be determined by quantitative mass spectrometry. The time course of the pulse-chase experiment can be analyzed to determine the binding rate of each protein. This technique was recently used to determine binding rates for 17 of the 20 ribosomal proteins in the small subunit (55). The binding rates of the ribosomal proteins are mapped onto the individual protein binding sites, as shown in **Figure 5**. In general, primary binding proteins bind more quickly than secondary or tertiary binding proteins, consistent with the hierarchical nature of assembly. Proteins in the 5′ domain generally bind more quickly overall than those in the central or 3′ domain, consistent with the idea of a 5′ to 3′ directionality for assembly (42). The slowly binding 5′ domain tertiary proteins S5 and S12 do not strictly follow the 5′ to 3′ directionality of assembly, but these two proteins bind at the intersection of all the domains near the decoding site of the 30S subunit.

TIME-RESOLVED HYDROXYL RADICAL FOOTPRINTING OF ASSEMBLY

The most recent work on 30S assembly uses time-resolved hydroxyl radical footprinting to obtain information on 16S rRNA folding in the presence of proteins and 30S subunit assembly (1). Synchrotron X-ray radiation can generate a sufficiently concentrated burst of hydroxyl radicals in a short pulse to give a footprint, allowing kinetic footprinting information to be gathered about the protection from hydroxyl radical cleavage of individual nucleotides, and is summarized in **Figure 6**. Most nucleotides exhibited protection in two phases, suggesting an initial fast RNA folding or protein binding event followed by a second, slower event. In these experiments, the rRNA was prefolded in the presence of magnesium ions, and much of the 16S secondary structure is likely preformed at the start of the experiment. However, initial protections may still be due to secondary structure formation as it reorganizes during the early stages of protein binding while the initial tertiary contacts form. The second, slower protections are likely to be due to the formation of RNA tertiary contacts from large conformational changes such as the RI to RI* transition, or the final stages of protein binding. Within the set of nucleotide protections due to interactions with a particular protein, two different rates of protection were often observed. This general observation suggests that protein binding itself may be a two-phase process, with an initial loose association with some fraction of the binding site followed by a second set of interactions formed upon conformational rearrangement of the protein and/or RNA. Perhaps the most striking observation in the context of previous work was the seeming lack of any 5′ to 3′ directionality in the fast folding events. Initial burst phases of protection were observed for nucleotides across the entire 16S rRNA, suggesting that RNA folding nucleates from many different sites spread across the entire molecule.

Figure 4

Hybrid representations of the 16S rRNA annotated with information about changes in accessibility to chemical modification during different steps in 30S subunit assembly (23, 24). Both decreased accessibility (protection) and increased accessibility (enhancement) are shown. The extent of protection or enhancement is indicated by the size of the circle used to annotate the nucleotides. (*a*) Changes to the 16S rRNA during the formation of a reconstitution intermediate (RI). (*b*) Changes to the 16S rRNA during the RI to RI* transition. (*c*) Changes to the 16S rRNA while a complete 30S subunit is formed from RI*. The RI to RI* transition in particular has many enhancements to modification suggestive of a large refolding of the RNA that exposes several nucleotides.

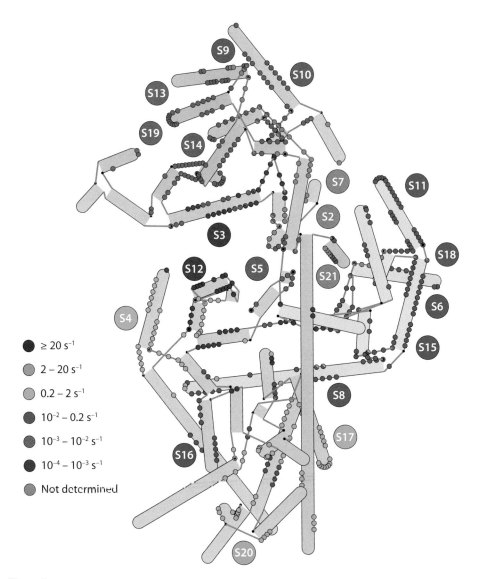

Figure 5

A hybrid representation of the 16S rRNA annotated with rate constants for protein binding as determined by pulse-chase quantitative mass spectrometry (55). Nucleotides that make protein contacts are color-coded according to the binding rate of the protein that they contact. In the case in which two proteins are contacted by a single nucleotide, semicircles are drawn. Nucleotides that contact proteins S2, S7, and S21 are marked in gray, as no rate constant was obtained for these proteins. Rates generally cluster by domain, with the fastest binding rates observed in the 5′ domain and the slowest in the 3′ domain, the exception being the 5′ domain protein S12, which is among the slowest binders (indicated by the brown circles in the 5′ domain).

DISCUSSION

The view of assembly that emerges from consideration of the sum of the body of experimental data is rich with detail, but not without apparent discrepancies. Chemical footprinting (23, 24, 44) and pulse-chase experiments measuring the kinetics of protein binding using pulse-chase quantitative mass spectrometry

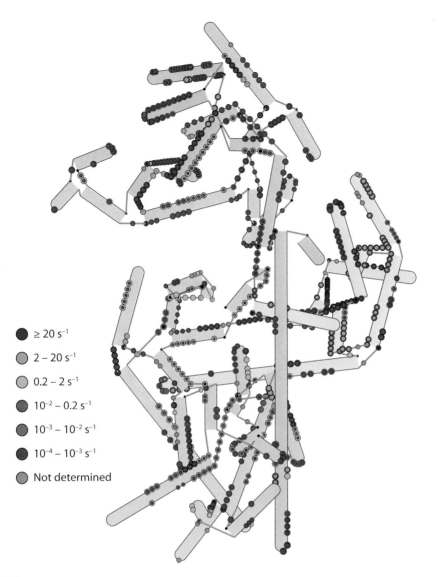

≥ 20 s⁻¹

2 – 20 s⁻¹

0.2 – 2 s⁻¹

10^{-2} – 0.2 s⁻¹

10^{-3} – 10^{-2} s⁻¹

10^{-4} – 10^{-3} s⁻¹

Not determined

Figure 6

A hybrid representation of the 16S rRNA annotated with rate constants for protection from hydroxyl radical cleavage (1). In the cases in which two rate constants are calculated for two phases of protection, concentric circles are displayed, with the inner circle colored according to the rate constant of the initial burst of protection and the outer circle colored according to the second, slower protection. In all cases the area displayed is proportional to the amplitude of the protection, so smaller amplitudes are reflected by smaller circles. Fast rates are evident across the entire 16S, suggesting multiple nucleation sites for assembly. Whereas the initial burst rates are in almost all cases faster than the rates observed for protein binding (**Figure 5**), the second slower rates are on par with those observed for protein binding in many cases.

(PC/QMS) (55) indicate a 5′ to 3′ directionality of assembly, but this was not observed in the time-resolved hydroxyl radical footprinting data (1). The key to reconciling these data sets is the observation that nucleotides that contact a particular protein do not always experience the same rate of protection. In the PC/QMS experiments, only tightly bound proteins are observed, as proteins that are loosely bound during the pulse can exchange during the chase. In contrast, the footprinting data can capture occupancy by both a weakly binding protein and a stably associated protein. It is likely that the faster phase of RNA protections correspond to thermodynamically favorable, but kinetically labile interactions of an initial encounter complex. The agreement of the measured rates for the slow phase in the RNA footprinting data and the rates from pulse-chase experiments is quite good. The general picture that emerges is consistent with the idea of alternating protein binding and RNA conformational changes that emerged from the studies of central domain fragments.

Most ribosomal protein binding sites have some fast fractional protection for at least part of the binding site. This indicates that some portion of each protein binding site on the 16S rRNA is preorganized in such a way to permit rapid protein binding, and that most of the RNA structure is capable of forming at least transiently in the absence of proteins. The hierarchy observed in the original Nomura map, as well as the kinetics observed by PC/QMS, indicates that protein binding to many of these sites is kinetically labile. Weakly bound proteins can dissociate during the ultracentrifugation step under nonequilibrium conditions after reconstitution or during the chase step in PC/QMS. The observed order of stable protein binding provides clues to the order of RNA conformational changes that lock down initial encounter complexes.

Furthermore, while the chemical footprinting data on assembly intermediates RI and RI* also suggest a 5′ to 3′ directionality, these data report on a specific set of steps during assembly. Indeed, the conversion of RI* to a complete 30S subunit gives rise to changes in nucleotide protections heavily concentrated in the 3′ domain, but this is a preordained result because four of the five proteins missing in the reconstituted RI* particle are 3′ domain proteins. The data on the transition between naked 16S rRNA and the RI particle show nucleotide protections across the entire subunit, owing to the addition of proteins from all three domains, and the RI to RI* transition describes a single, albeit significant step in the assembly process rather than a global rearrangement of the entire RNA, so its localization does not directly imply a 5′ to 3′ directionality.

There is also the question of whether RI represents a true assembly intermediate. Clearly RI is a prominent feature of in vitro reconstitution reactions at low temperature. It is possible to form and isolate an RI particle from reconstitutions, but this does not demand that all assembly trajectories necessarily converge on RI. The temperature dependence as shown by Arrhenius plots of the protein binding rates measured by pulse-chase mass spectrometry was linear over the measured range (15–40°C) (55). This indicates that the rate-determining step for each protein is the same at high and low temperatures, which is not consistent with a single temperature-dependent rate-limiting step such as the RI to RI* transition. The data indicate a series of multiple parallel folding pathways whose point of convergence is the final 30S subunit and not a single intermediate. The 30S subunit assembles over a complicated assembly landscape, and not a unique trajectory (55). The time-resolved hydroxyl radical footprinting data also suggest that many portions of the RNA are folding in parallel. The intermediate RI is most likely a misfolded RNA structure in which portions of the 3′ domain, and possibly the central domain, are mispaired, and the RI to RI* transition is the rate-limiting refolding of this stable kinetic trap.

The striking observation remains that 5′ domain proteins generally bind more quickly than 3′ domain proteins do, even if only in their final tight association. This is a curious result especially in the face of data showing that each domain can assemble independently. Reconciliation of all the in vitro data may ultimately be achieved by understanding what happens in vivo, where assembly occurs cotranscriptionally. The 5′ domain of the RNA is transcribed first, thereby having the opportunity to fold first into appropriate binding domains for the proteins. Because the 5′ domain proteins are present first in practice, binding of the central domain proteins and 3′ domain proteins may have evolved to be maximally efficient in their presence. Each domain is capable of assembling independently but may assemble most efficiently in the context of the entire subunit. In addition, cotranscriptional folding and assembly of the 5′ and central domains is likely to avoid formation of the RI kinetic trap.

The structure of the 16S rRNA is fairly monolithic, and it is possible to imagine that much of the RNA secondary and tertiary structure can form in the absence of the proteins dotting its periphery. The 5′ domain is highly folded in the absence of proteins (2), but it is not stable without the complement of ribosomal proteins. The body of work on central domain assembly suggests that protein binding does not so much induce a conformational change as it stabilizes an unstable RNA conformation. Transient RNA folding and unstable conformations may be populated enough to influence the footprinting data, but this does not imply a single stable conformation.

The ribosomal proteins serve to consolidate RNA folding gains as assembly proceeds. The three-helix junction binding site of S15 folds significantly in the presence of magnesium ions, and subsequent binding of S6:S18 is possible but greatly reduced in the absence of S15. S15 is nonessential for growth (9), which is consistent with a primary role for S15 in ribosome assembly rather than ribosome function. The hierarchy evident in the Nomura map is apparently not absolute in vivo. The binding sites for S6:S18 must be able to fold sufficiently well for transient protein binding. There is reciprocity of the thermodynamics of cooperativity, meaning that if S6:S18 binding stabilizes subsequent RNA tertiary structure, then that same tertiary structure can stabilize S6:S18 binding if it can

form in the absence of S15. The parallel nature of assembly makes it possible and likely that such defects can be overcome.

The overall impression of 30S subunit assembly is that of a dynamic and fluid process. Proteins do not simply lock into place in a strict processive manner; they work their way into the network of RNA helices in a multiphasic manner. The 16S rRNA begins folding on its own but reacts and adapts to the influx of ribosomal proteins, adopting new local conformations around them and stabilizing others. There are many parallel folding and binding pathways, each of them proceeding simultaneously toward the final product of a complete 30S subunit.

The assembly process for ribosomes has been highly evolved for efficiency and accuracy, and it appears that some combination of parallel and sequential folding and binding events has been selected. Although the efficient in vitro assembly of 30S subunits has been instrumental in allowing biophysical characterization of specific steps in assembly, many features of the 30S assembly mechanism remain to be understood in vivo. The assembly of the 50S subunit is much more complex, and although it also forms a monolithic RNA tertiary structure, it is much more topologically intricate. The cotranscriptional nature of assembly and the participation of tens of exogenous cofactors and chaperones are critical for rapid and efficient assembly of the 50S subunit. The next frontier in ribosome assembly is tackling this formidable problem, which will no doubt require experimental innovation for both in vitro and in vivo methodologies. Ten years ago, it was difficult to imagine the "movie" of the process of translation, but extraordinary efforts in structural biology have made this possible. Similar efforts will hopefully generate the ribosome assembly movie in the coming decade.

SUMMARY POINTS

1. Assembly of the 30S subunit occurs via multiple parallel folding pathways
2. The incorporation of proteins into the 30S subunit occurs in a hierarchical manner.

3. The 16S rRNA determines the overall fold of the 30S subunit.

4. Whereas the final tight binding of proteins occurs in a 5′ to 3′ manner, folding and assembly nucleate from several points in the ribosome independent of the domain structure. The directionality observed in vitro may be a side effect of optimization for in vivo assembly.

5. The three domains are capable of assembling independently.

6. Some ribosomal proteins help drive large conformational rearrangements of the RNA.

FUTURE ISSUES

1. An unambiguous description of the precise events that confer protection against hydroxyl radical cleavage of the 16S rRNA is needed.

2. A statistical view of the simultaneous folding pathways that nucleate from different sites along the 16S rRNA is important to understanding the parallel nature of ribosome assembly.

3. The 3D structures of important assembly intermediates need to be determined.

4. A complete kinetic and energetic description of each step of the assembly trajectory are necessary for the construction of a true mechanism for assembly.

5. The specific roles of assembly cofactors need to be determined.

6. The assembly process in vivo needs to be fully explored and compared to the process in vitro.

DISCLOSURE STATEMENT

The authors are not aware of any biases that might be perceived as affecting the objectivity of this review.

ACKNOWLEDGMENTS

This work was supported by grants from the NIH (F32-GM083510 to M.T.S. and R37-GM053757 to J.R.W.). The authors thank Drs. Gloria Culver and Sarah Woodson for critical comments on the manuscript.

LITERATURE CITED

1. A high-resolution kinetic study of RNA footprinting during assembly.

2. A study of the 5′ domain reveals native structure in the absence of proteins.

1. Adilakshmi T, Bellur DL, Woodson SA. 2008. Concurrent nucleation of 16S folding and induced fit in 30S ribosome assembly. *Nature* 455:1268–72
2. Adilakshmi T, Ramaswamy P, Woodson SA. 2005. Protein-independent folding pathway of the 16S rRNA 5′ domain. *J. Mol. Biol.* 351:508–19
3. Agalarov SC, Sridhar Prasad G, Funke PM, Stout CD, Williamson JR. 2000. Structure of the S15, S6, S18-rRNA complex: assembly of the 30S ribosome central domain. *Science* 288:107–13
4. Agalarov SC, Zheleznyakova EN, Selivanova OM, Zheleznaya LA, Matvienko NI, et al. 1998. In vitro assembly of a ribonucleoprotein particle corresponding to the platform domain of the 30S ribosomal subunit. *Proc. Natl. Acad. Sci. USA* 95:999–1003

5. Arnold RJ, Reilly JP. 1999. Observation of *Escherichia coli* ribosomal proteins and their posttranslational modifications by mass spectrometry. *Anal. Biochem.* 269:105–12

6. Ban N, Nissen P, Hansen J, Moore PB, Steitz TA. 2000. The complete atomic structure of the large ribosomal subunit at 2.4 A resolution. *Science* 289:905–20

7. Batey RT, Williamson JR. 1996. Interaction of the *Bacillus stearothermophilus* ribosomal protein S15 with 16S rRNA: I. Defining the minimal RNA site. *J. Mol. Biol.* 261:536–49

8. Batey RT, Williamson JR. 1996. Interaction of the *Bacillus stearothermophilus* ribosomal protein S15 with 16S rRNA: II. Specificity determinants of RNA-protein recognition. *J. Mol. Biol.* 261:550–67

9. Bubunenko M, Korepanov A, Court DL, Jagannathan I, Dickinson D, et al. 2006. 30S ribosomal subunits can be assembled in vivo without primary binding ribosomal protein S15. *RNA* 12:1229–39

10. Cannone JJ, Subramanian S, Schnare MN, Collett JR, D'Souza LM, et al. 2002. The comparative RNA web (CRW) site: an online database of comparative sequence and structure information for ribosomal, intron, and other RNAs. *BMC Bioinform.* 3:2

11. Chow CS, Lamichhane TN, Mahto SK. 2007. Expanding the nucleotide repertoire of the ribosome with post-transcriptional modifications. *ACS Chem. Biol.* 2:610–19

12. de Narvaez CC, Schaup HW. 1979. In vivo transcriptionally coupled assembly of *Escherichia coli* ribosomal subunits. *J. Mol. Biol.* 134:1–22

13. DeLano WL. 2002. *The PyMOL molecular graphics system.* **http://www.pymol.org/**

14. Deras ML, Brenowitz M, Ralston CY, Chance MR, Woodson SA. 2000. Folding mechanism of the *Tetrahymena* ribozyme P4-P6 domain. *Biochemistry* 39:10975–85

15. Dutcă LM, Jagannathan I, Grondek JF, Culver GM. 2007. Temperature-dependent RNP conformational rearrangements: analysis of binary complexes of primary binding proteins with 16S rRNA. *J. Mol. Biol.* 368:853–69

16. Graphviz: *Graph Visualization Software.* **http://www.graphviz.org/**

17. Green R, Noller HF. 1999. Reconstitution of functional 50S ribosomes from in vitro transcripts of *Bacillus stearothermophilus* 23S rRNA. *Biochemistry* 38:1772–79

18. Grondek JF, Culver GM. 2004. Assembly of the 30S ribosomal subunit: positioning ribosomal protein S13 in the S7 assembly branch. *RNA* 10:1861–66

19. Heilek GM, Marusak R, Meares CF, Noller HF. 1995. Directed hydroxyl radical probing of 16S rRNA using Fe(II) tethered to ribosomal protein S4. *Proc. Natl. Acad. Sci. USA* 92:1113–16

20. Heilek GM, Noller HF. 1996. Site-directed hydroxyl radical probing of the rRNA neighborhood of ribosomal protein S5. *Science* 272:1659–62

21. Held WA, Ballou B, Mizushima S, Nomura M. 1974. Assembly mapping of 30S ribosomal proteins from *Escherichia coli*. Further studies. *J. Biol. Chem.* 249:3103–11

22. Herold M, Nierhaus KH. 1987. Incorporation of six additional proteins to complete the assembly map of the 50S subunit from *Escherichia coli* ribosomes. *J. Biol. Chem.* 262:8826–33

23. Holmes KL, Culver GM. 2004. Mapping structural differences between 30S ribosomal subunit assembly intermediates. *Nat. Struct. Mol. Biol.* 11:179–86

24. Holmes KL, Culver GM. 2005. Analysis of conformational changes in 16S rRNA during the course of 30S subunit assembly. *J. Mol. Biol.* 354:340–57

25. Jagannathan I, Culver GM. 2003. Assembly of the central domain of the 30S ribosomal subunit: roles for the primary binding ribosomal proteins S15 and S8. *J. Mol. Biol.* 330:373–83

26. Jagannathan I, Culver GM. 2004. Ribosomal protein-dependent orientation of the 16S rRNA environment of S15. *J. Mol. Biol.* 335:1173–85

27. Kaczanowska M, Rydén-Aulin M. 2007. Ribosome biogenesis and the translation process in *Escherichia coli*. *Microbiol. Mol. Biol. Rev.* 71:477–94

28. Laederach A, Shcherbakova I, Jonikas MA, Altman RB, Brenowitz M. 2007. Distinct contribution of electrostatics, initial conformational ensemble, and macromolecular stability in RNA folding. *Proc. Natl. Acad. Sci. USA* 104:7045–50

29. Lescoute A, Westhof E. 2006. The interaction networks of structured RNAs. *Nucleic Acids Res.* 34:6587–604

30. Lewicki BT, Margus T, Remme J, Nierhaus KH. 1993. Coupling of rRNA transcription and ribosomal assembly in vivo. Formation of active ribosomal subunits in *Escherichia coli* requires transcription of rRNA genes by host RNA polymerase which cannot be replaced by bacteriophage T7 RNA polymerase. *J. Mol. Biol.* 231:581–93

31. The first appearance of an assembly map for the 30S ribosomal subunit.

31. Mizushima S, Nomura M. 1970. Assembly mapping of 30S ribosomal proteins from *E. coli*. *Nature* 226:1214–18

32. Moazed D, Stern S, Noller HF. 1986. Rapid chemical probing of conformation in 16S ribosomal RNA and 30S ribosomal subunits using primer extension. *J. Mol. Biol.* 187:399–416

33. Nomura M. 1973. Assembly of bacterial ribosomes. *Science* 179:864–73

34. Orr JW, Hagerman PJ, Williamson JR. 1998. Protein and Mg$^{(2+)}$-induced conformational changes in the S15 binding site of 16S ribosomal RNA. *J. Mol. Biol.* 275:453–64

35. Pan J, Deras ML, Woodson SA. 2000. Fast folding of a ribozyme by stabilizing core interactions: evidence for multiple folding pathways in RNA. *J. Mol. Biol.* 296:133–44

36. Pan J, Thirumalai D, Woodson SA. 1997. Folding of RNA involves parallel pathways. *J. Mol. Biol.* 273:7–13

37. Pan J, Woodson SA. 1998. Folding intermediates of a self-splicing RNA: mispairing of the catalytic core. *J. Mol. Biol.* 280:597–609

38. Pan T, Sosnick T. 2006. RNA folding during transcription. *Annu. Rev. Biophys. Biomol. Struct.* 35:161–75

39. Paul BJ, Ross W, Gaal T, Gourse RL. 2004. rRNA transcription in *Escherichia coli*. *Annu. Rev. Genet.* 38:749–70

40. Petry S, Brodersen DE, Murphy FV, Dunham CM, Selmer M, et al. 2005. Crystal structures of the ribosome in complex with release factors RF1 and RF2 bound to a cognate stop codon. *Cell* 123:1255–66

41. Polevoda B, Sherman F. 2007. Methylation of proteins involved in translation. *Mol. Microbiol.* 65:590–606

42. Powers T, Daubresse G, Noller HF. 1993. Dynamics of in vitro assembly of 16S rRNA into 30S ribosomal subunits. *J. Mol. Biol.* 232:362–74

43. Powers T, Noller HF. 1995. A temperature-dependent conformational rearrangement in the ribosomal protein S4.16S rRNA complex. *J. Biol. Chem.* 270:1238–42

44. Powers T, Noller HF. 1995. Hydroxyl radical footprinting of ribosomal proteins on 16S rRNA. *RNA* 1:194–209

45. Recht MI, Williamson JR. 2004. RNA tertiary structure and cooperative assembly of a large ribonucleoprotein complex. *J. Mol. Biol.* 344:395–407

46. Rook MS, Treiber DK, Williamson JR. 1998. Fast folding mutants of the *Tetrahymena* group I ribozyme reveal a rugged folding energy landscape. *J. Mol. Biol.* 281:609–20

47. Rook MS, Treiber DK, Williamson JR. 1999. An optimal Mg$^{(2+)}$ concentration for kinetic folding of the *Tetrahymena* ribozyme. *Proc. Natl. Acad. Sci. USA* 96:12471–76

48. Samaha RR, O'Brien B, O'Brien TW, Noller HF. 1994. Independent in vitro assembly of a ribonucleoprotein particle containing the 3′ domain of 16S rRNA. *Proc. Natl. Acad. Sci. USA* 91:7884–88

49. Schuwirth BS, Borovinskaya MA, Hau CW, Zhang W, Vila-Sanjurjo A, et al. 2005. Structures of the bacterial ribosome at 3.5 A resolution. *Science* 310:827–34

50. Selmer M, Dunham CM, Murphy FV, Weixlbaumer A, Petry S, et al. 2006. Structure of the 70S ribosome complexed with mRNA and tRNA. *Science* 313:1935–42

51. Silverman SK, Deras ML, Woodson SA, Scaringe SA, Cech TR. 2000. Multiple folding pathways for the P4-P6 RNA domain. *Biochemistry* 39:12465–75

52. Srivastava AK, Schlessinger D. 1990. Mechanism and regulation of bacterial ribosomal RNA processing. *Annu. Rev. Microbiol.* 44:105–29

53. Svensson P, Changchien LM, Craven GR, Noller HF. 1988. Interaction of ribosomal proteins, S6, S8, S15 and S18 with the central domain of 16S ribosomal RNA. *J. Mol. Biol.* 200:301–8

54. Takamoto K, Das R, He Q, Doniach S, Brenowitz M, et al. 2004. Principles of RNA compaction: insights from the equilibrium folding pathway of the P4-P6 RNA domain in monovalent cations. *J. Mol. Biol.* 343:1195–206

55. A nearly complete set of kinetic data on protein binding during assembly.

55. Talkington MW, Siuzdak G, Williamson JR. 2005. An assembly landscape for the 30S ribosomal subunit. *Nature* 438:628–32

56. An early description of the assembly of 30S subunits via RI.

56. Traub P, Nomura M. 1969. Structure and function of *Escherichia coli* ribosomes. VI. Mechanism of assembly of 30S ribosomes studied in vitro. *J. Mol. Biol.* 40:391–413

57. Treiber DK, Rook MS, Zarrinkar PP, Williamson JR. 1998. Kinetic intermediates trapped by native interactions in RNA folding. *Science* 279:1943–46

58. Treiber DK, Williamson JR. 2001. Beyond kinetic traps in RNA folding. *Curr. Opin. Struct. Biol.* 11:309–14

59. Treiber DK, Williamson JR. 2001. Concerted kinetic folding of a multidomain ribozyme with a disrupted loop-receptor interaction. *J. Mol. Biol.* 305:11–21

60. Weitzmann CJ, Cunningham PR, Nurse K, Ofengand J. 1993. Chemical evidence for domain assembly of the *Escherichia coli* 30S ribosome. *FASEB J.* 7:177–80

61. Wilson DN, Nierhaus KH. 2005. Ribosomal proteins in the spotlight. *Crit. Rev. Biochem. Mol. Biol.* 40:243–67

62. Wilson DN, Nierhaus KH. 2007. The weird and wonderful world of bacterial ribosome regulation. *Crit. Rev. Biochem. Mol. Biol.* 42:187–219

63. Wimberly BT, Brodersen DE, Clemons WM, Morgan-Warren RJ, Carter AP, et al. 2000. Structure of the 30S ribosomal subunit. *Nature* 407:327–39

64. Zengel JM, Lindahl L. 1994. Diverse mechanisms for regulating ribosomal protein synthesis in *Escherichia coli*. *Prog. Nucleic Acid Res. Mol. Biol.* 47:331–70

Mechanical Signaling in Networks of Motor and Cytoskeletal Proteins

Jonathon Howard

Max Planck Institute of Molecular Cell Biology and Genetics, 01307 Dresden, Germany;
email: howard@mpi-cbg.de

Annu. Rev. Biophys. 2009. 38:217–34

First published online as a Review in Advance on February 5, 2009

The *Annual Review of Biophysics* is online at biophys.annualreviews.org

This article's doi:
10.1146/annurev.biophys.050708.133732

Key Words

cell motility, spontaneous oscillations, symmetry breaking, axonemal beat, bidirectional motion

Abstract

The motions of cells and organelles are highly coordinated. They are driven by motor proteins moving along cytoskeletal filaments, and by the dynamic growth and shrinkage of the filaments themselves. The initiation of cellular motility is triggered by biochemical signaling pathways, but the coordination of motility at different locations or times is not well understood. In this review I discuss a new hypothesis that motility is coordinated through mechanical signals passing between and regulating the activity of motors and filaments. The signals are carried by forces and sensed through the acceleration of protein-protein dissociation rates. Mechanical signaling can lead to spontaneous symmetry breaking, switching, and oscillations, and it can account for a wide range of cell motions such as the flagellar beat, mitotic spindle movements, and bidirectional organelle transport. Because forces can propagate quickly, mechanical signaling is ideal for coordinating motion over large distances.

Contents

Motor protein: an enzyme that converts chemical energy derived from the hydrolysis of ATP into mechanical work

CELL MOVEMENTS ARE HIGHLY COORDINATED

The motions of cells and organelles are highly coordinated in space and time. A dramatic example is the serpentine movement of animal spermatozoa, whose motile organelle is the axoneme, a microtubule-based structure that extends from the head to the end of the tail (**Figure 1a**). On one side of the axoneme, motor proteins in the dynein family switch on and drive microtubule sliding to generate a bend in one direction; these motors then switch off and motor proteins on the other side switch on to generate a bend in the other direction. This alternating pattern of activity then travels as a wave from head to tail to generate the undulations that push the cell through the fluid.

Cells exhibit a wondrous variety of movements that they use to locomote, change shape, and divide. Keratocytes, skin cells in fish and amphibia that quickly move to sites of injury, have a characteristic shape that is preserved as the cell advances (**Figure 1b**); the protrusion of the actin-based lamella at the front must be coordinated with the retraction at the rear, otherwise the cell would stretch out or scrunch up (35). During mitosis, the duplicated chromosomes undergo a highly choreographed sequence of movements. They are captured by microtubules of the mitotic spindle, move to the center of the bipolar spindle (**Figure 1c**), and then, following separation, the sister chromatids move to different poles prior to cell division. This elaborate maneuver involves several checkpoints that ensure that the next phase begins only after the last phase has been completed, thus increasing the reliability of the partitioning of the genomes to the two daughter cells (44). Even the transport of membrane-bounded organelles, especially prominent in the long axons of neurons, must be coordinated because the outward-moving kinesin carries the inward-moving cytoplasmic dynein: Both motors cannot be active at the same time, or else the transport vesicles would get caught up in a local tug-of-war rather than participate in an orderly circulation of proteins, lipids, and metabolites from the cell body to the synapse and back. Coordination ensures that motors work with each other and not against each other, that motors move cells and not pull them apart, and that events take place in the correct sequence so that errors are not made.

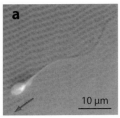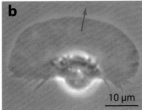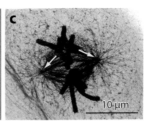

Figure 1

Cell locomotion. (*a*) The serpentine motion of human spermatozoa is driven by the propagation of bending waves along the axoneme, the microtubule-based motile organelle in the sperm's tail (image courtesy of Ingmar Riedel-Kruse). (*b*) The keratocyte migrates by protruding its actin-rich lamella (upward in the image) while retracting its rear (35) (courtesy of Julie Theriot). (*c*) Chromosomes lined up in the middle of the spindle at metaphase prior to their segregation to the two poles from which the microtubules emanate (courtesy Richard McIntosh). The arrows indicate direction of motion.

How are such mechanical processes regulated? There is no doubt that cell signaling pathways play crucial roles. For example, mammalian sperm are quiescent in the epididymis and become motile only after being triggered by bicarbonate anion present in reproductive fluids through a pathway that involves cAMP and calcium (9). Sea urchin sperm chemotax: They target eggs in the open water by swimming up a concentration gradient of a small peptide that they detect through a sensitive guanyly-cyclase receptor pathway (34). The protrusion of actin at the leading edge of moving cells is regulated by members of the Rho family of G-proteins (31). The cell cycle and mitosis are timed by phosphorylation pathways (44). These examples make clear that cell motion is turned on and off through the signaling pathways that are universal regulators of cellular activities.

But to what extent do biochemical signaling processes control the detailed movements of the motor proteins and the cytoskeleton that form the "nuts and bolts" of cell motility? Are the individual motors and filaments micromanaged like marionettes by a small group of cellular puppeteers (signaling pathways) that ensure that the players work together? Or is there a certain degree of autonomy of action of collectives of motors and filaments that allow them to self-organize and coordinate? If so, what is the nature of this coordination—how do the motors and cytoskeleton communicate with each other, and what are the sizes of the collectives that can work without centralized instructions?

In this review I describe recent progress on understanding how a certain type of mechanical communication between motor proteins can lead to coordinated cellular motions such as switching the direction of motion or spontaneous oscillation. The communication involves the load-dependent acceleration of protein-protein dissociation. The idea originated in a pair of theoretical papers by Jülicher & Prost (32, 33), though its essence is contained in an earlier paper by Brokaw (7). To understand how this can lead to coordination within a network of motors and filaments, we must develop a new way of thinking about the process of force generation by motor proteins in which velocity rather than force is the independent variable. With this shift in perspective, load-accelerated dissociation can be built into mechanical signaling networks, which can then be analyzed using the same control theoretic approaches that have been so successful in understanding electrical signaling networks in the nervous system (27) and chemical signaling networks in cells in general (2, 61). In this way, load-accelerated protein-protein dissociation has recently accounted for experimentally observed switching and oscillations that take place during the flagellar beat (8, 49), during spindle positioning in mitosis (21, 47), and in other cellular processes

such as bidirectional organelle movement
(5, 43).

FORCE DEPENDENCE
OF MOTOR PROTEINS

Cell movements are driven by motor proteins that hydrolyze ATP as they move along cytoskeletal filaments. In addition, the filaments themselves hydrolyze nucleotides, ATP by actin and GTP by tubulin, through reactions coupled to polymerization and depolymerization; this coupling enables the filaments to do mechanical work as they push and pull intracellular structures (28). I first consider force generation by motor proteins. The concepts generalize to the cytoskeleton, and indeed any system in which a change in physical position—such as motion along a filament or change in length of a polymer—is coupled to an energetically favorable chemical reaction.

Our understanding of how motor proteins work has increased profoundly over the past 20 years (28). This has been mainly due to two developments: Single-molecule techniques allow one to follow and manipulate individual motors as they move on their cytoskeletal tracks (54), and crystallographic and EM studies have provided atomic structures of many motor and cytoskeletal proteins (41, 42). A fundamental insight from these studies is that load forces—forces directed parallel to the filament but in a direction opposite of unloaded motion—couple to the ATP hydrolysis reaction by slowing down steps that involve forward movement against the load. In this view, the ATP hydrolysis reaction is such a highly energetically favorable process that, if it is tightly coupled to vectorial motion such as movement along a filament, the reaction (and the motion) will proceed even in the presence of an external load force. This leads naturally to the question of how the force influences the velocity. The load dependence of velocity has been measured using viscous loads and elastic loads exerted by flexible glass fibers and by optical tweezers, among others (28 and references therein) (**Figure 2a**). A velocity-force curve for kinesin-1 is shown in **Figure 2b**. Without load, the kinesin-1 moves at ~800 nm s^{-1}, corresponding to about 100 8-nm steps per second (8 nm is the distance from one tubulin dimer to the next along the protofilament of the microtubule). As the load force is increased, the speed decreases until the movement stalls at ~7 pN. At stall, there are low rates of forward and backward stepping that are balanced such that the net displacement is zero. When the load force exceeds the stall force, the motor goes backward: The stall force is also a reversal force (10).

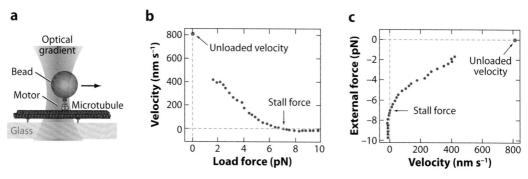

Figure 2

The velocity-force and force-velocity curves. (*a*) A two-headed motor protein (*green*) attached to a bead moving in the direction of the arrow out of an optical gradient (*yellow*), which generates a force directed toward the focus of the laser beam. (*b*) Plot of velocity as a function of the load force (points from Reference 60; original data from Reference 10). (*c*) Replot of panel *b* with external force as a function of the velocity.

A NEW VIEW OF MOTOR PROTEINS: FORCE GENERATORS WITH INTERNAL DAMPING

This chemical view of how force influences the velocity of motors is not useful for understanding how motor proteins interact with each other and work together to produce movement in cells. The reason is that our goal is to explain the motion rather than the force, the latter playing the role of an invisible hand that guides and coordinates.

However, there is an alternative, mechanical view of a motor protein that is useful: The motor comprises a force generator together with an internal damping element. The force generator tries to keep going and going, like a wind-up soldier or a walking toy. The internal damping element is needed to account for the surprisingly low speeds of unloaded motor proteins: Because the viscous drag force from the surrounding fluid on a nanometer-sized motor moving at only one micron per second is tiny compared with the stall force of several piconewtons (28), the unloaded speed is not limited by external friction. There must therefore be internal friction in the workings of the engine. This friction originates in the tight coupling between chemical and mechanical transitions—large chemical activation barriers lead to slow transitions in the associated mechanical states so that motion is slow even in the presence of external force. The motor is analogous to a watch with an intricate arrangement of cogs and wheels that limit how quickly one can force the hands to turn.

This view accounts, in a simple way, for the effects of external loads: The speed is simply the net force—the external force plus that of the force-generator—divided by the drag coefficient of the internal damping element. If we pull back on the motor, it slows down; if we push it, it speeds up. This is analogous to a walking toy slowing down when it moves uphill and speeding up when it moves downhill; the incline acts as an external force, which can be assisting (downhill) or loading (uphill). In the simplest case, the force-velocity curve is linear, i.e., the drag coefficient is constant.

The important thing about this new view of a motor protein comprising a force generator and an internal drag element is that we do not need to consider the chemical and mechanical mechanisms in any more detail. This is because our mechanical picture is sufficient to explain how the motor works in the presence of external forces, for example, those generated by other molecular machines.

In this new picture, it is natural to plot the external force as a function of the velocity (**Figure 2c**) because this allows us to estimate the drag of the motor as the slope of the curve. In this representation, we treat the velocity as the independent variable, which makes sense because ultimately we want to know how a network of motors generates motion.

Motors can interact with each other through the forces that they generate. For example, if two equally strong motors pull against each other, they will stop. But the outcomes of the interactions between motors comprising a force generator and internal friction are relatively uninteresting: A collection of such motors behaves simply as the superposition of the force-velocity curves with the appropriate sign to denote whether they are pulling together or pushing against each other. To use an electrical analogy, a motor is like a battery (force generator) that drives current flow (motion) through a resistor. Although the motors are active elements in that they are out of equilibrium and drive motion, they act in a passive manner in that the movements are always downhill. Networks of coupled motors with force-velocity curves display no complex behavior such as switching or oscillation that is characteristic of active electrical circuits. For motors to do more, they need to have additional properties. But what are they?

LOAD-ACCELERATED PROTEIN-PROTEIN DISSOCIATION

There is a simple property that endows mechanical networks of motors with active properties that can drive switching and oscillations. This property is load-accelerated dissociation: the increase induced by a load force of the rate

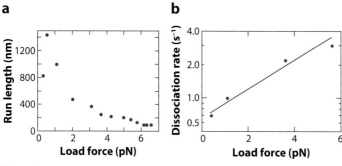

a

Run length (nm)

1200

800

400

0

0   2   4   6

Load force (pN)

b

Dissociation rate (s⁻¹)

4.0

2.0

1.0

0.5

0   2   4   6

Load force (pN)

Figure 3

Load-accelerated dissociation of kinesin-1. (*a*) Decrease in run length as a function of load force (53). (*b*) Taking into account the relation between force and speed, the curve in panel *a* has been replotted as dissociation rate, showing that kinesin has a load-accelerated rate of dissociation. The characteristic force, the load force that increases the dissociation rate e-fold, is 3.3 pN (*red line* in the log-linear plot).

Thermal energy (*kT*): the Boltzmann constant (*k*) multiplied by the absolute temperature (*T*) measured in Kelvin. At room temperature $kT \sim 4 \times 10^{-21}$ J

Positive feedback: a situation in which a change leads to an augmentation of the agent that caused the change

of detachment of the motor from the filament (or the motor's tail from its anchor point in the cell).

For a processive motor that takes many steps along a filament before detaching, load-accelerated dissociation implies that its run along the microtubule is briefer in the presence of a load force. Such a load-dependent decrease in the run length has been measured for kinesin-1 (**Figure 3*a***) (53). At zero load, the run length is about 1 μm, corresponding to about 120 steps. As the load increases, the run length decreases and approaches ~200 nm near stall. Because the velocity also decreases as the load increases, this experiment does not indicate that the dissociation rate is also increasing with increasing load force. However, if one takes into account the force-velocity curve, then the dissociation rate does indeed increase with load (**Figure 3*b***). Importantly, the attached time does not go to zero at stall, or else it would not be possible to make a motor go backward. Instead, as the force increases, the attached time decreases approximately exponentially with a characteristic force of ~3 pN: The dissociation rate increases e-fold per ~3 pN. At stall the attached time is ~0.2 s, and even at superstall loads there is still time to make several backsteps (10).

What is the physical mechanism by which force influences dissociation? The prevailing

hypothesis is that dissociation occurs via a transition state that is displaced in the direction toward which the load force is pulling (17). The characteristic force of 3 pN for kinesin-1 implies a displacement of 1.3 nm (*kT*/3 pN, where *kT* is thermal energy), which indicates that the motor domain is highly strained prior to dissociation.

Load-accelerated protein-protein dissociation is well established outside the motor field. For example, force accelerates dissociation of antibodies from antigens, integrins from extracellular matrix molecules such as collagen and fibronectin, cadherins from each other, and P-selectin from glycoproteins (reviewed in Reference 25). For motor proteins, load-dependent dissociation is not a general phenomenon; load has no effect on the run length of myosin-V (11), and yeast cytoplasmic dynein has a long-lived high-force state (18), though this is not found in metazoan dyneins (58). Thus, the load-accelerated dissociation found in some motors may be a specific adaptation for mechanical signaling and may not be advantageous in all systems. However, there is a caveat to the interpretation of run-length measurements that have been made with optical tweezers. Because of the geometry of a bead in an optical trap, a force directed along the filament also has a component pulling the motor away from the surface of the filament and it is not clear which component of the force—load or lift—causes detachment. Experiments in which the direction of the load is more precisely controlled are therefore needed to fully establish the principle of load-accelerated dissociation for motor proteins. Nevertheless, the evidence is still strong enough to explore its implications as a mechanism of mechanical communication between proteins.

IMPLICATIONS OF LOAD-DEPENDENT DISSOCIATION: POSITIVE FEEDBACK AND NEGATIVE DAMPING

Load-accelerated dissociation of a motor can lead to positive feedback. This can be best

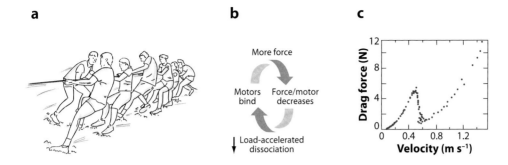

a

b

More force

Motors bind — Force/motor decreases

↓ Load-accelerated dissociation

c

Drag force (N) vs Velocity (m s⁻¹)

Figure 4

Positive feedback when a motor has load-accelerated dissociation (*a*) Tug-of-war. (*b*) The positive feedback cycle associated with load-accelerated dissociation. (*c*) The drag force on a sphere as a function of air velocity showing a region of negative slope at velocities around ~0.5 m s⁻¹, corresponding to a Reynolds number of ~350,000, where the flow switches from laminar to turbulent. Data from Reference 1, with parameters for a sphere of diameter 0.1 m at 20°C and a pressure of 1 bar.

appreciated by considering a tug-of-war (**Figure 4*a***). Imagine that both sides are evenly matched (symmetric) and that the probability of one person letting go (or slipping on the grass) increases with the load that he or she is carrying. If someone on one side accidentally lets go, then the load is shared among fewer people on that side and the load per person increases. This increases the likelihood that another person lets go, leading to a catastrophic release on that side. Consideration of the other side leads to the same result: If another person attaches, then the load per person is decreased so that even more people can take hold, leading to an ever-increasing force (**Figure 4*b***). The tug-of-war illustrates that positive feedback associated with load-dependent detachment can lead to spontaneous symmetry breaking.

Positive feedback in a mechanical system can be understood using the concept of negative damping. This is an unusual concept about which many people initially feel uncomfortable. To understand it, I remind the reader about positive damping. Positive damping refers to the increased resistance to movement that a body experiences as its speed increases. For example, to move a spoon faster in a jar of honey requires more force. Negative damping corresponds to the situation in which the resisting force decreases as the speed increases. A classic example

from the fluid mechanics literature is the transition from laminar to turbulent flow past a sphere: As the speed increases, a critical region is reached where the drag force decreases and the drag coefficient (the slope of the force-velocity curve) is less than zero (**Figure 4*c***). If a spoon in a jar of honey were to suffer such viscosity breakdown, a critical force would be reached in which the spoon would suddenly start accelerating in the jar—it would lurch forward. A loose analogy is quicksand that has high resistance when stationary but lower resistance when agitated. Another example is the phenomenon that the coefficient of moving friction is usually smaller than the coefficient of stationary friction (i.e., a velocity-dependent friction coefficient): This phenomenon is associated with the skidding of a wheel and can lead to slip-stick oscillations, such as during the bowing of a violin string (3, 55).

The reason that negative damping is a useful concept is that it leads to instability. For example, consider a linear oscillator described by the equation $ma + \gamma v + \kappa x = 0$ where m is the mass, a is the acceleration, γ is the drag coefficient that provides damping, v is the velocity, κ is the stiffness, and x is the position. If γ is positive, the oscillations will die out; if γ is negative, they will build up, and we call this spontaneous oscillation. The build-up requires

Spontaneous symmetry breaking: occurs when a spatially symmetric system is not stable with respect to small perturbations and rather than returning to its initial position it moves one way or the other and becomes asymmetric

Negative damping (or drag): a force that augments motion and whose magnitude increases with velocity

Positive damping (or drag): a force that opposes motion and whose magnitude increases with velocity

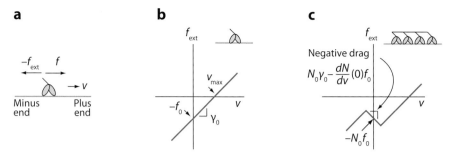

Figure 5

Positive and negative slopes of the force-velocity curve. (*a*) Balance of the motor force (f) and the external force (f_{ext}). (*b*) Linear force-velocity curve for a plus-end-directed motor. (*c*) For an ensemble of motors with load-accelerated dissociation, the slope of the force-velocity curve, the drag coefficient, can be negative if the load-dependence is strong enough.

energy, which enters the system through the damping element: The power dissipated to the surroundings by a moving particle is $1/2\gamma \langle v^2 \rangle$, where $\langle v^2 \rangle$ is the mean squared velocity, and if γ is negative, then the particle takes up energy from the surroundings.

FORCE BALANCE OF MOTOR PROTEINS: POSITIVE AND NEGATIVE DRAG

I now formalize these ideas of positive and negative damping. Imagine a single motor protein moving an object at constant velocity (**Figure 5a**). If an external force f_{ext} is also applied to the object, then there will be a force balance:

$$f + f_{ext} = 0, \qquad 1.$$

where f is the force of the motor. We adopt the following sign convention for motor proteins: The direction toward the plus end of the filament is positive. Thus, kinesin-1 and myosin-V have positive unloaded velocities, and minus-end-directed motors like dynein and myosin-VI have negative unloaded velocities. Usually, we show positive to the right. A load force is directed opposite the unloaded direction of motion. Thus, for a plus-end-directed motor a load is in the negative direction (i.e., $f_{ext} < 0$). Usually, the velocity is assumed to be small because we are usually interested in the force-limited regime, in which the motors have built up high

tension in the network (20). It is under these conditions, such as when there is a tug-of-war, that the network shows interesting behavior; importantly, at the onset of instability the amplitude of the velocity is still small. The force-velocity curve can be written as

$$f_{ext}(v) = -f(v) \approx -f_0 + \gamma_0 v, \qquad 2.$$

where v is the velocity, f_0 is the stall force (the force associated with the motor's force-generating element), and γ_0 is the slope, at stall, of the force-velocity curve (it is the drag coefficient associated with the motor's internal friction element). For a plus-end-directed motor like kinesin-1, f_0 is positive. For all motors characterized so far, the drag coefficient at stall is positive, corresponding to a drag force that opposes the motion. A linear force-velocity curve for a plus-end-directed motor is shown in **Figure 5b**. For a given external force, f_{ext}, Equation 2 can be used to calculate the speed as $v = (f_{ext} + f_0)/\gamma_0$.

Suppose now that there is an ensemble of several motors in which on average N of them are attached and therefore active (**Figure 5c**). The force balance is $f_{ext} = -Nf$. The number of attached motors depends on the association rate constant (k_{on}) and the dissociation rate constant [$k^{off}(f_{ext})$, which depends on the external force] according to $N = k_{on}(k_{on}+k^{off})^{-1} N_{tot}$, where N_{tot} is the total number of motors. Note that it does not make

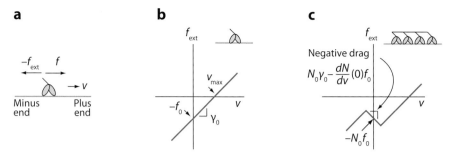

sense to have a load-dependent on-rate because the detached motor can feel no force. Differentiation of this equation shows that N varies with the dissociation rate constant according to $dN/k^{off} = -N(1 - N/N_{tot})/k^{off}$. As a consequence of load-accelerated dissociation, $dk^{off}/df_{ext} = -k_0^{off}/f_{off}$, where k_0^{off} is the dissociation rate at the stall force, $f_{off} > 0$ is the sensitivity of the dissociation rate to the load (an increase in the load by f_{off} increases the off-rate e-fold), and the sign is negative because increasing the load means making the external force more negative. Application of the chain rule—$dN/dv = (dN/dk^{off}) \cdot (dk^{off}/df_{ext}) \cdot (df_{ext}/df)$—for forces near the stall force gives

$$f_{ext}(v) \approx$$
$$- N_0 f_0 + N_0 \gamma_0 \left[1 - \frac{f_0}{f_{off}}(1 - N_0/N_{tot}) \right] v. \quad 3.$$

This reduces to Equation 2 if there is no load-accelerated dissociation ($f_{off} = \infty$), and the ensemble damping is positive due to the stabilizing effect of the force-velocity curve. However, if the load-dependent dissociation is strong enough (f_{off} is small enough), then the ensemble drag coefficient can be negative, corresponding to negative damping and consequently to positive feedback (similar to tug-of-war). Note that negative damping also relies on the motors being dynamic in that they can fluctuate between on and off states. If they are almost always on ($k_{on} \gg k^{off}$ and $N_0 \approx N_{tot}$), then load-dependent detachment will not significantly change N_0. For kinesin-1, $f_0 \approx 7$ pN and $f_{off} \approx 3$ pN, so the load-accelerated dissociation is strong enough, in principle, to give instability: Of course, this will not be seen in single-molecule conditions because after completion of a run, reassociation is slow; however, instabilities or oscillations might be observable at higher densities, especially in a tug-of-war scenario.

It is useful to describe our situation in the language of nonlinear dynamics. For an ensemble of motors at stall, a net drag coefficient equal to zero corresponds to a critical point. Decreasing the drag coefficient through zero causes the system to transition (i.e., to switch) from stability (positive damping) to instability (negative damping) and to start moving spontaneously. There are several potential control variables. For example, the motors could be modulated (e.g., by a biochemical signaling pathway) to have a higher stall force (f_0) or a stronger load dependence (f_{off} smaller). The number of active motors can also act as a control variable: If there is another source of positive damping (such as a filament array; 29) that stabilizes an otherwise unstable system, then increasing the number of active motors by a signaling pathway might eventually make the system unstable.

Critical point: the value of a parameter at which the system changes its behavior

Control variable: a parameter whose change can alter a system in a qualitative way, e.g., from being stationary to undergoing oscillations

DELAYS AND CHEMICAL INERTIA

Load-accelerated dissociation leads to oscillations in a natural way: Even though the dissociation rate constant changes as soon as the load is changed, the actual number of attached motors changes only slowly as the system relaxes to its new state of association appropriate for the new load. Thus, there is a delay between the application of the load and the response of the motors to that load. The system responds sluggishly to a change in force: It has inertia. The inertia is not due to mass, but rather it is a chemical inertia due to the finite time that it takes the motors to detach from their filaments. A central result from the theory of oscillators (3) is that for a system to undergo spontaneous oscillations, it must have a negative damping coefficient (to actively drive the motion), a positive stiffness to define a zero point (about which the motion oscillates), and a positive inertia that causes an overshoot of the zero point rather than a gradual relaxation to the zero point. Load-dependent detachment provides a positive inertia in a natural way.

The magnitude of the inertia associated with load-dependent detachment can be calculated by considering the kinetics of motor association and dissociation. The simplest situation is when these processes are first order and the number of attached motors changes with time according to

$$\frac{dN}{dt}(t) = -k^{off}(f_{ext}) \cdot N(t) + k_{on}[N_{tot} - N(t)]. \quad 4.$$

MTOC:
microtubule-
organizing
center

Cortex: the proteins
associated with the
inner surface of the
plasma membrane

This equation has a correlation time $\tau = (k_{on} + k_{off})^{-1}$ that determines how quickly the number of motors will respond to a changed load. If the ensemble is near stall and accelerating slowly $|dv/dt| \ll v_{max}/\tau_0$, where dv/dt is the acceleration and τ_0 is the correlation time at stall, then the number of attached motors satisfies $N \approx N_0 + (dN/dv)_0 v - \tau_0 (dN/dv)_0 dv/dt$ (47, 49). The total force is now

$$f_{ext} \approx -N_0 f_0 + [N_0 \gamma_0 - (dN/dv)_0 f_0]v$$
$$+ [\tau_0 (dN/dv)_0 f_0]\frac{dv}{dt}. \qquad 5.$$

The important thing is that the sign of the inertial term is positive, and the system has the capacity to oscillate.

DYNAMICS OF FILAMENT ARRAYS

The phenomena of switching and spontaneous oscillations can also be driven by the polymerization and depolymerization of cytoskeletal filaments. Indeed, they can occur whenever a motion-producing process is sensitive to the load or the velocity. For cytoskeletal filaments, cargos attached to the plus end, the more dynamic end, are pushed during growth and pulled during shrinkage. Examples of pushing include actin polymerization that drives cell protrusion (35) and some types of membrane

traffic (38), microtubule pushing that positions the nucleus and spindle in fission yeast (59), and probably the microtubule-organizing center (MTOC) in metazoan cells (50). Examples of pulling include microtubule depolymerization that asymmetrically positions the mitotic spindle (19) and moves chromosomes during mitosis (51). Although these movements require additional plus-end-binding proteins (such as the ARP2/3 complex and formins, which promote actin filament growth) (39) and anchoring proteins (such as the DAM1 complex that can bind to the ends of depolymerizing microtubules) (4, 65), the energy for these motions comes from the hydrolysis of ATP by actin or GTP by tubulin. Arrays of dynamic filaments display a rich variety of mechanical properties: elasticity, drag, and inertia.

Stiffness

A confined array of dynamic filaments can have stiffness (29). Consider the situation shown in **Figure 6a** in which microtubules grow out from an MTOC and push against the cell cortex. This centers the MTOC because if it is displaced from the cell center, then the microtubules will spend less time growing to the cortex on the nearer side than to that on the farther side. Therefore, the fraction of time that the microtubules spend pushing on the nearer

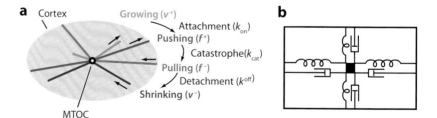

Figure 6

Stiffness and damping associated with a dynamic array of microtubules. (*a*) Microtubules grow out from the microtubule-organizing center (MTOC) and push on the cortex. After catastrophe, they pull on the cortex and then detach and shrink. If the MTOC is displaced to the right, then the time spent growing and shrinking to the right cortex will decrease and so the fraction of time that a microtubule spends pushing from the right (and therefore the total number of pushing microtubules on the right) will increase, leading to a restoring force that tends to push the MTOC back to the cell center. (*b*) Equivalent mechanical circuit showing the stiffness elements (*springs*) and damping elements (*dashpots*) in the *x* and *y* axes. Modified with permission from Reference 29.

cortex will be increased, leading to an increased force that opposes the displacement. Centration is therefore a form of collective stiffness because it corresponds to a force that opposes displacement away from the cell center, just as a spring opposes displacement away from its resting length. The further the MTOC is displaced, the greater the net forces pushing it back to the center, just like a spring. Pulling microtubules have an anticentering effect: The closer the MTOC gets to one cortex, the more microtubules will pull from that side. The net result of pushing and pulling is a force on the MTOC that depends on its displacement, x, from the cell center according to

$$f_{ext} \approx -2\left(N_0^+ f_0^+ - N_0^- f_0^-\right)\frac{x}{R}, \qquad 6.$$

where $N_0^+ (N_0^-)$ is the number of pushing (pulling) microtubules on each side when the MTOC is centered, $f_0^+ (f_0^-)$ is the stall force for a pushing (pulling) microtubule, and R is the radius of the cell (29). Thus, the dynamic array confers stiffness $2(N_0^+ f_0^+ - N_0^- f_0^-)/R$, which can be denoted as a spring in a mechanical circuit diagram (**Figure 6b**). Pushing microtubules confer a positive stiffness—they are stabilizing and drive centration of the pole. Pulling microtubules confer a negative stiffness—they are destabilizing. If the net stiffness is negative, then a small perturbation of the spindle from the center will lead to an ever-increasing force that moves the pole toward the cortex.

Damping

In addition to stiffness, a dynamic array of confined filaments can also give rise to damping (29). The net damping can be positive or negative. Positive drag comes from the force-velocity curves of the growing and shrinking microtubules, by the same argument that the force-velocity curve of a motor protein contributes positive drag. The force-velocity curve associated with pushing has been measured for growing microtubules (14) and actin filaments (36). Positive drag also results from the growth of a microtubule toward the cortex:

The faster the movement of the MTOC toward the cortex, the higher the net rate at which the microtubule plus end approaches the cortex, and the higher the attachment rate and so the larger the number of pushing microtubules. Negative drag can arise from three sources: a load-accelerated catastrophe rate that increases the number of shrinking (pulling) microtubules, the growth of the microtubule toward the cortex (a higher attachment rate also leads to a higher number of pulling microtubules), and a load-accelerated dissociation of shrinking (pulling) microtubules. Assuming that only a small fraction of the microtubules are pulling and pushing, the force on the pole then depends on MTOC velocity, v, according to

$$f_{ext}(v) \approx \left[N_0^+\left(\gamma_0^+ + \frac{f_0^+}{v_+} - \gamma_0^+ \frac{f_0^+}{f_{cat}}\right) \right. $$
$$\left. + N_0^-\left(\gamma_0^- - \frac{f_0^-}{v_+} - \gamma_0^- \frac{f_0^-}{f_{off}}\right)\right]v, \qquad 7.$$

where $\gamma_0^+ (\gamma_0^-)$ is the slope of the force-velocity curve associated with pushing (pulling), v^+ is the growth rate, f_{cat} is the compressive force required to increase the catastrophe rate constant e-fold (14), and f_{off} is the tensile force required to increase the detachment rate e-fold. In addition, there is an inertial term (proportional to dv/dt), but we do not calculate it here.

EXAMPLES OF MECHANICAL SIGNALING NETWORKS

The Flagellar Beat

Cilia and flagella are slender cellular appendages of eukaryotic cells. They are motile structures that drive fluid flow and motion through fluids. Their central core is an evolutionarily conserved organelle, the axoneme, that comprises a cylindrical arrangement of nine doublet microtubules surrounding a pair of singlet microtubules (45). Dynein motors located between adjacent doublet microtubules (**Figure 7a,b**) generate shear forces that cause sliding between the doublet microtubules. Passive components such as nexin

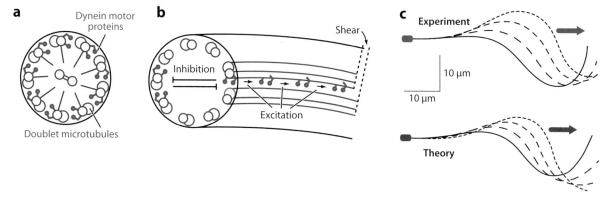

Figure 7

The flagellar beat. (*a*) The dynein motors (*blue*) are anchored at the A-fibers (*open circles*) of the nine doublet microtubules (*red*) and interact with the B-fiber (*eclipsed open circle*) via their motor domains (*blue circles*). The view is from the head. (*b*) Proposed mechanical network underlying the flagellar beat. Dyneins on opposite sides of the cross-section inhibit each other via load-accelerated unbinding. This leads to switching of the activity of the motors from one side to the other, bending the axoneme one way and then the other. Interactions along the axoneme mediated by sliding forces lead to bend propagation. (*c*) Comparison between the measured (*top*) and predicted (*bottom*) beats. The different curves show snapshots at different times (solid lines occur earlier than dashed lines). The arrows show the direction of propagation of the beat; the sperm moves to the left. Modified with permission from Reference 49.

links and constraints at the base convert sliding between adjacent microtubules into bending deformations.

Although the mechanism of bending is known, how the motors generate the regular, oscillatory beat patterns is not. Motor coordination manifests itself in two ways. First, if motors on opposite sides of the axoneme were equally active, the forces would cancel and there would be no bending. The observed bending therefore implies that the motor activity periodically varies from being higher on one side of the axoneme to being higher on the other side. There is a switch. Second, for the generation of a wave-like propagation of bends, the activity of motors must be coordinated along the length of the axoneme. The dynein activity cannot be coordinated by a traveling biochemical or electrical signal because demembranated axonemes can still generate regular beat patterns.

Recently, we have shown that load-accelerated dissociation of axonemal dynein from the microtubule can account for the beat (7, 49). Load-accelerated dissociation leads to (*a*) a negative drag that causes switching of the activity of the dyneins on either side of the axoneme and (*b*) an inertial delay. In combination with a stiffness contributed by the flexural rigidity of the microtubules that tends to return the axoneme to its straight conformation, the negative drag and positive inertia lead to spontaneous oscillations. The oscillations propagate because activation of motors at one place causes shearing of the microtubules, which in turn activates dyneins in adjacent locations along the axoneme. The shape of the beat of bull sperm accords with this theory (**Figure 7c**).

The flagellar beat corresponds to a mode of excitation whose amplitude builds up over many oscillation cycles owing to the negative damping contributed by the dyneins. In this respect, the flagellar beat is analogous to the resonance of a string: It builds up slowly, and the mechanical properties of the two ends—the boundary conditions—are important. The beat is a delocalized, collective phenomenon. This delocalization has several interesting consequences. First, the compliance at the base determines whether the beat propagates from base to tip (as it does for most flagella) or from tip to base (as it does for many cilia such as *Chlamydomonas* and *Paramecia* when swimming forward). Therefore, the change in the direction of propagation of the beat seen during avoidance

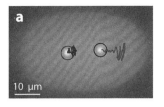
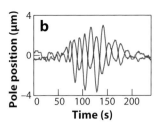
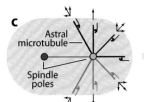

Figure 8

Spindle oscillations in the one-cell *Caenorhabditis elegans* embryo. (*a*) Motion of the anterior (*red*) and posterior (*blue*) poles of the mitotic spindle superimposed on a micrograph showing γ-tubulin localization. (*b*) Plot of the pole positions over time showing the out-of-phase oscillations of the two poles. (*c*) Model showing cortical force generators pulling on astral microtubules. (*d*) A mechanical model in which the springs correspond to a centering stiffness, the dashpots correspond to a positive drag contributed by the cytoskeleton, and the blue elements represent the load dependence of the force generators. (*e*) Equivalent mechanical circuit in which the force generators contribute negative drag (*blue dashpots*) and inertia (*blue square*). Modified with permission from Reference 47.

behavior in these and other organisms could be due to a change in the basal boundary conditions (49) rather than a modulation of the activity of the motors all along the length. Another consequence of the inherently collective nature of the axonemal beat is that for helically beating cilia, there exist modes that beat with opposite handedness, even though the location of the motors with respect to the doublet microtubules (**Figure 7*a***) confers a fixed handedness on the axoneme (26).

Oscillations of the Mitotic Spindle

Oscillations are also pronounced during mitosis, especially during asymmetric cell division. During displacement of the mitotic spindle into the posterior half of the cell, a transverse oscillation, a rocking of the spindle about a pivot point on the cell axis, is sometimes seen. It is prominent in the one-cell embryos of nematode worms (**Figure 8*a,b***), but it is also observed during asymmetric cell division in other tissues such as the developing mammalian cortex (23). The spindle movements are due to cortical force generators located on the cell cortex that attach to and pull on the astral microtubules that grow outward from the two spindle poles (**Figure 8*c***); the pulling forces generate a tension that is dramatically revealed when the spindle is cut with a laser, causing the two poles to fly apart (19). The molecular identity of the force generators is not known. Cytoplasmic dynein

plays an important role in both posterior displacement (12) and transverse oscillations (47); perhaps the force is generated by the dynein ATPase that pulls the pole toward the cortex. Alternatively, the force may be generated during depolymerization by the GTPase activity of tubulin.

Load-dependent dissociation of opposing cortical force generators located off the anterior-posterior axis can account for the oscillations (47), a conclusion confirmed by numerical simulations (37). As in the axoneme, load-accelerated dissociation leads to negative drag, and the delays lead to inertia. In addition, a stiffness element exists that precisely centers the spindle prior to anaphase (perhaps through a microtubule pushing mechanism discussed above). Together, these elements lead to oscillations. Why do the oscillations build up and then die down? The tug-of-war model gives a simple explanation. The pulling activity of the cortical force generators is a control variable: As it increases, a critical point is reached beyond which the net drag coefficient is negative and the oscillations start to build up. As most of the cortical force generators become active, the negative drag decreases again because negative drag relies on a dynamic equilibrium of force generators (Equation 3), and the oscillations die down again. Thus, a monotonic increase in the number of cortical force generations can lead to a monotonic posterior displacement but a transient oscillation.

Insect Flight Muscle

Another classic example of mechanical oscillation is insect flight muscle (48). In some insects such as flies, mosquitoes, and giant water bugs, the oscillation of the wings is not controlled neuronally, as it is for other insects and for most other animals with undulatory locomotion such as worms, lampreys (22), fish, and snakes. Instead the oscillation is mechanical. The neuronal input is not timed with the beat cycle, but rather the beat is self-oscillating. Experiments on isolated insect flight muscle show that it is activated by rapid stretches. Muscle fibers from other species also oscillate under some circumstances, though this might be a different mechanism (52). It will be interesting to determine, perhaps using single-molecule assays, whether load-accelerated dissociation of insect muscle myosin from the actin filaments underlies insect flight muscle oscillation.

Axonal Transport, Bidirectional Transport, and Chromosome Oscillations

The examples so far have been of oscillating systems. However, load-accelerated dissociation of motors may also play important switching roles. There are many examples of organelles that bind oppositely directed motors and have the capacity to move in both directions (reviewed in Reference 64). Examples include membrane-bounded vesicles in axons in which the anterograde kinesins carry the retrograde dynein to the distal end of the axoneme; oil droplets in *Drosophila* oocytes; melanosomes in melanocytes that move outward to darken the skin or inward to lighten it; and the kinetochores of chromosomes, which have both dynein and plus-end-directed kinesins (44). Often these systems display bidirectionality: persistent motion in one direction and then abrupt switching to persistent motion in the other direction (64). An important question is how the activities of the motors are regulated to ensure that both sets of motors are not activated at the same time: If this occurred, a futile battle

between motors might impede the orderly circulation of material from the cell center to the periphery and back. A theoretical analysis suggests that load-accelerated dissociation of motors from the microtubules can account for bidirectional movement (5, 43). Thus, even though biochemical signaling pathways play key roles in regulating organelle transport (64) and chromosome movements, the motors themselves have the capacity to self-organize.

Other Mechanical Forms of Intracellular and Intercellular Communication

It would be wrong to claim that load-dependent dissociation is the only mechanism underlying coordinated cellular movements. Mechanical oscillations of the hair bundle, the mechanoreceptive organelle of sensory hair cells in the inner ear, are thought to be driven by complex feedback loops involving motor proteins (57). Mechanical interactions of cells with the substrate (63) via focal adhesions (6) can control the position of the mitotic spindle (56), cell and tissue shape (30), and cell fate (16). Cortical tension regulates cell division (15), which may involve spontaneous symmetry breaking (46). In addition, cells may use motors to measure filament length (62) to control the length of organelles such as the mitotic spindle and axoneme (40). Load-accelerated dissociation is one of several processes by which cells can sense and respond to mechanical forces.

ANALOGIES TO ELECTRICAL AND CHEMICAL SIGNALING

It is instructive to compare mechanical signaling through load-dependent dissociation with electrical and chemical signaling mechanisms. The electrical analog of a motor protein is an ion channel in a membrane. The electrochemical gradient that drives ion flow through the channel is analogous to the phosphorylation potential of the ATP hydrolysis reaction that drives motor movement. Load-accelerated dissociation of a motor has an ion-channel analogy: voltage-activated channel opening. For example, the probability of a sodium channel being open increases as the voltage is increased. If the voltage sensitivity is sufficiently strong and a threshold voltage is reached, then there is positive feedback such that channel opening allows positively charged sodium ions into the cell, making the cell more positive and in turn opening more channels. This so-called Hodgkin cycle underlies the all-or-nothing action potential. Voltage-activated opening confers negative conductance to an ensemble of sodium channels that is analogous to the negative drag coefficient of an ensemble of motors. The lag between the change in voltage and the change in open probability of a channel gives rise to inductance, the electrical analog to inertia; this can lead to voltage oscillations when the capacitance of the cell (analogous to stiffness in a mechanical system) is taken into account (13). Electrical signaling has its origin in the transistor-like properties of ion channels; a motor with load-accelerated dissociation acts as a mechanical transistor.

However, there are important differences between electrical systems and the mechanical systems that I have been discussing. Although there is a superficial similarity between the action potential of nerve cells and the flagellar beat—they are traveling waves of electrical and mechanical activity, respectively—there is a fundamental difference. The actin potential is an all-or-nothing opening of sodium channels triggered by a local electrical disturbance and whose propagation is not influenced by the electrical properties remote from the site of excitation. By contrast, the flagellar beat is a mode of excitation whose amplitude builds up slowly over many oscillation cycles, and the mechanical properties of the two ends—the boundary conditions—are important.

There are also analogies between mechanical and chemical signaling networks. Autocatalysis is analogous to negative damping. Delays associated with phosphorylation or transcription lead to chemical inertia, and stiffness is conferred by the regulation of protein levels by degradation or dephosphorylation. These biochemical properties are the building blocks of systems biology (2, 61), and many of the circuit principles underlying biochemical signaling pathways also apply to mechanical systems. An important advantage of mechanical signaling over chemical signaling is that the former is fast: Mechanical deformations propagate quickly (at the speed of sound in the protein material) and so will outpace chemical signals that rely on diffusion to propagate (24).

SUMMARY POINTS

1. Cell movements are coordinated in space and time. This coordination may be due, at least in part, to direct mechanical coupling between motor proteins.

2. Mechanical coupling in the form of load-accelerated protein-protein dissociation can lead to positive feedback and switching between attached and detached states.

3. Load-accelerated protein-protein dissociation naturally leads to delays, which in conjunction with positive feedback can lead to oscillations.

4. Several examples of coordinated cellular motions such as the flagellar beat, spindle oscillations, insect flight muscle, axonal transport, and bidirectional motility can be understood in terms of positive feedback provided by load-accelerated dissociation.

5. There are close analogies between mechanical signaling mechanisms and those found in electrical and chemical networks.

1. Load-accelerated dissociation needs to be examined experimentally for other motor proteins (in addition to kinesin) and for protein-cytoskeletal interactions.

2. The direction of load needs to be controlled more precisely to determine which component of the force vector accelerates dissociation.

3. Bistability and oscillations need to be reconstituted with purified proteins.

4. More theory and experiment are needed for the study of nonprocessive motors.

5. Better techniques for studying the mechanics of organelles and whole cells are needed.

6. Do other cellular mechanochemical processes such as transcription and translation and endocytosis and exocytosis involve coordination through mechanical signaling?

DISCLOSURE STATEMENT

The author is not aware of any biases that might be perceived as affecting the objectivity of this review.

ACKNOWLEDGMENTS

I apologize to all the investigators who could not be appropriately cited owing to space limitations. I thank members of the Howard laboratory for thoughtful comments on the manuscript and my colleagues in Dresden with whom these ideas have been developed, especially Stefan Diez, Stephan Grill, Tony Hyman, Frank Jülicher, and Iva Tolić.

LITERATURE CITED

1. Achenbach E. 1972. Experiments on the flow past spheres at very high Reynolds numbers. *J. Fluid Mech.* 54:565–75

2. Alon U. 2007. *An Introduction to Systems Biology*. Boca Raton, FL: Chapman & Hall/CRC

3. Andronov AA, Vitt AA, Kaikin SE. 1987. *Theory of Oscillators*. Mineola, NY: Dover. 815 pp.

4. Asbury CL, Gestaut DR, Powers AF, Franck AD, Davis TN. 2006. The Dam1 kinetochore complex harnesses microtubule dynamics to produce force and movement. *Proc. Natl. Acad. Sci. USA* 103:9873–78

5. Badoual M, Jülicher F, Prost J. 2002. Bidirectional cooperative motion of molecular motors. *Proc. Natl. Acad. Sci. USA* 99:6696–701

6. Balaban NQ, Schwarz US, Riveline D, Goichberg P, Tzuer G, et al. 2001. Force and focal adhesion assembly: a close relationship studied using elastic micropatterned substrates. *Nat. Cell Biol.* 3:466–72

7. **Brokaw CJ. 1975. Molecular mechanism for oscillation in flagella and muscle. *Proc. Natl. Acad. Sci. USA* 72:3102–6**

8. Camalet S, Jülicher F. 2000. Generic aspects of axonemal beating. *N. J. Phys.* 2:1–23

9. Carlson AE, Hille B, Babcock DF. 2007. External Ca^{2+} acts upstream of adenylyl cyclase SACY in the bicarbonate signaled activation of sperm motility. *Dev. Biol.* 312:183–92

10. **Carter NJ, Cross RA. 2005. Mechanics of the kinesin step. *Nature* 435:308–12**

11. Clemen AE, Vilfan M, Jaud J, Zhang J, Bärmann M, Rief M. 2005. Force-dependent stepping kinetics of myosin-V. *Biophys. J.* 88:4402–10

12. Couwenbergs C, Labbé JC, Goulding M, Marty T, Bowerman B, Gotta M. 2007. Heterotrimeric G protein signaling functions with dynein to promote spindle positioning in *C. elegans*. *J. Cell Biol.* 179:15–22

7. Early paper that has the essence of the concept that load-accelerated motor dissociation can lead to spontaneous switching and oscillations.

10. The stall force of kinesin-1 is also a reversal force.

13. Detwiler PB, Hodgkin AL, McNaughton PA. 1980. Temporal and spatial characteristics of the voltage response of rods in the retina of the snapping turtle. *J. Physiol.* 300:213–50

14. Dogterom M, Kerssemakers JW, Romet-Lemonne G, Janson ME. 2005. Force generation by dynamic microtubules. *Curr. Opin. Cell Biol.* 17:67–74

15. Effler JC, Iglesias PA, Robinson DN. 2007. A mechanosensory system controls cell shape changes during mitosis. *Cell Cycle* 6:30–35

16. Engler AJ, Sen S, Sweeney HL, Discher DE. 2006. Matrix elasticity directs stem cell lineage specification. *Cell* 126:677–89

17. Evans E. 2001. Probing the relation between force–lifetime–and chemistry in single molecular bonds. *Annu. Rev. Biophys. Biomol. Struct.* 30:105–28

18. Gennerich A, Carter AP, Reck-Peterson SL, Vale RD. 2007. Force-induced bidirectional stepping of cytoplasmic dynein. *Cell* 131:952–65

19. Grill SW, Gonczy P, Stelzer EH, Hyman AA. 2001. Polarity controls forces governing asymmetric spindle positioning in the *Caenorhabditis elegans* embryo. *Nature* 409:630–33

20. Grill SW, Howard J, Schaeffer E, Stelzer E, Hyman AA. 2003. Forces controlling spindle position in mitosis. *Science* 301:518–21

21. Grill SW, Kruse K, Jülicher F. 2005. Theory of mitotic spindle oscillations. *Phys. Rev. Lett.* 94:108104

22. Grillner S. 1996. Neural networks for vertebrate locomotion. *Sci. Am.* 274:64–69

23. Haydar TF, Ang EJ, Rakic P. 2003. Mitotic spindle rotation and mode of cell division in the developing telencephalon. *Proc. Natl. Acad. Sci. USA* 100:2890–95

24. Helenius J, Brouhard G, Kalaidzidis Y, Diez S, Howard J. 2006. The depolymerizing kinesin MCAK uses lattice diffusion to rapidly target microtubule ends. *Nature* 441:115–19

25. Helenius J, Heisenberg CP, Gaub HE, Muller DJ. 2008. Single-cell force spectroscopy. *J. Cell Sci.* 121:1785–91

26. Hilfinger A, Jülicher F. 2008. The chirality of ciliary beats. *Phys. Biol.* 5:16003

27. Hopfield JJ, Tank DW. 1985. "Neural" computation of decisions in optimization problems. *Biol. Cybern.* 52:141–52

28. Howard J. 2001. *Mechanics of Motor Proteins and the Cytoskeleton*. Sunderland, MA: Sinauer. 367 pp.

29. **Howard J. 2006. Elastic and damping forces generated by confined arrays of dynamic microtubules. *Phys. Biol.* 3:54–66**

30. Ingber DE. 2006. Mechanical control of tissue morphogenesis during embryological development. *Int. J. Dev. Biol.* 50:255–66

31. Jaffe AB, Hall A. 2005. Rho GTPases: biochemistry and biology. *Annu. Rev. Cell Dev. Biol.* 21:247–69

32. **Jülicher F, Prost J. 1995. Cooperative molecular motors. *Phys. Rev. Lett.* 75:2618–21**

33. **Jülicher F, Prost J. 1997. Spontaneous oscillations of collective molecular motors. *Phys. Rev. Lett.* 78:4510–13**

34. Kaupp UB, Kashikar ND, Weyand I. 2008. Mechanisms of sperm chemotaxis. *Annu. Rev. Physiol.* 70:93–117

35. Keren K, Pincus Z, Allen GM, Barnhart EL, Marriott G, et al. 2008. Mechanism of shape determination in motile cells. *Nature* 453:475–80

36. Kovar DR, Pollard TD. 2004. Insertional assembly of actin filament barbed ends in association with formins produces piconewton forces. *Proc. Natl. Acad. Sci. USA* 101:14725–30

37. Kozlowski C, Srayko M, Nedelec F. 2007. Cortical microtubule contacts position the spindle in *C. elegans* embryos. *Cell* 129:499–510

38. Lanzetti L. 2007. Actin in membrane trafficking. *Curr. Opin. Cell Biol.* 19:453–58

39. Le Clainche C, Carlier MF. 2008. Regulation of actin assembly associated with protrusion and adhesion in cell migration. *Physiol. Rev.* 88:489–513

40. Marshall WF. 2004. Cellular length control systems. *Annu. Rev. Cell Dev. Biol.* 20:677–93

41. Marx A, Müller J, Mandelkow E. 2005. The structure of microtubule motor proteins. *Adv. Protein Chem.* 71:299–344

42. Ménétrey J, Llinas P, Mukherjea M, Sweeney HL, Houdusse A. 2007. The structural basis for the large powerstroke of myosin VI. *Cell* 131:300–8

29. Dynamic arrays of microtubules have stiffness and drag.

32. Coupled motor proteins can lead to spontaneous symmetry breaking.

33. Coupled motor proteins can lead to spontaneous oscillations.

43. Müller MJ, Klumpp S, Lipowsky R. 2008. Tug-of-war as a cooperative mechanism for bidirectional cargo transport by molecular motors. *Proc. Natl. Acad. Sci. USA* 105:4609–14

44. Musacchio A, Salmon ED. 2007. The spindle-assembly checkpoint in space and time. *Nat. Rev. Mol. Cell Biol.* 8:379–93

45. Nicastro D, Schwartz C, Pierson J, Gaudette R, Porter ME, McIntosh JR. 2006. The molecular architecture of axonemes revealed by cryoelectron tomography. *Science* 313:944–48

46. Paluch E, van der Gucht J, Sykes C. 2006. Cracking up: symmetry breaking in cellular systems. *J. Cell Biol.* 175:687–92

47. Pecreaux J, Röper JC, Kruse K, Jülicher F, Hyman AA, et al. 2006. Spindle oscillations during asymmetric cell division require a threshold number of active cortical force generators. *Curr. Biol.* 16:2111–22

48. Pringle JW. 1978. The Croonian Lecture, 1977. Stretch activation of muscle: function and mechanism. *Proc. R. Soc. London Ser. B* 201:107–30

49. Riedel-Kruse IH, Hilfinger A, Howard J, Jülicher F. 2007. How molecular motors shape the flagellar beat. *HFSP J.* 1:192–208

50. Rodionov VI, Borisy GG. 1997. Self-centring activity of cytoplasm. *Nature* 386:170–73

51. Rogers GC, Rogers SL, Schwimmer TA, Ems-McClung SC, Walczak CE, et al. 2004. Two mitotic kinesins cooperate to drive sister chromatid separation during anaphase. *Nature* 427:364–70

52. Sasaki D, Fujita H, Fukuda N, Kurihara S, Ishiwata S. 2005. Auto-oscillations of skinned myocardium correlating with heartbeat. *J. Muscle Res. Cell Motil.* 26:93–101

53. Schnitzer MJ, Visscher K, Block SM. 2000. Force production by single kinesin motors. *Nat. Cell Biol.* 2:718–23

54. Selvin HA, Ha T. 2008. *Single Molecule Techniques*. New York: Cold Spring Harbor Lab. Press

55. Strogatz SH. 1994. *Nonlinear Dynamics and Chaos*. Reading, MA: Perseus Books

56. Théry M, Jiménez-Dalmaroni A, Racine V, Bornens M, Jülicher F. 2007. Experimental and theoretical study of mitotic spindle orientation. *Nature* 447:493–96

57. Tinevez JY, Jülicher F, Martin P. 2007. Unifying the various incarnations of active hair-bundle motility by the vertebrate hair cell. *Biophys. J.* 93:4053–67

58. Toba S, Watanabe TM, Yamaguchi-Okimoto L, Toyoshima YY, Higuchi H. 2006. Overlapping hand-over-hand mechanism of single molecular motility of cytoplasmic dynein. *Proc. Natl. Acad. Sci. USA* 103:5741–45

59. Tolić-Nørrelykke IM. 2008. Push-me-pull-you: how microtubules organize the cell interior. *Eur. Biophys. J.* 37:1271–78

60. Tsygankov D, Fisher ME. 2007. Mechanoenzymes under superstall and large assisting loads reveal structural features. *Proc. Natl. Acad. Sci. USA* 104:19321–26

61. Tyson JJ, Chen KC, Novak B. 2003. Sniffers, buzzers, toggles and blinkers: dynamics of regulatory and signaling pathways in the cell. *Curr. Opin. Cell Biol.* 15:221–31

62. Varga V, Helenius J, Tanaka K, Hyman AA, Tanaka TU, Howard J. 2006. Yeast kinesin-8 depolymerizes microtubules in a length-dependent manner. *Nat. Cell Biol.* 8:957–62

63. Vogel V, Sheetz M. 2006. Local force and geometry sensing regulate cell functions. *Nat. Rev. Mol. Cell Biol.* 7:265–75

64. Welte MA. 2004. Bidirectional transport along microtubules. *Curr. Biol.* 14:R525–37

65. Westermann S, Wang HW, Avila-Sakar A, Drubin DG, Nogales E, Barnes G. 2006. The Dam1 kinetochore ring complex moves processively on depolymerizing microtubule ends. *Nature* 440:565–69

47. Load-accelerated dissociation accounts for the transverse oscillations of the mitotic spindle in the one-cell *C. elegans* embryo.

49. Load-accelerated dissociation accounts for the shape of the flagellar beat.

53. The processivity of kinesin-1 is decreased by the load force.

Biochemical and Structural Properties of the Integrin-Associated Cytoskeletal Protein Talin

David R. Critchley

Department of Biochemistry, University of Leicester, Leicester LE1 9HN, United Kingdom; email: drc@le.ac.uk

Annu. Rev. Biophys. 2009. 38:235–54

First published online as a Review in Advance on February 5, 2009

The *Annual Review of Biophysics* is online at biophys.annualreviews.org

This article's doi:
10.1146/annurev.biophys.050708.133744

Key Words

vinculin, actin, PIP2, Rap1, focal adhesions, cytoskeleton

Abstract

Interaction of cells with the extracellular matrix is fundamental to a wide variety of biological processes, such as cell proliferation, cell migration, embryogenesis, and organization of cells into tissues, and defects in cell-matrix interactions are an important element in many diseases. Cell-matrix interactions are frequently mediated by the integrin family of cell adhesion molecules, transmembrane $\alpha\beta$-heterodimers that are typically linked to the actin cytoskeleton by one of a number of adaptor proteins including talin, α-actinin, filamin, tensin, integrin-linked kinase, melusin, and skelemin. The focus of this review is talin, which appears unique among these proteins in that it also induces a conformational change in integrins that is propagated across the membrane, and increases the affinity of the extracellular domain for ligand. Particular emphasis is given to recent progress on the structure of talin, its interaction with binding partners, and its mode of regulation.

Contents

INTRODUCTION

Talin was initially purified from human platelets and chicken gizzard, and key early discoveries were that talin is localized in the adhesion junctions formed between cells and the extracellular matrix (focal adhesions, FAs) (12) and that it binds two other FA proteins, namely vinculin (13) and integrins (44). The primary amino acid sequence of talin was deduced from mouse (89) and chicken (43) cDNAs, and although this identified a single talin isoform, it is now clear that there are two vertebrate talin genes (*Tln1* and *Tln2*) (70, 95) encoding similar proteins (74% identical). *Tln2* is the ancestral gene and is larger and more complex than *Tln1*. To date, most studies have focused on talin1, but talin2 shares the same domain structure and binds many of the same ligands. However, *Dictyostelium* has two talin genes with distinct functions (104).

FA: focal adhesion

PTB domain: phosphotyrosine-binding domain

DOMAIN STRUCTURE OF TALIN

Talin ($\sim$270 kDa) comprises a globular N-terminal head ($\sim$50 kDa) and a large flexible C-terminal rod domain ($>$200 kDa) (**Figure 1a**) and reportedly exists in equilibrium between monomeric and dimeric forms (69). The talin head contains a FERM domain (residues 86–400) homologous to that found in the band 4.1/ezrin/radixin/moesin family of cytoskeletal proteins, and composed of F1, F2, and F3 domains. The X-ray structure of F2F3 showed that the F3 domain has a phosphotyrosine-binding (PTB)-like fold (33), but attempts to crystallize the entire FERM domain have been unsuccessful probably because the F1 domain (a ubiquitin-like fold) contains a unique 30-amino-acid insert that likely forms a large unstructured loop. The loop contains two of the most abundant phosphorylation sites (T144/T150) in platelet talin (88). The 85 amino acids preceding F1 have been largely ignored, but our recent NMR studies show that it also is a ubiquitin-like fold, referred to here as the F0 domain (B.T. Goult & D.R. Critchley, unpublished data). The talin head and rod are readily liberated from the intact protein by a variety of proteases, including calpain-II, that cleave within the linker region (residues 401–481) between the two domains (89). This region, which is predicted to be largely though not totally unstructured, contains numerous phosphorylation sites (88), and it will be important to determine the effects of phosphorylation on its structure, dynamics, and protease sensitivity. In broad agreement with secondary structure predictions (65), the talin rod contains 62 helices that are organized into a series of helical bundles (36), followed by a single C-terminal helix that forms an antiparallel homodimer (34), although the relative orientation of the two subunits within the dimer is uncertain (**Figure 1b**). The N-terminal region of the rod starts with a five-helix bundle (residues 482–655) (81), and the C-terminal half of the rod is also made up of a series of five-helix bundles (34) (**Figure 1c,d**). However, the tertiary structure of the helices between these regions

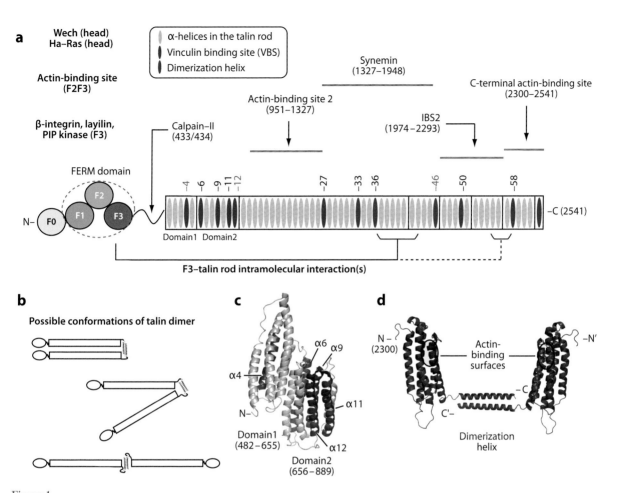

a

Wech (head)
Ha–Ras (head)

Actin-binding site
(F2F3)

β-integrin, layilin,
PIP kinase (F3)

- α-helices in the talin rod
- Vinculin binding site (VBS)
- Dimerization helix

Synemin
(1327–1948)

C-terminal actin-binding site
(2300–2541)

Actin-binding site 2
(951–1327)

IBS2
(1974–2293)

FERM domain

Calpain–II
(433/434)

−4 −6 −9 −11 −12   −27 −33 −36   −46 −50   −58

F2

F1 F3

N– F0   −C (2541)

Domain1 Domain2

F3–talin rod intramolecular interaction(s)

b

Possible conformations of talin dimer

c

α6 α9

α4

N–

α11

Domain1
(482–655)

α12

Domain2
(656–889)

d

N –
(2300)

Actin-
binding
surfaces

–N′

–C

C′–

Dimerization
helix

Figure 1

Domain structure of talin. (*a*) The N-terminal talin head (1–400) contains a FERM domain comprising F1, F2, and F3 domains preceded by the F0 domain. The talin rod (482–2541) contains 62 amphipathic α-helices, the most C-terminal of which is required for talin dimerization (*blue*). The position of the various ligand-binding sites is indicated, including the intramolecular interaction between F3 and the talin rod. Helices containing VBSs are numbered—helices 4, 12, and 46 contain VBS1–3 (4). (*b*) Possible orientations of the subunits in the talin dimer. (*c*) Model of domains 1 and 2 at the N-terminal region of the talin rod (PDB 1xwx). Vinculin-binding helices are in red. (*d*) Model of the dimeric C-terminal THATCH domain (2300–2541) indicating the position of the actin-binding surface (PDB 2jsw and 2qdq).

has proved difficult to establish. The nature of the interactions between bundles also requires further investigation. The X-ray crystal structure of talin 482–789 shows that residues 656–789 form a four-helix bundle that packs tightly against the talin 482–655 five-helix bundle in a staggered fashion via an extensive hydrophobic interface (81). However, there are no equivalent hydrophobic surfaces on the other bundles for which we have structural information.

TALIN-BINDING PARTNERS

The talin F3 FERM domain binds the cytoplasmic domains of β-integrins (15) and the hyaluronan receptor layilin (8, 107), as well as the C-terminal region of PIPK1γ90 (29, 59), a splice variant of phosphatidylinositol-4-phosphate 5 kinase type Iγ that is thought to regulate the assembly of FAs (**Figure 1a**). The talin head also contains an F-actin-binding site (53), and recent proteomic and

PIPK1γ90: a splice
variant of phos-
phatidylinositol(4)
phosphate 5 kinase
type Iγ

genetic studies have identified Ha-Ras (38) and the *Drosophila* protein Wech (61) (mouse homologue mLIN41), respectively, as additional FERM domain binding partners. Because Wech also binds integrin-linked kinase, this may serve to cross-link different integrin complexes (58). Pulldown assays indicate that the FERM domain also binds focal adhesion kinase (FAK) (19), a protein tyrosine kinase that regulates FA dynamics, although this interaction is probably indirect (S. Lorenz & M. Noble, personal communication). Finally, the talin head binds acidic phospholipids and inserts into lipid bilayers (28, 40, 42, 76, 79). The talin rod contains an additional integrin-binding site, IBS2 (68, 90, 103, 110), and at least two actin-binding sites (ABSs) (43), the best characterized of which is at the C terminus (34, 64, 94, 96), a region homologous to the Sla2p/HIP1R family of actin-binding proteins. It also contains multiple binding sites for vinculin (36), which itself has numerous binding partners including F-actin (113). A binding site for the muscle-specific intermediate filament protein α-synemin has also recently been identified in the talin rod, providing a mechanism for linking integrin/talin/actin complexes to the intermediate filament protein network (98). Finally, yeast two-hybrid and proteomic screens have identified several additional talin-binding proteins (111).

STUDIES ON TALIN AT THE CELLULAR AND ORGANISMAL LEVEL

Apart from linking integrins to the actin cytoskeleton, talin plays a key role in the energy-dependent activation of αIIbβ3-, αvβ3-, and β1-integrins. Thus, the talin head induced integrin activation in Chinese hamster ovary (CHO) cells expressing αIIbβ3, whereas shRNA-mediated talin knockdown reduced activation (100). Integrin activation is thought to involve disruption of a salt bridge between the α- and β-subunits, resulting in separation of their cytoplasmic tails (14, 16), and a role for the talin head in this process is supported by exper-

iments in which it reduced the FRET efficiency between the tails of αL-CFP and β2-YFP integrin subunits expressed in K562 cells (51). However, a number of recent studies show that integrin activation by talin requires members of the kindlin family of FERM domain proteins, although kindlins themselves do not activate integrins (62, 72, 73). Kindlins bind to the more C-terminal of the NPxY motifs in β-integrin tails that are not involved in talin binding, and it been suggested that kindlins localized to the membrane via their PH domain might reduce the flexibility of the β-integrin tails, thereby facilitating talin binding (62).

Disruption of both *Tln1* alleles in mouse embryonic stem cells confirms previous studies that talin1 is required for efficient cell spreading and FA assembly (87). Optical trap experiments using *Tln1*(−/−) cells and fibronectin-coated (type III repeats 7–10) silica beads show that talin1 is required for a characteristic 2pN slip bond or clutch between fibronectin-integrin complexes and the actin cytoskeleton, and for strengthening the integrin-cytoskeleton linkage (49). Problematically, *Tln1*(−/−) cells upregulate talin2, which compensates for loss of talin1, and Zhang et al. (112) therefore used an shRNA plasmid to suppress talin2. Surprisingly, talin depletion did not affect matrix-induced activation of Src or the initiation of cell spreading, but the cells were unable to (*a*) support integrin or FAK activation, (*b*) maintain the spread phenotype, (*c*) assemble FAs, or (*d*) exert traction force on the substrate. Expression of the talin head but not the rod rescued integrin activation and cell spreading, but not FAK activation or FA assembly, indicating that only full-length talin can provide the linkage between the matrix and the actin cytoskeleton required for FA assembly and signaling. The talin head also promotes integrin clustering (21) and thus might also induce the avidity changes required for efficient ligand binding, and cleavage of the talin head from the rod by calpain-II is required for FA turnover and cell migration (31).

Gene disruption studies in worms (23), flies (101), and mice (71) confirm that talin is essential for a variety of integrin-mediated developmental events, although a totally unexpected role for talin as a transcriptional suppressor of cadherin gene expression has emerged in flies (5). Disruption of the *Tln1* gene in mice is embryonic lethal due to arrested gastrulation (71), whereas mice that are homozygous for a *Tln2* gene trap allele are viable and fertile (20). More recently, a conditional allele has been used to investigate the role of talin1 in specific tissues. Importantly, such studies prove that talin1 is essential for activation of platelet integrins in vivo (78, 84) and for the integrity of the myotendinous junction (22), and that both talin1 and PIPK1γ90 (although not other PIPK1 isoforms) are required to strengthen the membrane cytoskeletal interface in megakaryocytes (105). Remarkably, however, neither integrins nor talin is required for the migration of dendritic cells from the skin into lymph nodes (52).

Structural Studies on the Integrin-Binding Site in the Talin Head

Although evidence for an interaction between integrins and talin was presented over 20 years ago (44), it was not until recombinant integrin tails became available that this observation was confirmed (85). Key discoveries were the mapping of the β-integrin-binding site to the talin head, and subsequently the F3 PTB-like domain, and the demonstration that the membrane proximal NPxY motif in the β-integrin tails was a key determinant of binding (16). The PTB domain of ICAP-1 competes with talin for binding to this same region of the β1-integrin tail and in so doing reduces the affinity of integrins for fibronectin (67).

The structure of the F3 PTB-like domain bound to residues [739]WDTANNPLYDEA[750] of the β3-integrin tail shows that the integrin interacts predominantly with the hydrophobic surface on strand S5 of the F3 domain, and mutations in S5 markedly reduce integrin binding (33) (**Figure 2*a,b***). W739 in the β3-tail occupies a hydrophobic pocket in F3 created by R358, A360, and Y377, and a R358A

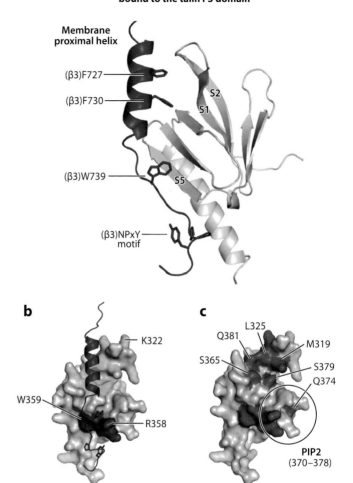

a Structure of the β3-integrin tail bound to the talin F3 domain

Membrane proximal helix

(β3)F727

(β3)F730

S2
S1

(β3)W739

S5

(β3)NPxY motif

b

K322

W359

R358

c

L325
Q381

M319

S365

S379
Q374

PIP2
(370–378)

Figure 2

Structure of the talin F3 domain. (*a*) Ribbon representation and (*b*) surface representation of F3 (*gray*) bound to the β3-integrin cytoplasmic tail (*blue*) (PDB 2h7d). (*b*) The binding sites for the membrane proximal helix of the β3-tail (*green*) and the NPxY motif (*red*) are indicated. (*c*) Magenta indicates the regions in F3 involved in binding the talin rod (residues 1654–2344). The PIP2-binding region in F3 is circled.

mutation markedly reduces integrin binding. The integrin [744]NPLY[747] sequence forms a β-turn, and Y747 projects into an acidic pocket in F3, whereas the equivalent region in those PTB domains that bind phosphotyrosine is strongly basic. Phosphorylation of Y747 would therefore be predicted to inhibit talin binding, and this has been confirmed by NMR using ^{15}N-labeled F3 domain and a β3-integrin tail pY747-phosphopeptide, although the affinity was only reduced 2-fold compared with a 400-fold increase in affinity for the Dok1 PTB domain (80). The WVENEIYY sequence in the layilin cytoplasmic domain (107) and the WVYSPLHY sequence at the C terminus of PIPK1γ90 (2, 24) interact with F3 in a broadly similar manner. More recently, NMR studies have shown that F3 also interacts with the membrane proximal helix of the β3-tail, and that F727 and F730, which are on the same face of the membrane proximal helix, bind to a hydrophobic pocket in F3 made up of the flexible loop between β-strands S1 and S2 (108) (**Figure 2a**). Significantly, mutation of either F727 or F730 in β3-integrin, or the interacting residues in F3 (notably L325), markedly reduced activation of αIIbβ3-integrin expressed in CHO cells. Other PTB domain proteins that bind β-integrin tails lack this flexible loop and do not activate integrins. The authors propose a model in which talin F3 initially binds to the β3-integrin NPxY motif and subsequently engages the membrane proximal helix, triggering separation of the α- and β-integrin tails and leading to integrin activation (108). This hypothesis is supported by the results of recent in-cell biomolecular fluorescence complementation studies in which a talin L325R mutant was recruited to αIIbβ3-integrin tails but was unable to activate the integrin (106). The integrin-talin complex may be further stabilized by an interaction between talin K322 (in the loop between β-strands S1 and S2 of talin F3) and acidic membrane lipids (108). Whether other regions of the talin head contribute to integrin binding has not been investigated, but recent studies show that talin F3 alone is not sufficient to activate β1-integrins and that the F0 and F1 domains are also required (10).

Characterization of Integrin-Binding Site 2 in the Talin Rod

Early indications that integrins might also bind to the talin rod (44) were eventually confirmed by Xing et al. (110), who showed that purified platelet αIIbβ3-integrin bound in a dose-dependent manner to microtiter wells containing the C-terminal region (residues 1984–2541) of the talin rod. More recently, FRET studies have shown that this same talin rod fragment is recruited to β1-integrin tails in cells plated on fibronectin, although no FRET signal was observed between β1-integrin and the talin head, possibly reflecting the transient nature of this interaction (82). The αIIbβ3-integrin-binding site in the talin rod was further defined to residues 1984–2113, and as for the talin head, binding was dependent on the integrin NPxY sequence, although a talin 1984–2344 polypeptide was unable to activate αIIbβ3-integrin expressed in CHO cells (103). Subsequently, Moes et al. (68) have shown that H50, a single helix in the talin rod, binds GST-β3-integrin tails in a blot assay, and they have recently shown that mutation of K2085 and K2089 on the hydrophilic surface of helix 50 abolished binding of talin 1843–2108 to GST-β3-integrin tails on a surface plasmon resonance chip (90). Moreover, a polyclonal antibody against H50 and H51 inhibited integrin binding. The authors also showed that mutation of two membrane proximal residues, E726 and E733, in the β3-integrin tail blocked binding to talin 1843–2108, and they propose an attractive model in which these acidic residues interact with K2085/K2089 on H50 of the talin rod. E726 and E733 are on the same side of the membrane proximal helix of the β3-integrin tail as D726, which forms the salt bridge with R998 in the αIIb-subunit that keeps the integrin in a low-affinity state (16). Therefore, E726/E733 might be masked by the α-subunit in the inactive form of the integrin, suggesting that IBS2 might only interact with activated integrins (90). The crystal structure of talin 1974–2293 (which contains IBS2) shows two five-helix bundles, the more N-terminal of which contains H50. However, β-integrin tail–binding assays show that both domains are required for high-affinity integrin binding, suggesting that the interaction may be more complex (A.R. Gingras, M. H. Ginsberg & D.R. Critchley, unpublished data).

INTERACTIONS OF TALIN WITH ACTIN

The literature on the interaction between talin and actin is confusing, in part because assays have been performed at different salt concentrations and pH levels. Indeed, initial studies with chicken talin failed to detect any interaction (13). Subsequently, chicken talin was shown to bind to both G- and F-actin, to stimulate polymerization of G-actin (74), and to augment the gelation of actin induced by α-actinin (75). Platelet talin was also reported to stimulate filament nucleation, although these experiments were conducted at a 1:1 molar ratio (50), and subsequent studies show that talin is much less efficient at actin nucleation than gelsolin (64). Using fluorescently labeled G-actin, researchers have determined a binding stoichiometry of 1:3 talin:G-actin for both gizzard (39) and platelet talin (50). The K_d value for gizzard talin was 0.3 μM and the association rate constant determined by stopped flow was $7 \times 10^6 M^{-1} \times s^{-1}$ with a calculated dissociation rate constant of 2–3 s^{-1}. Studies with highly purified gizzard talin have shown that it binds to F-actin more efficiently below pH 7.0 and in low salt buffers at 37°C, and under these conditions talin induces both cross-linking and bundling of actin filaments as judged by viscosity and electron microscopy (EM) assays (92). The number and location of the ABSs in talin have also been controversial. Several reports indicate that the talin rod but not the head binds to F-actin (74, 76). However, using GST-talin fusion proteins, Hemmings et al. (43) showed that the talin head as well as two distinct regions of the talin rod (**Figure 1**) cosedimented with F-actin. Subsequent studies confirmed that the talin head binds to F-actin, and EM image reconstruction experiments showed that the F2F3 domain bound to domain 4 of actin (53). Therefore, talin has a number of ABSs spread along the length of the molecule, a finding consistent with a binding stoichiometry of

1:3 (39, 50) and reminiscent of recent studies on filamin (77).

We have recently used NMR and X-ray crystallography to determine the structure of the C-terminal ABS of talin1 (34). It comprises a five-helix bundle followed by the C-terminal helix required for dimer formation (**Figure 1d**), and it has a structure similar to the homologous region of the HIP1R protein determined by Brett et al. (11). This domain is now referred to as a THATCH domain (talin/HIP1R/Sla2p actin-tethering C-terminal homology). The ABS maps to a conserved hydrophobic surface on helices 3 and 4 of the five-helix bundle that is flanked by basic residues. Helix 1, which packs against the opposite face of the bundle, negatively regulates actin binding (94), and deletion of this helix or mutation of the hydrophobic residues involved in helix 1 packing significantly increases actin binding (34, 94). NMR studies show that removing helix 1 causes a substantial conformational change in the THATCH domain, suggesting that it stabilizes the domain in a low-affinity state. S2338 in helix 1 is phosphorylated in platelet talin (88), although whether this regulates actin binding has not been evaluated. The X-ray structure of the C-terminal helix shows that it forms an antiparallel coiled-coil dimer that is stabilized by intermolecular nonpolar interactions and inter- and intramolecular salt bridges (34). F-actin only binds efficiently to the THATCH dimer (34, 96), and mutagenesis of highly conserved solvent-exposed residues on one face of the dimerization helix suggests that they may also contribute to actin binding (34). Small-angle X-ray scattering (SAXS) experiments indicate that the talin THATCH dimer adopts an elongated structure (124 Å long) in solution, and the high-resolution structure of the five-helix bundle and dimerization domain can be readily modeled within the SAXS envelope. EM and image reconstruction show that the THATCH dimer binds to three actin monomers along the long pitch of the same actin filament, and it does not cross-link F-actin (34). Presumably, the actin-bundling activity of talin is explained by the presence of the other ABSs in talin (43).

EM: electron microscopy

Actin binding to the talin THATCH domain is markedly reduced above pH 6.9, whereas binding of the talin head to F-actin is much less pH sensitive (53). The possible significance of this observation has recently emerged from the studies of Barber and colleagues (26), who have previously shown that proton efflux mediated by the Na/H antiport is required for FA turnover and cell migration. On the basis of computer simulations and NMR data, they suggest a model in which protonation of His2418 in the talin THATCH domain allosterically regulates actin binding (97). Therefore, a localized increase in pH may weaken the linkage between integrin/talin complexes and F-actin and facilitate FA turnover. Indeed, mutation of His2418Phe increased the stability of FAs and reduced cell migration.

INTERACTIONS OF TALIN WITH VINCULIN

Initial studies identified at least three vinculin-binding sites (VBSs) in the talin rod, and these were further defined to three short peptide sequences (VBS1–3), each corresponding to a single amphipathic α-helix (4) (**Figure 1a**). Conversely, the talin-binding site in vinculin was localized to residues 1–258 within the globular vinculin head (3). Much of cellular vinculin exists in an autoinhibited state, and recent structural studies show how the talin-binding site in vinculin is masked by a vinculin head/tail interaction (1, 7, 47). Vinculin 1–258 (Vinculin domain 1, Vd1) is composed of seven α-helices organized into two four-helix bundles bridged by a common long helix (α-helix4), and it binds the vinculin tail (Vt) with high affinity (K_d 50–80 nM). The structure of the Vd1/Vt complex shows that Vt binds to α-helices 1 and 4 of Vd1 and in so doing buries about 1300 Å of the solvent-accessible surface of Vd1 (**Figure 3a**). Binding of the talin VBS3 peptide induces a marked conformational change in Vd1 that distorts the Vt-binding surface, displacing Vt, and the VBS3 peptide itself is engulfed in a hydrophobic groove formed predominantly by α-helices 1 and 2 in Vd1 (47) (**Figure 3b**). Similar results have been reported for Vd1 in complex with other VBSs (36, 81). However, structures of the entire vinculin molecule (1, 7) show that Vt makes additional contacts with other vinculin head domains, which further stabilizes the autoinhibited form of the molecule. This has led to the suggestion that vinculin activation involves a combinatorial mechanism and only occurs when two or more of its binding partners are in close proximity (1, 17), although others have suggested that exposure of the VBS sequences in talin alone may be sufficient to activate vinculin (6, 48). Vd1 contains a second surface-exposed hydrophobic pocket that can bind VBS-like peptides (102), and perhaps vinculin can bind talin by more than one mechanism. This may explain why FRET studies indicate that vinculin can localize to FAs even in the autoinhibited state (18).

At the same time, the structure of the N-terminal part of the talin rod (residues 482–655) was determined (81), a five-helix bundle in which α-helix4 is equivalent to VBS1 (4) (**Figure 1c**). Peptide arrays of VBS1 alanine mutants were used to identify the key vinculin-binding determinants, five hydrophobic residues (L608, A612, L615, V619, and L623) on one face of the VBS1 amphipathic α-helix (81). These residues are buried within the hydrophobic core of the five-helix bundle and contribute to the interactions that stabilize the fold. Indeed the five-helix bundle binds Vd1 only weakly, but removal of α-helix 5 resulted in a marked increase in binding. Remarkably, comparison of the structures of the Vd1/VBS1 complex with talin 482–655 shows that Vd1 α-helices 1–4 occupy the equivalent positions in relation to the VBS1 helix as α-helices 1, 5, 2, and 3 in the talin five-helix bundle, and key hydrophobic contacts are maintained by interactions with similar side chains in Vd1. The result is that the talin VBS1 helix is extracted from its own five-helix bundle and forms an equivalent five-helix bundle with the four helices present in the N-terminal region of Vd1 (81) (**Figure 3b**).

As the major determinants of vinculin binding are hydrophobics on one surface of an amphipathic helix, it was perhaps unsurprising

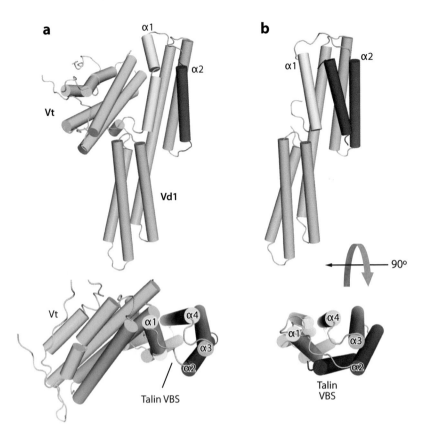

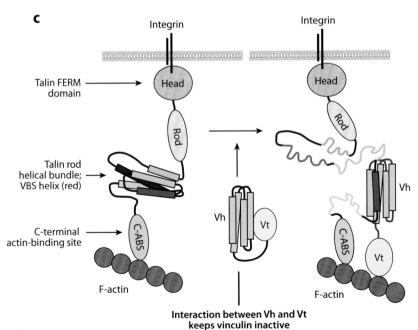

a

α1
α2
Vt
Vd1

Vt
α1 α4
α3
α2
Talin VBS

b

α1 α2

90°

α4
α1 α3
α2
Talin
VBS

c

Integrin

Talin FERM
domain → Head

Rod

Talin rod
helical bundle;
VBS helix (red) →

C-terminal
actin-binding site → C-ABS

F-actin

Vh Vt

**Interaction between Vh and Vt
keeps vinculin inactive**

Integrin

Head

Rod

Vh

C-ABS

Vt

F-actin

Figure 3

Interaction of talin with vinculin. (*a*) Structure of the complex between vinculin Vd1 (residues 1–258) and Vt (*green*) (PDB 1rke) and (*b*) Vd1 and the talin VBS3 helix (*red*) (PDB 1xwj). Note the change in orientation of the α1- and α2-helices in the two complexes. (*c*) Cartoon of the integrin/talin/actin complex (*left*). Vinculin is shown in the inactive form owing to an intramolecular interaction between the Vh (*light blue*) and the vinculin tail Vt (*yellow*). Only the first four N-terminal helices of Vh are shown. Once vinculin is activated, Vh extracts a vinculin-binding helix (*red*) from a five-helix bundle in the talin rod, forming a new five-helix bundle. Vt is suggested to cross-link talin to F-actin. Talin helices that do not bind vinculin are suggested to unfold (*right*). Abbreviations: ABS, actin-binding site; VBS, vinculin-binding site; Vd1, vinculin domain 1; Vh, vinculin head; Vt, vinculin tail.

that eight additional helices were identified that bound strongly to vinculin, plus an equivalent number of weaker binders (36). The nature of the binding specificity was determined using mutant peptides, and the consensus for vinculin binding was 1-LxxAAxxVAxxVxxLLxxA-19. Position 8 on turn 3 of the helix is the most restrictive and only allows Ileu, Leu, or Val, whereas positions 1, 8, 12, 15, and 16 require an aliphatic side chain and will not tolerate Ala. Positions 4, 5, and 9 are less restrictive, and position 11 will allow a charged side chain. However, no one position is absolutely critical for binding and there is little sequence identity across the 11 strong VBSs. Indeed, the affinities of individual VBSs for vinculin appear to vary (48). Helices that bind weakly or not at all generally violate the consensus sequence in one or more positions, and the common occurrence of alanine residues probably explains why most helices within the rod do not bind vinculin.

What regulates the VBSs in talin remains to be explored, but several factors require consideration. First, talin can adopt a globular conformation owing to an interaction between the talin head and rod (69, 109), and this appears to inhibit vinculin binding. Thus, the K_d value for Vd1 binding to intact talin determined by isothermal titration calorimetry confirms that the VBSs are in a low-affinity state ($K_d \sim 8.9 \pm 3.4$ μM) (83). The binding affinity is slightly increased by PIP2 (K_d 4.7 $\pm$ 1.6 μM), which inhibits the talin head/rod interaction (37), and the isolated rod domain showed a significant increase in affinity (K_d 2.4 $\pm$ 1.2 μM) over intact talin. Second, the stability of the helical bundles containing the VBS appears important. For example, domain 1 of the talin rod (residues 482–655) that contains VBS1 does not bind Vd1 efficiently at 37°C unless helix 5 is removed. Differential scanning calorimetry shows that the T_m values of the five-helix and four-helix bundles are 60.2°C and 53.9°C, respectively. Together with the higher unfolding enthalpy of the five-helix bundle, this suggests that helix 5 stabilizes the bundle, making VBS1 in helix 4 unavailable for binding (83). Domain 2 of the talin rod is unique in containing a cluster

of four VBSs, although the C-terminal boundary of this domain remains uncertain. Initial crystallographic studies on a 482–789 polypeptide demonstrated that residues 656–789 form a four-helix bundle containing two of the four VBSs (81). The T_m of this polypeptide (70.2°C) is greater than the preceding five-helix bundle alone (60.2°C), indicating that interactions between domains can affect the stability of individual bundles. This increased stability inhibits vinculin binding to the talin 482–789 polypeptide, and it binds Vd1 weakly even at elevated temperatures (83). We have also determined the NMR structure of a talin 755–889 polypeptide, a four-helix bundle that overlaps the 482–789 polypeptide by one helix, and on the basis of these two structures, we have modeled domain 2 (residues 656–889) as a seven-helix bundle (30) (**Figure 1c**). Remarkably, a polypeptide spanning domains 1 and 2 (residues 482–889) binds Vd1 with high affinity (K_d 0.14 $\pm$ 0.1 μM) (83), and perhaps the presence of four VBSs within domain 2 somehow facilitates vinculin binding. Support for this idea comes from studies on the talin 755–889 four-helix bundle that contains three of the four VBSs in domain 2. This has a stable fold (T_m 74.2°C), yet surprisingly, it binds three Vd1 molecules even at room temperature. An unusual feature of this fold is the presence of two threonine pairs (T775/T809 and T833/T867) within the hydrophobic core of the bundle (30), and substitution of these with more hydrophobic residues dramatically increased bundle stability and markedly reduced Vd1 binding (83).

Intriguingly, NMR studies show that whereas the three VBS α-helices in talin 755–889 become immobilized in the complex with Vd1, α-helix 2 unfolds (30). Similar results were obtained with a talin 1843–1973 four-helix bundle that contains a single VBS (35), and Vd1-induced unfolding of this bundle was confirmed by electron paramagnetic resonance (35) (**Figure 3c**). This renders the bundle exquisitely sensitive to proteolysis, and we used this to identify where vinculin binds in full-length talin. On its own, trypsin cleaves talin into head and rod fragments, but in the presence of Vd1 a previously cryptic tryptic cleavage site between residues 888/899 in the talin rod is exposed. This is just C-terminal to the cluster of four VBSs in domain 2 of the rod. Moreover the liberated N-terminal rod fragment (residues 482–888) was almost completely degraded, indicating that Vd1 induces a major conformational change in this region of the molecule (83). Overall, these experiments indicate that in full-length talin, vinculin binds preferentially to the cluster of four VBSs toward the N terminus of the rod, whereas the other sites are cryptic.

Molecular Dynamic Studies on the Talin-Vinculin Interaction

Vinculin recruitment to FAs is induced by applied force (32, 113), and we have previously suggested that force exerted on talin by actomyosin contraction might destabilize the helical bundles in the talin rod and facilitate vinculin binding (113). Indeed, recent studies using magnetic tweezers have shown that physiological forces applied to a talin rod polypeptide spanning domains 1 and 2 (**Figure 1a**) caused stretching and increased vinculin binding (25). Because the α-helical bundles have different stabilities, the force required to activate each VBS may vary, providing a mechanism in which vinculin recruitment is linked to applied force. These ideas have been explored using molecular dynamics, initially on the first five-helix bundle (residues 482–655) in the talin rod that contains VBS1 (α-helix 4). Force induced a conformational change in the bundle, and charged and polar residues on VBS1 that interact strongly with α-helix1 and more weakly with α-helix5 were the site of force transmission. The weaker VBS1-helix 5 interaction breaks and the resulting torque on VBS1 disrupts hydrophobic contacts, with helices 3 and 5 leading to rotation of VBS1, exposing its hydrophobic residues which gradually penetrate the hydrophobic core of Vd1 (55, 56). A similar approach has been used to investigate Vd1 binding to the first 12 helices of the talin rod (45), which starts with the 482–655 five-helix bundle and is thought

to be followed by a seven-helix bundle (656–889), although the latter is a model (**Figure 1c**), not a determined structure (30). This domain is unique in that it contains a cluster of four VBS helices (helices 6, 9, 11, and 12). Force resulted in fragmentation of these two domains into smaller helical bundles by rupturing the hydrophobic interfaces between bundles. The first split occurred between helices 1–8 and helices 9–12, after which helices 1–8 fragmented into helices 1–5 and helices 6–8. Water penetration into the strained helical bundles loosens up the fracture plane, leading to the sequential exposure of the VBS helices, each of which has a unique activation point. The VBSs in helices 6, 9, and 12 are located at the interfaces between bundles and are the first to be exposed, whereas helix 4, which is buried within a five-helix bundle, is only exposed later.

Interaction of Talin with Lipids and Phosphoinositides

Differential scanning calorimetry and lipid monolayer film balance studies provided the first indications that talin partially penetrates the hydrophobic region of membranes (28, 42). This was confirmed using liposomes containing a photoactivatable radio-labeled phospholipid probe in which the reactive group was on the apolar region of the molecule (40), and it was the talin head, not the rod, that was labeled (79). Similarly, the talin head but not the rod bound to phosphatidylserine-containing vesicles (76). The fact that talin is surface active is also consistent with the finding that it creates holes in liposomes (91). Using secondary structure predictions, Seelig et al. (93) identified three short amphipathic regions in the talin head as potential membrane anchors. Analysis of the interaction of the equivalent synthetic peptides with lipid monolayers or unilamellar vesicles using circular dichroism, isothermal titration calorimetry, and monolayer expansion measurements showed that residues 385–406 spanning the C-terminal boundary of the talin F3 FERM domain penetrated negatively charged membranes with a partition coefficient of

$K_{app} = 1.1 \pm 0.2 \times 10^5 \text{ M}^{-1}$. Membrane penetration was largely entropy driven with a free energy of binding of $\Delta G_0 = -9.4 \text{ kcal mol}^{-1}$. This is comparable to that observed with myristoylated proteins, consistent with the proposal that this region has the properties of a membrane anchor. The 385–406 peptide is also fusogenic and promotes membrane destabilization (46). Calculations based on the asymmetric hydrophobic potential along the 385–406 α-helix suggest that it inserts into the membrane at an angle of 18°, a value consistent with the penetration area of 160 Å determined by the monolayer expansion technique. The authors speculate that talin may relieve membrane tension at the leading edge, facilitating the actin polymerization required for membrane protrusion, while its fusogenic potential may support incorporation of membrane vesicles.

TALIN CAN ADOPT A GLOBULAR CONFORMATION

Sedimentation velocity, gel filtration, and rotary shadowing EM studies show that talin is more compact in low salt buffers, whereas it is a flexible elongated molecule (~60 nm in length) in high salt buffers (69). In contrast, the isolated talin rod was elongated under both conditions, indicating that the head is required for talin to adopt a globular conformation. More recent studies on negatively stained talin confirmed that it adopts a compact structure in low salt, whereas in 0.15 M KCl it appeared U-shaped (56 nm in length), with 10–12 globular domains (~3.8 nm in diameter), each with a calculated molecular mass of ~24 kDa (109). This fits reasonably well with the mass of five-helix bundles (18 kDa). The isolated talin head contains two such domains, and our recent SAXS studies show that the talin head adopts a more open structure than the typical globular FERM domain, possibly because it contains an additional F0 domain (B.T. Goult & D.R. Critchley, unpublished data).

The recent elegant studies of Goksoy et al. (37) have now provided new insights into the nature of the interaction between the talin head and rod. Using NMR, they showed that ^{15}N-labeled talin F3 binds to a talin rod fragment spanning residues 1654–2344, an interaction confirmed by surface plasmon resonance (K_d 0.57 μM). Within this region, two nonoverlapping binding sites for F3 were identified, one (residues 1654–1848) with a K_d of 3.6 μM and the other (residues 1984–2344) with lower affinity (K_d of ~78 μM). These results establish a biochemical basis for the talin head/rod interaction, although the full extent of the intramolecular interactions that allow talin to adopt a globular conformation requires further investigation. NMR chemical shift data show that the talin rod–binding site in F3 overlaps that for the membrane proximal helix of the β3-integrin tail, but not the region that interacts with the membrane proximal NPxY motif (**Figure 2b,c**). In agreement with this result, a β-integrin tail polypeptide containing the membrane proximal helix displaced the rod from F3, whereas a PIPK1γ90 peptide, which binds to the same region in F3 as the β-integrin tail NPxY motif, failed to inhibit binding. Thus, the interaction of the rod with F3 partially masks the integrin-binding site in F3 while that for PIPK1γ90 remains exposed. The F3-binding site in the rod is close to IBS2 (**Figure 1a**), and it will be important to establish whether IBS2 is also masked by the intramolecular talin head/rod interaction. That such interactions do indeed regulate talin activity was confirmed with a M319A talin F3 mutant that abolishes rod binding. Full-length talin carrying this mutation was almost twice as effective at activating αIIbβ3-integrin expressed in CHO as wild-type talin.

Pathways that Activate Talin

Mechanisms that regulate the talin head/rod interaction are therefore of much current interest, and significant breakthroughs have recently been made in this regard. The small GTPase Rap1A plays a key role in integrin activation (9), and it has recently emerged that the effects of Rap1A are mediated at least in part by talin. Thus, Han et al. (41) showed

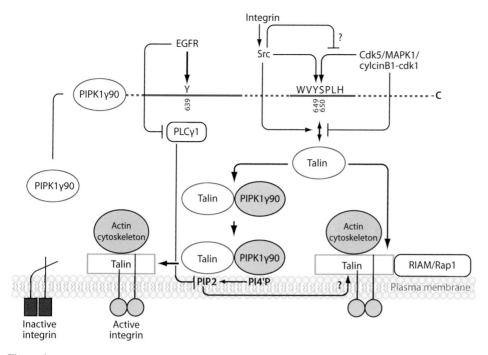

Figure 4

Signaling pathways that regulate talin activation. (*Top*) PIPK1γ90 (*oval*) with the C-terminal region shown as a line. The minimal talin F3-binding motif (WVYSPLH) is indicated along with phosphorylation sites for kinases that regulate either talin or phospholipase Cγ1 (PLCγ1) binding. (*Bottom*) Integrin signaling via Src promotes phosphorylation of PIPK1γ90 on Y649. This facilitates binding of inactive talin to PIPK1γ90 and results in activation of the enzyme. The talin/PIPK1γ90 complex then translocates to the plasma membrane, where the PIP2 synthesized by PIPK1γ90 activates talin. The scheme then indicates that talin activates integrins and couples them to the actin cytoskeleton. However, binding of talin to integrins and PIPK1γ90 is mutually exclusive, and PIPK1γ90 would have to be displaced for this to occur. Talin can also be recruited to the membrane and activated by a Rap1-GTP/RIAM complex. The possibility that PIP2 synthesized by PIPK1γ90 contributes to talin activation by Rap1/RIAM has not been investigated. In all cases, red and green indicate the inactive and active forms of the molecule, respectively.

that reconstituting αIIbβ3-integrin activation in CHO cells was dependent on the expression of both PKCα and talin at the high levels found in platelets. PKCα acted upstream of Rap1A, while talin is downstream of Rap1A, since constitutively active Rap1A(G12V) failed to activate integrins in cells expressing low levels of talin. Moreover, the Rap1-interacting adaptor molecule (RIAM) was sufficient, even in the absence of Rap1A(G12V), to activate αIIbβ3-integrin, implicating RIAM directly in talin activation (**Figure 4**). Indeed RIAM coimmunoprecipitates with talin in this system and in human platelets treated with PAR1 agonists. Recent biomolecular fluorescence complemen-

tation studies have also demonstrated that Rap1A and RIAM promote talin recruitment to the cytoplasmic domain of αIIbβ3-integrin, as did the RIAM-related protein lamellipodin, although the latter was not Rap1A dependent, implicating alternative signaling pathways in integrin activation (106). The talin binding sites in RIAM and lamellipodin have now been mapped to a short N-terminal amphipathic α-helix, and a chimeric protein containing this region fused to the Rap1A membrane targeting sequence was sufficient to activate integrins (54). However, studies in flies show that there is not a simple 1:1 relationship between integrins and talin in cell-matrix junctions, and there must be additional mechanisms that recruit talin to these sites (27).

PIP2 also activates integrin binding by talin (63), and PIP2 levels increase in response to cell adhesion to fibronectin (66). Recent studies show that PIP2 binds talin F3 (**Figure 2c**) and inhibits its association with the rod (37), thereby exposing the integrin-binding site in F3 and perhaps also IBS2 in the talin rod. The likely source of PIP2 involved in talin activation is PIPK1γ90 because (*a*) PIPK1γ90 binds talin, (*b*) binding leads to activation of PIPK1γ90 (29, 59), (*c*) integrin signaling via FAK and Src promotes their association and translocation to the plasma membrane, and (*d*) PIPK1γ90 kinase activity is required for assembly of talin-containing FA during cell spreading (60). The C-terminal region of PIPK1γ90 contains two phosphorylation sites within the WVYSPLH talin F3 recognition motif that regulate talin binding (**Figure 4**). Human S650 is substrate for cdk5, MAPK1, and cyclinB1-cdk1, and phosphorylation inhibits talin binding (57). Tyr 649 (Y644 in mouse) is a substrate for Src (60), and phosphorylation increases talin binding probably by suppressing phosphorylation of S650 (24, 57), although it also directly increases the affinity of binding (60). Y634 (mouse) flanking the talin recognition motif is directly phosphorylated by the EGFR, and this is required for EGF-induced talin recruitment to FA and for cell migration (99). This region of PIPK1γ90 binds PLCγ1, and phosphorylation of Y634 negatively regulates PLCγ1 binding. PLCγ1 is also important in cell migration (86) and likely works together with PIPK1γ90 to regulate local PIP2 levels at the leading edge of cells. Indeed, overexpression of PIPK1γ90 disrupts FAs (29, 59). The data suggest a model (**Figure 4**) in which integrin signaling via Src leads to PIPK1γ90 phosphorylation and its association with inactive talin. Importantly, the PIPK1γ90-binding site in talin F3 is exposed even in the autoinhibited form of talin (37). PIPK1γ90, which is activated by talin, then drives translocation of the complex to the membrane, leading to localized PIP2 synthesis. PIP2 then relieves the talin head/rod interaction (37), exposing the binding site in F3 for the membrane proximal helix of the β-integrin tail, leading to integrin activation and ultimately integrin clustering and the formation of the higher-order complexes required for FA formation and FAK signaling. One conundrum is that the binding affinity of talin F3 for PIPK1γ90 is greater than that for β-integrin tails (2, 107), raising the question whether mechanisms exist to displace the PIPK1γ90 from F3. In conclusion, talin has emerged as a key player in integrin activation, although it acts in concert with the kindlin family of FERM domain proteins. In turn, the activity of talin is tightly regulated by both Rap1/RIAM and PIPK1γ90/PIP2, and it will be important to establish whether there is synergy between these two pathways.

SUMMARY POINTS

1. Structural studies show that the talin F3 FERM domain binds to both the membrane proximal helix and the NPxY motif of the β3-integrin cytoplasmic domain, an interaction that increases the affinity of the integrin extracellular domain for ligand.

2. Talin exists in a globular autoinhibited form due to an interaction between the F3 domain and the large flexible talin rod that masks the β3-integrin-binding site in F3, and Src, PIP2, and Rap1 have been implicated in talin activation.

3. Structures of parts of the talin rod show that it is made up of a series of amphipathic helical bundles. The C-terminal five-helix bundle, which contains an ABS whose activity is inhibited by helix 1, is followed by a single helix that is required for both talin dimerization and actin binding.

4. The talin rod contains multiple VBSs, the key determinants of which are hydrophobic residues that are normally buried within the helical bundles. The affinity of these sites for vinculin is determined both by the talin F3-rod interaction and by the stability of the helical bundles in which they are embedded. Vinculin recruitment to FAs is force dependent, suggesting that the VBSs in talin might be exposed by conformational changes induced by actomyosin contraction.

FUTURE ISSUES

1. The full extent of the intramolecular interactions between the talin head and rod domains that result in the globular autoinhibited state requires further investigation. How many of the various ligand-binding sites are buried in this form of the molecule? Does talin exist as a constitutive dimer or is the autoinhibited form monomeric?

2. The signaling pathways that activate talin require further investigation. What is the mechanism of action of PKCα, Rap1, and RIAM, and how does this relate to the role of PIPK1γ90 and PIP2? What role do the multiple potential phosphorylation sites in talin play in regulating its activity?

3. How do kindlins cooperate with talin in activating integrins, and what is the role of IBS2 in the rod domain?

4. What role does mechanical force play in regulating the exposure of the VBSs in talin and what are the consequences of vinculin binding on the structure of the rest of the rod? Do the hydrophobic residues in the helices that unfold insert into the membrane or do these regions refold and form new helical bundles? Are the vinculin-induced conformational changes in talin reversible?

5. What is the role of talin2 in cells and tissues? Is it coexpressed with talin1 or is it the major isoform in certain tissues such as brain? Does it bind some or all of the same ligands as talin1 and does it have unique binding partners? Does it form heterodimers with talin1?

DISCLOSURE STATEMENT

The author is not aware of any biases that might be perceived as affecting the objectivity of this review.

ACKNOWLEDGMENTS

I am very grateful to Dr. Alexandre R. Gingras for help with the figures. I apologize to all those who have contributed to the extensive literature on talin but whose work has not been cited due to space limitations. The work in the author's laboratory is supported by the Wellcome Trust, Cancer Research U.K., and an NIH Cell Migration Consortium Grant U54 GM64346 from the National Institute of General Medical Sciences (NIGMS).

LITERATURE CITED

1. Bakolitsa C, Cohen DM, Bankston LA, Bobkov AA, Cadwell GW, et al. 2004. Structural basis for vinculin activation at sites of cell adhesion. *Nature* 430:583–86

2. Barsukov IL, Prescot A, Bate N, Patel B, Floyd DN, et al. 2003. Phosphatidylinositol phosphate kinase type 1gamma and beta1-integrin cytoplasmic domain bind to the same region in the talin FERM domain. *J. Biol. Chem.* 278:31202–9

3. Bass MD, Patel B, Barsukov IG, Fillingham IJ, Mason R, et al. 2002. Further characterization of the interaction between the cytoskeletal proteins talin and vinculin. *Biochem. J.* 362:761–68

4. Bass MD, Smith BJ, Prigent SA, Critchley DR. 1999. Talin contains three similar vinculin-binding sites predicted to form an amphipathic helix. *Biochem. J.* 341:257–63

5. Becam IE, Tanentzapf G, Lepesant JA, Brown NH, Huynh JR. 2005. Integrin-independent repression of cadherin transcription by talin during axis formation in *Drosophila*. *Nat. Cell Biol.* 7:510–16

6. Bois PR, O'Hara BP, Nietlispach D, Kirkpatrick J, Izard T. 2006. The vinculin binding sites of talin and alpha-actinin are sufficient to activate vinculin. *J. Biol. Chem.* 281:7228–36

7. Borgon RA, Vonrhein C, Bricogne G, Bois PR, Izard T. 2004. Crystal structure of human vinculin. *Structure* 12:1189–97

8. Borowsky ML, Hynes RO. 1998. Layilin, a novel talin-binding transmembrane protein homologous with C-type lectins, is localized in membrane ruffles. *J. Cell Biol.* 143:429–42

9. Bos JL. 2005. Linking Rap to cell adhesion. *Curr. Opin. Cell Biol.* 17:123–28

10. Bouaouina M, Lad Y, Calderwood DA. 2008. The N-terminal domains of talin cooperate with the phosphotyrosine binding-like domain to activate beta1 and beta3 integrins. *J. Biol. Chem.* 283:6118–25

11. Brett TJ, Legendre-Guillemin V, McPherson PS, Fremont DH. 2006. Structural definition of the F-actin-binding THATCH domain from HIP1R. *Nat. Struct. Mol. Biol.* 13:121–30

12. Burridge K, Connell L. 1983. A new protein of adhesion plaques and ruffling membranes. *J. Cell Biol.* 97:359–67

13. Burridge K, Mangeat P. 1984. An interaction between vinculin and talin. *Nature* 308:744–46

14. Calderwood DA. 2004. Integrin activation. *J. Cell Sci.* 117:657–66

15. Calderwood DA, Yan B, de Pereda JM, Alvarez BG, Fujioka Y, et al. 2002. The phosphotyrosine binding-like domain of talin activates integrins. *J. Biol. Chem.* 277:21749–58

16. Campbell ID, Ginsberg MH. 2004. The talin-tail interaction places integrin activation on FERM ground. *Trends Biochem. Sci.* 29:429–35

17. Chen H, Choudhury DM, Craig SW. 2006. Coincidence of actin filaments and talin is required to activate vinculin. *J. Biol. Chem.* 281:40389–98

18. Chen H, Cohen DM, Choudhury DM, Kioka N, Craig SW. 2005. Spatial distribution and functional significance of activated vinculin in living cells. *J. Cell Biol.* 169:459–70

19. Chen HC, Appeddu PA, Parsons JT, Hildebrand JD, Schaller MD, Guan JL. 1995. Interaction of focal adhesion kinase with the cytoskeletal protein talin. *J. Biol. Chem.* 270:16995–99

20. Chen NT, Lo SH. 2005. The N-terminal half of talin2 is sufficient for mouse development and survival. *Biochem. Biophys. Res. Commun.* 337:670–76

21. Cluzel C, Saltel F, Lussi J, Paulhe F, Imhof BA, Wehrle-Haller B. 2005. The mechanisms and dynamics of (alpha)v(beta)3 integrin clustering in living cells. *J. Cell Biol.* 171:383–92

22. Conti FJ, Felder A, Monkley S, Schwander M, Wood MR, et al. 2008. Progressive myopathy and defects in the maintenance of myotendinous junctions in mice that lack talin 1 in skeletal muscle. *Development* 135:2043–53

23. Cram EJ, Clark SG, Schwarzbauer JE. 2003. Talin loss-of-function uncovers roles in cell contractility and migration in *C. elegans*. *J. Cell Sci.* 116:3871–78

24. de Pereda JM, Wegener KL, Santelli E, Bate N, Ginsberg MH, et al. 2005. Structural basis for phosphatidylinositol phosphate kinase type Igamma binding to talin at focal adhesions. *J. Biol. Chem.* 280:8381–86

25. del Rio A, Perez-Jimenez R, Liu R, Roca-Cusachs P, Fernandez JM, Sheetz MP. 2009. Stretching single talin rod molecules activates vinculin binding. *Science* 323:638–41

26. Denker SP, Barber DL. 2002. Cell migration requires both ion translocation and cytoskeletal anchoring by the Na-H exchanger NHE1. *J. Cell Biol.* 159:1087–96

27. Devenport D, Bunch TA, Bloor JW, Brower DL, Brown NH. 2007. Mutations in the *Drosophila* alphaPS2 integrin subunit uncover new features of adhesion site assembly. *Dev. Biol.* 308:294–308

28. Dietrich C, Goldmann WH, Sackmann E, Isenberg G. 1993. Interaction of NBD-talin with lipid monolayers—a film balance study. *FEBS Lett.* 324:37–40

29. Di Paolo G, Pellegrini L, Letinic K, Cestra G, Zoncu R, et al. 2002. Recruitment and regulation of phosphatidylinositol phosphate kinase type 1 gamma by the FERM domain of talin. *Nature* 420:85–89

30. Fillingham I, Gingras AR, Papagrigoriou E, Patel B, Emsley J, et al. 2005. A vinculin binding domain from the talin rod unfolds to form a complex with the vinculin head. *Structure* 13:65–74

31. Franco SJ, Rodgers MA, Perrin BJ, Han J, Bennin DA, et al. 2004. Calpain-mediated proteolysis of talin regulates adhesion dynamics. *Nat. Cell Biol.* 6:977–83

32. Galbraith CG, Yamada KM, Sheetz MP. 2002. The relationship between force and focal complex development. *J. Cell Biol.* 159:695–705

33. García-Alvarez B, de Pereda JM, Calderwood DA, Ulmer TS, Critchley D, et al. 2003. Structural determinants of integrin recognition by talin. *Mol. Cell* 11:49–58

34. Gingras AR, Bate N, Goult BT, Hazelwood L, Canestrelli I, et al. 2008. The structure of the C-terminal actin-binding domain of talin. *EMBO J.* 27:458–69

35. Gingras AR, Vogel KP, Steinhoff HJ, Ziegler WH, Patel B, et al. 2006. Structural and dynamic characterization of a vinculin binding site in the talin rod. *Biochemistry* 45:1805–17

36. Gingras AR, Ziegler WH, Frank R, Barsukov IL, Roberts GC, et al. 2005. Mapping and consensus sequence identification for multiple vinculin binding sites within the talin rod. *J. Biol. Chem.* 280:37217–24

> 36. Identifies multiple VBSs in the talin rod and determines the basis of binding specificity.

37. Goksoy E, Ma YQ, Wang X, Kong X, Perera D, et al. 2008. Structural basis for the autoinhibition of talin in regulating integrin activation. *Mol. Cell* 31:124–33

> 37. Provides the first structural insights into the interaction between the talin FERM domain and talin rod, and how PIP2 regulates this interaction and the integrin-binding site in the FERM domain.

38. Goldfinger LE, Ptak C, Jeffery ED, Shabanowitz J, Han J, et al. 2007. An experimentally derived database of candidate Ras-interacting proteins. *J. Proteome Res.* 6:1806–11

39. Goldmann WH, Isenberg G. 1991. Kinetic determination of talin-actin binding. *Biochem. Biophys. Res. Commun.* 178:718–23

40. Goldmann WH, Niggli V, Kaufmann S, Isenberg G. 1992. Probing actin and liposome interaction of talin and talin vinculin complexes—a kinetic, thermodynamic and lipid labeling study. *Biochemistry* 31:7665–71

41. Han J, Lim CJ, Watanabe N, Soriani A, Ratnikov B, et al. 2006. Reconstructing and deconstructing agonist-induced activation of integrin alphaIIbbeta3. *Curr. Biol.* 16:1796–806

> 41. Defines a pathway by which PKCα, Rap1, and RIAM regulate talin and thereby integrin activation.

42. Heise H, Bayerl T, Isenberg G, Sackmann E. 1991. Human platelet p-235, a talin-like actin binding-protein, binds selectively to mixed lipid bilayers. *Biochim. Biophys. Acta* 1061:121–31

43. Hemmings L, Rees DJG, Ohanian V, Bolton SJ, Gilmore AP, et al. 1996. Talin contains three actin-binding sites each of which is adjacent to a vinculin-binding site. *J. Cell Sci.* 109:2715–26

44. Horwitz A, Duggan K, Buck C, Beckerle MC, Burridge K. 1986. Interaction of plasma-membrane fibronectin receptor with talin—a transmembrane linkage. *Nature* 320:531–33

45. Hytonen VP, Vogel V. 2008. How force might activate talin's vinculin binding sites: SMD reveals a structural mechanism. *PLoS Comput. Biol.* 4:e24

46. Isenberg G, Doerhoefer S, Hoekstra D, Goldmann WH. 2002. Membrane fusion induced by the major lipid-binding domain of the cytoskeletal protein talin. *Biochem. Biophys. Res. Commun.* 295:636–43

47. Izard T, Evans G, Borgon RA, Rush CL, Bricogne G, Bois PR. 2004. Vinculin activation by talin through helical bundle conversion. *Nature* 427:171–75

> 47. Describes the conformational changes required for Vd1 to bind the talin VBS3 peptide, and how this is regulated by the Vd1-Vt interaction.

48. Izard T, Vonrhein C. 2004. Structural basis for amplifying vinculin activation by talin. *J. Biol. Chem.* 279:27667–78

49. Jiang G, Giannone G, Critchley DR, Fukumoto E, Sheetz MP. 2003. Two-piconewton slip bond between fibronectin and the cytoskeleton depends on talin. *Nature* 424:334–37

50. Kaufmann S, Piekenbrock T, Goldmann WH, Barmann M, Isenberg G. 1991. Talin binds to actin and promotes filament nucleation. *FEBS Lett.* 284:187–91

51. Kim M, Carman CV, Springer TA. 2003. Bidirectional transmembrane signaling by cytoplasmic domain separation in integrins. *Science* 301:1720–25

52. Lammermann T, Bader BL, Monkley SJ, Worbs T, Wedlich-Soldner R, et al. 2008. Rapid leukocyte migration by integrin-independent flowing and squeezing. *Nature* 453:51–55

53. Lee HS, Bellin RM, Walker DL, Patel B, Powers P, et al. 2004. Characterization of an actin-binding site within the talin FERM domain. *J. Mol. Biol.* 343:771–84

54. Lee HS, Lim CJ, Puzon-McLaughlin W, Shattil SJ, Ginsberg MH. 2009. RIAM activates integrins by linking talin to Ras GTPase membrane-targeting sequences. *J. Biol. Chem.* 284:5119–27

55. Lee SE, Chunsrivirot S, Kamm RD, Mofrad MR. 2008. Molecular dynamics study of talin-vinculin binding. *Biophys. J.* 95:2027–36

56. Lee SE, Kamm RD, Mofrad MR. 2007. Force-induced activation of talin and its possible role in focal adhesion mechanotransduction. *J. Biomech.* 40:2096–106

57. Lee SY, Voronov S, Letinic K, Nairn AC, Di Paolo G, De Camilli P. 2005. Regulation of the interaction between PIPKI gamma and talin by proline-directed protein kinases. *J. Cell Biol.* 168:789–99

58. Legate KR, Montanez E, Kudlacek O, Fassler R. 2006. ILK, PINCH and parvin: the tIPP of integrin signalling. *Nat. Rev. Mol. Cell Biol.* 7:20–31

59. Ling K, Doughman RL, Firestone AJ, Bunce MW, Anderson RA. 2002. Type I gamma phosphatidylinositol phosphate kinase targets and regulates focal adhesions. *Nature* 420:89–93

60. Provides the first insights into how the interaction between talin and PIPK1γ90 is regulated by phosphorylation.

60. Ling K, Doughman RL, Iyer VV, Firestone AJ, Bairstow SF, et al. 2003. Tyrosine phosphorylation of type Igamma phosphatidylinositol phosphate kinase by Src regulates an integrin-talin switch. *J. Cell Biol.* 163:1339–49

61. Loer B, Bauer R, Bornheim R, Grell J, Kremmer E, et al. 2008. The NHL-domain protein Wech is crucial for the integrin-cytoskeleton link. *Nat. Cell Biol.* 10:422–28

62. Ma YQ, Qin J, Wu C, Plow EF. 2008. Kindlin-2 (Mig-2): a coactivator of beta3 integrins. *J. Cell Biol.* 181:439–46

63. Martel V, Racaud-Sultan C, Dupe S, Marie C, Paulhe F, et al. 2001. Conformation, localization, and integrin binding of talin depend on its interaction with phosphoinositides. *J. Biol. Chem.* 276:21217–27

64. McCann RO, Craig SW. 1997. The I/LWEQ module: a conserved sequence that signifies F-actin binding in functionally diverse proteins from yeast to mammals. *Proc. Natl. Acad. Sci. USA* 94:5679–84

65. McLachlan AD, Stewart M, Hynes RO, Rees DJG. 1994. Analysis of repeated motifs in the talin rod. *J. Mol. Biol.* 235:1278–90

66. McNamee HP, Ingber DE, Schwartz MA. 1993. Adhesion to fibronectin stimulates inositol lipid synthesis and enhances PDGF-induced inositol lipid breakdown. *J. Cell Biol.* 121:673–78

67. Millon-Frémillon A, Bouvard D, Grichine A, Manet-Dupé S, Block MR, Albiges-Rizo C. 2008. Cell adaptive response to extracellular matrix density is controlled by ICAP-1-dependent beta1-integrin affinity. *J. Cell Biol.* 180:427–41

68. Moes M, Rodius S, Coleman SJ, Monkley SJ, Goormaghtigh E, et al. 2007. The integrin binding site 2 (IBS2) in the talin rod domain is essential for linking integrin beta subunits to the cytoskeleton. *J. Biol. Chem.* 282:17280–88

69. Molony L, McCaslin D, Abernethy J, Paschal B, Burridge K. 1987. Properties of talin from chicken gizzard smooth muscle. *J. Biol. Chem.* 262:7790–95

70. Monkley SJ, Pritchard CA, Critchley DR. 2001. Analysis of the mammalian talin2 gene TLN2. *Biochem. Biophys. Res. Commun.* 286:880–85

71. Monkley SJ, Zho X-H, Kinston SJ, Giblett SM, Hemmings L, et al. 2000. Disruption of the talin gene arrests mouse development at the gastrulation stage. *Dev. Dyn.* 219:560–74

72. Montanez E, Ussar S, Schifferer M, Bosl M, Zent R, et al. 2008. Kindlin-2 controls bidirectional signaling of integrins. *Genes Dev.* 22:1325–30

73. Moser M, Nieswandt B, Ussar S, Pozgajova M, Fassler R. 2008. Kindlin-3 is essential for integrin activation and platelet aggregation. *Nat. Med.* 14:325–30

74. Muguruma M, Matsumura S, Fukazawa T. 1990. Direct interactions between talin and actin. *Biochem. Biophys. Res. Commun.* 171:1217–23

75. Muguruma M, Matsumura S, Fukazawa T. 1992. Augmentation of alpha-actinin-induced gelation of actin by talin. *J. Biol. Chem.* 267:5621–24

76. Muguruma M, Nishimuta S, Tomisaka Y, Ito T, Matsumara S. 1995. Organisation of the functional domains in membrane cytoskeletal protein talin. *J. Biochem.* 117:1036–42

77. Nakamura F, Osborn TM, Hartemink CA, Hartwig JH, Stossel TP. 2007. Structural basis of filamin A functions. *J. Cell Biol.* 179:1011–25

78. Nieswandt B, Moser M, Pleines I, Varga-Szabo D, Monkley S, et al. 2007. Loss of talin1 in platelets abrogates integrin activation, platelet aggregation, and thrombus formation in vitro and in vivo. *J. Exp. Med.* 204:3113–18

79. Niggli V, Kaufmann S, Goldmann WH, Weber T, Isenberg G. 1994. Identification of functional domains in the cytoskeletal protein talin. *Eur. J. Biochem.* 224(3):951–57

80. Oxley CL, Anthis NJ, Lowe ED, Vakonakis I, Campbell ID, Wegener KL. 2008. An integrin phosphorylation switch: the effect of beta3 integrin tail phosphorylation on Dok1 and talin binding. *J. Biol. Chem.* 283:5420–26

81. Papagrigoriou E, Gingras AR, Barsukov IL, Bate N, Fillingham IJ, et al. 2004. Activation of a vinculin-binding site in the talin rod involves rearrangement of a five-helix bundle. *EMBO J.* 23:2942–51

82. Parsons M, Messent AJ, Humphries JD, Deakin NO, Humphries MJ. 2008. Quantification of integrin receptor agonism by fluorescence lifetime imaging. *J. Cell Sci.* 121:265–71

83. Patel B, Gingras AR, Bobkov AA, Fujimoto LM, Zhang M, et al. 2006. The activity of the vinculin binding sites in talin is influenced by the stability of the helical bundles that make up the talin rod. *J. Biol. Chem.* 281:7458–67

84. Petrich BG, Marchese P, Ruggeri ZM, Spiess S, Weichert RA, et al. 2007. Talin is required for integrin-mediated platelet function in hemostasis and thrombosis. *J. Exp. Med.* 204:3103–11

85. Pfaff M, Liu S, Erle DJ, Ginsberg MH. 1998. Integrin beta cytoplasmic domains differentially bind to cytoskeletal proteins. *J. Biol. Chem.* 273:6104–9

86. Piccolo E, Innominato PF, Mariggio MA, Maffucci T, Iacobelli S, Falasca M. 2002. The mechanism involved in the regulation of phospholipase Cgamma1 activity in cell migration. *Oncogene* 21:6520–29

87. Priddle H, Hemmings L, Monkley S, Woods A, Patel B, et al. 1998. Disruption of the talin gene compromises focal adhesion assembly in undifferentiated but not differentiated ES cells. *J. Cell Biol.* 142:1121–33

88. Ratnikov B, Ptak C, Han J, Shabanowitz J, Hunt DF, Ginsberg MH. 2005. Talin phosphorylation sites mapped by mass spectrometry. *J. Cell Sci.* 118:4921–23

89. Rees DJG, Ades SE, Singer SJ, Hynes RO. 1990. Sequence and domain-structure of talin. *Nature* 347:685–89

90. Rodius S, Chaloin O, Moes M, Schaffner-Reckinger E, Landrieu I, et al. 2008. The talin rod IBS2 alpha-helix interacts with the beta3 integrin cytoplasmic tail membrane proximal helix by establishing charge complementary salt bridges. *J. Biol. Chem.* 283:24212–23

91. Saitoh A, Takiguchi K, Tanaka Y, Hotani H. 1998. Opening-up of liposomal membranes by talin. *Proc. Natl. Acad. Sci. USA* 95:1026–31

92. Schmidt JM, Zhang J, Lee H-S, Stromer MH, Robson RM. 1999. Interaction of talin with actin: sensitive modulation of filament crosslinking activity. *Arch. Biochem. Biophys.* 366:139–50

93. Seelig A, Blatter XL, Frentzel A, Isenberg G. 2000. Phospholipid binding of synthetic talin peptides provides evidence for an intrinsic membrane anchor of talin. *J. Biol. Chem.* 275:17954–61

94. Senetar MA, Foster SJ, McCann RO. 2004. Intrasteric inhibition mediates the interaction of the I/LWEQ module proteins Talin1, Talin2, Hip1, and Hip12 with actin. *Biochemistry* 43:15418–28

95. Senetar MA, McCann RO. 2005. Gene duplication and functional divergence during evolution of the cytoskeletal linker protein talin. *Gene* 362:141–52

96. Smith SJ, McCann RO. 2007. A C-terminal dimerization motif is required for focal adhesion targeting of Talin1 and the interaction of the Talin1 I/LWEQ module with F-actin. *Biochemistry* 46:10886–98

97. Srivastava J, Barreiro G, Groscurth S, Gingras AR, Goult BT, et al. 2008. Structural model and functional significance of pH-dependent talin-actin binding for focal adhesion remodelling. *Proc. Natl. Acad. Sci. USA* 105:14436–41

98. Sun N, Critchley DR, Paulin D, Li Z, Robson RM. 2008. Identification of a repeated domain within mammalian alpha-synemin that interacts directly with talin. *Exp. Cell Res.* 314:1839–49

81. Shows that the N-terminal part of the talin rod is a five-helix bundle in which the VBS1 is buried in the hydrophobic core of the bundle.

99. Sun Y, Ling K, Wagoner MP, Anderson RA. 2007. Type I gamma phosphatidylinositol phosphate kinase is required for EGF-stimulated directional cell migration. *J. Cell Biol.* 178:297–308

100. **Tadokoro S, Shattil SJ, Eto K, Tai V, Liddington RC, et al. 2003. Talin binding to integrin beta tails: a final common step in integrin activation. *Science* 302:103–6**

101. Tanentzapf G, Martin-Bermudo MD, Hicks MS, Brown NH. 2006. Multiple factors contribute to integrin-talin interactions in vivo. *J. Cell Sci.* 119:1632–44

102. Tran Van Nhieu G, Izard T. 2007. Vinculin binding in its closed conformation by a helix addition mechanism. *EMBO J.* 26:4588–96

103. Tremuth L, Kreis S, Melchior C, Hoebeke J, Ronde P, et al. 2004. A fluorescence cell biology approach to map the second integrin-binding site of talin to a 130-amino acid sequence within the rod domain. *J. Biol. Chem.* 279:22258–66

104. Tsujioka M, Yoshida K, Nagasaki A, Yonemura S, Müller-Taubenberger A, Uyeda TQ. 2008. Overlapping functions of the two talin homologues in *Dictyostelium*. *Eukaryot. Cell* 7:906–16

105. Wang Y, Litvinov RI, Chen X, Bach TL, Lian L, et al. 2008. Loss of PIP5KIgamma, unlike other PIP5KI isoforms, impairs the integrity of the membrane cytoskeleton in murine megakaryocytes. *J. Clin. Invest.* 118:812–19

106. Watanabe N, Bodin L, Pandey M, Krause M, Coughlin S, et al. 2008. Mechanisms and consequences of agonist-induced talin recruitment to platelet integrin alphaIIbbeta3. *J. Cell Biol.* 181:1211–22

107. Wegener KL, Basran J, Bagshaw CR, Campbell ID, Roberts GC, et al. 2008. Structural basis for the interaction between the cytoplasmic domain of the hyaluronate receptor layilin and the talin F3 subdomain. *J. Mol. Biol.* 382:112–26

108. **Wegener KL, Partridge AW, Han J, Pickford AR, Liddington RC, et al. 2007. Structural basis of integrin activation by talin. *Cell* 128:171–82**

109. Winkler J, Lunsdorf H, Jockusch BM. 1997. Energy-filtered electron microscopy reveals that talin is a highly flexible protein composed of a series of globular domains. *Eur. J. Biochem.* 243:430–36

110. Xing B, Jedsadayanmata A, Lam SC. 2001. Localization of an integrin binding site to the C terminus of talin. *J. Biol. Chem.* 276:44373–78

111. Zaidel-Bar R, Itzkovitz S, Ma'ayan A, Iyengar R, Geiger B. 2007. Functional atlas of the integrin adhesome. *Nat. Cell Biol.* 9:858–67

112. **Zhang X, Jiang G, Cai Y, Monkley SJ, Critchley DR, Sheetz MP. 2008. Talin depletion reveals independence of initial cell spreading from integrin activation and traction. *Nat. Cell Biol.* 10:1062–68**

113. Ziegler WH, Liddington RC, Critchley DR. 2006. The structure and regulation of vinculin. *Trends Cell Biol.* 16:453–60

100. The definitive study showing that the talin head is required for integrin activation.

108. A definitive structural study showing how the talin F3 FERM domain binds both the membrane proximal helix and the NPxY motif in the β3-integrin cytoplasmic domain.

112. Talin is not required for the initial phase of integrin-mediated cell spreading or for Src activation, but it is required for integrin activation, sustained spreading, FAK signaling, FA assembly, and substrate traction.

Single-Molecule Approaches to Stochastic Gene Expression

Arjun Raj and Alexander van Oudenaarden

Department of Physics, Massachusetts Institute of Technology, Cambridge, Massachusetts 02139; email: avano@mit.edu

Annu. Rev. Biophys. 2009. 38:255–70

First published online as a Review in Advance on February 5, 2009

The *Annual Review of Biophysics* is online at biophys.annualreviews.org

This article's doi: 10.1146/annurev.biophys.37.032807.125928

Key Words

single-molecule mRNA protein, gene expression noise

Abstract

Both the transcription of mRNAs from genes and their subsequent translation into proteins are inherently stochastic biochemical events, and this randomness can lead to substantial cell-to-cell variability in mRNA and protein numbers in otherwise identical cells. Recently, a number of studies have greatly enhanced our understanding of stochastic processes in gene expression by utilizing new methods capable of counting individual mRNAs and proteins in cells. In this review, we examine the insights that these studies have yielded in the field of stochastic gene expression. In particular, we discuss how these studies have played in understanding the properties of bursts in gene expression. We also compare the array of different methods that have arisen for single mRNA and protein detection, highlighting their relative strengths and weaknesses. In conclusion, we point out further areas where single-molecule techniques applied to gene expression may lead to new discoveries.

Contents

INTRODUCTION

Until relatively recently, scientists studying gene expression have measured the properties of gene expression on populations of cells rather than in individual cells, largely because of the technical challenges involved in making single-cell measurements. However, the advent of simple and accurate measurements of gene expression in individual cells has led researchers to find that the numbers of mRNAs and proteins can vary, sometimes dramatically, from cell to cell and that this variability is caused by the fundamentally stochastic nature of the biochemical events involved in gene expression.

Primary among these technical advances is the use of fluorescent proteins, such as GFP, whose importance in the field of stochastic gene expression is hard to overstate. Of course, even before fluorescent proteins were available, a few researchers still showed that gene expression was highly variable; such efforts include the pioneering work of Novick & Weiner (26), who used serial dilution and amplification of individual bacteria, and Ko et al. (20), who used single-cell enzymatic assays to show that levels of β-galactosidase expression varied significantly in individual mammalian cells. Yet, while these studies and others (36, 51) established the phenomenon, the ease with which GFP can be used to measure gene expression in individual cells led to an explosion in experimental work in stochastic gene expression that continues to this day, beginning with the seminal studies of Elowitz et al. (11) and Ozbudak et al. (27). These studies and the ones that have followed have shed light on many of the mechanisms that result in cell-to-cell variability in gene expression by using GFP and its variants in combination with time-lapse imaging, flow cytometry, and microscopy.

However, as researchers probe ever more deeply into the stochastic processes underlying gene expression variability, the limitations of GFP are becoming more and more apparent. One of the most serious limitations is sensitivity: When using conventional microscopy or flow cytometry, it is difficult to detect small numbers of fluorescent proteins. Given that stochastic effects are more prevalent at these low molecule numbers, sensitivity issues may make GFP an inappropriate choice of assay in some situations. Another problem is that GFP is typically measured in arbitrary fluorescence units rather than molecular units [with the notable exceptions of Rosenfeld et al. (35) and Gregor et al. (17)], thus limiting the ability to quantitatively evaluate increasingly sophisticated models of stochastic gene expression.

Ultimately, the ideal way to study stochastic gene expression would be to monitor the production, degradation, and functional states of individual biomolecules in real time in living cells. While such a goal may seem almost laughably unrealistic at first glance, the work highlighted in this review shows that researchers have made remarkable progress toward these seemingly unattainable ends. We begin by examining some recent work demonstrating the ability to count individual mRNAs within single cells, and then discuss developments in counting individual proteins. One of the key benefits of counting individual molecules is that it provides rigorous tests for stochastic models of gene expression, and we examine these connections, focusing in particular on the notion of bursts in transcription and translation, in which the production of mRNAs and proteins occurs in a pulsatile rather than continuous fashion. We conclude with some speculations about potential new areas in which single-molecule detection may drive the field of stochastic gene expression forward.

SINGLE-mRNA DETECTION

The detection of individual molecules of mRNA in single cells has the potential to dramatically enhance our understanding of transcription, not only in terms of its effects on cell-to-cell variability in gene expression but also in providing insights into the biochemical mechanisms involved. Using a variety of experimental methods, researchers have begun to understand some of these mechanisms, perhaps the most dramatic of which is transcriptional bursting.

Initially, stochastic models of gene expression assumed that mRNAs are produced and degraded according to the statistics of a Poisson process (43); that is, while the production and degradation happen at random, the probability of a transcript produced within any given time period is a constant that does not change in time (**Figure 1**). If one looks across a population of cells that are transcribing in this fashion, then one would expect to see a Poisson distribution of mRNA per cell:

$$P(m) = \frac{\bar{m}^m}{m!} e^{-\bar{m}},$$

where m is the number of mRNA molecules per cell and $\bar{m}$ denotes the average mRNA number. The situation becomes more complex, however, when considering models in which mRNA production does not occur with a constant probability in time but rather occurs with much greater likelihood at some time periods than others (19, 29, 31). These transcriptionally active time periods are often referred to as transcriptional bursts. (In this review, we explicitly refer to bursts as being either transcriptional or translational to avoid confusion.)

One important consequence of transcriptional bursts is that they result in much higher variability in gene expression than the Poisson model predicts. However, experimentally distinguishing bursty transcription from nonbursty Poissonian transcription requires that measures of mRNA number per cell be made in molecular units. This requirement arises from the way in which variability scales as a function of mean mRNA number. For instance, in the Poisson model, as the mean increases, the relative variability about that mean should decrease, meaning that for large means, variability should be essentially negligible.

Transcriptional bursting, however, can lead to high variability even with high mean expression levels. In principle, it should thus be easy to tell the difference between these two situations, but the problem is that in the absence of molecular units, it is difficult to say whether an observation of high variability is the result of bursting or simply due to low levels of Poissonian transcription. Mathematically, one can encapsulate this argument through the use of the Fano factor, defined as the ratio of the variance to the mean. When measured in molecular units, the Fano factor for a Poisson distribution is exactly 1, whereas transcriptional bursts can result in Fano factors much larger than 1. (Some intuition can be gained from the fact that the Fano factor is approximately equal to the average number of transcripts produced during a burst, often referred to as the burst size.) However, when measured in arbitrary fluorescence units, the Fano factor contains an arbitrary scaling factor that makes such absolute numerical comparisons impossible (28), providing a strong rationale for counting the actual numbers of transcripts in individual cells.

It was against this theoretical backdrop that Golding et al. (16) began their beautiful study of the kinetics of transcription in *Escherichia coli*. Their main tool was the MS2 mRNA detection technique, developed simultaneously by Bloom and colleagues (4) and Singer and colleagues (6), which can be sensitive enough to visualize single mRNA molecules (15). In the variant of the method used by Golding et al. (16), a gene is engineered to transcribe an mRNA containing 96 copies of a specific RNA hairpin in its untranslated region, each of which binds tightly to the coat protein of the bacteriophage MS2. This gene is then expressed in a cell that already expresses the MS2 coat protein fused to GFP. When 96 of the MS2-GFP proteins bind to an individual mRNA, enough fluorescent signal is generated that the individual mRNAs are detectable as diffraction-limited spots by

Fano factor: mathematically defined as the variance of a distribution divided by the mean

MS2: a bacteriophage whose coat protein binds strongly with a particular RNA hairpin

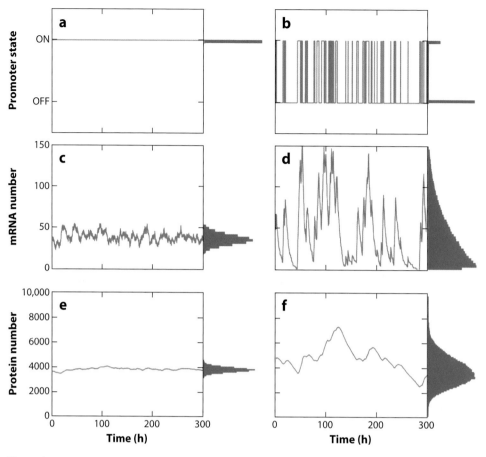

Figure 1

(*a*) Promoter dynamics for a gene that is always in the active state (i.e., nonbursting) versus (*b*) promoter dynamics for a gene that switches between active and inactive states (i.e., bursty dynamics). (*c*) mRNA dynamics for nonbursting and (*d*) bursting genes. In the nonbursting case, one obtains a Poisson distribution of mRNAs per cell across the population, as shown in the marginal histogram, whereas the distribution of mRNAs per cell in the bursting case is much wider than a Poisson distribution despite having the same mean. Protein dynamics for (*e*) nonbursting and (*f*) bursting genes, again with the same mean. Although the underlying gene expression dynamics are bursty, the relatively long half-life of the protein results in a wide but Gaussian-looking population distribution, pointing out the need for single-molecule mRNA-counting approaches when studying bursty gene expression. The marginal histograms on the right of the time courses show the distribution of the promoter states, mRNAs, and proteins across a population.

conventional fluorescence microscopy. One can thus count the number of mRNAs in single cells by counting spots or, if the spots contain multiple mRNAs, by integrating the fluorescence in each spot. Upon performing this counting across an entire population of cells, they measured a Fano factor of roughly 4, which, being greater than 1, provided strong evidence for transcriptional bursting.

Impressively, they went even further by measuring transcriptional activity in real time by using time-lapse microscopy. The authors found that transcription did indeed occur in bursts, with the gene itself switching randomly between transcriptionally active and inactive states. These switching events appeared to happen at exponentially distributed times, indicating that gene activation and inactivation were themselves Poisson processes, justifying the assumptions made in many models of transcriptional bursts (19, 29, 31). Moreover, using the temporal statistics of the switching events, they used a model of transcriptional bursts to predict the statistics of the population snapshots that they had measured experimentally, which showed a fairly good match between the two.

Another method by which one can count single molecules of mRNAs in individual cells is fluorescence in situ hybridization (FISH) (12, 32). In this method, samples are fixed and then a hybridization is performed using a set of fluorescently labeled oligonucleotides, each complementary to a unique portion of the target mRNA. As with the MS2 method, the presence of sufficiently large numbers of fluorophores bound to an individual mRNA renders the molecule sufficiently fluorescent to be detected by fluorescence microscopy. One recent application of this method to the study of stochastic gene expression in bacteria was to the phenomenon of competence in *Bacillus subtilis* (23). *B. subtilis* has the remarkable property of being naturally competent (i.e., it takes up foreign DNA from the environment). This property only manifests itself, though, at the beginning of stationary phase, and only a small percentage (~15%) of the total population actually becomes competent. Maamar et al. (23) showed that noise in the expression of the transcription factor primarily responsible for competence, *comK*, underlies the stochastic decision to become competent or not: Occasionally, a stochastic accumulation of ComK protein will become large enough to allow the ComK protein to bind to its own promoter, dramatically upregulating its expression and resulting in cell competence. One difficulty in studying the expression of *comK* in noncompetent cells, however, is that ComK is lowly expressed, making it impossible to measure gene expression from the *comK* promoter with fluorescent proteins. Instead, the authors used single-molecule FISH to count the numbers of *comK* mRNAs in individual bacteria. They found that *comK* was indeed expressed at a low level in noncompetent cells (less than 1 transcript per cell), and

that this expression level was modulated over time, resulting in a concomitant modulation in frequency of transition to competence. The authors also measured some of the statistical properties of fluctuations in mRNA numbers throughout the population. They found that the Fano factor was relatively close to 1 for the mRNAs they measured, indicating that for this gene bursting is not likely to be a significant source of variability.

Owing to the increased complexity of eukaryotic transcription, one might expect cell-to-cell variability in eukaryotic transcription to have stochastic properties different from those in prokaryotic transcription. Also utilizing a single-molecule FISH assay, Raj et al. (31) found that transcription in mammalian cells was extremely bursty, with short, infrequent bursts resulting in large variations in mRNA numbers from cell to cell, leading to Fano factors of 40 or higher. Their assay also showed a single, intensely bright spot in some cells (but not others) resulting from mRNAs that had not yet diffused away from an active site of transcription. Moreover, cells with these active transcription sites exhibited a larger percentage of nuclear mRNA than those without active transcription sites. Together, these facts present a picture of transcription in which short transcriptional bursts cause the production of large quantities of mRNAs, which are exported from the nucleus to the cytoplasm, where they slowly decay, highlighting the benefits of measuring the spatial locations of single mRNAs in individual cells. Further, the authors used multicolor FISH to visualize simultaneously two different mRNA transcripts, showing that transcription from genes located far apart from each other on the genome were expressed in uncorrelated transcriptional bursts, whereas those located near each other were expressed in strongly correlated transcriptional bursts.

Another completely different approach to counting the number of particular mRNAs within single cells is the use of single-cell quantitative reverse transcriptase polymerase chain reaction (RT-PCR). Bengtsson et al. (5) used such a method to show that gene expression

FISH: fluorescence in situ hybridization

RT-PCR: reverse transcriptase polymerase chain reaction

in individual cells isolated from mouse pancreatic islets is subject to large fluctuations. Their assay, which involves isolating individual cells and performing RT-PCR on each cell, can yield absolute measures of transcript numbers with appropriate controls and standardization. The authors found that most population histograms of the numbers of mRNA per cell were close to lognormal distributions, which are distributions that appear Gaussian when a histogram is made of the log of the mRNA number. Although such distributions appear Gaussian in logarithmic coordinates, they can exhibit long tails in nonlogarithmic coordinates, similar to those observed by Raj et al. (31) and Warren et al. (50).

Another advantage of their assay is the ability to detect simultaneously the levels of five different target genes through the use of multiplex PCR. In their assay, Bengtsson et al. found that two related genes, *Ins1* and *Ins2*, showed highly correlated expression between cells, whereas the other pairs of genes exhibited no significant correlations. Such correlations may arise from a number of sources, including fluctuations in common upstream gene expression factors. Thus, the analysis of correlations has the potential to uncover previously hidden regulatory connections between genes. Traditionally, the way to check if the expression of two genes is related would be to use an external trigger (such as a signaling molecule or some environmental change) and check if the mean levels of the two genes change concurrently. However, this presupposes the existence of such an external trigger, which might be available for the genes in question. By looking for correlations in cell-to-cell variations between the transcript levels of two different mRNAs, one might effectively perform a coexpression analysis without requiring any such trigger.

One interesting extension of the single-cell RT-PCR technique is so-called digital RT-PCR (50), which is a variation on digital PCR (48). In this assay, cDNA obtained from reverse-transcribing mRNA from a single cell is partitioned into many (potentially thousands) of individual PCR reactions. The result of this massive dilution of the cDNA is that each PCR reaction will contain either 0 or 1 cDNA molecules as a template, and the presence or absence of a single cDNA is then detected by the PCR itself in digital fashion. To facilitate the large amount of liquid handling required, the reactions are typically performed with a microfluidic device that fractionates the reactions into appropriately sized volumes. By providing a digital readout of gene expression, one can sidestep the need for the many careful controls necessary for quantifying mRNA counts by conventional single-cell RT-PCR. Warren et al. (50) used digital RT-PCR to examine variability in the expression of the transcriptional factor PU.1, which plays a central role in the process of hematopoiesis, the process by which blood stem cells differentiate into different blood cell types. Cell fate decisions in this process are thought to have a significant stochastic component, thus motivating measurements of variability in the expression of PU.1.

The authors found that PU.1 does indeed display a large variability in all the different blood cell types examined, although the mean expression level was different in the various lineages. The authors also performed an experiment in which they presorted common myeloid progenitors according to whether they displayed high or low levels of the cytokine receptor flk2, which has been correlated with differential functionality of common myeloid progenitors. They found that cells with high levels of flk2 displayed high expression of PU.1, whereas cells with low levels of flk2 showed low expression of PU.1. This discovery showed that variability in PU.1 expression is indeed correlated with functional distinctions between otherwise identical cells, a finding that has recently been extended by using microarrays (8).

From a methodological standpoint, each of the techniques used in these single-mRNA detection studies has various advantages and disadvantages (**Table 1**). For the MS2 technique, one major advantage is the ability to measure mRNA levels in real time—all the other methods except for molecular beacons (described

Table 1 Comparison of different single mRNA detection methods

Method	Endogenous mRNA detection?	Real-time measurements?	Detection of multiple mRNA species at once?	Other advantages	Other disadvantages	Reference(s)
MS2	No	Yes	No	No need for external interventions (e.g., microinjection), yields spatial information	mRNA tend to form clumps, requires long UTR sequence elements	(4, 6, 16)
FISH	Yes	No	Up to 3	Yields spatial information	Imaging can be difficult in small organisms	(12, 31, 32)
Single-cell RT-PCR	Yes	No	Up to 5	Simple to perform with a large dynamic range	Requires careful standardization, questions about efficiency and sensitivity at low numbers, no spatial information	(5)
Digital single-cell RT-PCR	Yes	No	Up to 2	Easily interpretable signals, sensitive at low numbers of molecules	Requires microfluidics, questions about RT efficiency, no spatial information	(50)
Molecular beacons	No	Yes	Yes	No clumping of transcripts, yields spatial information	Requires microinjection or other invasive delivery methods, requires long UTR sequence elements	(46, 47)

Abbreviations: FISH, fluorescence in situ hybridization; RT, reverse transcriptase; RT-PCR, reverse transcriptase polymerase chain reaction; UTR, untranslated region.

below) require the use of fixed or lysed samples. Moreover, it yields spatial information on the locations of the individual mRNAs, which could prove invaluable in developmental studies in which positional information is critical (for an example of the use of MS2 in developmental systems, see Reference 13). The main problem, however, is that one must generate transgenes with large untranslated regions that may affect mRNA dynamics; for instance, Golding et al. (16) found that the incorporation of 96 protein-bound hairpins into the untranslated region of mRNAs rendered the mRNAs resistant to cellular nucleases. Also, the tendency of the MS2 coat protein to multimerize requires that one make a careful estimation of the total fluorescence within individual spots to determine the number of mRNAs contained therein.

Another method for the real-time detection of individual mRNAs is in vivo hybridization of target mRNAs with molecular beacons, which are single-stranded nucleic acid probes that only fluoresce upon hybridization to a target molecule (46, 47). The most comparable of the above methods is the MS2 technique. One advantage that molecular beacons possess is that they have no tendency to multimerize, thus simplifying the image analysis. One

Molecular beacon: a single-stranded hairpin-shaped nucleic acid probe with a fluorophore and a quencher that fluoresces upon hybridization to a single-stranded target nucleic acid

Steady-state
distribution: the
distribution of mRNA
per cell across a
population that is
equilibrated in the
sense that the
distribution will not
change over time

downside, however, is the delivery of the molecular beacons to the cell itself. The most commonly used methods are microinjection (47) and listeriolysin-O (34, 49), which may result in irregular dosages and decreased cell viability.

For FISH, the primary advantages in comparison to the MS2 method are the ability to detect endogenous transcripts, obviating the need for genetic manipulations that are often difficult to perform in many organisms, and the ability to detect simultaneously at least three separate transcripts (32). Meanwhile, FISH shares with MS2 the ability to provide spatial information. However, both FISH and MS2 also share the difficulty of counting transcripts when the mRNA density is high: If many mRNAs are in close spatial proximity (in bacteria, for instance), it is hard to distinguish individual fluorescent spots using conventional microscopy, although it is possible that the use of sophisticated subdiffraction limit microscopy techniques can alleviate these problem (38, 39).

The RT-PCR-based methods are notable both for their potentially higher throughput and possibly simpler setup compared with FISH and MS2, and the data are less prone to subjective decisions in quantification than the fluorescence spot-finding algorithms required for FISH and MS2. Also, Bengtsson et al. (5) detected five different transcripts simultaneously within single cells, a feat difficult to perform with FISH. The two RT-PCR methods suffer, however, from uncertainties about the efficiency of the reverse transcriptase enzyme itself. Upon comparison, the single-cell RT-PCR experiments of Bengtsson et al. (5) are simpler to perform than the digital RT-PCR experiments of Warren et al. (50), which require the use of microfluidic devices to manage the large number of individual reactions. However, Bentgsson et al. (5) also note that their method is unable to detect transcripts at numbers below 10–20 copies per cell, whereas Warren et al. (50) counted mRNAs in individual cells at arbitrarily low copy numbers.

The studies described above have also contributed greatly to evaluating models of burst-like stochastic gene expression. The most common model was that first analyzed by Peccoud & Ycart (29) in which the gene itself transitions randomly between transcriptionally active and inactive states (**Figure 2**). Such a model contains four parameters: λ, the rate at which the gene transitions from the inactive to the active state; γ, the rate at which the gene transitions from the active to the inactive state; μ, the rate of transcription when the gene is in the active state; and δ, the rate of mRNA degradation. Peccoud & Ycart solved this model for the moments of steady-state distribution (29), which was extended to a complete analytic expression for the distribution by Raj et al. (31). This distribution can then be used to extract parameters from mRNA-counting experiments, potentially revealing new information about what parameters are subjected to regulation. For instance, Raj et al. (31) used this model to show that modulating the amount of transcription factor resulted in a modulation of the average burst size (μ/γ) while leaving the burst frequency fixed; more generally, it is possible for transcriptional regulation to occur through a change of any one (or combination) of the parameters μ, λ, and γ.

One important parameter regime of this model is that of instantaneous bursts, which occur when the rate of gene inactivation γ is larger than both the rate of mRNA degradation δ and the rate of gene activation λ. Intuitively, the former condition allows one to effectively ignore mRNA degradation during the burst itself and the latter condition ensures that individual activation events are infrequent enough that their appearance is a Poisson process, thus allowing one to make the approximation that all the mRNAs are synthesized at the same time. The number of parameters is thus reduced by 1: The model now consists only of λ, which can be interpreted as the burst frequency, and μ/γ, which is the average burst size, with δ unchanged. The steady-state distribution of this reduced model can also be solved approximately (14, 31), and this model appears to apply well for certain situations (31). There are situations in which this model cannot apply, though, the most notable being cases of bimodal mRNA

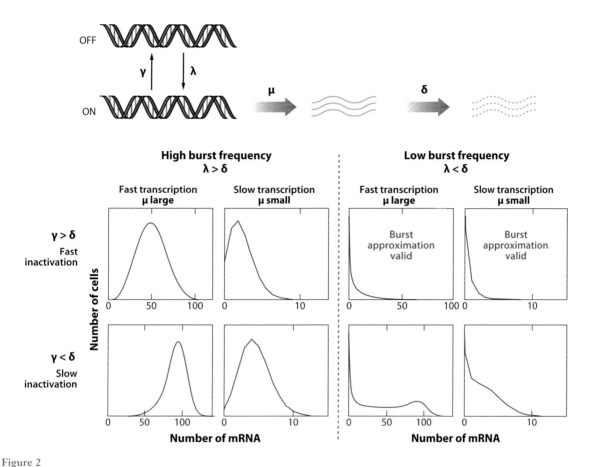

Figure 2

Distributions resulting from different values of the parameters in the gene activation/inactivation model of Peccoud & Ycart (29). The top row corresponds to the parameter γ being larger than the mRNA decay rate δ. The left side of the figure corresponds to high burst frequency compared with δ, whereas the right side corresponds to low burst frequency. The transcription rate γ was also altered as indicated. As mentioned in the text, the burst approximation is only valid when the burst frequency is low and the inactivation rate is faster than the mRNA decay rate. In particular, the bimodal expression pattern that appears with high μ and small λ, and γ cannot appear when one uses the burst approximation.

distributions, which are the result of long transcriptional bursts during which the mRNA level approaches a new steady state.

Implicit in this model is the assumption that the gene activation and inactivation events are random Poisson processes, in which case the time between events would be exponentially distributed. Recent theoretical work by Pedraza & Paulsson (30) showed that one would obtain similar (i.e., experimentally indistinguishable) distributions even if the times between the gene activation and inactivation events were not close to Poisson. This raises the possibility that

parameters extracted from steady-state measurements do not correspond to anything physical, an option that cannot be excluded using snapshot data such as those obtained from fixed or lysed cells. However, real-time imaging of transcription in living cells has shown that, at least for certain genes in *Escherichia coli* (16) and *Dictyostelium discoideum* (9), the distributions of these events are indeed exponential. Nevertheless, owing to the complexities of transcriptional regulation in higher eukaryotes, researchers will have to obtain real-time observations of transcription in those organisms to specify exactly what sorts of models are applicable.

Yet while there is now a growing body of evidence supporting transcriptional bursts, their biological origins remain unclear. In prokaryotes, the results of Golding et al. (16) convincingly show that transcriptional bursts do indeed occur in *E. coli*, and the authors proposed a host of possible causes, including simple mechanisms such as transcription factor binding and unbinding as well as more complex processes such as DNA conformational changes and sigma factor retention resulting in pulsatile reinitiation of transcription. Indeed, the very presence of prokaryotic transcriptional bursts themselves may be gene specific, since Maamar et al. (23) found that the Fano factor for the mRNA distributions they measured were close to 1, thus arguing against transcriptional bursts in that particular case. Only further experimentation can provide answers to these questions.

Meanwhile, in eukaryotes in general and higher eukaryotes in particular, it seems as though transcriptional bursting is most certainly the norm, with most if not all noise studies in the field providing some evidence for pulsatile transcription. One early candidate for the cause of transcriptional bursts was chromatin remodeling. Eukaryotic genes are wrapped around histone proteins that form chromatin fibers, and chromatin can be remodeled from a tightly bound, transcriptionally inert structure to a more loosely bound, transcriptionally active conformation through the action of various chromatin-remodeling enzymes. Thus, random events of chromatin remodeling could result in random bursts of transcription. Yet, despite the clarity of this hypothesis, it has yet to be decisively proven or disproven; so far, the only studies providing any hints are those of Raser and O'Shea (33), in which the alteration of chromatin-remodeling enzymes resulted in changes in stochastic gene expression, and Raj et al. (31), in which genomic position (and thus chromatin context) appeared to have a strong effect on covariation in bursting between multiple genes. A conclusive test of the connection between chromatin remodeling and transcriptional bursting will also require single-molecule techniques, this time directed at the gene itself. Given that experiments monitoring chromatin remodeling in real time have already been carried out (45), a suitable combination of these different single-molecule techniques will likely settle the question.

Another consideration is the propagation of fluctuations in mRNA levels compared to those of proteins. Although no study has yet combined single-molecule mRNA detection with single-molecule protein detection, two of the single-molecule mRNA studies highlighted here have used conventional fluorescent proteins to examine these problems. Golding et al. (16) found that the mRNA and protein levels exhibited a linear correlation in single cells. They also found that the correlation was weakest in the time just following cell division, which they ascribed to the randomizing effects of the binomial partitioning of mRNAs and proteins upon cell division. Raj et al. (31) tried to examine the relationship between protein degradation rates and the correlation between mRNA and protein levels. They found that mRNA and protein levels correlated strongly when protein lifetime was short, but that this correlation decreased when protein lifetime was long, a finding also born out in models of stochastic protein and mRNA production. The authors found it generally difficult, however, to detect small numbers of protein molecules in individual eukaryotic cells owing to their large cellular volumes, making the development of single-molecule techniques to count the number of proteins in individual cells important. We outline some recent efforts toward this goal in the next section.

SINGLE-PROTEIN DETECTION

Ultimately, much of cellular function is carried out by the proteins encoded for by the mRNAs, and hence the enumeration of individual proteins is essential to a complete understanding of stochastic effects in gene expression. Unfortunately, achieving the required probe specificity is far more difficult with proteins

Table 2   Comparison of different single protein detection methods

Method	Endogenous protein detection?	Real-time measurements?	Detection of multiple proteins species at once?	Other advantages	Other disadvantages	Reference
β-galactosidase microfluidics	No	Yes	No	Works in a variety of cell types	Requires microfluidics, cells must be permeabilized, not as effective with large molecule numbers, no spatial information	(7)
Single fluorescent protein imaging	No	Yes	Potentially	Yields spatial information	Unlikely to work in present form in organisms larger than bacteria	(53)
Single-protein antibody labeling	Yes	No	Potentially	Simple to perform with a large dynamic range, works in many organisms	Requires microfluidics and complex optics, questions about antibody efficiency and sensitivity at low numbers, no spatial information	(18)

than with nucleic acids. Nevertheless, new techniques are emerging that are giving researchers a glimpse into the protein content of individual cells (**Table 2**).

Recently, two exciting studies from the laboratory of X. Sunney Xie have detailed their efforts to detect individual protein molecules in living cells. Although both studies reached strikingly similar biological conclusions, their approaches were rather different. In Cai et al. (7), the authors combined the high efficiency of the β-galactosidase enzyme with microfluidics to count protein numbers by measuring enzymatic activity. β-galactosidase is efficient at cleaving substrates, and several reagents produce easily detectable substances upon enzymatic activity. However, such product molecules are usually quickly exported from the cell itself, thus diffusing the signal greatly, which is why such assays are typically performed on populations rather than single cells. To circumvent this problem, the authors confined each cell to a defined small volume using a

microfluidic device, an approach based on previously described methods used to detect the activity of single enzymes (37). The concentration of the fluorescent product from a single enzyme is made high enough so that the fluorescent signal is easily detectable. Because the nonfluorescent substrate is present in saturating quantities, the increase in signal is linear, with the slope directly proportional to the number of β-galactosidase enzymes present. Thus, by measuring changes in the signal slope, the authors detected the formation of single enzymes.

Yu et al. (53) used fluorescent proteins in a manner that allowed for the detection of single-protein molecules. The authors noted that individual fluorescent proteins generate enough fluorescence for detection given a long enough exposure time, but the problem is that they diffuse too rapidly to produce a localized signal within such time periods. To solve this problem, they fused a bright, fast-folding variant of yellow fluorescent protein (Venus YFP) to a peptide sequence that anchors itself to the membrane, thus dramatically reducing the mobility of the YFP molecules. Once anchored in this fashion, they directly imaged the molecules using a standard fluorescence microscope coupled with laser illumination.

Impressively, although these methods are different in character, the results obtained from both studies were almost identical. The main finding was that proteins were produced in short but infrequent bursts, presumably occurring during the lifetime of single, infrequently transcribed mRNAs. They parameterized their data using a model with two parameters: a, referring to the burst frequency; and b, referring to the average burst size. (This model is in principle similar to models used for describing mRNA distributions arising from short transcriptional bursts.) The main assumption is that individual burst events are short compared with the protein lifetime, which, mathematically speaking, means that the mRNA lifetime is much shorter than the protein lifetime, likely a valid approximation for a significant number of genes. During the lifetime of the mRNA, the number of proteins produced is taken from a geometric distribution, which has a nice biological interpretation: Once the mRNA is transcribed, it will be continuously translated into protein by ribosomes. The presence of the ribosomes also confers protection from various ribonucleases. However, every time a ribosome finishes translating and thus unbinds from the mRNA, there is a certain probability that a ribonuclease will bind rather than another ribosome. This process leads to the geometric distribution of burst sizes of mean size b (24), and when one combines this burst size distribution with the random appearances of bursts (parameterized by a), Cai et al. (7) found that the distribution of proteins across a population is given by the γ distribution. This model has been applied to small gene networks such as genetic autoregulation and transcriptional cascades by Friedman et al. (14), and a recent study has shown how to extend this work to find distributions in the presence of transcriptional bursts (41). In terms of the underlying rates of transcription, translation, and mRNA and protein degradation, parameter a is the rate of transcription and parameter b is the ratio of the translation rate to the mRNA degradation rate (43).

Yet, although these two techniques are undeniably elegant, it is unclear how well they will translate to other types of organisms in which the protein count is much higher and the cellular volume is much larger. Moreover, the use of various reporter gene constructs presupposes the ability to perform genetic manipulations, which are often difficult or impossible to perform in many organisms. To circumvent these problems, Huang et al. (18) used a combination of microfluidics, immunofluorescent labeling, and new optics to count the number of endogenous protein molecules in organisms both large and small. Their approach was to lyse cells in small microfluidic chambers and then use fluorescently conjugated antibodies to label the target protein. They then flowed the now fluorescent proteins over a confocal microscope to image the individual proteins. The imaging step is one of the principal difficulties in this type of method, since the field of illumination is usually much smaller than the channels through which the proteins flow, thus making it hard to detect all the proteins as they pass by the objective. The authors solved this problem by utilizing cylindrical optics, thereby illuminating the entire cross section of the protein channel.

They then used their system to measure the number of β-adrenergic receptors in individual insect cells and found that the numbers of proteins fluctuated wildly from cell to cell, with numbers as low as 2000 and as high as 60,000. They also measured the numbers of the constituents of the phycobilisome (i.e., the complex responsible for harvesting light energy from the sun) in individual cyanobacteria. Huang et al. found that expression of this complex was much more variable in nitrogen-starved conditions than in nitrogen-rich conditions. Some caveats to this method include the limits of its sensitivity (which the authors estimated to be around seven molecules in their cyanobacteria experiments) and its throughput, but it is nevertheless a promising and general methodology for measuring cell-to-cell variability in endogenous protein levels.

FUTURE DIRECTIONS

The application of single-molecule techniques to the measurement of gene expression in single cells has provided many new insights into the field of stochastic gene expression, and the utilization of these and other methods not yet invented for progressively more complex biological problems will undoubtedly lead to further discoveries. Such research may move toward studying all the individual stochastic biochemical reactions involved in gene expression, rather than just counting and monitoring mRNA and protein numbers. Some work has already been done along these lines, with the study of Elf et al. (10) examining the kinetics of individual transcription factors searching for their DNA binding site in living *E. coli*; further work may reveal the contribution of these random binding and unbinding events to stochastic expression of the target genes. In higher eukaryotes, the notion that chromatin remodeling is responsible for transcriptional bursting has still not been proven directly and remains a ripe target for combining single-molecule DNA measurements with transcriptional measurements. More generally, virtually all the enzymatic activities involved in gene expression can be subjected to single molecule scrutiny to determine exactly which individual processes are the most important in making gene expression stochastic. Candidates include the stepping behavior of individual RNA polymerases; splicing and other posttranscriptional mRNA processing; the nuclear export of mRNAs, including gene translocation to the nuclear periphery (40); the activity(ies) of ribosomes, ribonucleases, and proteases. These are but a smattering of the many important elements involved in gene expression, and studying how these individual molecules function in vivo will almost certainly change our conception of stochastic gene expression.

Another avenue of inquiry in which single-molecule techniques may provide fresh insights is the biological consequences of noise. Thus far, the field has focused primarily on cases in which noise in gene expression can be beneficial, providing useful phenotypic variability in genetically identical populations (2, 21, 22, 44, 52). Often this variability is enhanced by thresholding and amplification of noise by genetic feedback loops (1, 23, 42). Maamar et al. (23) used single-mRNA detection to try to infer thresholding behavior at the protein level, but it is likely that direct observation of individual protein molecules in real time will be required to truly observe the actions of such feedbacks. Less well studied (but perhaps more important in general) are instances in which noise is detrimental to robust function, an example of which is development in multicellular organisms. In such cases single-molecule detection may be of primary importance in detecting low numbers of important biomolecules during developmental processes (32), thus allowing researchers to gauge the extent to which noise is tolerated in such systems.

Biological insights may also arise from parallelization of single-molecule gene expression measurements to a genomic scale, i.e., measuring the detailed stochastic properties of gene expression (such as mRNA and protein burst frequency and size) for most genes in an organism. These sorts of measurements can lead to insights into the nature and consequences of noise, as demonstrated by using GFP in yeast (3, 25). Single-molecule measurements allow for the detection of many potentially interesting genes whose expression levels are below the GFP detection limit, and would allow for more careful measurements of important parameters that can only be inferred by GFP measurements. It remains to be seen how easily such methods can be parallelized to facilitate such studies, but the capacity for new insights is great.

In conclusion, we feel that these trailblazing single-molecule stochastic gene expression experiments are pointing in the direction to which the rest of the field will head. The ability of these methods to yield quantitative data raises several exciting possibilities that seemed impossible only a few years ago. We look forward to expecting the unexpected as the combination of single-molecule detection and molecular biology breathe new life into the still-young field of stochastic gene expression.

DISCLOSURE STATEMENT

The authors are not aware of any biases that might be perceived as affecting the objectivity of this review.

ACKNOWLEDGMENTS

We would like to thank Ido Golding for many helpful comments on the manuscript. We also apologize to any authors whose work we were unable to mention due to space constraints. A.v.O. was supported by NSF grant PHY-0548484 and NIH grants R01-GM068957 and R01-GM077183. A.R. was supported by NSF Fellowship DMS-0603392 and a Burroughs Wellcome Fund Career Award at the Scientific Interface.

LITERATURE CITED

1. Acar M, Becskei A, van Oudenaarden A. 2005. Enhancement of cellular memory by reducing stochastic transitions. *Nature* 435:228–32
2. Acar M, Mettetal JT, van Oudenaarden A. 2008. Stochastic switching as a survival strategy in fluctuating environments. *Nat. Genet.* 40:471–75
3. Bar-Even A, Paulsson J, Maheshri N, Carmi M, O'Shea E, et al. 2006. Noise in protein expression scales with natural protein abundance. *Nat. Genet.* 38:636–43
4. Beach DL, Salmon ED, Bloom K. 1999. Localization and anchoring of mRNA in budding yeast. *Curr. Biol.* 9:569–78
5. Bengtsson M, Stahlberg A, Rorsman P, Kubista M. 2005. Gene expression profiling in single cells from the pancreatic islets of Langerhans reveals lognormal distribution of mRNA levels. *Genome Res.* 15:1388–92
6. Bertrand E, Chartrand P, Schaefer M, Shenoy SM, Singer RH, Long RM. 1998. Localization of ASH1 mRNA particles in living yeast. *Mol. Cell* 2:437–45
7. Cai L, Friedman N, Xie XS. 2006. Stochastic protein expression in individual cells at the single molecule level. *Nature* 440:358–62
8. Chang HH, Hemberg M, Barahona M, Ingber DE, Huang S. 2008. Transcriptome-wide noise controls lineage choice in mammalian progenitor cells. *Nature* 453:544–47
9. Chubb JR, Trcek T, Shenoy SM, Singer RH. 2006. Transcriptional pulsing of a developmental gene. *Curr. Biol.* 16:1018–25
10. Elf J, Li GW, Xie XS. 2007. Probing transcription factor dynamics at the single-molecule level in a living cell. *Science* 316:1191–94
11. Elowitz MB, Levine AJ, Siggia ED, Swain PS. 2002. Stochastic gene expression in a single cell. *Science* 297:1183–86
12. Femino AM, Fay FS, Fogarty K, Singer RH. 1998. Visualization of single RNA transcripts in situ. *Science* 280:585–90
13. Forrest KM, Gavis ER. 2003. Live imaging of endogenous RNA reveals a diffusion and entrapment mechanism for nanos mRNA localization in *Drosophila*. *Curr. Biol.* 13:1159–68
14. Friedman N, Cai L, Xie XS. 2006. Linking stochastic dynamics to population distribution: an analytical framework of gene expression. *Phys. Rev. Lett.* 97:168302
15. Fusco D, Accornero N, Lavoie B, Shenoy SM, Blanchard JM, et al. 2003. Single mRNA molecules demonstrate probabilistic movement in living cells. *Curr. Biol.* 13:161–67
16. Golding I, Paulsson J, Zawilski SM, Cox EC. 2005. Real-time kinetics of gene activity in individual bacteria. *Cell* 123:1025–36

17. Gregor T, Tank DW, Wieschaus EF, Bialek W. 2007. Probing the limits to positional information. *Cell* 130:153–64

18. Huang B, Wu H, Bhaya D, Grossman A, Granier S, et al. 2007. Counting low-copy number proteins in a single cell. *Science* 315:81–84

19. Kepler TB, Elston TC. 2001. Stochasticity in transcriptional regulation: origins, consequences, and mathematical representations. *Biophys. J.* 81:3116–36

20. Ko MS, Nakauchi H, Takahashi N. 1990. The dose dependence of glucocorticoid-inducible gene expression results from changes in the number of transcriptionally active templates. *EMBO J.* 9:2835–42

21. Kussell E, Leibler S. 2005. Phenotypic diversity, population growth, and information in fluctuating environments. *Science* 309:2075–78

22. Losick R, Desplan C. 2008. Stochasticity and cell fate. *Science* 320:65–68

23. Maamar H, Raj A, Dubnau D. 2007. Noise in gene expression determines cell fate in *Bacillus subtilis*. *Science* 317:526–29

24. McAdams HH, Arkin A. 1997. Stochastic mechanisms in gene expression. *Proc. Natl. Acad. Sci. USA* 94:814–19

25. Newman JR, Ghaemmaghami S, Ihmels J, Breslow DK, Noble M, et al. 2006. Single-cell proteomic analysis of *S. cerevisiae* reveals the architecture of biological noise. *Nature* 441:840–46

26. Novick A, Weiner M. 1957. Enzyme induction as an all-or-none phenomenon. *Proc. Natl. Acad. Sci. USA* 43:553–66

27. Ozbudak EM, Thattai M, Kurtser I, Grossman AD, van Oudenaarden A. 2002. Regulation of noise in the expression of a single gene. *Nat. Genet.* 31:69–73

28. Paulsson J. 2004. Summing up the noise in gene networks. *Nature* 427:415–18

29. Peccoud J, Ycart B. 1995. Markovian modelling of gene product synthesis. *Theor. Popul. Biol.* 48:222–34

30. Pedraza JM, Paulsson J. 2008. Effects of molecular memory and bursting on fluctuations in gene expression. *Science* 319:339–43

31. Raj A, Peskin CS, Tranchina D, Vargas DY, Tyagi S. 2006. Stochastic mRNA synthesis in mammalian cells. *PLoS Biol.* 4:e309

32. Raj A, Van Den Bogaard P, Rifkin SA, van Oudenaarden A, Tyagi S. 2008. Imaging individual mRNA molecules using multiple singly labeled probes. *Nat. Methods* 5:877–79

33. Raser JM, O'Shea EK. 2004. Control of stochasticity in eukaryotic gene expression. *Science* 304:1811–14

34. Rhee WJ, Santangelo PJ, Jo H, Bao G. 2008. Target accessibility and signal specificity in live-cell detection of BMP-4 mRNA using molecular beacons. *Nucleic Acids Res.* 36:e30

35. Rosenfeld N, Young JW, Alon U, Swain PS, Elowitz MB. 2005. Gene regulation at the single-cell level. *Science* 307:1962–65

36. Ross IL, Browne CM, Hume DA. 1994. Transcription of individual genes in eukaryotic cells occurs randomly and infrequently. *Immunol. Cell Biol.* 72:177–85

37. Rotman B. 1961. Measurement of activity of single molecules of beta-D-galactosidase. *Proc. Natl. Acad. Sci. USA* 47:1981–91

38. Rust MJ, Bates M, Zhuang X. 2006. Sub-diffraction-limit imaging by stochastic optical reconstruction microscopy (STORM). *Nat. Methods* 3:793–95

39. Schermelleh L, Carlton PM, Haase S, Shao L, Winoto L, et al. 2008. Subdiffraction multicolor imaging of the nuclear periphery with 3D structured illumination microscopy. *Science* 320:1332–36

40. Sexton T, Schober H, Fraser P, Gasser SM. 2007. Gene regulation through nuclear organization. *Nat. Struct. Mol. Biol.* 14:1049–55

41. Shahrezaei V, Swain PS. 2008. Analytical distributions for stochastic gene expression. *Proc. Natl. Acad. Sci. USA* 105:17256–61

42. Suel GM, Kulkarni RP, Dworkin J, Garcia-Ojalvo J, Elowitz MB. 2007. Tunability and noise dependence in differentiation dynamics. *Science* 315:1716–19

43. Thattai M, van Oudenaarden A. 2001. Intrinsic noise in gene regulatory networks. *Proc. Natl. Acad. Sci. USA* 98:8614–19

44. Thattai M, van Oudenaarden A. 2004. Stochastic gene expression in fluctuating environments. *Genetics* 167:523–30

45. Tumbar T, Sudlow G, Belmont AS. 1999. Large-scale chromatin unfolding and remodeling induced by VP16 acidic activation domain. *J. Cell Biol.* 145:1341–54
46. Tyagi S, Kramer FR. 1996. Molecular beacons: probes that fluoresce upon hybridization. *Nat. Biotechnol.* 14:303–8
47. Vargas DY, Raj A, Marras SA, Kramer FR, Tyagi S. 2005. Mechanism of mRNA transport in the nucleus. *Proc. Natl. Acad. Sci. USA* 102:17008–13
48. Vogelstein B, Kinzler KW. 1999. Digital PCR. *Proc. Natl. Acad. Sci. USA* 96:9236–41
49. Wang W, Cui ZQ, Han H, Zhang ZP, Wei HP, et al. 2008. Imaging and characterizing influenza A virus mRNA transport in living cells. *Nucleic Acids Res.* 36:4913–28
50. Warren L, Bryder D, Weissman IL, Quake SR. 2006. Transcription factor profiling in individual hematopoietic progenitors by digital RT-PCR. *Proc. Natl. Acad. Sci. USA* 103:17807–12
51. White MR, Masuko M, Amet L, Elliott G, Braddock M, et al. 1995. Real-time analysis of the transcriptional regulation of HIV and hCMV promoters in single mammalian cells. *J. Cell Sci.* 108(Pt. 2):441–55
52. Wolf DM, Vazirani VV, Arkin AP. 2005. Diversity in times of adversity: probabilistic strategies in microbial survival games. *J. Theor. Biol.* 234:227–53
53. Yu J, Xiao J, Ren X, Lao K, Xie XS. 2006. Probing gene expression in live cells, one protein molecule at a time. *Science* 311:1600–3

Comparative Enzymology and Structural Biology of RNA Self-Cleavage

Martha J. Fedor

Department of Chemical Physiology, Department of Molecular Biology and The Skaggs Institute for Chemical Biology, The Scripps Research Institute, La Jolla, California 92037; email: mfedor@scripps.edu

Annu. Rev. Biophys. 2009. 38:271–99

First published online as a Review in Advance on February 20, 2009

The *Annual Review of Biophysics* is online at biophys.annualreviews.org

This article's doi: 10.1146/annurev.biophys.050708.133710

Key Words

ribozyme, catalytic RNA, RNA structure, RNA folding

Abstract

Self-cleaving hammerhead, hairpin, hepatitis delta virus, and *glmS* ribozymes comprise a family of small catalytic RNA motifs that catalyze the same reversible phosphodiester cleavage reaction, but each motif adopts a unique structure and displays a unique array of biochemical properties. Recent structural, biochemical, and biophysical studies of these self-cleaving RNAs have begun to reveal how active site nucleotides exploit general acid-base catalysis, electrostatic stabilization, substrate destabilization, and positioning and orientation to reduce the free energy barrier to catalysis. Insights into the variety of catalytic strategies available to these model RNA enzymes are likely to have important implications for understanding more complex RNA-catalyzed reactions fundamental to RNA processing and protein synthesis.

Contents

INTRODUCTION

Recent X-ray crystal structures of hammerhead, hairpin, hepatitis delta virus (HDV), and *glmS* ribozymes highlight differences in global and active site structures among these self-cleaving RNAs and have fueled renewed interest in the enzymology of RNA self-cleavage. These self-cleaving RNAs mediate the same reversible phosphodiester cleavage reaction that involves nucleophilic attack of a $2'$ oxygen on the adjacent phosphodiester and breaking of a $5'$ oxygen-phosphorus bond to give cleavage products with $2',3'$-cyclic and $5'$-hydroxyl termini (18, 50, 67) (**Figure 1**). For hammerhead and hairpin ribozymes, cleavage of a chiral R_P phosphorothioate linkage occurs with inversion of stereochemical configuration, implicating an $S_N2(P)$ in-line attack mechanism in which the $2'$-oxygen nucleophile and the $5'$-oxygen leaving group occupy apical positions in a trigonal

HDV: hepatitis delta virus

Transition state: a high-energy, short-lived chemical state that presents the highest free energy barrier in a reaction pathway between substrates and products

General acid-base catalysis: a process in which the free energy barrier to a reaction is reduced by proton transfer that is mediated by a species other than water

bipyramidal transition state during cleavage and reverse roles during ligation (77, 157, 166).

The fundamental chemistry and enzymology of phosphoryl transesterification reactions have been the subject of considerable interest since investigations in the 1950s of alkaline hydrolysis of RNA, which is the uncatalyzed version of the same reaction (for recent reviews see References 26 and 88). These studies suggest that cleavage occurs through an associative mechanism in which the bonds between phosphorus and the $2'$ and $5'$ oxygens are partially formed in the transition state. Therefore, there is considerable negative charge accumulation on both nucleophilic and leaving group oxygens. Several kinds of catalytic strategies could be used to accomplish the rate accelerations on the order of 10^6- to 10^8-fold that are typical of self-cleaving ribozymes. An active site scaffold could lower the energy barrier to catalysis by fixing the reactive phosphodiester in the trigonal bipyramidal geometry optimal for an $S_N2(P)$ in-line attack mechanism. Removal of a proton from a hydroxyl to generate an oxyanion could activate nucleophilic attack on phosphorus, and protonation of an oxyanion leaving group could facilitate breaking the oxygen-phosphorus bond. Proton transfer could be mediated through general acid-base catalysis by an active site functional group, by buffer or a bound cofactor, through specific acid-base catalysis by hydroxide or hydronium ions, or by metal-bound water. Positively charged active site residues, metal cations, or hydrogen bonding interactions also could provide electrostatic stabilization of the negative charge that develops in the pentacovalent transition state. Factors that destabilize the substrate ground state, such as an unfavorable geometry or charge distribution, also could lower the energy barrier to catalysis.

RNase A mediates the same chemical reaction as the self-cleaving ribozymes and provides a textbook example of concerted general acid-base catalysis (132). In the RNase A mechanism, His12 serves as a general base catalyst to remove a proton from the attacking $2'$-oxygen nucleophile and His119 acts as a general acid

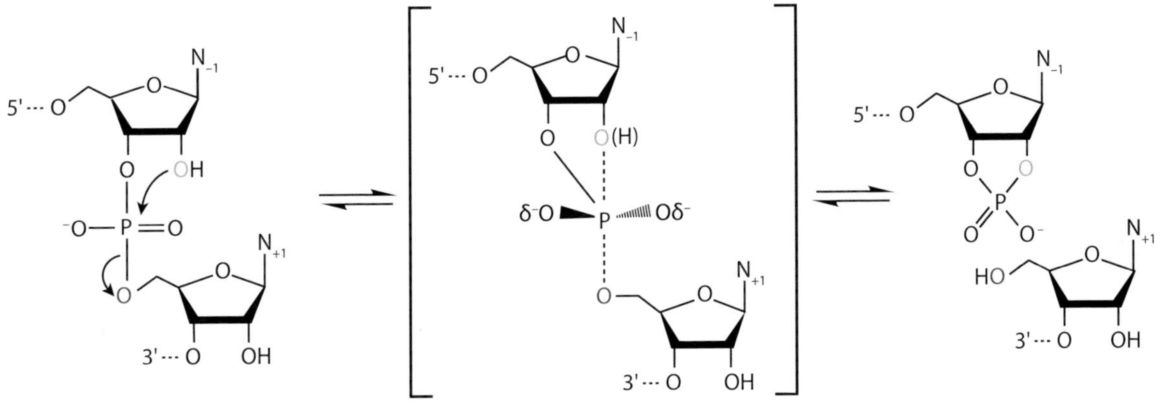

Figure 1

Chemical mechanism of RNA self-cleavage. The vicinal 2′ hydroxyl (*blue*) is the nucleophile during cleavage and the leaving group during ligation. The 5′ oxygen (*red*) is the leaving group during cleavage and the nucleophile during ligation. The nucleotide upstream of the reactive phosphate that contributes the 2′ hydroxyl is numbered −1, and the downstream nucleotide that contributes the 5′ oxygen is +1.

catalyst to protonate the 5′-oxygen leaving group. Hydrogen bonding between the ε-amino side chain of Lys41 and the *pro-R*$_P$ oxygen provides electrostatic stabilization to the transition state. The imidazole side chain of histidine ionizes with a pK$_a$ value near 6, making histidine residues particularly well suited to catalysis of proton transfer because significant fractions of histidine are present in the appropriate ionization states at neutral pH to accept and donate protons. Likewise, the positive charge on the ε-amino group of lysine enables these residues to contribute electrostatic stabilization to a pentacovalent transition state with five electronegative oxygens.

The concerted general acid-base mechanism of RNase A provides a framework for investigating RNA self-cleavage, but the comparison of protein and RNA enzymes is not at all straightforward due to the difference between amino acids and nucleotides in chemical properties. Ribonucleotide bases display pK$_a$ values above 9.2 or below 4.3, at least for nucleotides free in solution, and no RNA functional groups are positively charged at neutral pH. Like proteins, however, RNAs do fold into precise 3D structures. Thus, an active site composed of RNA could fix a phosphodiester in the trigonal bipyramidal geometry to facilitate an S$_N$2(P)

in-line attack mechanism and position functional groups that mediate catalytic chemistry. RNAs also exhibit specific, high-affinity binding of metal cations and small ligands that could serve as catalytic cofactors. Each of the self-cleaving RNA motifs exhibits a unique constellation of pH, metal cation, and functional group dependencies, suggesting that they exploit distinct catalytic mechanisms. Nevertheless, a complete understanding of the catalytic mechanism of any self-cleaving RNA remains elusive. This review aims to present the current understanding of the catalytic mechanisms of hammerhead, hairpin, HDV, and *glmS* ribozymes, informed by striking advances in structural biology. (For recent related reviews, see References 38, 66, 147, 150, and 156.) The active site structure of the *Neurospora* Varkud satellite self-cleaving RNA has not yet been determined crystallographically and it is not included in the review, but it has been reviewed elsewhere (29, 86).

HAMMERHEAD RIBOZYME

The hammerhead ribozyme was the first self-cleaving RNA to be discovered in plant satellite RNAs, where self-cleavage generates monomers from concatameric genomes

Specific acid-base catalysis: a process in which the free energy barrier to a reaction is reduced by the addition or removal of protons by solvent ions

Electrostatic stabilization: a mechanism for lowering the energy barrier to a reaction by neutralizing charge accumulation in the transition state

produced by rolling circle replication (17, 67, 127). The minimal hammerhead ribozyme that was defined by phylogenetic comparison, deletion, truncation, and mutagenesis experiments has three base-paired helices of variable composition surrounding a 15-nucleotide core containing 11 strongly conserved nucleotides that are positioned precisely with respect to the reactive phosphodiester (18, 49, 50, 67) (**Figure 2a**). Hammerhead ribozyme motifs also have been identified in newt (42), schistosome (46), cricket (135), *Arabidopsis thaliana* (128) and, recently, mammalian (93) genomes. The frequent emergence of hammerhead motifs during in vitro evolution experiments suggests that this simple self-cleaving domain could have evolved repeatedly (143).

Minimal and Extended Forms of the Hammerhead Ribozyme

In X-ray crystal structures of minimal hammerhead ribozymes (126, 148), helices I and II form the short arms in a global Y-shaped conformation, with the core nucleotides falling into two distinct domains rather than a single active site, and many functional groups that are critical for catalysis form no obvious interactions (11, 56, 97, 164, 175) (**Table 1**) (**Figure 2a**). A similar Y-shaped global conformation was inferred from the results of a variety of spectroscopic and electrophoretic studies of minimal hammerheads arguing that crystal structures represent the same conformation that predominates in solution (4, 6, 7, 56, 65, 163). However,

the lack of agreement between the arrangement of functional groups in the structures and the effects of nucleotide modifications on activity made it difficult to arrive at a consistent model of the catalytic mechanism. Because functional studies monitor interactions in the transition state and physical studies probe ground state structures, these discrepancies suggest that a major conformational rearrangement is needed to create the transition state from the ground state crystal structures.

It was not until extended hammerhead motifs were characterized that structure-function relationships began to come into clear focus (66, 110, 111, 129). Natural forms of hammerhead ribozymes include additional sequences that are not essential for catalysis but promote assembly of the functional structure through interhelical interactions (36, 72). These interactions can occur between hairpin loops, between hairpin loops and internal bulges, and between two internal bulges, and loops and bulges can vary in size and position within the conserved helices I and II (134, 151). Unlike minimal ribozymes, extended hammerhead ribozymes exhibit robust self-cleavage activity in vivo and in vitro in reactions that approximate intracellular ionic conditions. Extended hammerhead ribozymes can exhibit fast cleavage kinetics, with one sequence emerging through in vitro selection cleaving 1000-fold faster than minimal ribozymes under similar conditions (20, 21, 34, 36, 72, 73, 109, 115, 117, 137, 151).

Extended hammerhead ribozymes also display enhanced ligation activity relative to

---

Figure 2

Hammerhead ribozyme structure and mechanism. Ribozyme and substrate strands are blue and green, respectively, and the two nucleotides upstream (C_{17}) and downstream ($C_{1.1}$) of the reactive phosphodiester are gold. (*a*) Secondary structure of a minimal hammerhead ribozyme with a sequence derived from the hammerhead motif of *Schistosoma mansoni* arranged to reflect the structure of minimal ribozymes. (*b*) Sequence and secondary structure of the extended hammerhead ribozyme sequence from *S. mansoni* (95) (PDB 2GOZ). Red lines indicate interhelical hydrogen bonding interactions. (*c*) Global architecture of the hammerhead ribozyme (95) (PDB 2GOZ). (*d*) Active site of an extended hammerhead ribozyme from *S. mansoni* with cleavage blocked by a 2′-methoxyl modification of the nucleophilic oxygen (95) (PDB 2GOZ). (*e*) Hypothetical cleavage mechanism in which G_{12} serves as a general base catalyst by deprotonating the 2′ hydroxyl. Alternatively, G_{12} could participate indirectly in proton shuttling through water. The 2′ hydroxyl of G_8 has been proposed to play a role as a general acid during cleavage by protonating the 5′-oxygen-leaving group. A metal cation could interact with the *pro-R*$_P$ oxygen as a metal ligand to provide electrostatic stabilization. Alternatively, or in addition, metal-bound water could mediate proton transfer between leaving group and nucleophilic oxygens.

minimal ribozymes (21, 109, 115). Minimal hammerhead ribozymes catalyze ligation with rate constants below 0.01 min^{-1} and favor cleavage over ligation by more than 100-fold (63, 64). Extended hammerhead ribozymes exhibit ligation rate constants as much as 2000-fold faster and favor cleavage over ligation by only two- to threefold under some conditions

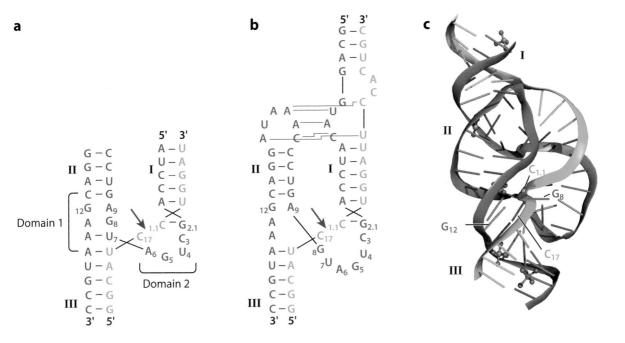

Table 1 X-ray crystal structures of hammerhead ribozymes

RNA construct	Salt	pH	Resolution (Å)	PDB ID	Reference
Minimal, RNA-DNA hybrid	(NH$_4$)$_2$SO$_4$, Li$_2$SO$_4$, NH$_4$(CH$_3$)$_2$AsO$_2$, spermine	6.0	2.6	1HMH	(126)
Minimal, 2'-OCH$_3$ ribose at N$_{-1}$	NH$_4$(CH$_3$)$_2$AsO$_2$, NH$_4$COOCH$_3$, Mg(COOCH$_3$)$_2$, spermine	6.5	3.1	1MME	(148)
Minimal, freeze-trapped RNA	Li$_2$SO$_4$, NH$_4$(CH$_3$)$_2$AsO$_2$	6.0	3.0	299D	(149)
Minimal, freeze-trapped RNA	Li$_2$SO$_4$, MnSO$_4$	5.0	3.0	300D	(149)
Minimal, freeze-trapped RNA	Li$_2$SO$_4$, MgSO$_4$	8.5	3.0	301D	(149)
Minimal, *talo*-5'-*C*-methyl at N$_{-1}$	Na(CH$_3$)$_2$AsO$_2$, Li$_2$SO$_4$, NaCOOCH$_3$	6.0	3.0	379D	(103)
Minimal, lattice-trapped mixture of unmodified RNA and RNA with 2' OCH$_3$ at N$_{-1}$	NaCOOCH$_3$, Li$_2$SO$_4$	6.5	3.1	488D	(102)
Minimal, tethered between helices I and II	NaCOOCH$_3$, Li$_2$SO$_4$	5.0	2.85	1NYI	(40)
Minimal, tethered between helices I and II	NaCOOCH$_3$, Li$_2$SO$_4$	5.0	2.99	1Q29	(40)
Schistosoma mansoni, 2' OCH$_3$ at N$_{-1}$	MES, (NH$_4$)$_2$SO$_4$, MgCl$_2$	5.5	2.2	2GOZ	(95)
Schistosoma mansoni, 2' OCH$_3$ at N$_{-1}$	MES, (NH$_4$)$_2$SO$_4$, MgCl$_2$, MnCl$_2$	5.5	2.0	2OEU	(94)

(21, 109). The ability of interhelical loop interactions to increase ligation activity of extended ribozymes is consistent with earlier observations that covalent cross-links between helix I and helix II enhance the ligation efficiency of minimal ribozymes (10, 40, 158).

Hammerhead Ribozyme Active Site Structure

Two X-ray crystal structures of the extended hammerhead ribozyme have been solved at 2.2 Å and 2.0 Å using the hammerhead sequence from *Schistosoma mansoni*, in which cleavage was blocked by methylation of the C$_{17}$ ribose upstream of the reactive phosphate (94, 95) (**Figure 2b–d**). The structures reveal a novel catalytic core that reconciles many of the discrepancies between biochemical and structural features of minimal ribozymes (66, 110, 111, 129). The effects of functional group modifications on the activity of minimal ribozymes agree well with the extended ribozyme structure, and reactions catalyzed by both forms of the hammerhead respond similarly to modifications of

core nucleotides (95, 110–112, 130, 146). These findings show that reactions catalyzed by both forms of the hammerhead proceed through the same transition state. The agreement between structural, and functional studies indicates that the active site structure in the *Schistosoma* ribozyme crystals more closely resembles the transition state, and provides valuable insights into the catalytic mechanism.

A striking of feature of the active site architecture is the near in-line geometry of the nucleophilic and leaving group oxygens (**Figure 2d**). In contrast to minimal ribozyme structures, in which the angle between the 2' and 5' oxygens diverged from the angle expected for an in-line attack mechanism by about 60°, the angle between the attacking 2' oxygen and phosphorus is just 17° away from perfect 180° in-line geometry. An extensive network of stacking and hydrogen bonding interactions fix the C$_{17}$ and C$_{1.1}$ nucleotides in an extended geometry. Consistent with cross-links that place G$_5$, G$_8$ and G$_{12}$ near the reactive phosphodiester in a functional ribozyme (62, 81), the G$_5$ purine stacks on A$_6$ to form a wedge between the C_{17} and $C_{1.1}$ nucleotides flanking the reactive phosphate, which are fixed in position by a G_8-C_3 base pair. The functional importance of the G_8-C_3 pair has been confirmed through compensating base changes that simultaneously eliminate and restore base pairing and activity (95, 112, 130). Previous biochemical studies had shown that substitutions of G_8 with guanosine analogs that vary in the pK_a value for ionization at the N1 ring nitrogen alter the pH dependence of cleavage activity, leading to the proposal that the guanine nucleobase participates in proton transfer (60). However, this effect on pH dependence can now be understood in light of the crystal structure as a requirement for hydrogen bond donation by N1 of G_8 to N3 of C_3 to maintain the Watson-Crick pairing that is essential to active site architecture.

Metal Cation Dependence of Hammerhead Folding and Catalysis

Molar concentrations of monovalent cations support the activity of both minimal and extended forms of the ribozyme, excluding an absolute requirement for divalent cations as catalytic cofactors (32, 101, 109, 114). Although divalent cations are not obligatory catalytic cofactors, divalent cations still are needed to reach maximum cleavage rates. Micromolar concentrations of magnesium support global folding of extended ribozymes, but activity continues to increase with increasing magnesium concentrations through the 100 mM range (20, 73, 109, 117), suggesting that a separate class of low-affinity magnesium interactions also contributes to folding or catalysis.

Changes in divalent cation specificity that accompany substitution of sulfur for oxygen ligands are commonly used to distinguish specific metal interactions from the diffuse, nonspecific interactions with counterions that are needed to stabilize compact RNA structures. Hammerhead ribozymes exhibit a change in cation specificity when the pro-R_P nonbridging oxygens of the $C_{1.1}$ and A_9 phosphates are substituted with sulfur. Cleavage of a phosphorothioate linkage

is reduced in reactions with magnesium, which binds sulfur ligands weakly, but is restored in reactions with more thiophilic cations such as Cd^{2+} or Mn^{2+}. This change in cation specificity is interpreted as evidence that a Mg^{2+} associated with pro-R_P oxygen normally stimulates phosphodiester cleavage, a conclusion confirmed by spectroscopic studies (34, 37, 76, 77, 92, 115, 118, 138, 145, 157, 167, 174). Quantitative thiophilic cation titrations identified pro-R_P oxygens of $P_{1.1}$ and P_9 as ligands of the same divalent metal cation (174), a result that also is consistent with molecular dynamics simulations (83, 84). No metal was detected at this position in the initial structure obtained from crystals formed in high concentrations of $(NH_4)_2SO_4$ (95), but a Mn^{2+} cation subsequently was found near the pro-R_P oxygen of A_9 through difference Fourier analyses of crystals soaked briefly in $MnCl_2$ (94). The second ligand of the metal cation is N_7 of $G_{10.1}$ in the crystal structure, but molecular dynamics simulations suggest that a minor conformational rearrangement could allow the same cation to bind the $P_{1.1}$ pro-R_P oxygen (84).

Divalent cations are estimated to contribute 20- to 100-fold catalytic rate enhancement from the difference between hammerhead cleavage rate constants measured in reactions with divalent and monovalent salts (32, 101, 109, 114). This rate acceleration could arise from a direct contribution to catalysis through electrostatic stabilization of negative charge that accumulates in the pentacovalent transition state, analogous to the role of the lysine side chain in the mechanism of RNase A, or through indirect effects on the ionization of bound water or another functional group involved in proton transfer. Divalent metals also might enhance activity without participating directly in catalytic chemistry. Bound divalent metals might contribute to positioning and orientation of reactive groups. Owing to their smaller size and higher charge density, divalent cations might be more effective than monovalent cations in stabilizing compact RNA structures through diffuse interactions (25, 39, 156).

The Hammerhead Proton-Transfer Mechanism

pH-rate profile: a plot of reaction rate versus pH

Hammerhead ribozymes display a log-linear increase in cleavage rates with increasing pH (20, 33, 64, 109). The difficulty of interpreting pH-rate profiles in terms of specific functional groups that donate or accept protons during ribozyme catalysis is widely recognized (9, 45, 108). This log-linear pH dependence is consistent with a rate-determining step that depends on the deprotonated form of a functional group with a high pK_a value, such as nucleophilic attack of a 2′ oxyanion. Candidate functional groups with high pK_a values include the 2′ hydroxyl of G_8, with the pK_a value for an internucleotidic 2′ hydroxyl expected to be greater than 12 (2, 90); Mg^{2+}-bound water, with a pK_a value of 11.4 (85); the imine N1 of guanosine, with a pK_a value near 10 (1); and specific acid-base catalysis by water itself, which ionizes with a pK_a value of 14.

High-resolution views of the hammerhead active site poised for catalysis provide an invaluable foundation for efforts to gain a more precise understanding of the proton-transfer mechanism. The X-ray crystal structure of the *Schistosoma* ribozyme places the 2′ hydroxyl of G_8 and the N1 ring nitrogen of G_{12} near the nucleophilic and leaving group oxygens, where they have been proposed to participate in general acid-base catalysis (60, 94, 95) (**Figure 2d,e**). The 2′ hydroxyl of G_8 is near the 5′ oxygen of $C_{1.1}$, where it might play a role in protonating the 5′ oxyanion leaving group as the 5′-oxygen-phosphorus bond breaks during cleavage. N1 of G_{12} lies near the 2′ hydroxyl of C_{17}, where it could facilitate nucleophilic attack by deprotonating the 2′ oxygen during cleavage. More than 200 solvent molecules also were identified in the most recent 2.0 Å X-ray crystal structure, including a water molecule, or NH_4^+, associated with both the reactive phosphate and the 2′ hydroxyl of G_8 (94). Molecular dynamics simulations point to a potential role for water in proton transfer between nucleophilic and leaving group oxygens through a hydrogen bonding network that includes the di-

valent cation associated with P_9 and $P_{1.1}$, the 2′ hydroxyl of G_8, and the guanine nucleobase of G_{12} (83, 84, 94). Each of these models could be consistent with the observed pH dependence.

HAIRPIN RIBOZYME

The hairpin ribozyme motif was discovered in the opposite strand of the *Tobacco ringspot virus* satellite RNA, which harbors a hammerhead ribozyme (15, 16, 18, 166) where both self-cleaving motifs process replication intermediates produced by rolling circle transcription (14, 19, 24, 165). Differences between hammerhead and hairpin ribozyme reactions gave an early indication that RNA enzymes make use of different strategies to catalyze the same chemical reaction. In contrast to hammerhead ribozymes that require thiophilic cations to cleave a phosphorothioate linkage efficiently, hairpin ribozyme rate constants are virtually the same for cleavage of phosphodiester and phosphorothioate linkages. Furthermore, hairpin ribozyme reactions with phosphorothioate substrates do not exhibit the changes in cation specificity that would be expected if nonbridging oxygens interact directly with divalent cations (57, 113, 180). The absence of significant phosphorothioate effects argues that direct metal cation coordination to nonbridging oxygens makes no contribution to hairpin ribozyme activity. Furthermore, submillimolar concentrations of cobalt hexammine support full hairpin ribozyme activity in the absence of any other salt (57, 113, 180). Cobalt hexammine resembles hexahydrated magnesium in size and geometry, but unlike magnesium-bound water, the amine ligands of cobalt are inert and unable to exchange with other ligands on the timescale of ribozyme reactions (31). Therefore, the ability of hairpin ribozymes to function in cobalt hexammine excludes a requirement for direct metal cation coordination to any ribose, phosphate, or water oxygens. Until the recent discovery that *glmS* ribozymes also function in cobalt hexammine and cleave phosphorothioate linkages efficiently in the absence of thiophilic cations (136), the hairpin ribozyme was the only catalytic RNA known to exhibit these properties.

Kinetic Mechanisms of Minimal and Natural Forms of the Hairpin Ribozyme

A minimal hairpin ribozyme defined by truncation and deletion experiments includes two helix-loop-helix domains termed A and B that bend sharply at the interdomain junction to bring loop nucleotides from each domain together to form the active site (44) (**Figure 3a**). Under standard reaction conditions, the ribozyme partitions almost equally between inactive extended, and active docked conformations (172). The natural form of the hairpin ribozyme in viral satellite RNAs contains two additional helices inserted between the A and B domains, which allow the ribozyme to assemble in the context of a four-way junction (4WJ) (58) (**Figure 3b**). Similar to the peripheral sequences in natural forms of hammerhead ribozymes, the C and D arms of the junction are not essential for activity but promote interhelical docking under less favorable ionic conditions (100, 168–170, 172) and support rapid cleavage and ligation under physiological salt conditions in vitro and in vivo (179). Evidently, natural forms of both hammerhead and hairpin ribozymes include peripheral elements that compensate for unfavorable intracellular ionic conditions by stabilizing tertiary structures. A similar principle might underlie the larger size and complexity of riboswitches (the RNA regulatory elements that bind small ligands in noncoding regions of bacterial mRNAs) compared to their simple aptamer counterparts, which evolved to bind small ligands under in vitro conditions that typically include high concentrations of divalent cations (41, 98, 131, 150).

Minimal hairpin ribozyme cleavage rate constants are similar to those of other ribozymes, on the order of 0.5 min^{-1}, but ligation is favored relative to cleavage by almost 10-fold (44). The 4WJ form of the ribozyme exhibits cleavage rate constants similar to those of minimal ribozymes but catalyzes ligation sixfold faster under the same conditions (43). An equilibrium favoring ligation is consistent with an enthalpic advantage for 3′,5′ phosphodiesters relative to 2′,3′-cyclic phosphates that presumably reflects steric strain in the 2′,3′-cyclic phosphate (51). The finding that natural forms of both hammerhead and hairpin ribozymes exhibit enhanced ligation activity supports the notion that tertiary structure stability influences the balance between cleavage and ligation, perhaps by fixing cleavage product termini in an orientation that is favorable for religation.

Hairpin Ribozyme Structure

Almost 30 X-ray crystal structures are available for hairpin ribozymes (**Table 2**). Structures have been solved for 4WJ forms of the ribozyme in complexes with a substrate analog in which cleavage is blocked by a 2′-OCH$_3$ modification of the −1 nucleotide, with a vanadate mimic of the trigonal bipyramidal transition state, and with cleavage product RNAs that have 2′,3′-cyclic and 5′-hydroxyl termini (139) (**Figure 3b–d**). Wedekind and coworkers have focused on structures of minimal hairpin ribozymes assembled from shorter oligonucleotides that are amenable to chemical synthesis, which allowed analysis of active site nucleotide modifications and a 2′,5′ linkage that mimics the transition state in regard to the positions of the nonbridging oxygens and the 2′-oxygen-phosphorus bond (3, 55, 91, 144, 161, 162).

Hairpin ribozyme active site structures agree well with each other and with biochemical data regarding the functional consequences of nucleotide modifications. The reactive phosphodiester adopts the in-line geometry consistent with an S$_N$2(P)-type in-line mechanism, evidence that the ground state structure in the crystals resembles the transition state (**Figure 3d**). A Watson-Crick base pair between G$_{+1}$ in loop A and C$_{25}$ in loop B is the key feature of the network of hydrogen bonding and stacking interactions that dock the A and B domains and align the reactive phosphodiester,

4WJ: four-way junction

S$_N$2(P)-type in-line mechanism: the mechanism of a class of nucleophilic substitution reactions that involves concerted nucleophilic attack on phosphorus and leaving group departure

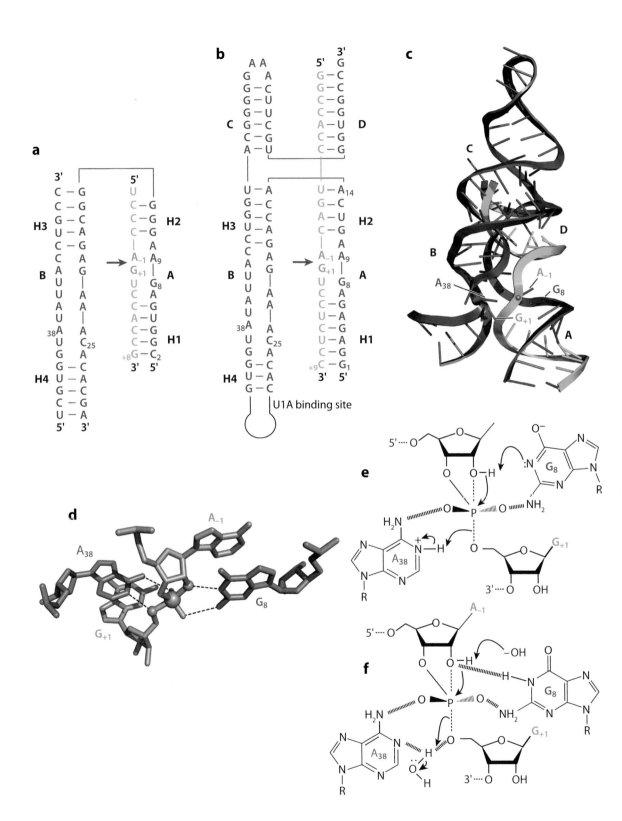

as predicted from the functional consequences of eliminating and then restoring complementary bases at these positions (125). The G_{+1} nucleotide inserts into a pocket in domain B that presents interacting partners for virtually every element of the G_{+1} nucleotide (**Figure 3a–c**). The loss of activity that accompanies modifications of these interactions highlights the contribution of positioning and orientation to catalytic rate enhancement (30, 54, 82, 155). Metal cation binding sites were identified in domain B in both minimal and 4WJ ribozymes, and a cobalt hexammine binding site also was observed in the major groove of H2 in minimal ribozymes (3). However, no cations were observed near the active site, as expected from the absence of a divalent cation requirement for catalysis (57, 101, 113, 180).

Nucleobase Contributions to Catalytic Chemistry

The active site structures place the Watson-Crick faces of G_8 and A_{38} within hydrogen bonding distance of the reactive phosphate (**Figure 3d**). The ring N1 of G_8 lies near the 2′ oxygen of A_{-1}, and its exocyclic N2 amine lies within hydrogen bonding distance of the *pro-R*$_P$ oxygen. The exocyclic N6 amine of A_{38} interacts with the *pro-S*$_P$ nonbridging oxygen, and N1 of A_{38} is near the 5′ oxygen of G_{+1}. High-resolution structures also identified four ordered water molecules in the active site, including two that are near the 2′ oxygen of A_{-1} (144).

A resemblance between the active site structures of the hairpin ribozyme and RNase A suggests that both enzymes might use a similar concerted general acid-base catalysis mechanism, supposing that G_8 and A_{38} nucleobases in the hairpin ribozyme active site perform the same roles in proton transfer as do the active site histidines in RNase A (139) (**Figure 3e**). Consistent with this model was the observation that the activity of minimal hairpin ribozymes exhibits a shallow, bell-shaped pH-rate profile, which could signify dependence on the ionization states of functional groups that ionize at high and low pH extremes (8, 113, 124). In this model, the peak in activity at neutral pH would reflect an optimal combination of the deprotonated form of a general base catalyst, such as N1 of G_8, that ionizes with a high pK_a value near 10, and the protonated form of a general acid catalyst, such as N1 of A_{38}, that ionizes with a pK_a below 4 (8, 141). If these active site nucleotides have pK_a values below 4 or above 10, just a small fraction of A_{38} or G_8 would be in the appropriate protonation state at neutral pH. However, sufficient amounts could be available at neutral pH to support the moderate rate acceleration characteristic of hairpin ribozymes (8). Biochemical studies of ribozyme variants in which G_8 and A_{38} are replaced by nucleobase analogs that have different ionization constants for N1 ionization show corresponding changes in pH-rate profiles, evidence that the protonation state of the nucleobases at these positions is important for activity (79, 80, 82, 87, 124, 177).

Figure 3

Hairpin ribozyme structure and mechanism. (*a*) Sequence and secondary structure of a minimal hairpin ribozyme optimized for crystallization (161). The essential A and B domains are joined by a tri(ethylene glycol) linker in place of the A_{14} nucleotide. (*b*) Sequence and secondary structure of a four-way junction (4WJ) hairpin ribozyme sequence optimized for crystallization (139). The 4WJ ribozyme contains two helices, termed C and D, in addition to the essential A and B domains. (*c*) Global architecture of the hairpin ribozyme complex with a vanadate mimic of the transition state (PDB 1M5O) (140). (*d*) Active site in the X-ray crystal structure of the hairpin ribozyme complex with a vanadate mimic of the transition state (PDB 1M5O) (140). (*e*) Hypothetical mechanism in which G_8 and A_{38} act as general acid-base catalysts, with N1 of G_8 withdrawing a proton from the 2′ hydroxyl during cleavage and donating a proton to the 2′ oxyanion during ligation and with N1 of A_{38} donating a proton to the 5′-oxyanion-leaving group during cleavage and withdrawing a proton from the 5′ hydroxyl during ligation. (*f*) Hypothetical mechanism in which hydrogen bond formation with the amidine groups of G_8 and A_{38} contributes to positioning and orientation and provides electrostatic stabilization to the pentacovalent transition state, and in which proton transfer occurs through specific acid-base catalysis.

Table 2   X-ray crystal structures of hairpin ribozymes

RNA construct	Salts	pH	Resolution (Å)	PDB ID	Reference
4WJ, 2′ OCH$_3$ at N$_{-1}$	NH$_4$Cl, MgCl$_2$, CaCl$_2$, spermine		2.4	1HP6	(139)
4WJ, 2′ OCH$_3$ at N$_{-1}$	NH$_4$Cl, MgCl$_2$, CaCl$_2$, spermine		2.4	1M5K	(140)
4WJ, vanadate complex	NH$_4$Cl, MgCl$_2$, CaCl$_2$, NH$_4$VO$_3$, spermine		2.2	1M5O	(140)
4WJ, product complex	NH$_4$Cl, MgCl$_2$, CaCl$_2$, spermine		2.4	1M5V	(140)
4WJ, 3′ product with 5′-Cl terminus	NH$_4$Cl, MgCl$_2$, CaCl$_2$, spermine		2.6	1M5P	(140)
Minimal, 2′ OCH$_3$ at N$_{-1}$	(NH$_4$)$_2$SO$_4$, Na(CH$_3$)$_2$AsO$_2$, Li$_2$SO$_4$, Co(NH$_3$)$_6$Cl$_3$, spermidine	6.0	2.65	2D2K	(3)
Minimal, 2′ OCH$_3$ at N$_{-1}$	(NH$_4$)$_2$SO$_4$, Na(CH$_3$)$_2$AsO$_2$, Li$_2$SO$_4$, Co(NH$_3$)$_6$Cl$_3$, spermidine	6.0	2.65	1X9K	(3)
Minimal, U39C mutant, 2′ OCH$_3$ at N$_{-1}$	(NH$_4$)$_2$SO$_4$, Na(CH$_3$)$_2$AsO$_2$, Li$_2$SO$_4$, Co(NH$_3$)$_6$Cl$_3$, spermidine	6.0	2.19	1X9C	(3)
Minimal, U39 with a propyl (C3) linker, 2′ OCH$_3$ at N$_{-1}$	Li$_2$SO$_4$, Co(NH$_3$)$_6$Cl$_3$, NaHEPES, spermidine	7.6	2.50	2D2L	(3)
Minimal, 2′ OCH$_3$ at N$_{-1}$	Na(CH$_3$)$_2$AsO$_2$, Li$_2$SO$_4$, Co(NH$_3$)$_6$Cl$_3$, spermidine	6.0	2.05	1ZFR	(144)
Minimal, G$_8$I, 2′ OCH$_3$ at N$_{-1}$	NaHEPES, Li$_2$SO$_4$, Co(NH$_3$)$_6$Cl$_3$, spermidine	7.2	2.33	1ZFT	(144)
Minimal, G$_8$I, 2′ deoxy at N$_{-1}$	NaHEPES, Li$_2$SO$_4$, Co(NH$_3$)$_6$Cl$_3$, spermidine	6.0	2.40	2BCZ	(144)
Minimal, G$_8$2AP	TrisHCl, Li$_2$SO$_4$, Co(NH$_3$)$_6$Cl$_3$, spermidine	8.8	2.70	2BCY	(144)
Minimal, G$_8$DAP	TrisHCl, Li$_2$SO$_4$, Co(NH$_3$)$_6$Cl$_3$, spermidine	8.6	2.40	2FGP	(144)
Minimal, G$_8$A	Na(CH$_3$)$_2$AsO$_2$, Li$_2$SO$_4$, Co(NH$_3$)$_6$Cl$_3$, spermidine	6.0	2.40	1ZFV	(144)
Minimal, G$_8$U	Na(CH$_3$)$_2$AsO$_2$, Li$_2$SO$_4$, Co(NH$_3$)$_6$Cl$_3$, spermidine	6.0	2.38	1ZFX	(144)
Minimal, vanadate complex	Li$_2$SO$_4$, Co(NH$_3$)$_6$Cl$_3$, NH$_4$VO$_3$, Na(CH$_3$)$_2$AsO$_2$, spermidine	6.0	2.05	2P7E	(161)
Minimal, 3′-deoxy, 2′,5′ phosphodiester	Li$_2$SO$_4$, Co(NH$_3$)$_6$Cl$_3$, NaHEPES, spermidine, nicotinic acid	7.0	2.35	2P7F	(161)
Minimal, product complex with cis-diol terminus	Li$_2$SO$_4$, Co(NH$_3$)$_6$Cl$_3$, Na(CH$_3$)$_2$AsO$_2$, spermidine	6.0	2.25	2P7D	(161)
Minimal, 3′-deoxy, 2′,5′ phosphodiester	Li$_2$SO$_4$, Co(NH$_3$)$_6$Cl$_3$, Na(CH$_3$)$_2$AsO$_2$, spermidine	6.5	2.80	3CQS	(162)
Minimal, A$_{38}$DAP, 2′ OCH$_3$ at N$_{-1}$	Li$_2$SO$_4$, Co(NH$_3$)$_6$Cl$_3$, HEPES, spermidine	7.0	2.25	3B5S	(91)
Minimal, A$_{38}$AP, 2′ OCH$_3$ at N$_{-1}$	Li$_2$SO$_4$, Co(NH$_3$)$_6$Cl$_3$, HEPES, spermidine	7.6	2.35	3BBI	(91)
Minimal, A$_{38}$C, 2′ OCH$_3$ at N$_{-1}$	Li$_2$SO$_4$, Co(NH$_3$)$_6$Cl$_3$, Na(CH$_3$)$_2$AsO$_2$, spermidine	6.5	2.65	3CR1	(91)
Minimal, A$_{38}$C, 2′ OCH$_3$ at N$_{-1}$	Li$_2$SO$_4$, MgCl$_2$, Na(CH$_3$)$_2$AsO$_2$, spermidine	6.0	2.25	3CR1	(91)
Minimal, A$_{38}$G, 2′ OCH$_3$ at N$_{-1}$	Li$_2$SO$_4$, Co(NH$_3$)$_6$Cl$_3$, Tris, spermidine	8.6	2.35	3B5A	(91)

(Continued)

Table 2 *(Continued)*

RNA construct	Salts	pH	Resolution (Å)	PDB ID	Reference
Minimal, A_{38} DAP, 3'-deoxy, 2',5' phosphodiester	Li_2SO_4, $Co(NH_3)_6Cl_3$, $Na(CH_3)_2AsO_2$, spermidine	6.8	2.70	3B5F	(91)
Minimal, A_{38} AP, 3'-deoxy, 2',5' phosphodiester	Li_2SO_4, $Co(NH_3)_6Cl_3$, $Na(CH_3)_2AsO_2$, spermidine	6.5	2.75	3B91	(91)
Minimal, A_{38} C, 3'-deoxy, 2',5' phosphodiester	Li_2SO_4, $Co(NH_3)_6Cl_3$, $Na(CH_3)_2AsO_2$, spermidine	6.5	2.75	3BBK	(91)
Minimal, A_{38} G, 3'-deoxy, 2',5' phosphodiester	Li_2SO_4, $Co(NH_3)_6Cl_3$, HEPES, spermidine	7.8	2.65	3B58	(91)

Abbreviations: 2AP, 2-aminopurine; 4WJ, four-way junction; DAP, 2,6-diamonopurine.

However, this interpretation of pH dependence in terms of a general acid-base catalysis model is complicated by the dependence of RNA structure stability on the ionization state of hydrogen bond donors and acceptors, particularly for minimal ribozymes without stabilizing tertiary interactions. The 4WJ form of the hairpin ribozyme exhibits a pH-rate profile different from that exhibited by the minimal ribozyme (79, 80). For 4WJ ribozymes, cleavage and ligation rate constants continue to increase with increasing pH until pH values are above 10, the pH at which RNA secondary structures undergo denaturation due to deprotonation of all guanosine and uridine hydrogen bond donors. The pH-rate profile for cleavage and ligation rate constants of a 4WJ form of the ribozyme exhibits two pH-dependent transitions in activity, one occurring near pH 6 and another that falls above the experimentally accessible pH range (79, 80).

The interaction between G_8 and the reactive phosphate observed in the X-ray crystal structures is important for activity, because an abasic substitution of G_8 reduces cleavage and ligation rate constants by 850- and 1600-fold, respectively. However, variants lacking G_8 display the same pH-rate profile as unmodified ribozymes, indicating that the pH-dependent step in the reaction pathway does not involve G_8 (80). In contrast, an abasic substitution of A_{38} not only reduces cleavage and ligation rate constants more dramatically, by 14,000- and 370,00-fold, respectively, but also eliminates the pH-dependent transition in activity that

occurs near pH 6 in unmodified ribozymes (79). These results show that an ionization event directly or indirectly associated with A_{38} is responsible for a pH-dependent step in the reaction pathway. Although the apparent pK_a value of 6 for hairpin ribozyme activity is more than two units above the pK_a value for adenosine ionization in solution, there are ample precedents for shifts in the ionization equilibria of nucleobases in the context of structured RNAs (9, 99, 108, 159).

Certain exogenous nucleobase analogs restore activity to ribozymes with abasic substitutions of G_8 or A_{38} (79, 80, 82). Each of the nucleobases capable of rescue is a planar heterocycle with an amidine group, that is, an amino group in α position to a ring nitrogen, similar to the Watson-Crick hydrogen bonding face of the adenine or guanine nucleobases they replace. Simple purine, with no exocyclic amine, inhibits rescue with a K_i value similar to the apparent K_d values near 20 mM that are obtained from the concentration dependence of exogenous nucleobase rescue. The similarity between apparent K_d and K_i values suggests that the amidine group of the rescuing nucleobases interacts with the transition state to promote catalysis without contributing to ground state binding energy.

The absence of an effect of G_8 deletion on pH-rate profiles does not exclude the possibility that G_8 mediates general acid-base catalysis. However, changes in pH-rate profiles that accompany substitutions of G_8 with nucleobase analogs do not provide convincing support of a role for G_8 in proton transfer, since a change in the protonation state of G_8 could affect activity through positioning and orientation or electrostatic stabilization of the transition state. Structural studies show that modifications of G_8 that eliminate the N2 exocyclic amine are associated with changes in the angle between phosphorus and the nucleophilic and leaving group oxygens, evidence that hydrogen bond donation from the exocyclic amine to the *pro-R_P* oxygen promotes the productive in-line geometry (144) (**Figure 3f**). Computational studies suggest that hydrogen bonding between the amidine face of G_8 and the 2′ oxygen and *pro-R_P* oxygen could also contribute electrostatic stabilization to the pentacovalent transition state (107, 116, 133, 144, 171) (**Figure 3f**). The identification of bound water near the 2′ hydroxyl of A_{-1} suggests that proton transfer could occur through specific acid-base catalysis, a model that also agrees with computational studies.

The effects of an abasic substitution of A_{38} on pH-rate profiles could be consistent with a role in general acid-base catalysis, with A_{38} in the protonated form donating a proton to the departing 5′ oxyanion during cleavage and A_{38} in the deprotonated form acting as a general base to activate the 5′-oxygen nucleophile during ligation (**Figure 3e**). In an alternative scenario, unprotonated adenosine might accept a hydrogen bond from a 5′ hydroxyl to provide electrostatic stabilization to the transition state and contribute to positioning and orientation (**Figure 3f**). Water or hydronium ion would then contribute the proton to the 5′ hydroxyl, although no candidate water bound at this location has yet been identified in crystal structures. Either scenario is consistent with X-ray crystal structures that place N1 of A_{38} within hydrogen bonding distance of the 5′ oxygen. Structural characterizations of a hairpin ribozyme with a 2-aminopurine substitution of A_{38} showed that loss of the N6 exocyclic amine that interacts with the *pro-S_P* oxygen interfered with positioning of the reactive phosphate (91). Replacement of A_{38} with cytosine or guanine caused significant perturbations of the in-line geometry, which could explain the loss of activity that results from these mutations (91). A more detailed understanding of the roles of these active site nucleobases in catalytic chemistry requires more information about the protonation states of A_{38} and G_8.

HEPATITIS DELTA VIRUS RIBOZYME

HDV self-cleaving motifs are found in a small pathogenic satellite of hepatitis B virus, where they function in replication by cleaving concatameric intermediates of rolling circle transcription (154). Both genomic and antigenomic strands of HDV RNA harbor self-cleaving sequences that fold into similar, but not identical, secondary structures that include five base-paired helices joined by unpaired loops (**Figure 4a**). A self-cleaving sequence similar to the HDV ribozyme recently was identified in an intron of a human gene involved in regulating polyadenylation, but its biological function is not yet known (142).

HDV ribozymes can cleave with rate constants as high as 60 min^{-1} under favorable conditions (13). In contrast to hammerhead and hairpin ribozymes, HDV ribozymes exhibit little ligation activity, in part because

---

Figure 4

Hepatitis delta virus (HDV) ribozyme structure and mechanism. (*a*) Sequence and secondary structure of the genomic HDV ribozyme (71). (*b*) Global architecture of the 3′ product of genomic HDV ribozyme cleavage (47) (PDB 1DRZ). (*c*) Global architecture of the uncleaved HDV ribozyme with cleavage blocked by a uridine substitution of the essential active site C_{75} residue (71). (*d*) Active site in the X-ray crystal structure of the HDV 3′ cleavage product (47) (PDB 1DRZ). (*e*) Hypothetical cleavage mechanism in which C_{75} and hydrated Mg^{2+} mediate general base and general acid catalysis, respectively. (*f*) Active site in the X-ray crystal structure of the uncleaved HDV ribozyme with cleavage blocked by a $C_{75}U$ mutation (71) (PDB 1SJ3). (*g*) Hypothetical cleavage mechanism in which C_{75} and hydrated Mg^{2+} mediate general acid and general base catalysis, respectively.

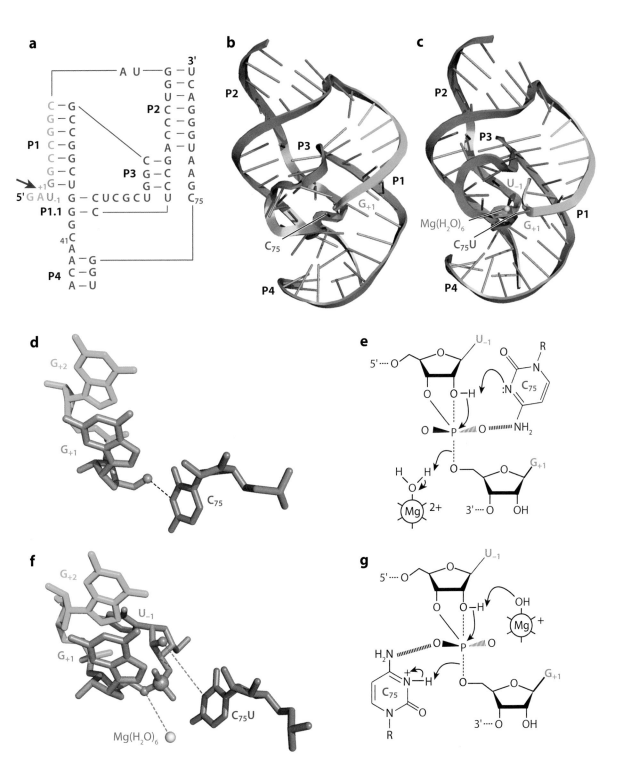

the low affinity and rapid dissociation of the 5′ cleavage product limit assembly of a stable ribozyme-product ternary complex (152). Local and global conformational changes coincident with cleavage that might also interfere with binding and religation of cleavage products have been detected (61, 119, 173). Sequences upstream of minimal ribozyme sequences in HDV genomic and antigenomic RNAs that are not essential for catalysis affect cleavage rates, extents, and salt dependence (12, 13, 22, 23). These effects are reminiscent of the effects of peripheral elements on the activity of natural forms of hammerhead and hairpin ribozymes. However, in the case of HDV ribozymes, peripheral sequences seem to influence partitioning among alternative functional and nonfunctional structures during assembly rather than stabilize functional structures directly.

The first X-ray crystal structure of an HDV ribozyme was solved for the 3′ product of genomic ribozyme self-cleavage and provided the first view of a ribozyme active site devoid of metal cations (48) (**Figure 4b**) (**Table 3**). This structure revealed an active site cleft in which N3 of a catalytically essential cytosine residue, C_{75}, lies within hydrogen bonding distance of the 5′ hydroxyl (**Figure 4d**). This structure stimulated a flurry of experiments aimed at understanding how this cytosine contributes to catalysis. Self-cleavage activity of both genomic and antigenomic HDV ribozymes exhibits a bell-shaped pH-rate profile that is consistent with dependence on the protonation state of two functional groups that ionize with apparent

Table 3 X-ray crystal structures of hepatitis delta virus RNAs

RNA construct	Salt	pH	Resolution (Å)	PDB ID	Reference
Genomic, 3′ cleavage product	Li_2SO_4, $MgCl_2$, TrisHCl, $Co(NH_3)_6Cl_3$, spermine	7.0	2.3	1DRZ	(48)
Genomic, $C_{75}U$	TrisHCl, NaCl, $MgCl_2$, $Na(CH_3)_2AsO_2$, spermidine	7.5	2.2	1SJ3	(71)
Genomic, $C_{75}U$	TrisHCl, NaCl, $MgCl_2$, $MnCl_2$, $Na(CH_3)_2AsO_2$, spermidine	7.5	2.8	1VBY	(71)
Genomic, $C_{75}U$	TrisHCl, NaCl, $MgCl_2$, $Co(NH_3)_6Cl_3$, $Na(CH_3)_2AsO_2$, spermidine	7.5	2.7	1SJF	(71)
Genomic, $C_{75}U$	TrisHCl, NaCl, $MgCl_2$, $CuCl_2$, $Na(CH_3)_2AsO_2$, spermidine	7.5	2.7	1SJ4	(71)
Genomic, $C_{75}U$	TrisHCl, NaCl, EDTA, $Na(CH_3)_2AsO_2$, spermidine	7.5	2.7	1VBX	(71)
Genomic, $C_{75}U$	TrisHCl, NaCl, $MgCl_2$, $BaCl_2$, $Na(CH_3)_2AsO_2$, spermidine	7.5	2.9	1VBZ	(71)
Genomic, $C_{75}U$	TrisHCl, NaCl, $MgCl_2$, $SrCl_2$, imidazole, $Na(CH_3)_2AsO_2$, spermidine	7.5	2.5	1VCO	(71)
Genomic, $C_{75}U$, 3′ cleavage product	TrisHCl, NaCl, $MgCl_2$, $Na(CH_3)_2AsO_2$, spermidine	7.5	2.8	1VC6	(71)
Genomic, $C_{75}U$	TrisHCl, NaCl, $MgCl_2$, $SrCl_2$, $Na(CH_3)_2AsO_2$, spermidine	7.5	2.45	1VC7	(71)
Genomic, 2′ deoxy at N_{-1}	TrisHCl, NaCl, $MgCl_2$, $Na(CH_3)_2AsO_2$, spermidine	7.5	3.4	1VC5	(71)
Genomic, $C_{75}U$	$Na(CH_3)_2AsO_2$, NaCl, $SrCl_2$, spermine, $TlCOOCH_3$	6.0	2.4	2OIH	(70)
Genomic, $C_{75}U$	$Na(CH_3)_2AsO_2$, NaCl, $SrCl_2$, spermine, $TlCOOCH_3$, $Co(NH_3)_6Cl_3$	6.0	2.4	2OJ3	(70)

pK$_a$ values of 6.5 and 9 (154). Changes in pH-rate profiles that result from C$_{75}$ mutations link C$_{75}$ to the pK$_a$ of 6.5.

Like abasic hairpin ribozymes, HDV ribozymes with inactivating C$_{75}$ deletions or mutations can be rescued by exogenous cytosine and nucleobase analogs (122, 123, 153). However, chemical rescue of HDV ribozymes seems to occur through a distinct mechanism. Robust rescue of mutant HDV ribozymes is observed with imidazole buffer and imidazole analogs, as well as with cytosine analogs (122, 123, 153). In contrast, hairpin ribozyme rescue requires nucleobase analogs that share the same Watson-Crick hydrogen bonding face as the guanine and adenosine bases they replace; no rescue is seen with imidazole buffer or imidazole analogs (79, 80, 82). Hairpin ribozymes with inactivating mutations or deletions are rescued much less efficiently than abasic ribozymes. In contrast, efficient HDV ribozyme rescue is observed with both mutations and deletions of C$_{75}$ and no saturation is observed at the highest imidazole buffer concentrations tested, suggesting that rescue does not require high-affinity binding (122, 153). Most importantly, comparison of a rescue by a series of imidazole analogs with different pK$_a$ values showed that reaction rates increase in a linear fashion with increasing pK$_a$ values of the analogs, after normalizing for the concentration of the ionized species. The correlation between the Brønsted base strength and rate acceleration provides strong evidence that imidazole, and by analogy C$_{75}$, participates directly in proton transfer in the HDV ribozyme transition state (122, 123, 153, 160). Biochemical features of hairpin ribozyme rescue, on the other hand, are equally consistent with other rescue mechanisms that depend on the ionization state of the nucleobase, such as electrostatic stabilization or stabilization of transition state geometry through hydrogen bonding (79, 80, 82).

The pH dependence of HDV ribozyme activity is equally consistent with a role for C$_{75}$ in general acid or base catalysis, because both mechanisms are associated with equivalent rate equations (69, 105, 153). C$_{75}$ initially was hypothesized to serve as a general base catalyst to deprotonate the 2′-hydroxyl nucleophile, even though information about its proximity to the 2′ hydroxyl of U$_{-1}$ was unavailable from the X-ray crystal structure of the 3′ cleavage product that was solved first (48, 122) (**Figure 4b,d,e**). Subsequently, a series of crystal structures of uncleaved ribozymes, most with an inactivating C$_{75}$U mutation, produced a different view of the active site (71) (**Table 3**) (**Figure 4c,e,f**). The full-length C$_{75}$U ribozyme adopts an active site structure distinctly different from that of the 3′ product RNA, consistent with evidence from fluorescence studies that a large conformational change accompanies cleavage (61, 119, 173). N3 of U$_{75}$ is no longer within hydrogen bonding distance of the 5′ hydroxyl, as in the 3′ product structure, and instead, a hydrated divalent metal cation is observed near the 5′-oxygen leaving group. The presence of a divalent metal cation in the precursor but not in the 3′ product RNA is consistent with recent biochemical, biophysical, and computational evidence that links changes in divalent cation specificity and concentration dependence to nucleotides upstream of the cleavage site (52, 78, 121). N3 of U$_{75}$ is 5.5 Å from the 2′-nucleophilic oxygen in the C$_{75}$U active site, but could be modeled close to a position expected for a general base catalyst after a slight rotation of the 3′-oxygen-phosphorus bond (71). The C$_{75}$U structure supports a model in which a hydrated metal cation protonates the 5′ oxygen during departure of the leaving group and C$_{75}$ mediates general base catalysis to activate the 2′ hydroxyl for nucleophilic attack. This model is also consistent with results of molecular dynamics calculations starting with the precursor structure (5, 71) (**Figure 4e**).

Recent biochemical and computational studies support an alternative model in which C$_{75}$ mediates general acid catalysis to facilitate departure of the 5′-oxygen leaving group and in which specific base catalysis, perhaps through metal hydroxide, activates the nucleophile (35, 105, 106, 176) (**Figure 4g**). Das & Piccirilli examined cleavage of 5′-phosphorothioate linkages, in which sulfur is the leaving group during

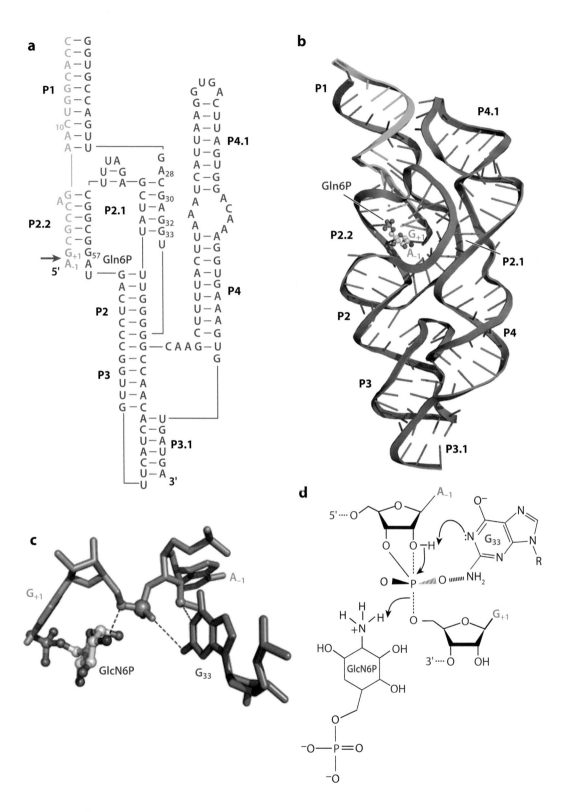

breaking of the 5'-sulfur-phosphorus bond (35) in a novel strategy to identify contributions to the bond-breaking step. Because sulfur is a better leaving group than oxygen is, factors that contribute to catalysis of the 5'-oxygen-phosphorus bond-breaking step should have little effect on rates of 5'-sulfur-oxygen bond cleavage in a phosphorothiolate substrate. A $C_{75}U$ mutation had virtually no effect on cleavage of the 5'-phosphorothioate linkage, consistent with a role for C_{75} in a facilitating leaving group departure, whereas other inactivating mutations interfered with cleavage of both phosphodiester and phosphorothiolate substrates as expected.

The apparent pK_a near 6.5 calculated from HDV ribozyme cleavage kinetics is significantly more basic than the microscopic pK_a value of 4.2 for ionization of cytosine in solution (141). Why might C_{75} undergo protonation more readily in the context of the ribozyme active site? A shift in the ionization equilibrium might be explained by interactions between the exocyclic amine of C_{75} and neighboring ribose or phosphate oxygens (48, 154). ^{13}C NMR studies did not provide direct evidence for a shifted pK_a value for C_{75} in the context of the 3'-product RNA, but these experiments with the product RNA did not exclude the possibility of a pK_a shift in the precursor ribozyme that includes the reactive phosphate (89). Raman spectroscopy recently was used to measure the C_{75} pK_a value directly in single crystals of the HDV ribozyme (53). The microscopic pK_a value of 6.4 measured for C_{75} in single crystals of an HDV ribozyme with an inactivating 2'-OCH_3 modification at the −1 position agrees well with the pK_a value of 6.5 calculated from the kinetics of HDV ribozyme cleavage under similar conditions (53, 105). This striking correlation between the apparent pK_a value determined from

the pH dependence of cleavage kinetics and the microscopic pK_a value for protonation of C_{75} in the ribozyme confirms that the ionization equilibria for a nucleotide base can shift toward neutrality in the context of a ribozyme active site.

Like other self-cleaving RNAs, HDV ribozymes retain catalytic activity in reactions with high concentrations of monovalent salts (101, 120). Similar to the case with hammerhead ribozymes, however, Mg^{2+} is more effective in promoting HDV ribozyme catalysis than monovalent cations are; rate constants with Mg^{2+} are 25- to 100-fold higher than rate constants in reactions with monovalent salts (104, 106, 120, 154). The molecular basis of this modest rate enhancement by divalent cations is not yet understood. Exchange-inert cobalt hexammine inhibits HDV ribozyme activity (70, 71, 154). It remains unclear whether this inhibition reflects competition for a directly coordinated metal cation or a subtle difference between hexahydrated magnesium and cobalt hexammine in the ability to stabilize the functional structure. Divalent metal cations could contribute directly to catalysis through electrostatic stabilization of the transition state or by shifting ionization equilibria of bound water or a nucleotide functional group to facilitate general acid-base catalysis (104, 154). Alternatively, divalent metal cations might simply be more effective at stabilizing functional ribozyme structures than monovalent cations are because of their smaller size and higher charge density.

glmS RIBOZYME

The *glmS* ribozyme was discovered in the 5'-untranslated region of the bacterial mRNA that encodes glucosamine-6-phosphate (GlcN6P) synthase by using a bioinformatics approach designed to identify ligand-binding RNA motifs,

---

Figure 5

glmS ribozyme structure and mechanism. (*a*) Sequence and secondary structure of the *glmS* ribozyme from *Bacillus anthracis* (27). (*b*) Global architecture of the *glmS* ribozyme complex with the glucosamine-6-phosphate (GlcN6P) cofactor (27) (PDB 2NZ4). (*c*) Active site structure of the *glmS* ribozyme complex with GlcN6P (27) (PDB 2NZ4). (*d*) Hypothetical cleavage mechanism in which G_{33} mediates general base catalysis and the amine of GlcN6P mediates general acid catalysis.

Table 4   X-ray crystal structures of *glmS* ribozymes

RNA construct	Cofactor	Salt	pH	Resolution (Å)	PDB ID	Reference
Thermoanaerobacter tengcongensis, 2′ deoxy at N_{-1}	Glc6P	MgCl$_2$, HEPES-KOH, spermine-HCL, LiCl, MES-NaOH	5.2–7.5	2.7	2HOZ	(75)
T. tengcongensis, 2′ amino at N_{-1}	GlcN6P (not seen)	MgCl$_2$, HEPES-KOH, spermine-HCL, LiCl, MES-NaOH	5.2–7.5	2.1	2GCS	(75)
T. tengcongensis, 2′ deoxy at N_{-1}	GlcN6P (not seen)	MgCl$_2$, HEPES-KOH, spermine-HCL, LiCl, MES-NaOH	5.2–7.5	2.35	2HOS	(75)
T. tengcongensis	Glc6P	MgCl$_2$, HEPES-KOH, spermine-HCL, LiCl, MES-NaOH	5.2–7.5	2.90	2HO7	(75)
T. tengcongensis	GlcN6P (not seen)	MgCl$_2$, HEPES-KOH, spermine-HCL, L-Cl, MES-NaOH	5.2–7.5	2.80	2HO6	(75)
T. tengcongensis	None	MgCl$_2$, HEPES-KOH, spermine-HCL, LiCl, MES-NaOH	5.2–7.5	2.3	2HOX	(75)
T. tengcongensis	GlcN6P (not seen)	MgCl$_2$, HEPES-KOH, spermine-HCL, LiCl, MES-NaOH	5.2–7.5	2.35	–	(75)
T. tengcongensis, cleaved, 5′ product not seen	GlcN6P (not seen)	MgCl$_2$, HEPES-KOH, spermine-HCL, LiCl, MES-NaOH	5.2–7.5	2.1	2GCV	(75)
T. tengcongensis, cleaved, 5′ product not seen	None	MgCl$_2$, HEPES-KOH, spermine-HCL, LiCl, MES-NaOH	5.2–7.5	2.40	2HOW	(75)
Bacillus anthracis	GlcN6P	MgCl$_2$, Na(CH$_3$)$_2$AsO$_2$, KCl	6.8	2.5	2NZ4	(27)
T. tengcongensis, G$_{40}$A	GlcN6P	MgCl$_2$, HEPES-KOH, spermine-HCL, LiCl, MES-NaOH	5.5	2.70	3B4A	(74)
T. tengcongensis, G$_{40}$A, 3′ deoxy at N_{-1} 2′,5′ phosphodiester	GlcN6P	MgCl$_2$, HEPES-KOH, spermine-HCL, LiCl, MES-NaOH	5.5	2.70	3B4B	(74)
T. tengcongensis, 3′ deoxy at N_{-1} 2′,5′ phosphodiester	GlcN6P	MgCl$_2$, HEPES-KOH, spermine-HCL, LiCl, MES-NaOH	5.5–8.5	3.00	3B4C	(74)

Abbreviations: Glc6P, glucose-6-phosphate; GlcN6P, glucosamine-6-phosphate.

known as riboswitches, that are novel regulators of gene expression in bacteria (178). The *glmS* ribozyme is distinct from other self-cleaving ribozymes in requiring a small ligand, GlcN6P, as a catalytic cofactor. Self-cleavage produces 2′,3′-cyclic phosphate and 5′-hydroxyl termini, as for other self-cleaving motifs, in a reaction that is accelerated ~10^5-fold by GlcN6P (96, 178). GlcN6P is the metabolic product of the enzyme encoded by the *glmS* gene. GlcN6P-dependent self-cleavage of *glmS* mRNA triggers degradation of the downstream coding sequences to serve as a negative-feedback mechanism regulating *glmS* expression (28).

Besides GlcN6P, other primary amines with a vicinal hydroxyl group, including Tris, also activate the ribozyme, whereas glucose-6-phosphate (Glc6P), with a hydroxyl group in place of the primary amine, is a potent inhibitor (96). Several features of the *glmS* ribozyme self-cleavage reaction resemble hairpin ribozyme reactions. Like the hairpin ribozyme (57, 113, 180), the *glmS* ribozyme remains functional in reactions with the exchange-inert $Co(NH_3)_6^{3+}$ cation in place of Mg^{2+} (136). Furthermore, a *pro-R*$_P$ sulfur substitution of the reactive phosphate, which inhibits hammerhead ribozyme reactions in Mg^{2+}, has only minor effects on hairpin (57, 113, 180) or *glmS* ribozyme activity (136). These findings indicate that *glmS* ribozyme catalysis, like hairpin ribozyme catalysis, does not require direct coordination of metal cations to phosphate or ribose oxygens.

Virtually no structural differences can be detected between *glmS* ribozymes alone or in complex with inhibitors or activators (27, 59, 68, 75, 178). Thus, in contrast to other riboswitches in which ligand binding is associated with major conformational changes, GlcN6P plays a biochemical rather than a structural role in *glmS* ribozyme function. High-resolution structures were recently reported for *glmS* ribozyme complexes formed with cleavable and noncleavable substrates and with bound Glc6P inhibitor (75) or GlcN6P activator (27) (**Figure 5**) (**Table 4**). Active site structures obtained from crystals that formed in the presence of GlcN6P or Glc6P are superimposable,

despite the 10^5-fold difference in catalytic activity between activator and inhibitor complexes, emphasizing that the amine functional group of GlcN6P is indispensible for catalysis. The amine of GlcN6P lies within hydrogen bonding distance of the 5′ oxygen, consistent with a role as a general acid catalyst to facilitate departure of the leaving group during bond breaking (**Figure 5c,d**). Two active site waters observed in the inhibitor complex also were suggested to participate with GlcN6P in a proton relay network for activation of the nucleophilic 2′ hydroxyl (75), but the same water molecules were not observed in the GlcN6P complex (27). A conserved guanine located with its N1 ring nitrogen within hydrogen bonding distance of the 2′ oxygen is an alternative candidate for the role of general base catalyst. A role in proton transfer is consistent with recent biochemical evidence that this guanine is essential for activity (74) (**Figure 5c**). A more detailed understanding of the cofactor-dependent mechanism of the *glmS* ribozyme awaits further biochemical and biophysical study.

CONCLUSION

Self-cleaving RNAs accomplish catalysis of the same chemical reaction through a remarkably diverse array of catalytic strategies, given the seeming paucity of chemical tools available from ribonucleotide functional groups. Structural, biochemical, and biophysical approaches have uncovered evidence for mechanisms including general acid-base catalysis, electrostatic stabilization, ground state destabilization, and positioning and orientation, with each catalytic motif adopting elements of each. pH-rate profiles are blunt tools for dissecting proton-transfer mechanisms, even when combined with nucleotide modifications that alter ionization equilibria or with information about active site structures. One challenge now is to develop methods that discriminate among alternative mechanisms with more rigor and precision to gain a detailed understanding of catalytic chemistry. Recent advances in spectroscopic approaches for precise

GlcN6P: glucosamine-6-phosphate

Glc6P: glucose-6-phosphate

localization of divalent metal interactions and nucleobase ionization states promise to provide higher-resolution views of active site chemistry while novel approaches for distinguishing bond-breaking steps from bond-making steps are applied more broadly. The explosion of high-resolution structures available for ribozyme active sites will continue to fuel mechanistic studies as more ribozymes with specific active site modifications are characterized structurally under varying pH and ionic conditions.

A second challenge is to understand the molecular basis for the striking rate enhancements that originate from peripheral sequences found in the natural forms of some ribozymes. Two broad classes of mechanisms can be envisioned for rate enhancements in the context of peripheral interactions remote from the active site. In one class of mechanisms, tertiary structure stabilization might lower the free energy of the transition state. Two potential mechanisms for stabilizing the transition state might be facilitating positioning and orientation of reactive groups, or organizing ligands in a binding site that captures a metal cation for electrostatic stabilization. A second class of possible mechanisms to lower the energy barrier to catalysis involves destabilizing the ground state and promoting an active site geometry closer to that of the transition state. Disentangling the contributions of interactions observed in ribozyme structures to catalytic rate enhancement remains an intriguing challenge in structure-function studies of self-cleaving ribozymes.

SUMMARY POINTS

1. Self-cleaving RNAs use different strategies to catalyze the same chemical reaction.

2. Active site nucleobases can contribute directly to catalytic chemistry through general acid-base catalysis, electrostatic stabilization, and positioning and orientation of reactive groups.

3. The contribution of divalent metal cations to catalytic rate accelerations is much smaller than once thought, and it is not yet clear if metal cations contribute directly to catalysis or facilitate catalysis indirectly by stabilizing ribozyme structures.

DISCLOSURE STATEMENT

The author is not aware of any biases that might be perceived as affecting the objectivity of this review.

ACKNOWLEDGMENTS

The author gratefully acknowledges Joseph Cottrell, Peter Watson, and Peg Engel for assistance with the manuscript. Research in the author's lab is supported by the National Institutes of Health and by The Skaggs Institute for Chemical Biology.

LITERATURE CITED

1. Acharya S, Barman J, Cheruku P, Chatterjee S, Acharya P, et al. 2004. Significant pK_a perturbation of nucleobases is an intrinsic property of the sequence context in DNA and RNA. *J. Am. Chem. Soc.* 126:8674–81

2. Acharya S, Földesi A, Chattopadhyaya J. 2003. The pK_a of the internucleotidic 2'-hydroxyl group in diribonucleoside (3'–>5') monophosphates. *J. Org. Chem.* 68:1906–10

3. Alam S, Grum-Tokars V, Krucinska J, Kundracik ML, Wedekind JE. 2005. Conformational heterogeneity at position U37 of an all-RNA hairpin ribozyme with implications for metal binding and the catalytic structure of the S-turn. *Biochemistry* 44:14396–408

4. Amiri KM, Hagerman PJ. 1994. Global conformation of a self-cleaving hammerhead RNA. *Biochemistry* 33:13172–77

5. Banáš P, Rulíšek L, Hánošová V, Svozil D, Walter NG, et al. 2008. General base catalysis for cleavage by the active-site cytosine of the hepatitis delta virus ribozyme: QM/MM calculations establish chemical feasibility. *J. Phys. Chem. B* 112:11177–87

6. Bassi GS, Møllegaard NE, Murchie AI, von Kitzing E, Lilley DM. 1995. Ionic interactions and the global conformations of the hammerhead ribozyme. *Nat. Struct. Biol.* 2:45–55

7. Bassi GS, Murchie AI, Walter F, Clegg RM, Lilley DM. 1997. Ion-induced folding of the hammerhead ribozyme: a fluorescence resonance energy transfer study. *EMBO J.* 16:7481–89

8. Bevilacqua PC. 2003. Mechanistic considerations for general acid-base catalysis by RNA: Revisiting the mechanism of the hairpin ribozyme. *Biochemistry* 42:2259–65

9. Bevilacqua PC, Brown TS, Nakano S, Yajima R. 2004. Catalytic roles for proton transfer and protonation in ribozymes. *Biopolymers* 73:90–109

10. Blount KF, Uhlenbeck OC. 2002. Internal equilibrium of the hammerhead ribozyme is altered by the length of certain covalent cross-links. *Biochemistry* 41:6834–41

11. Blount KF, Uhlenbeck OC. 2005. The structure-function dilemma for the hammerhead ribozyme. *Annu. Rev. Biophys. Biomol. Struct.* 34:415–40

12. Brown AL, Perrotta AT, Wadkins TS, Been MD. 2008. The poly(A) site sequence in HDV RNA alters both extent and rate of self-cleavage of the antigenomic ribozyme. *Nucleic Acids Res.* 36:2990–3000

13. Brown TS, Chadalavada DM, Bevilacqua PC. 2004. Design of a highly reactive HDV ribozyme sequence uncovers facilitation of RNA folding by alternative pairings and physiological ionic strength. *J. Mol. Biol.* 341:695–712

14. Bruening G, Passmore BK, van Tol H, Buzayan JM, Feldstein PA. 1991. Replication of a plant virus satellite RNA: Evidence favors transcription of circular templates of both polarities. *Mol. Plant-Microbe Interact.* 4:219–25

15. Buzayan JM, Feldstein PA, Bruening G, Eckstein F. 1988. RNA mediated formation of a phosphorothioate diester bond. *Biochem. Biophys. Res. Commun.* 156:340–47

16. Buzayan JM, Gerlach WL, Bruening G. 1986. Nonenzymatic cleavage and ligation of RNAs complementary to a plant virus satellite RNA. *Nature* 323:349–53

17. Buzayan JM, Gerlach WL, Bruening G. 1986. Satellite *Tobacco ringspot virus* RNA: A subset of the RNA sequence is sufficient for autolytic processing. *Proc. Natl. Acad. Sci. USA* 83:8859–62

18. Buzayan JM, Hampel A, Bruening G. 1986. Nucleotide sequence and newly formed phosphodiester bond of spontaneously ligated satellite *Tobacco ringspot virus* RNA. *Nucleic Acids Res.* 14:9729–43

19. Buzayan JM, van Tol H, Zalloua PA, Bruening G. 1995. Increase of satellite *Tobacco ringspot virus* RNA initiated by inoculating circular RNA. *Virology* 208:832–37

20. Canny MD, Jucker FM, Kellogg E, Khvorova A, Jayasena SD, Pardi A. 2004. Fast cleavage kinetics of a natural hammerhead ribozyme. *J. Am. Chem. Soc.* 126:10848–49

21. Canny MD, Jucker FM, Pardi A. 2007. Efficient ligation of the *Schistosoma* hammerhead ribozyme. *Biochemistry* 46:3826–34

22. Chadalavada DM, Cerrone-Szakal AL, Bevilacqua PC. 2007. Wild-type is the optimal sequence of the HDV ribozyme under cotranscriptional conditions. *RNA* 13:2189–201

23. Chadalavada DM, Senchak SE, Bevilacqua PC. 2002. The folding pathway of the genomic hepatitis delta virus ribozyme is dominated by slow folding of the pseudoknots. *J. Mol. Biol.* 317:559–75

24. Chay CA, Guan X, Bruening G. 1997. Formation of circular satellite *Tobacco ringspot virus* RNA in protoplasts transiently expressing the linear RNA. *Virology* 239:413–25

25. Chen SJ. 2008. RNA folding: conformational statistics, folding kinetics, and ion electrostatics. *Annu. Rev. Biophys.* 37:197–214

26. Cleland WW, Hengge AC. 2006. Enzymatic mechanisms of phosphate and sulfate transfer. *Chem. Rev.* 106:3252–78

27. X-ray crystal structure of the *glmS* ribozyme complex with the GlcN6P cofactor.

27. **Cochrane JC, Lipchock SV, Strobel SA. 2007. Structural investigation of the glmS ribozyme bound to its catalytic cofactor. *Chem. Biol.* 14:97–105**

28. Collins JA, Irnov I, Baker S, Winkler WC. 2007. Mechanism of mRNA destabilization by the glmS ribozyme. *Genes Dev.* 21:3356–68

29. Collins RA. 2002. The Neurospora Varkud satellite ribozyme. *Biochem. Soc. Trans.* 30:1122–26

30. Cottrell JW, Kuzmin YI, Fedor MJ. 2007. Functional analysis of hairpin ribozyme active site architecture. *J. Biol. Chem.* 282:13498–507

31. Cowan JA. 1993. Metallobiochemistry of RNA. $Co(NH_3)_6^{3+}$ as a probe for Mg^{2+}(aq) binding sites. *J. Inorg. Biochem.* 49:171–75

32. Curtis EA, Bartel DP. 2001. The hammerhead cleavage reaction in monovalent cations. *RNA* 7:546–52

33. Dahm SC, Derrick WB, Uhlenbeck OC. 1993. Evidence for the role of solvated metal hydroxide in the hammerhead cleavage mechanism. *Biochemistry* 32:13040–45

34. Dahm SC, Uhlenbeck OC. 1991. Role of divalent metal ions in the hammerhead RNA cleavage reaction. *Biochemistry* 30:9464–69

35. Comparison of phosphorothiolate and phosphodiester cleavage reactions implicates an active site cytosine as a general acid in HDV ribozyme cleavage.

35. **Das SR, Piccirilli JA. 2005. General acid catalysis by the hepatitis delta virus ribozyme. *Nat. Chem. Biol.* 1:45–52**

36. Discovery that natural forms of hammerhead ribozymes include peripheral sequences that stimulate activity.

36. **De la Pena M, Gago S, Flores R. 2003. Peripheral regions of natural hammerhead ribozymes greatly increase their self-cleavage activity. *EMBO J.* 22:5561–70**

37. Derrick WB, Greef CH, Caruthers MH, Uhlenbeck OC. 2000. Hammerhead cleavage of the phosphorodithioate linkage. *Biochemistry* 39:4947–54

38. Ditzler MA, Aleman EA, Rueda D, Walter NG. 2007. Focus on function: single molecule RNA enzymology. *Biopolymers* 87:302–16

39. Draper DE, Grilley D, Soto AM. 2005. Ions and RNA folding. *Annu. Rev. Biophys. Biomol. Struct.* 34:221–43

40. Dunham CM, Murray JB, Scott WG. 2003. A helical twist-induced conformational switch activates cleavage in the hammerhead ribozyme. *J. Mol. Biol.* 332:327–36

41. Edwards TE, Klein DJ, Ferré-D'Amaré AR. 2007. Riboswitches: small-molecule recognition by gene regulatory RNAs. *Curr. Opin. Struct. Biol.* 17:273–79

42. Epstein LM, Gall JG. 1987. Self-cleaving transcripts of satellite DNA from the newt. *Cell* 48:535–43

43. Fedor MJ. 1999. Tertiary structure stabilization promotes hairpin ribozyme ligation. *Biochemistry* 38:11040–50

44. Fedor MJ. 2000. Structure and function of the hairpin ribozyme. *J. Mol. Biol.* 297:269–91

45. Fedor MJ, Williamson JR. 2005. The catalytic diversity of RNAs. *Nat. Rev. Mol. Cell Biol.* 6:399–412

46. Ferbeyre G, Smith JM, Cedergren R. 1998. Schistosome satellite DNA encodes active hammerhead ribozymes. *Mol. Cell Biol.* 18:3880–88

47. Ferré-D'Amaré AR, Doudna JA. 2000. Crystallization and structure determination of a hepatitis delta virus ribozyme: use of the RNA-binding protein U1A as a crystallization module. *J. Mol. Biol.* 295:541–56

48. Ferré-D'Amaré AR, Zhou K, Doudna JA. 1998. Crystal structure of a hepatitis delta virus ribozyme. *Nature* 395:567–74

49. Forster AC, Symons RH. 1987. Self-cleavage of plus and minus RNAs of a virusoid and a structural model for the active sites. *Cell* 49:211–20

50. Forster AC, Symons RH. 1987. Self-cleavage of virusoid RNA is performed by the proposed 55-nucleotide active site. *Cell* 50:9–16

51. Gerlt JA, Westheimer FH, Sturtevant JM. 1975. The enthalpies of hydrolysis of acyclic, monocyclic, and glycoside cyclic phosphate diesters. *J. Biol. Chem.* 250:5059–67

52. Gondert ME, Tinsley RA, Rueda D, Walter NG. 2006. Catalytic core structure of the trans-acting HDV ribozyme is subtly influenced by sequence variation outside the core. *Biochemistry* 45:7563–73

53. The first application of single crystal Raman spectroscopy to determine the ionization equilibrium of an active site nucleobase.

53. **Gong B, Chen JH, Chase E, Chadalavada DM, Yajima R, et al. 2007. Direct measurement of a pK_a near neutrality for the catalytic cytosine in the genomic HDV ribozyme using Raman crystallography. *J. Am. Chem. Soc.* 129:13335–42**

54. Grasby JA, Mersmann K, Singh M, Gait MJ. 1995. Purine functional groups in essential residues of the hairpin ribozyme required for catalytic cleavage of RNA. *Biochemistry* 34:4068–76

55. Grum-Tokars V, Milovanovic M, Wedekind JE. 2003. Crystallization and X-ray diffraction analysis of an all-RNA U39C mutant of the minimal hairpin ribozyme. *Acta Crystallogr. D* 59:142–45

56. Hammann C, Lilley DM. 2002. Folding and activity of the hammerhead ribozyme. *ChemBioChem* 3:690–700

57. Hampel A, Cowan JA. 1997. A unique mechanism for RNA catalysis: the role of metal cofactors in hairpin ribozyme cleavage. *Chem. Biol.* 4:513–17

58. Hampel A, Tritz R. 1989. RNA catalytic properties of the minimum (-)sTRSV sequence. *Biochemistry* 28:4929–33

59. Hampel KJ, Tinsley MM. 2006. Evidence for preorganization of the glmS ribozyme ligand binding pocket. *Biochemistry* 45:7861–71

60. Han J, Burke JM. 2005. Model for general acid-base catalysis by the hammerhead ribozyme: pH-activity relationships of G8 and G12 variants at the putative active site. *Biochemistry* 44:7864–70

61. Harris DA, Rueda D, Walter NG. 2002. Local conformational changes in the catalytic core of the trans-acting hepatitis delta virus ribozyme accompany catalysis. *Biochemistry* 41:12051–61

62. Heckman JE, Lambert D, Burke JM. 2005. Photocrosslinking detects a compact, active structure of the hammerhead ribozyme. *Biochemistry* 44:4148–56

63. Hertel KJ, Herschlag D, Uhlenbeck OC. 1994. A kinetic and thermodynamic framework for the hammerhead ribozyme reaction. *Biochemistry* 33:3374–85

64. Hertel KJ, Uhlenbeck OC. 1995. The internal equilibrium of the hammerhead ribozyme reaction. *Biochemistry* 34:1744–49

65. Heus HA, Pardi A. 1991. Nuclear magnetic resonance studies of the hammerhead ribozyme domain. Secondary structure formation and magnesium ion dependence. *J. Mol. Biol.* 217:113–24

66. Hoogstraten CG, Sumita M. 2007. Structure-function relationships in RNA and RNP enzymes: recent advances. *Biopolymers* 87:317–28

67. Hutchins CJ, Rathjen PD, Forster AC, Symons RH. 1986. Self-cleavage of plus and minus transcripts of avocado sunblotch viroid. *Nucleic Acids Res.* 14:3627–40

68. Jansen JA, McCarthy TJ, Soukup GA, Soukup JK. 2006. Backbone and nucleobase contacts to glucosamine-6-phosphate in the glmS ribozyme. *Nat. Struct. Mol. Biol.* 13:517–23

69. Jencks WP. 1969. General acid-base catalysis. In *Catalysis in Chemistry and Enzymology*, pp. 163–242. New York: Dover

70. Ke A, Ding F, Batchelor JD, Doudna JA. 2007. Structural roles of monovalent cations in the HDV ribozyme. *Structure* 15:281–87

71. Ke A, Zhou K, Ding F, Cate JH, Doudna JA. 2004. A conformational switch controls hepatitis delta virus ribozyme catalysis. *Nature* 429:201–5

72. Khvorova A, Lescoute A, Westhof E, Jayasena SD. 2003. Sequence elements outside the hammerhead ribozyme catalytic core enable intracellular activity. *Nat. Struct. Biol.* 10:708–12

73. Kim NK, Murali A, DeRose VJ. 2005. Separate metal requirements for loop interactions and catalysis in the extended hammerhead ribozyme. *J. Am. Chem. Soc.* 127:14134–35

74. Klein DJ, Been MD, Ferré-D'Amaré AR. 2007. Essential role of an active-site guanine in glmS ribozyme catalysis. *J. Am. Chem. Soc.* 129:14858–59

75. Klein DJ, Ferré-D'Amaré AR. 2006. Structural basis of glmS ribozyme activation by glucosamine-6-phosphate. *Science* 313:1752–56

76. Knoll R, Bald R, Furste JP. 1997. Complete identification of nonbridging phosphate oxygens involved in hammerhead cleavage. *RNA* 3:132–40

77. Koizumi M, Ohtsuka E. 1991. Effects of phosphorothioate and 2-amino groups in hammerhead ribozymes on cleavage rates and Mg^{2+} binding. *Biochemistry* 30:5145–50

78. Krasovska MV, Sefcikova J, Reblova K, Schneider B, Walter NG, Sponer J. 2006. Cations and hydration in catalytic RNA: molecular dynamics of the hepatitis delta virus ribozyme. *Biophys. J.* 91:626–38

79. Kuzmin YI, Da Costa CP, Cottrell J, Fedor MJ. 2005. Contribution of an active site adenosine to hairpin ribozyme catalysis. *J. Mol. Biol.* 349:989–1010

80. Kuzmin YI, Da Costa CP, Fedor MJ. 2004. Role of an active site guanine in hairpin ribozyme catalysis probed by exogenous nucleobase rescue. *J. Mol. Biol.* 340:233–51

72. Discovery that natural forms of hammerhead ribozymes include peripheral sequences that stimulate activity by forming interhelical tertiary interactions.

75. X-ray crystal structures of the *glmS* ribozyme demonstrating virtually no structural differences between activator and inhibitor complexes.

81. Lambert D, Heckman JE, Burke JM. 2006. Three conserved guanosines approach the reaction site in native and minimal hammerhead ribozymes. *Biochemistry* 45:7140–47

82. Lebruska LL, Kuzmine II, Fedor MJ. 2002. Rescue of an abasic hairpin ribozyme by cationic nucleobases. Evidence for a novel mechanism of RNA catalysis. *Chem. Biol.* 9:465–73

83. Lee TS, Silva Lopez C, Giambasu GM, Martick M, Scott WG, York DM. 2008. Role of Mg^{2+} in hammerhead ribozyme catalysis from molecular simulation. *J. Am. Chem. Soc.* 130:3053–64

84. Lee TS, Silva-Lopez C, Martick M, Scott WG, York DM. 2007. Insight into the role of Mg^{2+} in hammerhead ribozyme catalysis from X-ray crystallography and molecular dynamics simulation. *J. Chem. Theory Comp.* 3:325–27

85. Lincoln SF, Richens DT, Sykes AG. 2004. Metal aqua ions. In *Comprehensive Coordination Chemistry II: From Biology to Nanotechnology*, ed. JA McCleverty, TJ Meyer, 1:515–55. Amsterdam: Elsevier

86. Lilley DM. 2004. The Varkud satellite ribozyme. *RNA* 10:151–58

87. Lilley DM. 2007. A chemo-genetic approach for the study of nucleobase participation in nucleolytic ribozymes. *Biol. Chem.* 388:699–704

88. Lönnberg T, Lönnberg H. 2005. Chemical models for ribozyme action. *Curr. Opin. Chem. Biol.* 9:665–73

89. Lupták A, Ferré-D'Amaré AR, Zhou K, Zilm KW, Doudna JA. 2001. Direct pK_a measurement of the active-site cytosine in a genomic hepatitis delta virus ribozyme. *J. Am. Chem. Soc.* 123:8447–52

90. Lyne PD, Karplus M. 2000. Determination of the pK_a of the 2'-hydroxyl group of a phosphorylated ribose: implications for the mechanism of hammerhead ribozyme catalysis. *J. Am. Chem. Soc.* 122:166–67

91. MacElrevey C, Salter JD, Krucinska J, Wedekind JE. 2008. Structural effects of nucleobase variations at key active site residue Ade38 in the hairpin ribozyme. *RNA* 14:1600–16

92. Maderia M, Hunsicker L, DeRose V. 2000. Metal-phosphate interactions in the hammerhead ribozyme observed by ^{31}P NMR and phosphorothioate substitutions. *Biochemistry* 39:12113–20

93. Martick M, Horan LH, Noller HF, Scott WG. 2008. A discontinuous hammerhead ribozyme embedded in a mammalian messenger RNA. *Nature* 454:899–902

94. Martick M, Lee TS, York DM, Scott WG. 2008. Solvent structure and hammerhead ribozyme catalysis. *Chem. Biol.* 15:332–42

95. The first X-ray crystal structure of an extended hammerhead ribozyme that includes interhelical interactions.

95. Martick M, Scott WG. 2006. Tertiary contacts distant from the active site prime a ribozyme for catalysis. *Cell* 126:309–20

96. McCarthy TJ, Plog MA, Floy SA, Jansen JA, Soukup JK, Soukup GA. 2005. Ligand requirements for glmS ribozyme self-cleavage. *Chem. Biol.* 12:1221–26

97. McKay DB. 1996. Structure and function of the hammerhead ribozyme: an unfinished story. *RNA* 2:395–403

98. Montange RK, Batey RT. 2008. Riboswitches: emerging themes in RNA structure and function. *Annu. Rev. Biophys.* 37:117–33

99. Moody EM, Lecomte JT, Bevilacqua PC. 2005. Linkage between proton binding and folding in RNA: a thermodynamic framework and its experimental application for investigating pK_a shifting. *RNA* 11:157–72

100. Murchie AI, Thomson JB, Walter F, Lilley DM. 1998. Folding of the hairpin ribozyme in its natural conformation achieves close physical proximity of the loops. *Mol. Cell* 1:873–81

101. Murray JB, Seyhan AA, Walter NG, Burke JM, Scott WG. 1998. The hammerhead, hairpin and VS ribozymes are catalytically proficient in monovalent cations alone. *Chem. Biol.* 5:587–95

102. Murray JB, Szoke H, Szoke A, Scott WG. 2000. Capture and visualization of a catalytic RNA enzyme-product complex using crystal lattice trapping and X-ray holographic reconstruction. *Mol. Cell* 5:279–87

103. Murray JB, Terwey DP, Maloney L, Karpeisky A, Usman N, et al. 1998. The structural basis of hammerhead ribozyme self-cleavage. *Cell* 92:665–73

104. Nakano S, Cerrone AL, Bevilacqua PC. 2003. Mechanistic characterization of the HDV genomic ribozyme: classifying the catalytic and structural metal ion sites within a multichannel reaction mechanism. *Biochemistry* 42:2982–94

105. Nakano S, Chadalavada DM, Bevilacqua PC. 2000. General acid-base catalysis in the mechanism of a hepatitis delta virus ribozyme. *Science* 287:1493–97

106. Nakano S, Proctor DJ, Bevilacqua PC. 2001. Mechanistic characterization of the HDV genomic ribozyme: assessing the catalytic and structural contributions of divalent metal ions within a multichannel reaction mechanism. *Biochemistry* 40:12022–38

107. Nam K, Gao J, York DM. 2008. Electrostatic interactions in the hairpin ribozyme account for the majority of the rate acceleration without chemical participation by nucleobases. *RNA* 14:1501–7

108. Narlikar GJ, Herschlag D. 1997. Mechanistic aspects of enzymatic catalysis: lessons from comparison of RNA and protein enzymes. *Annu. Rev. Biochem.* 66:19–59

109. Nelson JA, Shepotinovskaya I, Uhlenbeck OC. 2005. Hammerheads derived from sTRSV show enhanced cleavage and ligation rate constants. *Biochemistry* 44:14577–85

110. Nelson JA, Uhlenbeck OC. 2006. When to believe what you see. *Mol. Cell* 23:447–50

111. Nelson JA, Uhlenbeck OC. 2008. Hammerhead redux: Does the new structure fit the old biochemical data? *RNA* 14:605–15

112. Nelson JA, Uhlenbeck OC. 2008. Minimal and extended hammerheads utilize a similar dynamic reaction mechanism for catalysis. *RNA* 14:43–54

113. Nesbitt S, Hegg LA, Fedor MJ. 1997. An unusual pH-independent and metal-ion-independent mechanism for hairpin ribozyme catalysis. *Chem. Biol.* 4:619–30

114. O'Rear JL, Wang S, Feig AL, Beigelman L, Uhlenbeck OC, Herschlag D. 2001. Comparison of the hammerhead cleavage reactions stimulated by monovalent and divalent cations. *RNA* 7:537–45

115. Osborne EM, Schaak JE, DeRose VJ. 2005. Characterization of a native hammerhead ribozyme derived from schistosomes. *RNA* 11:187–96

116. Park H, Lee S. 2006. Role of solvent dynamics in stabilizing the transition state of RNA hydrolysis by hairpin ribozyme. *J. Chem. Theory Comp.* 2:858–62

117. Penedo JC, Wilson TJ, Jayasena SD, Khvorova A, Lilley DM. 2004. Folding of the natural hammerhead ribozyme is enhanced by interaction of auxiliary elements. *RNA* 10:880–88

118. Peracchi A, Beigelman L, Scott EC, Uhlenbeck OC, Herschlag D. 1997. Involvement of a specific metal ion in the transition of the hammerhead ribozyme to its catalytic conformation. *J. Biol. Chem.* 272:26822–26

119. Pereira MJ, Harris DA, Rueda D, Walter NG. 2002. Reaction pathway of the trans-acting hepatitis delta virus ribozyme: a conformational change accompanies catalysis. *Biochemistry* 41:730–40

120. Perrotta AT, Been MD. 2006. HDV ribozyme activity in monovalent cations. *Biochemistry* 45:11357–65

121. Perrotta AT, Been MD. 2007. A single nucleotide linked to a switch in metal ion reactivity preference in the HDV ribozymes. *Biochemistry* 46:5124–30

122. Perrotta AT, Shih I, Been MD. 1999. Imidazole rescue of a cytosine mutation in a self-cleaving ribozyme. *Science* 286:123–26

123. Perrotta AT, Wadkins TS, Been MD. 2006. Chemical rescue, multiple ionizable groups, and general acid-base catalysis in the HDV genomic ribozyme. *RNA* 12:1282–91

124. Pinard R, Hampel KJ, Heckman JE, Lambert D, Chan PA, et al. 2001. Functional involvement of G8 in the hairpin ribozyme cleavage mechanism. *EMBO J.* 20:6434–42

125. Pinard R, Lambert D, Walter NG, Heckman JE, Major F, Burke JM. 1999. Structural basis for the guanosine requirement of the hairpin ribozyme. *Biochemistry* 38:16035–39

126. Pley HW, Flaherty KM, McKay DB. 1994. Three-dimensional structure of a hammerhead ribozyme. *Nature* 372:68–74

127. Prody GA, Bakos JT, Buzayan JM, Schneider IR, Bruening G. 1986. Autolytic processing of dimeric plant virus satellite RNA. *Science* 231:1577–80

128. Przybilski R, Graf S, Lescoute A, Nellen W, Westhof E, et al. 2005. Functional hammerhead ribozymes naturally encoded in the genome of *Arabidopsis thaliana*. *Plant Cell* 17:1877–85

129. Przybilski R, Hammann C. 2006. The hammerhead ribozyme structure brought in line. *ChemBioChem* 7:1641–44

130. Przybilski R, Hammann C. 2007. The tolerance to exchanges of the Watson Crick base pair in the hammerhead ribozyme core is determined by surrounding elements. *RNA* 13:1625–30

131. Que-Gewirth NS, Sullenger BA. 2007. Gene therapy progress and prospects: RNA aptamers. *Gene Ther.* 14:283–91

132. Raines RT. 1998. Ribonuclease A. *Chem. Rev.* 98:1045–66

133. Rhodes MM, Reblova K, Sponer J, Walter NG. 2006. Trapped water molecules are essential to structural dynamics and function of a ribozyme. *Proc. Natl. Acad. Sci. USA* 103:13380–85

134. Rocheleau L, Pelchat M. 2006. The subviral RNA database: a toolbox for viroids, the hepatitis delta virus and satellite RNAs research. *BMC Microbiol.* 6:24

135. Rojas AA, Vazquez-Tello A, Ferbeyre G, Venanzetti F, Bachmann L, et al. 2000. Hammerhead-mediated processing of satellite pDo500 family transcripts from Dolichopoda cave crickets. *Nucleic Acids Res.* 28:4037–43

136. Roth A, Nahvi A, Lee M, Jona I, Breaker RR. 2006. Characteristics of the glmS ribozyme suggest only structural roles for divalent metal ions. *RNA* 12:607–19

137. Roychowdhury-Saha M, Burke DH. 2006. Extraordinary rates of transition metal ion-mediated ribozyme catalysis. *RNA* 12:1846–52

138. Ruffner DE, Uhlenbeck OC. 1990. Thiophosphate interference experiments locate phosphates important for the hammerhead RNA self-cleavage reaction. *Nucleic Acids Res.* 18:6025–29

139. Rupert PB, Ferré-D'Amaré AR. 2001. Crystal structure of a hairpin ribozyme-inhibitor complex with implications for catalysis. *Nature* 410:780–86

140. Rupert PB, Massey AP, Sigurdsson ST, Ferré-D'Amaré AR. 2002. Transition state stabilization by a catalytic RNA. *Science* 298:1421–24

141. Saenger W. 1984. 5.2 pK values of base, sugar, and phosphate groups: sites for nucleophilic attack. In *Principles of Nucleic Acid Structure*, pp. 107–10. New York: Springer-Verlag

142. Salehi-Ashtiani K, Luptak A, Litovchick A, Szostak JW. 2006. A genomewide search for ribozymes reveals an HDV-like sequence in the human CPEB3 gene. *Science* 313:1788–92

143. Salehi-Ashtiani K, Szostak JW. 2001. In vitro evolution suggests multiple origins for the hammerhead ribozyme. *Nature* 414:82–84

144. Salter J, Krucinska J, Alam S, Grum-Tokars V, Wedekind JE. 2006. Water in the active site of an all-RNA hairpin ribozyme and effects of Gua8 base variants on the geometry of phosphoryl transfer. *Biochemistry* 45:686–700

145. Scott EC, Uhlenbeck OC. 1999. A reinvestigation of the thio effect at the hammerhead cleavage site. *Nucleic Acids Res.* 27:479–84

146. Scott WG. 2007. Morphing the minimal and full-length hammerhead ribozymes: implications for the cleavage mechanism. *Biol. Chem.* 388:727–35

147. Scott WG. 2007. Ribozymes. *Curr. Opin. Struct. Biol.* 17:280–86

148. Scott WG, Finch JT, Klug A. 1995. The crystal structure of an all-RNA hammerhead ribozyme: a proposed mechanism for RNA catalytic cleavage. *Cell* 81:991–1002

149. Scott WG, Murray JB, Arnold JR, Stoddard BL, Klug A. 1996. Capturing the structure of a catalytic RNA intermediate: the hammerhead ribozyme. *Science* 274:2065–69

150. Serganov A, Patel DJ. 2007. Ribozymes, riboswitches and beyond: regulation of gene expression without proteins. *Nat. Rev. Genet.* 8:776–90

151. Shepotinovskaya IV, Uhlenbeck OC. 2008. Catalytic diversity of extended hammerhead ribozymes. *Biochemistry* 47:7034–42

152. Shih I, Been MD. 2000. Kinetic scheme for intermolecular RNA cleavage by a ribozyme derived from hepatitis delta virus RNA. *Biochemistry* 39:9055–66

153. Shih I, Been MD. 2001. Involvement of a cytosine side chain in proton transfer in the rate-determining step of ribozyme self-cleavage. *Proc. Natl. Acad. Sci. USA* 98:1489–94

154. Shih I, Been MD. 2002. Catalytic strategies of the hepatitis delta virus ribozymes. *Annu. Rev. Biochem.* 71:887–917

155. Shippy R, Siwkowski A, Hampel A. 1998. Mutational analysis of loops 1 and 5 of the hairpin ribozyme. *Biochemistry* 37:564–70

156. Sigel RKO, Pyle AM. 2007. Alternative roles for metal ions in enzyme catalysis and the implications for ribozyme chemistry. *Chem. Rev.* 107:97–113

157. Slim G, Gait MJ. 1991. Configurationally defined phosphorothioate-containing oligoribonucleotides in the study of the mechanism of cleavage of hammerhead ribozymes. *Nucleic Acids Res.* 19:1183–88

158. Stage-Zimmermann TK, Uhlenbeck OC. 2001. A covalent crosslink converts the hammerhead ribozyme from a ribonuclease to an RNA ligase. *Nat. Struct. Biol.* 8:863–67

159. Tang CL, Alexov E, Pyle AM, Honig B. 2007. Calculation of pK$_a$s in RNA: on the structural origins and functional roles of protonated nucleotides. *J. Mol. Biol.* 366:1475–96

160. Toney M, Kirsch J. 1989. Direct Brønsted analysis of the restoration of activity to a mutant enzyme by exogenous amines. *Science* 243:1485–88

161. Torelli AT, Krucinska J, Wedekind JE. 2007. A comparison of vanadate to a 2′-5′ linkage at the active site of a small ribozyme suggests a role for water in transition-state stabilization. *RNA* 13:1052–70

162. Torelli AT, Spitale RC, Krucinska J, Wedekind JE. 2008. Shared traits on the reaction coordinates of ribonuclease and an RNA enzyme. *Biochem. Biophys. Res. Commun.* 371:154–58

163. Tuschl T, Gohlke C, Jovin TM, Westhof E, Eckstein F. 1994. A three-dimensional model for the hammerhead ribozyme based on fluorescence measurements. *Science* 266:785–89

164. Vaish NK, Kore AR, Eckstein F. 1998. Recent developments in the hammerhead ribozyme field. *Nucleic Acids Res.* 26:5237–42

165. van Tol H, Buzayan JM, Bruening G. 1991. Evidence for spontaneous circle formation in the replication of the satellite RNA of *Tobacco ringspot virus*. *Virology* 180:23–30

166. van Tol H, Buzayan JM, Feldstein PA, Eckstein F, Bruening G. 1990. Two autolytic processing reactions of a satellite RNA proceed with inversion of configuration. *Nucleic Acids Res.* 18:1971–75

167. Vogt M, Lahiri S, Hoogstraten CG, Britt RD, DeRose VJ. 2006. Coordination environment of a site-bound metal ion in the hammerhead ribozyme determined by 15N and 2H ESEEM spectroscopy. *J. Am. Chem. Soc.* 128:16764–70

168. Walter F, Murchie AI, Duckett DR, Lilley DM. 1998. Global structure of four-way RNA junctions studied using fluorescence resonance energy transfer. *RNA* 4:719–28

169. Walter F, Murchie AIH, Lilley DMJ. 1998. Folding of the four-way RNA junction of the hairpin ribozyme. *Biochemistry* 37:17629–36

170. Walter F, Murchie AIH, Thomson JB, Lilley DMJ. 1998. Structure and activity of the hairpin ribozyme in its natural junction conformation: effect of metal ions. *Biochemistry* 37:14195–203

171. Walter NG. 2007. Ribozyme catalysis revisited: Is water involved? *Mol. Cell* 28:923–29

172. Walter NG, Burke JM, Millar DP. 1999. Stability of hairpin ribozyme tertiary structure is governed by the interdomain junction. *Nat. Struct. Biol.* 6:544–49

173. Walter NG, Harris DA, Pereira MJ, Rueda D. 2001. In the fluorescent spotlight: global and local conformational changes of small catalytic RNAs. *Biopolymers* 61:224–42

174. Wang S, Karbstein K, Peracchi A, Beigelman L, Herschlag D. 1999. Identification of the hammerhead ribozyme metal ion binding site responsible for rescue of the deleterious effect of a cleavage site phosphorothioate. *Biochemistry* 38:14363–78

175. Wedekind JE, McKay DB. 1998. Crystallographic structures of the hammerhead ribozyme: relationship to ribozyme folding and catalysis. *Annu. Rev. Biophys. Biomol. Struct.* 27:475–502

176. Wei K, Liu L, Cheng YH, Fu Y, Guo QX. 2007. Theoretical examination of two opposite mechanisms proposed for hepatitis delta virus ribozyme. *J. Phys. Chem. B* 111:1514–16

177. Wilson TJ, Ouellet J, Zhao ZY, Harusawa S, Araki L, et al. 2006. Nucleobase catalysis in the hairpin ribozyme. *RNA* 12:980–87

178. Winkler WC, Nahvi A, Roth A, Collins JA, Breaker RR. 2004. Control of gene expression by a natural metabolite-responsive ribozyme. *Nature* 428:281–86

179. Yadava RS, Choi AJ, Lebruska LL, Fedor MJ. 2001. Hairpin ribozymes with four-way helical junctions mediate intracellular RNA ligation. *J. Mol. Biol.* 309:893–902

180. Young KJ, Gill F, Grasby JA. 1997. Metal ions play a passive role in the hairpin ribozyme catalysed reaction. *Nucleic Acids Res.* 25:3760–66

178. Discovery of the *glmS* ribozyme.

Particle-Tracking Microrheology of Living Cells: Principles and Applications

Denis Wirtz

Department of Chemical and Biomolecular Engineering and Institute for NanoBioTechnology, Johns Hopkins University, Baltimore, Maryland 21218; email: wirtz@jhu.edu

Annu. Rev. Biophys. 2009. 38:301–26

The *Annual Review of Biophysics* is online at biophys.annualreviews.org

This article's doi: 10.1146/annurev.biophys.050708.133724

Key Words

cell mechanics, nanorheology, viscosity, elasticity, LINC complex, laminopathies, emerin

Abstract

A multitude of cellular and subcellular processes depend critically on the mechanical deformability of the cytoplasm. We have recently introduced the method of particle-tracking microrheology, which measures the viscoelastic properties of the cytoplasm locally and with high spatiotemporal resolution. Here we establish the basic principles of particle-tracking microrheology, describing the advantages of this approach over more conventional approaches to cell mechanics. We present basic concepts of molecular mechanics and polymer physics relevant to the microrheological response of cells. Particle-tracking microrheology can probe the mechanical properties of live cells in experimentally difficult, yet more physiological, environments, including cells embedded inside a 3D matrix, adherent cells subjected to shear flows, and cells inside a developing embryo. Particle-tracking microrheology can readily reveal the lost ability of diseased cells to resist shear forces.

Contents

INTRODUCTION

Many cellular and subcellular processes depend critically on the mechanical deformability of the cytoplasm. For example, the translocation of organelles (e.g., nucleus, mitochondria, and endoplasmic reticulum) within the cytoplasm is partly controlled by their frictional drag and therefore by the local viscoelastic properties of the cytoplasm (54, 55, 59, 68). Migrating cells at the edge of a wound significantly increase the stiffness of their cytoplasm to enable dendritic filamentous actin (F-actin) assemblies to produce net protruding forces against the plasma membrane (50, 80). Axonal elongation depends directly on the highly regulated intracellular viscosity of the growth cone (72). Cells need to adapt their intracellular physical properties to the physical properties of their extracellular milieu to grow, differentiate, and migrate (20, 22, 78). Moreover, changes in the mechanical properties of cells often correlate with disease state (14, 34, 58). For instance, cells derived from mouse models of progeria (premature aging) or muscular dystrophy display significantly softer (i.e., less elastic) cytoplasm (55) than wild-type controls. This affects the ability of these cells to resist shear forces and to migrate to the edge of a wound (60).

We have recently introduced the method of particle-tracking microrheology (102) to measure the viscoelastic properties of the cytoplasm locally and with high spatiotemporal resolution. In this approach, fluorescent beads of less than 1 μm in diameter are injected directly into the cytoplasm of live cells (**Figure 1**). These beads rapidly disperse throughout the cytoplasm and are subsequently tracked by fluorescence microscopy (102). The recorded movements of the beads are analyzed in terms of viscosity and elasticity of the cytoplasm. We and others have exploited particle-tracking microrheology to probe the viscoelastic properties of various types of cells in a wide range of conditions (12, 37, 45, 57, 60, 61, 67, 73, 77, 85, 92, 95, 103). These measurements have revealed important new mechanistic insights into how the physical properties of the cytoplasm adapt to various chemical and physical stimuli, how they can control basic cell functions, and how these properties can be significantly altered in diseased cells. Below, we briefly describe different cell biological questions, which have been recently addressed using particle-tracking microrheology.

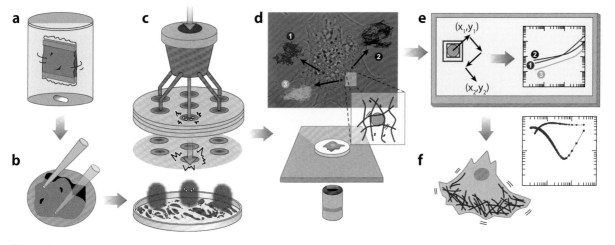

Figure 1

Particle-tracking microrheology. (*a*) Submicron fluorescent beads are dialyzed. (*b*, *c*) These beads are spread on a grid, which is placed inside a ballistic injection machine. After ballistic injection, the beads disperse rapidly within the cytoplasm. (*d*) The cells are placed under a high-magnification fluorescence microscope. The random spontaneous movements of the beads are monitored with high spatial and temporal resolution. The numerals 1, 2, and 3 refer to the order of steps in which particle-tracking is conducted. (*e*) Using the appropriate software, the time-dependent (x, y) coordinates of the beads are mathematically transformed into mean squared displacements (MSDs). (*f*) The time lag-dependent MSDs of the beads are subsequently transformed into local values of either the frequency-dependent viscoelastic moduli, $G'(\omega)$ and $G''(\omega)$, or the creep compliance, $\Gamma(t)$, of the cytoplasm. Modified with permission from Reference 76.

Particle-tracking microrheology (also called nanorheology) shows that the cytoplasm of adherent cells at rest, such as endothelial cells and fibroblasts on planar substrates, is typically more elastic than viscous (i.e., these cells show a rheological behavior which is akin to that of a solid) for timescales between 0.1 and ~10 s (102). However, at timescales greater than 10–20 s, these cells show a predominantly viscous response to shear forces: The cytoplasm behaves like a liquid (102). In general, the viscoelastic properties of adherent cells are dominated by the actin filament cytoskeleton. Indeed pharmacological treatment inducing the disassembly of actin filaments eliminates the deformability of the cytoplasm and renders the cytoplasm mostly viscous at all timescales (107, 116). Moreover, the level of elasticity in the cell correlates with the local concentration of F-actin present in the cytoplasm: The actin-rich cell periphery (i.e., the lamella) is significantly stiffer than the perinuclear region, which contains less actin. Serum-starved cells, which show little organized F-actin, have both a low viscosity and a low elasticity. However, when suddenly subjected to a shear flow similar to that present in blood vessels, the cytoplasm of serum-starved cells displays rapid assembly of actin filaments into organized structures (61, 76). Particle-tracking microrheology reveals that flow-induced actin filament assembly is accompanied by a rapid rise in cytoplasmic stiffness, followed by an equally rapid decrease in stiffness (61). This transient increase in cytoplasmic stiffness correlates with the transient activation of the Rho/Rho kinase (ROCK) pathway and the associated assembly and transient contraction of the actin filament network by myosin II. These results suggest that when an adherent cell is subjected to shear stresses, its first action is to prevent detachment from its substratum by greatly stiffening the cytoplasm through enhanced actin filament assembly and Rho-kinase-mediated contractility.

During migration at the edge of a wound, cells polarize their overall morphology and position the microtubule organizing center (MTOC) toward the cell's leading edge (31, 59). Particle-tracking microrheology indicates that the mechanical properties of the cytoplasm become spatially polarized as well: The cytoplasm is much stiffer at the leading edge than at the trailing edge of the migrating cell, i.e., the region near the nucleus that is positioned near the back of the cell (50). MTOC repositioning at the edge of the wound is abrogated in cells transfected with a dominant-negative mutant of the small GTPase Cdc42 (98). Similarly, cells with deactivated Cdc42 show no spatial polarization in their mechanical properties. These results suggest that a differential distribution of subcellular mechanical microenvironments is essential for directed cell migration and is coordinated through microtubules (50).

Particle-tracking microrheology has also shown that the elasticity of the cytoplasm is much lower than that of the nucleus in the same cell. The nucleus is also more elastic than viscous, which reveals that the intranuclear region displays an unexpectedly strong solid-like behavior (103). Indeed, when cells move though a dense extracellular matrix, a rate-limiting step in the 3D migration process is the squeezing of the nucleus by the pores of the matrix, not the deformation of the cytoplasm (6, 56). Measurements of the mean shear viscosity and the elasticity of the intranuclear region determine a lower bound of the propulsive forces (3–15 pN) required for nuclear organelles such as the promyelocytic leukemia body to undergo processive transport within the nucleus by overcoming the friction forces set by the intranuclear viscosity. Dynamic analysis of the spontaneous movements of submicron beads embedded in the nucleus also reveals the presence of transient nuclear microdomains of mean size 290 nm that are mostly absent in the cytoplasm (32, 103). The strong elastic character and the micro-organization of the intranuclear region revealed by particle-tracking analysis may help the nucleus preserve its structural coherence. This study highlights the difference between the low interstitial nucleoplasmic viscosity, which controls the transport of nuclear

proteins and molecules, and the much higher mesoscale viscosity, which affects the diffusion and the directed transport of nuclear organelles and reorganization of interphase chromosomes (103).

Finally, particle-tracking microrheology has shown that when human endothelial cells are placed inside a 3D matrix, their cytoplasm is much softer than when the same cells are placed on a thin planar layer of the same matrix (77). Vascular endothelial growth factor (VEGF) treatment, which enhances endothelial migration in the 3D matrix, increases the deformability of the cytoplasm of endothelial cells in the matrix. This VEGF-induced softening response of the cytoplasm is abrogated by specific ROCK inhibition. These results suggest that ROCK plays an essential role in the regulation of the intracellular mechanical response to VEGF of endothelial cells in a 3D matrix.

The above measurements could only have been achieved thanks to particle-tracking microrheology, which can reveal the mechanical properties of live cells in experimentally difficult—yet more physiological—environments. For instance, particle-tracking microrheology is the only cell mechanics method that can probe the mechanical properties of individual cells deeply embedded inside a 3D matrix and their mechanical response to agonists and/or drug treatments (77), monitor changes in cytoskeleton elasticity in cells subjected to shear flows (61), or measure in vivo the local viscoelastic properties of a *Caenorhabditis elegans* embryo embedded in an impenetrable shell (16). Cells in these more physiological environments cannot be probed by existing methods of cell mechanics, such as atomic force microscopy (AFM), because these cells are inaccessible to direct physical contact and can only be probed at a distance.

This review describes basic concepts of molecular mechanics and polymer physics applied to cells and introduces the fundamental principles underlying particle-tracking microrheology and its advantages over traditional cell mechanics methods. This review also shows how particle-tracking microrheology can readily reveal the lost ability of diseased cells to resist shear forces.

BASIC CONCEPTS OF MOLECULAR CELL MECHANICS

Working Definitions of Stress, Viscosity, Elasticity, and Compliance

Because different experimental methods measure different (but often related) rheological quantities, the intracellular mechanics of a living cell is best characterized by multiple rheological parameters, including viscosity, elasticity, and creep compliance. These rheological parameters simply describe the mechanical response of a material (such as the cell's cytoplasm) subjected to a force and how it measures the resulting deformation and, vice versa, the mechanical response of a material subjected to a deformation and how it measures the force required to produce the deformation. The rheological response of the cytoplasm can be either predominantly viscous or elastic, depending on the time of application and the magnitude of the force. These forces can be externally applied as in the case of endothelial cells subjected to blood flow (61), result from internal tension as in the case of actomyosin contractility, or both (49).

The shear viscosity of a liquid measures its propensity to flow under random or applied forces. The shear viscosity generates the drag forces that slow down the motion of organelles and protein complexes in the cytoplasm and nucleus. A simple way to measure the viscosity of a material or liquid is to use a falling-ball viscometer. Here the speed at which heavy metallic beads fall through the probed material depends on an effective viscosity. This method is highly approximate due to the uncontrolled interactions between the beads and the material, the inherent difficulty to measure the terminal velocity of the beads, and the assumptions that need to be made to compute this viscosity. Alternatively, the material can be subjected to a steady shear deformation of controlled rate using a rheometer. Here, the material is placed between two parallel plates or between a cone and plate. The viscosity is the ratio of the stress (force per unit area) induced in the material by the imposed deformation to the rate of shear deformation. In cells, the viscosity of the cytoplasm predominantly governs the transport and movements of subcellular organelles and cytoskeleton structures at long timescales. A material that is only viscous (and not elastic), such as water, cannot resist mechanical stresses; it can only slow down its deformation by the imposed mechanical stress. Upon cessation of the stress, the material or liquid has lost all memory of its original shape and location. Below, we show how the viscosity of a material or a liquid can be obtained by tracking the random movements of submicron beads embedded in the material.

The elasticity (also called the elastic modulus) of a material measures its stretchiness. Elasticity measures the ability of cytoplasmic structures to resist forces and store energy caused by deformation. A material that is only elastic (and not viscous) can deform under stress but cannot flow. As no viscous dissipation occurs during its deformation, the elastic material snaps back to its original shape upon cessation of the stress. Elasticity typically governs the response of the cytoplasm to mechanical stresses at short timescales.

Some materials, such as Silly Putty®, can be both viscous and elastic. Silly Putty can bounce as it deforms upon impact, but quickly regains its shape, which means it is elastic. It is also viscous, as Silly Putty rolled into a ball will partially flatten due to its own weight when left on a flat surface. These simple observations indicate that Silly Putty is elastic at short timescales (i.e., during impact) and viscous at long timescales. Similarly, the cytoplasm of living cells is both viscous and elastic, i.e., it is viscoelastic.

Instead of working in the time domain, rheologists tend to work in the frequency domain. The frequency-dependent elastic modulus of a material, $G'(\omega)$, can be obtained by subjecting it to oscillatory deformations of controlled frequency ω and constant (small) amplitude. When an oscillatory deformation is applied and is given by a sine function of time,

$\gamma(t) = \gamma_0 \sin \omega t$, the stress, τ, induced within the material by this deformation will typically have both sine and cosine components. Specifically, the stress can be decomposed into a sine (in-phase) component and a cosine (out-of-phase) component: $\tau(t) = \tau' \sin \omega t + \tau'' \cos \omega t = \gamma_0(G' \sin \omega t + G'' \cos \omega t)$. The elastic modulus of the material, G', is equal to the in-phase component of the frequency-dependent stress divided by the amplitude of the oscillatory deformation, i.e., $G'(\omega) = \tau'/\gamma_0$. The viscous modulus of the same material, $G''(\omega)$, can be obtained during the same measurement by extracting the out-of-phase component of the frequency-dependent stress and dividing by the amplitude of the oscillatory deformation, i.e., $G''(\omega) = \tau''/\gamma_0$ (24). If the material is an elastic solid (no or little viscosity), such as rubber, then the induced stress is exactly in phase with the input deformation and $\tau(t) = \tau' \sin \omega t$. In this case, $G'' = 0$. If the material is a viscous liquid (no elasticity), such as water or glycerol, then the induced stress is out of phase with the input deformation, and $\tau(t) = \tau'' \cos \omega t$. In this case, $G' = 0$.

Typical biological materials, such as cells and tissues, have rheological properties that depend on the rate of deformation, ω, i.e., both G' and G'' depend on ω. At low frequencies, the cytoplasm has the time to reorganize its cytoskeleton polymers and it can flow, behaving as a viscous liquid. At high frequencies, the cytoplasm does not have the time to reorganize and the cytoplasm behaves as an elastic solid, which resists the deformation. This underlies the importance of measuring the full frequency-dependent response of cells and tissues, which undergo both slow and rapid movements. However, the rheology of individual cells cannot be measured with a macroscopic device, such as a rheometer. Moreover, even if a microscopic rheometer existed, measuring the full frequency-dependent rheological response of the cytoplasm to oscillatory deformations would be tedious, as this response would have to be probed one frequency at a time over a wide range of frequencies. Below, we show that the frequency-dependent viscous and elastic moduli of a cell can be ob-

tained in ~10 s from the dynamic movements of submicron probe beads embedded in the cytoplasm, without subjecting the cell to any external force or deformation.

A material that is only viscous and has no elasticity, such as water, is a liquid and $G' = 0$. A material that is only elastic and has no or little viscosity, such as rubber or Jell-O, is a solid and $G'' \approx 0$. A material that is more viscous than elastic, i.e., $G''(\omega) > G'(\omega)$, is a viscoelastic liquid; a material that is more elastic than viscous, $G'(\omega) > G''(\omega)$, is a viscoelastic solid.

Finally, the creep compliance of the cytoplasm refers to its deformability. Experimentally, it is measured by the deformation of the cytoplasm resulting from an applied mechanical stress (applied force). A high compliance indicates that a material has a low propensity to resist mechanical deformation following application of a shear stress; a low compliance indicates that it can resist such stress. When the material is only viscous or only elastic and not both, the creep compliance of the material is inversely proportional to its viscosity or elasticity, respectively. Below, we show how the time-dependent creep compliance of the cytoplasm can be obtained directly from tracking the movements of submicron probe beads embedded in the cytoplasm, without subjecting the cell to any external force.

Viscoelastic moduli (G' and G'') and creep compliance of the cytoplasm are related. Indeed, it is possible to compute the time-dependent cytoplasmic creep compliance from the frequency-dependent viscoelastic parameters and, vice versa, the viscoelastic moduli of the cytoplasm can be computed from the time-dependent creep compliance. The shear viscosity can also be estimated from the frequency-dependent viscous and elastic moduli.

A Model System: A Solution of Actin Filaments

To obtain a more intuitive understanding of the rheological concepts defined above, we consider the simple (but not trivial) system of a concentrated solution of actin filaments. A key contributor to cytoplasmic stiffness is the cytoskeleton, which is composed of three major filamentous proteins: F-actin, microtubules, and intermediate filaments (92). The cytoskeleton provides the cytoplasm with its structure and shape. In the cell, actin monomers assemble into semiflexible F-actin polymers that readily form entangled networks (30, 97, 108) whose viscoelastic properties can be controlled by either altering the local density of actin filaments or the cross-linking/bundling activity of F-actin-binding proteins, such as α-actinin or filamin (43, 113). In most cells, F-actin and intermediate filaments are the primary contributors to cytoplasmic stiffness (92).

The cytoplasm of adherent cells is rheologically complex: It typically behaves as a viscoelastic liquid when sheared either slowly or for a long time, and as a viscoelastic solid when sheared either rapidly or for a short time. This rheological complexity can be captured by a solution of entangled actin filaments in vitro in the absence of motor or cross-linking proteins (3, 47). Below a threshold concentration of $\sim$0.2 mg/ml, actin filaments cannot resist shear deformations; their global rheological response is viscous, with a viscosity equal to that of buffer, with a small additional correction due to the presence of the filaments (17). However, past this threshold concentration, the actin filaments in suspension begin to form entanglements. Because these filaments are rigid (30), they form entanglements more readily than flexible polymers that have the same contour length. These entanglements form topological obstacles in the network that impede the lateral bending motion of the filaments, but not their longitudinal movements. These polymer entanglements render the solution of actin filaments elastic.

At short timescales, the filaments only undergo thermally driven lateral fluctuations

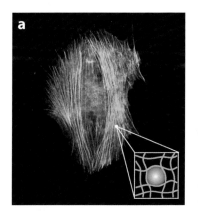 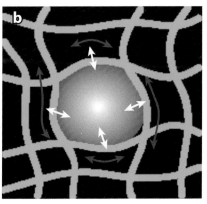 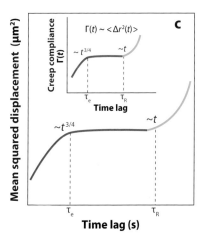

→← Slow longitudinal back-and-forth fluctuations ($\tau > \tau_e$)
⬛ Fast lateral fluctuations ($\tau > \tau_e$)
⌣ Cytoskeleton polymer

Figure 2

Physical origin of the displacements of beads in the cytoplasm. (*a*) Ballistically injected beads are lodged within the cytoskeleton. The size of the beads is larger than the average mesh size of the cytoskeleton network, which is $\sim$50 nm in fibroblasts. (*b*) The spontaneous movements of the cytoskeleton filaments that surround each bead induce displacements of the beads. At short timescales, the displacements of the beads are predominantly induced by the fast lateral bending fluctuations of the cytoskeleton filaments. At long times, the displacements of the beads are predominantly induced by the slow longitudinal back-and-forth lateral fluctuations of the cytoskeleton filaments. Finally, filaments move sufficiently to allow beads to escape the cage to move to the next cage. (*c*) Accordingly, the mean squared displacements (MSDs) of the beads show a $t^{3/4}$ power-law dependence at short timescales, a quasi plateau value at intermediate timescales between τ_e and τ_R, and a linear dependence at long times caused by the slow viscous diffusion of the beads. (*Inset*) The deformability of the cytoplasm [creep compliance, $\Gamma(t)$] is proportional to the MSD of the beads, $\langle \Delta r^2(t) \rangle$. See text for details.

(**Figure 2**). These fluctuations are fast, as they involve small sections of the filaments between network entanglements. A fast camera shows each actin filament undergoing lateral fluctuations within a confining tube-like region formed by the surrounding filaments (17, 18, 46). At intermediate timescales that are longer than the time for the lateral fluctuations to begin to hit the walls of the tube region, the back-and-forth longitudinal movements of the filament in the network can take place (**Figure 2**). These longitudinal motions are slow because they involve the entire filament. But these back-and-forth movements result in no net displacement of the filaments: The movements of the filaments are effectively curved 1D random walks. Finally, at timescales longer than a characteristic terminal relaxation time of the network, which depends on the length of the filament, the filament can finally escape the tube-like region (39, 70).

This description of the timescale-dependent movements of individual filaments in a dense network can be redescribed in rheological terms for the whole filament network following de Gennes' original concept (17). Consider an actin filament network subjected to a constant stress (force per unit area) of relatively small magnitude. At short timescales, the filaments can only relax the energetically unfavorable distortions, which are created in the polymer network by the externally applied stress, through lateral fluctuations (70). The creep compliance—or the deformation of the network—increases as a function of time as $t^{3/4}$ (114) (**Figure 2c**). The exponent 3/4 reflects the lateral bending fluctuations of the filaments (70, 74). At longer timescales, the network cannot relax anymore because no net filament motion occurs at these timescales. Accordingly, the creep compliance of the filament network becomes approximately constant (or a quasi plateau, **Figure 2c**) and the network behaves mostly as an elastic gel, as the elastic modulus becomes much larger than the viscous modulus. A stably cross-linked actin filament network behaves similarly, but at all timescales (44). At long timescales, the filaments can finally

diffuse out of their confining tubes, the network can relax the stress, and the creep compliance grows linearly with time, a proportionality that reflects viscous diffusion (21). The network becomes a viscoelastic liquid, for which the viscous modulus is much larger than the elastic modulus (24). If the actin filament network is stably cross-linked, the polymer cannot undergo this terminal relaxation and the plateau value of the compliance persists indefinitely. Often the transition from the plateau region to the terminal relaxation is not sharp because actin filaments have a wide distribution of lengths, which define a broad distribution of relaxation times.

A permanently cross-linked actin filament network provides cells with structural stability but does not allow them to readily soften to allow for cell shape changes that are, for instance, required during intravasation and extravasation of cancer cells in and out of blood vessels (110). To circumvent this problem and allow for a stiff but malleable cytoskeleton, cells have evolved an ingenious mechanism. This mechanism does not require dynamic disassembly and reassembly of the filaments themselves to modulate the mechanical properties of the network. Instead F-actin networks exploit dynamic cross-linking proteins, such as filamin or α-actinin (25, 99, 109, 110). When such a network is sheared more slowly than the lifetime of binding of the cross-linking protein, the network can flow and behaves mostly as a liquid. When sheared at a rate faster than the inverse binding lifetime of the cross-linking protein, the actin filament network behaves as an elastic gel and cannot flow (109, 110, 115).

BASIC PRINCIPLES OF PARTICLE-TRACKING MICRORHEOLOGY

Particle-Tracking Microrheology of a Viscous Liquid

To understand how the viscoelastic properties of the cytoplasm of cells can be obtained by tracking the movements of beads embedded in it, we consider first the simpler limit case of submicron beads suspended in a viscous liquid (no elasticity). These beads are smaller than 1 μm so that they undergo Brownian motion, as inertial forces (gravity) are negligible. If active transport of the nanoparticles is also negligible, only two types of forces act on the beads inside the cytoplasm:

- The small random force produced by the random bombardment of water molecules generated by the thermal energy $k_B T$ and the movements of other cytoplasmic structures, such as cytoskeleton filaments and other organelles (102).

- The counteracting frictional force, which results from the movement of the beads driven by thermal energy. The frictional force is proportional to the velocity of the bead and the bead's friction coefficient, which depends on the viscoelastic properties of the cytoplasm and the size of the bead.

Because we neglect inertia and directed motion, the mathematical equation describing the motion of a bead simply states that the sum of these two forces is zero. This resulting equation is not deterministic, but stochastic, because the force that powers the bead movements is random in amplitude and direction. The solution of this stochastic equation, i.e., the time-dependent coordinates of the bead, is the conventional random walk (5). Therefore, on average, the bead remains at the same position. However, the standard deviation of the displacements or the mean squared displacement (MSD) of the bead, $\langle \Delta r^2 \rangle$, is not zero.

In a liquid of unknown shear viscosity, η, the submicron bead undergoes Brownian motion driven by the thermal energy $k_B T$. Each time the bead takes a step in a random direction, it loses all memory of where it just came from. The next step occurs in an uncorrelated direction. Einstein showed that, in these conditions, the bead undergoes a random walk and its time-averaged MSD is simply $\langle \Delta r^2 \rangle = 4Dt$ (**Figure 3a**). Here t is time lag, $\langle ... \rangle$ indicates time-averaging, and D is the diffusion coefficient of the bead. The linear dependence of the MSD on time lag is a signature of pure viscous diffusion of the bead (81). Equivalently, the square-root dependence of the root MSD with time, $\sqrt{\langle \Delta r^2 \rangle} = \sqrt{4Dt} \sim t^{1/2}$, is a signature of viscous diffusion (5). For a spherical bead, the diffusion coefficient is given by the relation $D = k_B T/\xi$ (Stokes-Einstein relation), where $\xi = 6\pi \eta a$ is the friction coefficient of the bead in the liquid and a is the radius of the bead. After rearrangements, one finds that the

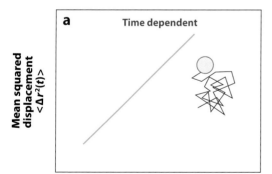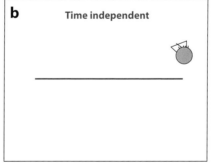

Figure 3

Particle-tracking microrheology of a viscous liquid and an elastic solid. (*a*) The mean squared displacement (MSD) of beads immersed in a viscous liquid (such as water or glycerol) grows linearly with time lag, with a slope inversely proportional to the viscosity of the liquid and the radius of the bead. (*b*) The MSD of beads embedded in an elastic solid (such as rubber) is independent of time lag. The plateau value is inversely proportional to the elasticity of the solid.

viscosity of the liquid can be estimated as soon as the MSD of the bead is measured (114):

$$\eta = \frac{2k_B T}{3\pi a} \frac{t}{\langle \Delta r^2(t) \rangle}.$$ 1.

This expression explains how one can estimate the viscosity of a liquid merely by suspending small beads in that liquid, subsequently tracking their thermally excited random motion using fluorescence light microscopy and appropriate particle-tracking software, and finally computing the MSD from the trajectories of the beads.

In practice, tens to hundreds of beads need to be tracked to ensure adequate statistical averaging. In this case, the ensemble-averaged MSD, $\langle\langle \Delta r^2(t) \rangle\rangle$, which is the means of all measured MSDs, is used instead of $\langle \Delta r^2 \rangle$. If the suspending fluid is indeed a liquid, then $t/\langle \Delta r^2(t) \rangle$ should be a constant independent of t (see Equation 1), which is a stringent test of viscous diffusion. If $t/\langle \Delta r^2(t) \rangle$ is nonconstant, it usually indicates that the suspending fluid is a viscoelastic material instead of a viscous liquid (65).

We consider a 100-nm-diameter bead suspended in corn syrup, which has a shear viscosity close to that of the cytoplasm of fibroblasts, $\eta = 20$ P $= 2$ Pa.s (typical values of viscosity are listed in **Table 1**). The thermal energy is $k_B T \approx 4.2$ pN.nm (where $T = 37°C$) and the diffusion coefficient of the bead is $D \approx 0.0022$ μm^2/s. By comparison, the diffusion of the same bead in water is 4.4 μm^2/s. Therefore, the value of the time-averaged MSD of the bead after $t = 0.1$ s, 1 s, and 10 s of tracking is ≈ 0.00088 μm^2 (or 88 nm^2), 0.0088 μm^2, and 0.088 μm^2, respectively. This result indicates that the average size of the region crisscrossed by the Brownian movements of the bead after the same times is $\sqrt{\langle \Delta r^2 \rangle} = \sqrt{4Dt} \approx 30$, 94, and 300 nm, respectively. This result suggests that high-resolution tracking is required to obtain good measurements of MSDs of 100-nm beads in a highly viscous liquid within a 10-s-long tracking time.

The prefactor 4 in the expression $\langle \Delta r^2 \rangle = 4Dt$ stems from the fact that microscopy probes the 2D projection of an intrinsically 3D movement of the bead. This is correct only if the material in the proximity of the bead is isotropic, i.e., this material or liquid has the same physical properties in all three (x, y, and z) directions. To test whether this is likely to be true without tracking the movements of the beads in all three directions (19), one can compute the MSDs of the bead in the x and y directions independently. Indeed, from the measured time-dependent coordinates of the bead, $x(t)$ and $y(t)$, the MSD is given by $\langle \Delta r^2(t) \rangle = \langle [x(t) - x(0)]^2 + [y(t) - y(0)]^2 \rangle$, where $x(0)$ and $y(0)$ correspond to the initial position of the bead. If the material is isotropic, then $\langle [x(t) - x(0)]^2 \rangle = \langle [y(t) - y(0)]^2 \rangle$ and therefore $\langle \Delta r^2(t) \rangle = 2\langle [x(t) - x(0)]^2 \rangle = 2\langle [y(t) - y(0)]^2 \rangle$. If the MSDs of the bead in the x and y directions are indeed Brownian (corresponding to two independent 1D random walks), then $\langle [x(t) - x(0)]^2 \rangle = \langle [y(t) - y(0)]^2 \rangle = 2Dt$. Hence, $\langle \Delta r^2(t) \rangle = 4Dt$, and therefore it is likely that the MSDs of the bead in the z direction are also Brownian and equal to those in the x and y directions, which indicates that the viscoelastic properties of the cytoplasm in the vicinity of the bead are isotropic (38).

In the case of a viscous liquid, the elastic modulus of the liquid is of course zero, $G' = 0$, and it can be shown that the viscous modulus is simply $G'' = \eta\omega$. Here ω is the frequency (or rate) of deformation in an oscillatory mode of deformation and is equal to the inverse of the time lag t, $\omega = 1/t$. This means that if this liquid were placed in a rheometer and subjected to oscillatory deformations, then the elastic modulus would be negligible and the viscous modulus would increase linearly with the input frequency ω. Equivalent to the linear dependence of the MSD with time, that the viscous modulus of a given material increases linearly with the rate of deformation indicates that this material is a viscous liquid. This rheological response, i.e., $G' \approx 0$ and $G'' = \eta\omega$, is for instance that of an unpolymerized G-actin solution in vitro or that of the cytoplasm of a cell treated with an actin-depolymerizing drug (e.g., latrunculin A or cytochalasin D).

Table 1 Elasticity and shear viscosity of the cytoplasm of different types of cells measured by particle-tracking microrheology[a]

Cell type and condition; values of viscoelastic parameters	Average viscosity (poise)	Average elasticity at 1 Hz (dyne/cm^2)	Reference
Serum-starved Swiss3T3 fibroblast[b]	10 ± 3	50 ± 20	(49)
Serum-starved Swiss 3T3 fibroblast treated with LPA[c]	95 ± 20	120 ± 30	(49)
Serum-starved Swiss3T3 fibroblast subjected to shear flow[d]	300 ± 40	600 ± 50	(61)
Swiss 3T3 fibroblast at the edge of a wound[e]	45 ± 15	330 ± 30	(50)
Swiss 3T3 fibroblast treated with bradykinin[f]	22 ± 13	90 ± 20	(50)
Swiss 3T3 fibroblast treated with PDGF[g]	24 ± 8	190 ± 30	(50)
Mouse embryonic fibroblast (*Lmna*$^{+/+}$ MEF)[h]	18 ± 2	140 ± 30	(60)
MEF treated with latrunculin B[i]	NA	80 ± 4	(60)
MEF treated with nocodazole[j]	NA	50 ± 4	(60)
MEF deficient in lamin A/C (*Lmna*$^{-/-}$ MEF)	8 ± 1	60 ± 4	(60)
HUVEC on a planar 2D peptide matrix[k]	17 ± 1	130 ± 10	P. Panorchan, J.S.H. Lee, D. Wirtz, unpublished data
HUVEC on a planar 2D peptide matrix and treated with VEGF[l]	8 ± 1	100 ± 5	P. Panorchan, J.S.H. Lee, D. Wirtz, unpublished data
HUVEC inside a 3D peptide matrix[k]	14 ± 1	55 ± 4	(77)
HUVEC inside a 3D peptide matrix treated with VEGF[l]	18 ± 1	40 ± 3	(77)
HUVEC inside a 3D fibronectin matrix	NA	58 ± 6	(117)
HUVEC inside a 3D fibronectin matrix treated with inhibiting FN peptide	NA	20 ± 4	(117)
Single-cell *C. elegans* embryo[m]	10 ± 1	Negligible	(16)
Interphase nucleus of Swiss 3T3 fibroblast[n]	520 ± 50	180 ± 30	(103)

[a]Particle-tracking microrheology was used to study the mechanical properties of many varieties of cells under a wide range of conditions. Unless stated, all values of viscosity and elasticity in the table are for the cytoplasm. The elasticity was evaluated at a shear frequency of 1 s^{-1} (1 s^{-1} = 1 Hz) and the shear viscosity was estimated as the product of plateau modulus and the relaxation time. The plateau modulus is the value of the elastic modulus at intermediate frequencies where it reaches a quasi plateau value. The relaxation time is the inverse of the frequency where elastic and viscous moduli are equal. All measurements are mean ± sem. Unit conversions are 1 dyne/cm^2 = 0.1 Pa = 0.1 N/m^2 = 0.1 pN/µm^2. Pa, Pascal; pN, piconewton; NA, not available.

[b]Cells placed on 50 µg/ml fibronectin deposited on glass were serum starved for 48 h before measurements.

[c]Serum-starved cells placed on 50 µg/ml fibronectin deposited on glass were treated with 1 µg/ml lysophosphatidic acid (LPA), which activates Rho-mediated actomyosin contractility. LPA was applied 15 min before measurements.

[d]Cells were grown on 20 µg/ml fibronectin for 24 h and exposed to shear flow (wall shear stress, 9.4 dyne/cm^2) for 40 min before measurements.

[e]Cells in complete medium and grown on 50 µg/ml fibronectin to confluence were wounded to induce migration. Measurements were conducted 4 h after wounding.

[f]Cells in complete medium and grown on 50 µg/ml fibronectin were treated with 100 ng/ml bradykinin for 10 min before measurements.

[g]Cells in complete medium and grown on 50 µg/ml fibronectin were treated with 10 ng/ml platelet-derived growth factor (PDGF) for 10 min before measurements.

[h]Cells in complete medium were grown on glass.

[i]*Lmna*$^{+/+}$ mouse embryonic fibroblasts (MEFs) in complete medium grown on glass were treated with 5 µg/ml actin-filament disassembly drug latrunculin B.

[j]*Lmna*$^{+/+}$ MEFs in complete medium and grown on glass were treated with 5 µg/ml microtubule disassembly drug nocodazole.

[k]Cells in complete medium were placed in a 0.5% puramatrix gel. HUVEC, human umbilical vein endothelial cell.

[l]Cells in complete medium were placed in a 0.5% puramatrix gel and treated with 4 ng/ml vascular endothelial growth factor (VEGF) for 24 h prior to the measurements.

[m]Young *Caenorhabditis elegans* eggs were obtained by cutting gravid hermaphrodites from worms in egg salts. The nanoparticles were then microinjected into the syncytial gonads of gravid hermaphrodites.

[n]Cells in complete medium.

If the probed system were in equilibrium and perfectly uniform, tracking one bead for a long time (e.g., 1000 s) and dividing this time span into 100 equal time spans of 10 s should be equivalent to tracking 100 beads each for 10 s. Although it is impractical to track a single bead for a long time even in in vitro systems because of drift problems, this equivalency is correct for liquids such as water and glycerol, a signature of ergodicity. But it is incorrect in suspensions of actin filaments in vitro and in live cells, which are systems far from equilibrium and highly spatially heterogeneous (100, 104). Indeed, beads well dispersed in the cytoplasm show a wide distribution in MSDs (102).

We note that individual trajectories of the beads, even in water, are typically highly asymmetric in shape (38). Even the overall shape of a computer-generated random walk is highly asymmetric (5, 87), i.e., it is not a circle. The shape of the object encompassing an individual trajectory of a single bead in a liquid is an ellipse in the 2D space (e.g., a protein diffusing within the plasma membrane) and an ellipsoid in the 3D space (e.g., GFP diffusing in the cytoplasm) (36). However, when these trajectories are ensemble-averaged and the trajectories are superimposed, the object encompassing these individual trajectories is a circle in 2D and a sphere in 3D.

Particle-Tracking Microrheology of an Elastic Solid

The second limit example involves the same bead, but this time embedded in a highly elastic material of negligible viscosity, such as rubber. Each time the bead is driven by the thermal energy in a random direction, the surrounding material pushes back instantaneously with equal force in the opposite direction. Therefore, the MSD of the bead is finite but constant: $\langle \Delta r^2 \rangle = K$ at all timescales (**Figure 3b**). The independence of the MSD with time is a signature of purely elastic solid behavior. The viscous modulus of this elastic solid is of course $G''(\omega) = 0$. The elastic modulus of an elastic solid is simply $G'(\omega) = 2k_B T/3\pi a \langle \Delta r^2 \rangle = 2k_B T/3\pi a K$,

which is a constant independent of ω, i.e., the elastic modulus of an elastic solid is inversely proportional to the MSD of the beads embedded in it. This expression shows that one can compute the elastic modulus of an elastic solid from particle-tracking measurements of $\langle \Delta r^2 \rangle$ without imposing any oscillatory deformation. The fact that the elastic modulus of a given material is independent of the rate of deformation or frequency, ω, is a signature of purely elastic behavior. This would be the rheological response of a permanently cross-linked actin filament network (75) or Jell-O, where G' is approximately a constant independent of frequency and G'' is much smaller than G'.

Let us consider a 100-nm-diameter bead suspended in a stiff elastic material of frequency-independent elasticity of $G' = 10$ Pa $= 100$ dyne/cm^2. Typical values of elasticity are given in **Table 1**. Then, the time-averaged MSD of the bead in this material at all timescales is $\langle \Delta r^2 \rangle = 2k_B T/3\pi a G' \approx 0.002$ μm^2, which can readily be measured by fluorescence microscopy and appropriate particle-tracking software.

These two examples—viscous liquid and elastic solid—illustrate how the time dependence of the MSD of beads inside a given material can help rapidly diagnose key physical properties of the material, including whether it is a liquid (then, $\langle \Delta r^2 \rangle \sim t$ or $\langle \Delta r^2 \rangle/t$ is a constant) or an elastic solid (then, $\langle \Delta r^2 \rangle$ is a constant).

The results above indicate that to estimate the viscosity of a liquid, a spatial resolution of at least 30 nm is required when tracking 100-nm-diameter beads for 10 s using a camera that collects images at a rate of 10 frames per second. In practice, the spatial resolution on the displacements of the beads must be an order of magnitude better, approximately 3 nm for a 100-nm-diameter fluorescent bead. Such a high, subpixel spatial resolution can be obtained by using either quadrant detection or fluorescence video microscopy. The advantage of quadrant detection is its superior detection capability (<1 nm) (93), but quadrant detection can only track one or two beads at a time (28, 29, 116). This is a serious problem when it is desirable to monitor simultaneously the viscoelastic properties of various locations within the cell. When using high-magnification, high-numerical-aperture lenses (11), video microscopy allows one to track hundreds of beads at the same time (13, 86, 102) but has a maximum resolution of the order of 3–10 nm, depending on the experimental setup. The spatial resolution of the measurements will vary from one instrument to another as it also depends on the stability of the microscope system used to track the beads. For a material that is essentially elastic and with a maximum spatial resolution of 3 nm, the highest elasticity that can still be measured with accuracy using particle-tracking microrheology is approximately $G'_{max} = 2k_B T/3\pi a(3 \text{ nm})^2 \approx 1,900 \text{ Pa} \approx 19,000 \text{ dyne/cm}^2$ (88, 89), which is much higher than typical values of the elasticity of the cytoplasm. Therefore, using video microscopy and high-resolution tracking allows for the measurements of highly elastic materials.

Particle-Tracking Microrheology of a Viscoelastic Material

The rheological behavior of the cytoplasm or a reconstituted actin filament network is intermediate between the two limit behaviors of a viscous liquid and an elastic solid, as described above. Indeed, the cytoplasm is predominantly viscous at long timescales ($t > \tau_R$) or low rates of shear, $G''(\omega) \gg G'(\omega)$ for $\omega < \omega_R$, and predominantly elastic at short timescales or high rates of shear, $G''(\omega) \ll G'(\omega)$ for $\omega > \omega_R$, where ω_R is the inverse of the relaxation time of the cytoskeleton network, τ_R (**Figure 4**). At high frequencies, $\omega > \omega_e$, both G' and G'' are proportional to $\omega^{3/4}$, a power-law dependence that reflects the bending lateral fluctuations of the semiflexible filaments in the network (40, 90). The exponent would be 1/2 if the polymers constituting the cytoskeleton were flexible (21), but intermediate filaments, actin filaments, and microtubules in live cells are semiflexible (1). The frequency ω_e is approximately equal to the

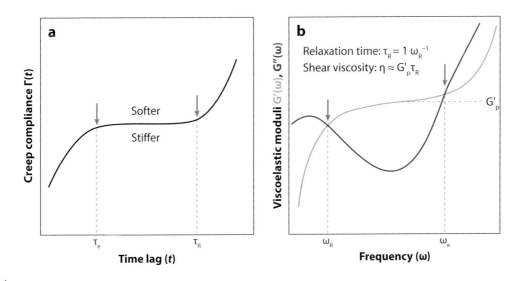

Figure 4

Creep compliance and viscoelastic moduli of the cytoplasm. (*a*) Typical creep compliance of the cytoplasm measured by the spontaneous displacements of beads embedded in the cytoplasm of a living cell (see also the inset in **Figure 2***c*). The creep compliance defines two characteristic timescales, τ_e and τ_R. (*b*) The frequency-dependent elastic (or storage) modulus, $G'(\omega)$, and viscous (or loss) modulus, $G''(\omega)$, of the cytoplasm can be approximately computed from the mean squared displacement of the beads or equivalently from the time-dependent creep compliance shown in panel *a*. In general, the cytoplasm behaves as a viscoelastic liquid at low frequencies (corresponding to long timescales, $t > \tau_R$): $G''(\omega) > G'(\omega)$ for $\omega < \omega_R$. The cytoplasm behaves as a viscoelastic solid at intermediate frequencies: $G'(\omega) > G''(\omega)$ for $\omega_e > \omega > \omega_R$. Finally, the cytoplasm behaves again as a viscous liquid at high frequencies: $\omega > \omega_e$, for which $G'(\omega) \sim G''(\omega) \sim \omega^{3/4}$.

inverse of the time required for lateral fluctuations to begin to touch the walls of the confining tube-like region (**Figure 4**).

In particle-tracking microrheology of living cells, the size of the probing beads is larger than the average mesh size of the cytoskeleton, typically 50 nm for fibroblasts (62, 63). Here again, without applying oscillatory deformations or mechanical stresses to the cytoplasm of the cells, the frequency-dependent elastic modulus and viscous modulus can be computed from the MSDs of embedded beads. In this case, the viscoelastic moduli can be obtained from the so-called complex modulus, $G^*(\omega)$, as (66)

$$G^*(\omega) = \frac{2k_B T}{3\pi a i \omega F_u[\Delta r^2(t)]}, \qquad 2.$$

where $F_u[\langle \Delta r^2(t) \rangle]$ is the unilateral Fourier transform of $\langle \Delta r^2(t) \rangle$. The above equation can be solved analytically (65), if we allow the frequency-dependent elastic modulus to be calculated algebraically using the following relationship:

$$G'(\omega) = |G^*(\omega)| \cos\left[\frac{\pi \alpha(\omega)}{2}\right], \qquad 3.$$

and the amplitude can be approximated as

$$|G^*(\omega)| \approx \frac{2k_B T}{3\pi a \langle \Delta r^2(1/\omega) \rangle \Gamma[1 + \alpha(\omega)]}. \qquad 4.$$

Here α is the local logarithmic slope of $\langle \Delta r^2(t) \rangle$ estimated at the frequency of interest and Γ is the gamma function.

Dynamic Viscosity Versus Shear Viscosity

The viscous modulus (also called the loss modulus), $G''(\omega)$, or alternatively the dynamic viscosity, $\eta''(\omega) = G''(\omega)/\omega$, and the shear viscosity, η, of a material ought not to be confused with each other. $G''(\omega)$ has units of pressure (Pa or dyne/cm^2) and represents the fraction of the energy induced by the imposed deformation, which is lost by viscous dissipation as opposed to stored elastically by the material. In a rheometer, it is computed by imposing an oscillatory deformation of controlled frequency and measuring the out-of-phase component of the

frequency-dependent stress induced in the material. In contrast, the shear viscosity η has units of Pa.s and is measured by imposing a steady deformation of constant rate. It is the ratio of the stress induced in the material to the rate of deformation.

Here again without applying any external forces to the cytoplasm, one can compute an approximate value of its shear viscosity η from the frequency-dependent viscoelastic moduli, $G'(\omega)$ and $G''(\omega)$:

$$\eta \approx G'_p \tau_R, \qquad 5.$$

where G'_p is the plateau value of the elastic modulus $G'(\omega)$ at intermediate frequencies (between ω_e and ω_R), and τ_R is the inverse of the frequency ω_R at which $G'(\omega)$ and $G''(\omega)$ are equal (see definitions in **Figure 4a,b**). Typical values of shear viscosity of the cytoplasm are given in **Table 1**.

Particle-tracking microrheology not only shows that the rheology of the cytoplasm is timescale dependent, but also that the frequency-dependent viscoelastic moduli define a single crossover frequency, ω_R, at low frequencies (**Figure 4b**). Below this frequency, the cytoplasm behaves essentially as a viscous liquid. Equivalently, in the time domain, if the cytoplasm were subjected to a step deformation, the time-dependent stress induced in the cytoplasm would adopt a plateau value at intermediate timescales, crossing over at a characteristic time τ_R ($= 1/\omega_R$) to a terminal viscous relaxation at long timescales (15).

Creep Compliance from Particle-Tracking Measurements

From a rheological point of view, the cytoplasm is a viscoelastic material. As suggested above, the mathematical transformation of the measured time-dependent MSD into frequency-dependent elastic and viscous moduli, $G'(\omega)$ and $G''(\omega)$, is not trivial (65, 66) (see Equations 2–4). This transformation involves Fourier/Laplace integrals that presume the knowledge of the MSD over an infinite range of timescales. Therefore, large errors can be introduced into the computation of $G'(\omega)$ and $G''(\omega)$ at low and high frequencies, corresponding to the maximum time of capture (at long timescales) and the rate of image capture of the camera for the MSD (at short timescales), respectively. However, the MSD of a bead is proportional to the creep compliance, $\Gamma(t)$, of the material in which the bead is embedded (114) (**Figures 2c** and **4a**):

$$\Gamma(t) = \frac{3\pi a}{2k_B T}\langle \Delta r^2(t)\rangle. \qquad 6.$$

The creep compliance of a material is its deformability and, in classical rheological measurements, is obtained by imposing a steady mechanical stress of constant magnitude and measuring the resulting deformation of the material. Here we can compute the creep compliance from particle-tracking measurements without imposing any deformations.

If we return to the examples of viscous liquid and elastic solid described above, then simple substitution of the expressions of $\langle \Delta r^2\rangle$, $\langle \Delta r^2\rangle = (2k_B T/3\pi a\eta)t$ and $\langle \Delta r^2\rangle = K$ in Equation 6 indicates that the creep compliance of a viscous liquid is $\Gamma(t) = t/\eta$, while the creep compliance of a highly elastic solid (negligible viscosity) is $\Gamma(t) = (3\pi a K/2k_B T) =$ constant. Therefore, the creep compliance of a viscous liquid is inversely proportional to the shear viscosity, $\Gamma(t) \sim 1/\eta$, and the creep compliance of an elastic solid is the inverse of the material's elasticity, $\Gamma = 1/G'$, where G' is a constant.

All the rheological information about the cytoplasm or the material in which the probing beads are embedded is contained in both the time dependence of creep compliance and its magnitude (94) (**Figure 4a**). To illustrate this, we return to the relatively simple example of a concentrated suspension of uncross-linked actin filaments, in which 100-nm-diameter beads have now been embedded. This bead is larger than the mesh size of the network. Hence the bead is in intimate physical contact with the filaments that, for a time, form a cage that confines the bead (**Figure 2a**). At short timescales, the motion of the bead is dictated by the lateral fluctuations of the filaments in contact with the bead (**Figure 2b**). Hence the MSD of the bead grows with time as $t^{3/4}$ (**Figure 2c**). An ultrafast camera would show that the bead undergoes random, small-magnitude, rapid displacements created by the lateral fluctuations of the confining filaments. At intermediate timescales, the bead does not undergo net motion anymore. A camera capturing images at that rate would not capture any net motion of the bead because, at intermediate timescales, the filaments surrounding it cannot get out of their confining tube-like region. Finally, at long timescales, the filaments surrounding the bead can escape their tube-like regions through large-magnitude longitudinal motion (**Figure 2b**). Accordingly, the filamentous cage that was confining the bead disappears because its constitutive filaments have moved out of the way. A slow camera would capture the bead going from cage to cage, undergoing a random walk—albeit at a slow pace set by the high viscosity of the actin filament solution. The MSD of the bead grows proportionally with time, a signature of viscous diffusion.

PARTICLE-TRACKING MICRORHEOLOGY OF CELLS

Interstitial Viscosity Versus Mesoscale Viscoelasticity of the Cytoplasm

The viscosity of the intracellular space depends critically on the length scale being probed. At small length scales, the viscosity of the cytoplasm is essentially that of water [more precisely, 1.2–1.4 times the viscosity of water (27)] or approximately 0.01 P (64, 91). It can be measured by monitoring the diffusion of GFP or small fluorescently labeled DNA fragments in the cell. This interstitial viscosity is a key parameter that controls the transport of small globular proteins and it varies slightly for various cell conditions.

At length scales larger than the effective mesh size of the cytoskeleton meshwork, the apparent viscosity is much higher, typically

>10 P, or 1000 times the viscosity of water. This effective mesh size can be measured by probing the diffusion of fluorescently labeled dextran or DNA inside the cell. The diffusion of dextran polymers with a gyration radius of <50 nm within the cytoplasm (of fibroblasts) is largely unhindered, whereas polymers with a gyration radius of >50 nm become immobilized (63, 64, 91). At length scales larger than the effective mesh size of the cytoplasm, the viscoelastic moduli and creep compliance computed from particle-tracking measurements should be independent of the size of the probe beads used in the experiments. This is readily seen from the expression in Equation 1 in the limit case of a viscous liquid: The creep compliance depends linearly on both bead size and MSD, which itself depends inversely on bead size.

The above discussion about how viscoelastic parameters of the cytoplasm can be obtained from particle-tracking measurements assumes that the probing beads are larger than the effective mesh size of the cytoplasm or the network being probed (**Figure 4a**). We call this viscosity mesoscale viscosity, as it describes the rheological properties at length scales intermediate between small globular proteins in the interstitial space and the whole cell. The mesoscale viscosity controls the rate of movements of mitochondria (53), nuclei (59), and phagosomes (96), as well as viruses, bacteria, and engineered drug delivery microcarriers (96), because these entities are larger than the effective mesh size of the cytoskeleton. The mesoscale viscosity of the cytoplasm also controls the rates of cell spreading and migration (35).

Active Versus Passive Microrheology

Direct injection of the beads into the cytoplasm (102), as opposed to passive engulfment of the beads inside the cytoplasm, circumvents the endocytic pathway and therefore the engulfment of the beads in vesicles tethered to cytoskeleton filaments by motor proteins (96). These vesicles move toward the nucleus; therefore, the probe beads undergo directed motion. Although interesting in their own right, such directed displacements prevent the computation of the viscoelastic parameters of the cytoplasm from MSDs.

Nevertheless, even in the absence of directed motion, actomyosin contractility could affect the movements of beads. Recent work with reconstituted actin filament networks containing myosin II has shown that in this in vitro system the movements of beads can be affected 10-fold, and even more at long timescales by the activity of motor proteins (69). Even without producing a net movement of the filaments, motor proteins enhance the random movements of the beads, akin to an effective >10-fold increase in the temperature of the system (69). These enhanced movements allow for faster relaxation of mechanical stresses. This result obtained with purified proteins suggests that particle-tracking measurements of the mechanical properties of the cytoplasm significantly underestimate the values of the viscoelastic properties of the cytoplasm in live cells.

However, our unpublished results suggest that reducing or eliminating actomyosin contractility in a live cell using either blebbistatin, which blocks the myosin heads in a complex with low actin affinity (52), or ML-7, which specifically inhibits MLCK (myosin light chain kinase), which normally phosphorylates myosin II, has no significant effect on the magnitude of the displacements of the beads compared with control cells. This absence of effect may result because myosin II molecules are mostly localized to the contractile actin stress fibers localized to the ventral side of the cell and the leading edge and are not present in large quantities in the body of the cell (108), where beads used for particle-tracking microrheology are lodged. Therefore, a direct extrapolation of results obtained in vitro with purified proteins, although often instructive of the more complex behavior of cytoskeleton in cells (79), can be misleading.

The movements of the cell itself can add an additional contribution to the overall movements of the probing beads. The speed of adherent cells (7) is typically much lower than the speed of the probing beads embedded in the cytoplasm. Therefore, the movements of the cell will be felt by the beads embedded in the cytoplasm only when they are monitored for a long time. The contribution of cell movements to the overall movements of the beads can readily be detected by the curvature of the MSDs of the embedded beads. To illustrate this effect, we consider the simple case of a viscous liquid. If an overall movement is added to the thermally driven random motion of the bead, then the MSD can be rewritten as (81)

$$\langle \Delta r^2 \rangle = 4Dt + v^2 t^2, \qquad 7.$$

where v is the overall speed of the suspending liquid. The first term in this expression corresponds to the random thermally driven motion of the bead, and the second term corresponds to the motion of the bead imposed by the overall movement of the liquid. At short timescales, the second term is negligible and we recover $\langle \Delta r^2 \rangle \approx 4Dt$, i.e., bead movements are dominated by viscous diffusion. At long timescales, the second term dominates, $\langle \Delta r^2 \rangle \approx v^2 t^2$, corresponding to a quadratic dependence on time, which is readily observed at long timescales in the MSD profile. The timescale at which the crossover between these two regimes occurs (when $4Dt = v^2 t^2$) is $t = 4D/v^2$. For instance, for a 100-nm-diameter bead in a 20 P viscous liquid, $D \approx 0.0022 \ \mu m^2/s$, moving at a constant speed of 5 μm/h (a typical speed for a cultured adherent cell), the crossover time is equal to $t = 4D/v^2 = 1.3$ s, which is readily detected when tracking a bead for 10 s.

Advantages of Particle-Tracking Microrheology Over Current Methods

Particle-tracking microrheology can measure directly the mechanical properties of the cytoplasm because of the intimate contact between the probing beads and subcellular structures (103). In contrast, most current single-cell mechanics methods rely on a direct contact between the cell surface and a physical probe. For example, AFM probes the mechanical properties of cells using soft cantilevers (41); magnetocytometry (106, 111) measures cell mechanics by subjecting large beads coated with extracellular matrix molecules bound to cell receptors to rotational movements. These methods cannot distinguish the contribution of the plasma membrane from the combined response of the nucleus and the cytoplasm without making drastic assumptions. If two elastic elements of different stiffness are connected to one another (here the nucleus and the cytoskeleton), their total response is dominated by the stiffer element. Therefore, methods such as AFM, magnetocytometry, or micropipette suction probe the combined response of the nucleus and cytoplasm, even when the probe is positioned far away from the nucleus. In contrast, particle-tracking microrheology, which relies on the smallest energy possible—thermal energy—is a highly localized measurement inside the cytoplasm.

Unlike most other approaches to cell mechanics, particle-tracking microrheology measures frequency-rate-dependent viscoelastic moduli, $G'(\omega)$ and $G''(\omega)$. This is particularly crucial for the cytoskeleton, which behaves like a liquid at long timescales (or low rates of shear) and like an elastic solid at short timescales (or high rates of shear). The frequency-dependent mechanical response of the cytoplasm of a live cell can be measured in about 10 s by particle-tracking microrheology, at least for timescales relevant to cell motility. In contrast, AFM measurements may take up to 1 h for high-resolution mechanical measurements (82). Such time-consuming measurements cannot capture fast subcellular dynamics and are unsuitable for motile cells.

Particle-tracking microrheology requires short times of data collection, typically 10–20 s, to measure frequency-dependent viscoelastic moduli over two decades in frequency (or timescales). This is not the case for other particle-tracking methods for which the correlated motion between two particles is used instead of the motion of individual particles as a probe of local cytoplasmic rheology (13). These methods are inappropriate for the measurement of the mechanical properties of live cells because they incorrectly assume that the intracellular milieu is homogeneous and static. Furthermore, to obtain statistically significant data, measurement times are on the order of 30–60 min, timescales for which migrating cells cannot be assumed to be stationary.

By tracking multiple beads simultaneously, microrheology can measure simultaneously the micromechanical responses to stimuli in various parts of the cell. This is particularly important because the viscoelastic properties can vary by more than 2 orders of magnitude within the same cell (51, 102, 103). By using video-based, multiple-particle tracking instead of laser-deflection particle tracking (65, 116), hundreds of beads embedded in the cytoplasm can be tracked at the same time.

Particle-tracking microrheology measurements of the cytoplasm of live cells are typically conducted using carboxylated polystyrene beads or polyethylene glycol (PEG)-coated beads. The rheology of reconstituted actin filament networks or DNA solutions measured by particle-tracking microrheology and using these beads is quantitatively similar to that measured by rheometry (75, 114). This indicates that the presence of the beads in polymer networks does not affect the global mechanical properties of the networks. Moreover, our unpublished results show that carboxylated polystyrene beads or PEG-coated beads yield the same frequency-dependent viscoelastic moduli for the cytoplasm of fibroblasts, but not for amino-modified beads (102). Finally, the proximity of some of the beads to the plasma membrane will not significantly affect the movements of the beads because hydrodynamic interactions caused by the movements of the beads and reflecting off the membrane are effectively screened by the cytoskeleton mesh (21, 116). Subcellular organelles can also be used as probing particles (40, 86, 116). However, because their interactions with subcellular structures are ill-defined, only the frequency dependence of the viscoelastic moduli is meaningful, at least at high frequency, not the absolute value of moduli.

Microrheological measurements are absolute and compare favorably with traditional rheometric measurements of standard fluids of known viscosity and elasticity (2, 65, 112, 114). This is not the case of some single-cell approaches that rely on a direct physical contact between the cell surface and the probe (such as a cantilever or a macroscopic bead). For instance, the apparent viscoelastic moduli measured by magnetocytometry and AFM depend greatly on the type of ligands coated on the magnetic beads or the AFM cantilever. Extracellular matrix ligands—including fibronectin or RGD peptide—coated onto magnetic beads and AFM cantilevers lead to vastly different values of (apparent) cell stiffness. Therefore, the measurements of viscoelastic properties of standard materials using these methods cannot be readily compared with those obtained using a rheometer. Furthermore, particle-tracking microrheology measures both elasticity and viscosity, whereas many other approaches cannot directly distinguish the elastic response from the viscous response of the cell.

Importantly, values of shear viscosity and elastic moduli of the cytoplasm in live cells measured by particle-tracking microrheology (**Table 1**) are similar to those measured in reconstituted actin filament networks in the presence of cross-linking/bundling proteins (**Table 2**). The elasticity of live cells, such as mouse embryonic fibroblasts or human umbilical vein endothelial cells (HUVECs), which are commonly used as models of cell biology, and of reconstituted actin filament networks are on the order of tens of Pascal (hundreds of dyne per cm^2). This is in striking contrast to values of cell elasticity obtained using AFM, which is of the order of hundreds and even thousands of Pascal (82). To our knowledge, no actin filament network reconstituted in vitro that contains a physiological concentration of actin polymerized in the presence of cross-linking and bundling proteins and under mechanical tension (which can further increase the network elasticity) reaches these high levels of elasticity (99). The discrepancy between values of cell elasticity measured by AFM and those obtained

Table 2 Viscosity and elasticity of common liquids and cytoskeletal filament networks measured by a cone-and-plate rheometer

Type of material; values of viscoelastic parameters	Average viscosity (poise)	Average elasticity at 1 rad/s (dyne/cm^2)	Reference
Water	0.01	0	–
Blood	0.1	Negligible	–
Glycerol	~1	Negligible	–
Corn syrup	~20	Negligible	–
Ketchup	~500	Negligible	–
Jell-O	Negligible	1,000	–
Polyacrylamide gel	–	500	(26)
F-actin network	–	8 ± 3	(113)
F-actin + filamin	–	450 ± 60	(99)
F-actin + α-actinin	–	120 ± 20	(113)
F-actin + Arp2/3 complex/WASp	–	60 ± 15	(105)
F-actin + fascin	–	80 ± 10	(101)
F-actin + fimbrin	–	300 ± 30	(48)
Vimentin network	–	14 ± 2	(23)

[a]Measurements are mean ± sem. The elasticity was measured at a shear amplitude of 1% and a shear frequency of 1 rad/s. Shear viscosity of the F-actin and vimentin networks was not measured because these filaments break under continuous shear.

[b]The concentrations of actin and vimentin solutions are 24 μM. The concentrations of α-actinin, fascin, fimbrin, and filamin in solutions are 0.24 μM. The concentration of the Arp2/3 complex is 0.12 μM and that of its activator WASp is 0.06 μM. The concentration of acrylamide and bis-acrylamide in the polyacrylamide gel is 0.04% and 0.05%, respectively. Unit conversions are 1 P = 0.1 Pa.s; 1 dyne/cm^2 = 0.1 Pa.

with actin filament networks by conventional rheometry or particle-tracking microrheology remains unexplained.

Applications of particle-tracking microrheology discussed in the Introduction were rendered possible thanks to the replacement of manual microinjection of beads into cells with ballistic injection (61, 76, 77). In a single ballistic injection, the number of injected cells amenable to measurements increases 1000-fold compared with conventional microinjection (61, 77). With a large sample size per condition (typical number of probed cells is ~30), particle-tracking microrheology results become more precise and significant. Ballistic transfer of beads to the cytoplasm coupled with particle tracking (named ballistic injection nanorheology, or BIN) provides a more precise and consistent value for global and local viscoelastic properties.

Illustrative Example of Particle-Tracking Microrheology of Living Cells

In typical measurements of cell mechanics (e.g., AFM) (82–84), it is assumed that the nucleus and the cytoplasm contribute independently to global cell stiffness. This assumption is an oversimplification that overlooks the existence of critical functional links between the nucleus and the cytoskeleton. These connections are established by specific linker proteins located at the nuclear envelope, including the LINC complex (10), and they physically connect the filamentous nuclear lamina underneath the nuclear envelope to the actin/microtubule/intermediate filament cytoskeleton in the cytoplasm. These physical connections between nucleus and cytoskeleton are as important to cell mechanics as the intrinsic mechanical properties of the nucleus and the cytoskeleton themselves. Indeed, when these functional links are disrupted, the elasticity of the cytoplasm is drastically reduced; the magnitude of this reduction is as important as when the actin cytoskeleton is completely disassembled with drugs (60).

Particle-tracking microrheology shows that the loss of the LINC complex dramatically affects the ability of cells to resist mechanical shear forces (95). Mutations scattered along *Lmna*, which encodes A-type lamins, have been associated with a broad range of human diseases collectively called laminopathies (4, 8, 9, 33, 42, 71). These diseases involve either specific or combined pathologies of neurons, muscle, and bone tissue (4, 42). Cytoplasmic fragility as measured by particle-tracking microrheology correlates with the loss of the LINC complex from the nuclear envelope in cells derived from mouse models of laminopathy (37, 60, 95) (**Figure 5**).

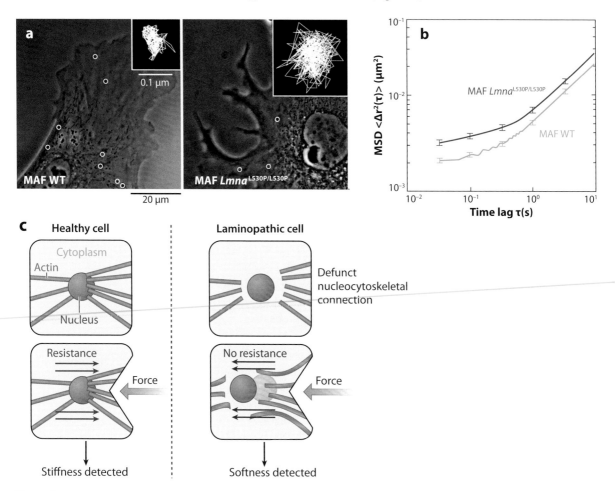

Figure 5

Intracellular microrheology of laminopathic fibroblasts. (*a*) Fluorescent 100-nm-diameter polystyrene beads are ballistically injected into wild-type (WT) (*left panel*) and *Lmna*[L530P/L530P] (*right panel*) mouse adult fibroblasts (MAFs), derived from a mouse model of progeria. Fluorescent micrographs of beads (*outlined in white circles*) are superimposed onto phase-contrast micrographs of cells. Representative trajectories of nanoparticles are shown at the top right of each micrograph (*inset*). (*b*) Ensemble-averaged mean squared displacements (MSDs) of the beads embedded in the cytoplasm of WT (*blue curve*) and *Lmna*[L530P/L530P] (*red curve*) MAFs are shown. (*c*) Model for the intracellular mechanics of healthy and laminopathic cells. Healthy cells in which the nucleocytoskeletal connections are intact can resist forces of large magnitude, whereas laminopathic cells cannot resist such forces because of defunct nucleocytoskeletal connections. Modified with permission from Reference 37.

SUMMARY POINTS

1. The viscoelastic properties of the cytoplasm and nucleoplasm control key cell functions, including cell motility, cell adhesion, and the regulated movements of organelles.

2. Particle-tracking microrheology can measure the viscoelastic properties of living cells without subjecting them to applied forces. In this approach, fluorescent submicron beads are injected directly into the cytoplasm of live cells. These beads rapidly disperse throughout the cytoplasm and are subsequently tracked by fluorescence microscopy. Analysis of the recorded movements of the beads transforms the MSDs of the beads in terms of either viscosity and elasticity or the compliance of the cytoplasm.

3. Particle-tracking microrheology can reveal the micromechanical properties of cells in more physiological environments, such as a 3D extracellular matrix or inside an embryo, better than traditional approaches to cell mechanics.

4. Particle-tracking microrheology measurements, which are fast, acknowledge that the mechanical properties of the cytoplasm are spatially variable and depend on timescale.

5. Particle-tracking microrheology reveals the reduced ability of cells from mouse models of laminopathies to resist mechanical shear forces.

FUTURE ISSUES

1. The levels of viscosity, elasticity, and compliance of the cytoplasm, as measured by particle-tracking microrheology, should be tested to determine whether they can serve as reliable diagnostic markers of disease, in particular cancer and laminopathies.

2. To circumvent the need of beads to probe cytoplasmic microrheology, exogenous beads should be rationally replaced with endogenous subcellular markers.

3. Researchers should assess quantitatively the role of motor proteins (e.g., myosins) on the movements of beads lodged in the cytoskeleton of living cells and therefore the computation of viscoelastic parameters.

4. Can the wide range of phenotypes associated with human laminopathy be explained by the differential mechanical response at the cellular level?

5. A high-throughput device based on particle-tracking microrheology would allow the method to become a diagnostic tool for human disease.

6. Particle-tracking microrheology should be extended to in vivo measurements.

DISCLOSURE STATEMENT

The author is not aware of any biases that might be perceived as affecting the objectivity of this review.

ACKNOWLEDGMENTS

The author thanks Yiider Tseng, Thomas P. Kole, Jerry S.H. Lee, Porntula Panorchan, Brian R. Daniels, and Melissa S. Thompson for key contributions to the development of particle-tracking microrheology and its applications in cell biology, as well as Didier Hodzic, Christopher M. Hale, Shyam B. Khatau, Philip Stewart-Hutchinson, and Colin L. Stewart, who contributed to the work on nucleocytoskeleton protein linkers. The author also thanks members of the Wirtz group for reading and editing the manuscript. Work in the author's laboratory is supported by the National Institutes of Health (NIGMS, NIBIB, and NCI), the Howard Hughes Medical Institute, and the American Heart Association.

LITERATURE CITED

1. Alberts B, Bray D, Lewis J, Raff M, Roberts K, Watson JD. 1994. *Molecular Biology of the Cell*. New York: Garland. 1408 pp.
2. Apgar J, Tseng Y, Federov E, Herwig MB, Almo SC, Wirtz D. 2000. Multiple-particle tracking measurements of heterogeneities in solutions of actin filaments and actin bundles. *Biophys. J.* 79:1095–106
3. Bausch AR, Kroy K. 2006. A bottom-up approach to cell mechanics. *Nat. Phys.* 2:231–38
4. Ben Yaou R, Muchir A, Arimura T, Massart C, Demay L, et al. 2005. Genetics of laminopathies. *Novartis Found. Symp.* 264:81–90
5. Berg HC. 1993. *Random Walks in Biology*. Princeton, NJ: Princeton Univ. Press. 152 pp.
6. Bloom RJ, George JP, Celedon A, Sun SX, Wirtz D. 2008. Mapping local matrix remodeling induced by a migrating tumor cell using three-dimensional multiple-particle tracking. *Biophys. J.* 95:4077–88
7. Bray D. 2001. *Cell Movements: From Molecules to Motility*. New York: Garland. 371 pp.
8. Broers JL, Hutchison CJ, Ramaekers FC. 2004. Laminopathies. *J. Pathol.* 204:478–88
9. Capell BC, Collins FS. 2006. Human laminopathies: Nuclei gone genetically awry. *Nat. Rev. Genet.* 7:940–52
10. Crisp M, Liu Q, Roux K, Rattner JB, Shanahan C, et al. 2006. Coupling of the nucleus and cytoplasm: role of the LINC complex. *J. Cell Biol.* 172:41–53
11. Crocker JC, Grier DG. 1996. Methods of digital video microscopy for colloidal studies. *J. Colloid Interface Sci.* 179:298–310
12. Crocker JC, Hoffman BD. 2007. Multiple-particle tracking and two-point microrheology in cells. *Methods Cell Biol.* 83:141–78
13. Crocker JC, Valentine MT, Weeks ER, Gisler T, Kaplan PD, et al. 2000. Two-point microrheology of inhomogeneous soft materials. *Phys. Rev. Lett.* 85:888–91
14. Cross SE, Jin YS, Rao J, Gimzewski JK. 2007. Nanomechanical analysis of cells from cancer patients. *Nat. Nanotechnol.* 2:780–83
15. Dahl KN, Engler AJ, Pajerowski JD, Discher DE. 2005. Power-law rheology of isolated nuclei with deformation mapping of nuclear substructures. *Biophys. J.* 89:2855–64
16. Daniels BR, Masi BC, Wirtz D. 2006. Probing single-cell micromechanics in vivo: the microrheology of *C. elegans* developing embryos. *Biophys. J.* 90:4712–19
17. de Gennes P-G. 1991. *Scaling Concepts in Polymer Physics*. Ithaca, NY: Cornell Univ. Press
18. de Gennes P-G, Leger L. 1982. Dynamics of entangled polymer chains. *Annu. Rev. Phys. Chem.* 33:49–61
19. Desai KV, Bishop TG, Vicci L, O'Brien ET Sr, Taylor RM 2nd, Superfine R. 2008. Agnostic particle tracking for three-dimensional motion of cellular granules and membrane-tethered bead dynamics. *Biophys. J.* 94:2374–84
20. Discher DE, Janmey P, Wang YL. 2005. Tissue cells feel and respond to the stiffness of their substrate. *Science* 310:1139–43
21. Doi M, Edwards SF. 1989. *The Theory of Polymer Dynamics*. Oxford: Clarendon. 391 pp.
22. Engler AJ, Sen A, Sweeney HL, Discher DE. 2006. Matrix elasticity directs stem cell lineage specification. *Cell* 126:677–89
23. Esue O, Carson AA, Tseng Y, Wirtz D. 2006. A direct interaction between actin and vimentin filaments mediated by the tail domain of vimentin. *J. Biol. Chem.* 281:30393–99
24. Ferry JD. 1980. *Viscoelastic Properties of Polymers*. New York: Wiley. 672 pp.
25. Flanagan LA, Chou J, Falet H, Neujahr R, Hartwig JH, Stossel TP. 2001. Filamin A, Arp2/3 complex, and the morphology and function of cortical actin filaments in human melanoma cells. *J. Cell Biol.* 155:511–18

26. Flanagan LA, Ju YE, Marg B, Osterfield M, Janmey PA. 2002. Neurite branching on deformable substrates. *NeuroReport* 13:2411–17

27. Fushimi K, Verkman AS. 1991. Low viscosity in the aqueous domain of cell cytoplasm measured by picosecond polarization microfluorimetry. *J. Cell Biol.* 112:719–25

28. Girard KD, Chaney C, Delannoy M, Kuo SC, Robinson DN. 2004. Dynacortin contributes to cortical viscoelasticity and helps define the shape changes of cytokinesis. *EMBO J.* 23:1536–46

29. Girard KD, Kuo SC, Robinson DN. 2006. *Dictyostelium* myosin II mechanochemistry promotes active behavior of the cortex on long time scales. *Proc. Natl. Acad. Sci. USA* 103:2103–8

30. Gittes F, Mickey B, Nettleton J, Howard J. 1993. Flexural rigidity of microtubules and actin filaments measured from thermal fluctuations in shape. *J. Cell Biol.* 120:923–34

31. Gomes ER, Jani S, Gundersen GG. 2005. Nuclear movement regulated by Cdc42, MRCK, myosin, and actin flow establishes MTOC polarization in migrating cells. *Cell* 121:451–63

32. Gorisch SM, Wachsmuth M, Ittrich C, Bacher CP, Rippe K, Lichter P. 2004. Nuclear body movement is determined by chromatin accessibility and dynamics. *Proc. Natl. Acad. Sci. USA* 101:13221–26

33. Gruenbaum Y, Margalit A, Goldman RD, Shumaker DK, Wilson KL. 2005. The nuclear lamina comes of age. *Nat. Rev. Mol. Cell. Biol.* 6:21–31

34. Guck J, Schinkinger S, Lincoln B, Wottawah F, Ebert S, et al. 2005. Optical deformability as an inherent cell marker for testing malignant transformation and metastatic competence. *Biophys. J.* 88:3689–98

35. Gupton SL, Anderson KL, Kole TP, Fischer RS, Ponti A, et al. 2005. Cell migration without a lamellipodium: translation of actin dynamics into cell movement mediated by tropomyosin. *J. Cell Biol.* 168:619–31

36. Haber C, Ruiz SA, Wirtz D. 2000. Shape anisotropy of a single random-walk polymer. *Proc. Natl. Acad. Sci. USA* 97:10792–95

37. Hale CM, Shrestha AL, Khatau SB, Stewart-Hutchinson PJ, Hernandez L, et al. 2008. Dysfunctional connections between the nucleus and the actin and microtubule networks in laminopathic models. *Biophys. J.* 95:5462–75

38. Hasnain IA, Donald AM. 2006. Microrheological characterization of anisotropic materials. *Phys. Rev. E* 73:031901

39. Hinner B, Tempel M, Sackmann E, Kroy K, Frey E. 1998. Entanglement, elasticity, and viscous relaxation of actin solutions. *Phys. Rev. Lett.* 81:2614–17

40. Hoffman BD, Massiera G, Van Citters KM, Crocker JC. 2006. The consensus mechanics of cultured mammalian cells. *Proc. Natl. Acad. Sci. USA* 103:10259–64

41. Hoh JH, Schoenenberger CA. 1994. Surface morphology and mechanical properties of MDCK monolayers by atomic force microscopy. *J. Cell Sci.* 107(Pt. 5):1105–14

42. Jacob KN, Garg A. 2006. Laminopathies: multisystem dystrophy syndromes. *Mol. Genet. Metab.* 87:289–302

43. Janmey PA. 1991. Mechanical properties of cytoskeletal polymers. *Curr. Opin. Cell Biol.* 2:4–11

44. Janmey PA, Hvidt S, Lamb J, Stossel TP. 1990. Resemblance of actin-binding protein/actin gels to covalently crosslinked networks. *Nature* 345:89–92

45. Jonas M, Huang H, Kamm RD, So PT. 2008. Fast fluorescence laser tracking microrheometry. II. Quantitative studies of cytoskeletal mechanotransduction. *Biophys. J.* 95:895–909

46. Kas J, Strey H, Sackmann E. 1994. Direct imaging of reptation for semiflexible actin filaments. *Nature* 368:226–29

47. Kas J, Strey H, Tang JX, Fanger D, Sackmann E, Janmey P. 1996. F-actin, a model polymer for semiflexible chains in dilute, semidilute, and liquid crystalline solutions. *Biophys. J.* 70:609–25

48. Klein MG, Shi W, Ramagopal U, Tseng Y, Wirtz D, et al. 2004. Structure of the actin crosslinking core of fimbrin. *Structure* 12:999–1013

49. Kole TP, Tseng Y, Huang L, Katz JL, Wirtz D. 2004. Rho kinase regulates the intracellular micromechanical response of adherent cells to rho activation. *Mol. Biol. Cell* 15:3475–84

50. Kole TP, Tseng Y, Jiang I, Katz JL, Wirtz D. 2005. Intracellular mechanics of migrating fibroblasts. *Mol. Biol. Cell* 16:328–38

51. Kole TP, Tseng Y, Wirtz D. 2004. Intracellular microrheology as a tool for the measurement of the local mechanical properties of live cells. *Methods Cell Biol.* 78:45–64

52. Kovacs M, Toth J, Hetenyi C, Malnasi-Csizmadia A, Sellers JR. 2004. Mechanism of blebbistatin inhibition of myosin II. *J. Biol. Chem.* 279:35557–63

53. Lacayo CI, Theriot JA. 2004. *Listeria* monocytogenes actin-based motility varies depending on subcellular location: a kinematic probe for cytoarchitecture. *Mol. Biol. Cell* 15:2164–75

54. Lammerding J, Lee RT. 2005. The nuclear membrane and mechanotransduction: impaired nuclear mechanics and mechanotransduction in lamin A/C deficient cells. *Novartis Found. Symp.* 264:264–73

55. Lammerding J, Schulze PC, Takahashi T, Kozlov S, Sullivan T, et al. 2004. Lamin A/C deficiency causes defective nuclear mechanics and mechanotransduction. *J. Clin. Invest.* 113:370–78

56. Lammermann T, Bader BL, Monkley SJ, Worbs T, Wedlich-Soldner R, et al. 2008. Rapid leukocyte migration by integrin-independent flowing and squeezing. *Nature* 453:51–55

57. Lau AW, Hoffman BD, Davies A, Crocker JC, Lubensky TC. 2003. Microrheology, stress fluctuations, and active behavior of living cells. *Phys. Rev. Lett.* 91:198101

58. Lee GY, Lim CT. 2007. Biomechanics approaches to studying human diseases. *Trends Biotechnol.* 25:111–18

59. Lee JS, Chang MI, Tseng Y, Wirtz D. 2005. Cdc42 mediates nucleus movement and MTOC polarization in Swiss 3T3 fibroblasts under mechanical shear stress. *Mol. Biol. Cell* 16:871–80

60. Lee JS, Hale CM, Panorchan P, Khatau SB, George JP, et al. 2007. Nuclear lamin A/C deficiency induces defects in cell mechanics, polarization, and migration. *Biophys. J.* 93:2542–52

61. Lee JS, Panorchan P, Hale CM, Khatau SB, Kole TP, et al. 2006. Ballistic intracellular nanorheology reveals ROCK-hard cytoplasmic stiffening response to fluid flow. *J. Cell Sci.* 119:1760–68

62. Luby-Phelps K, Castle PE, Taylor DL, Lanni F. 1987. Hindered diffusion of inert tracer particles in the cytoplasm of mouse 3T3 cells. *Proc. Natl. Acad. Sci. USA* 84:4910–13

63. Luby-Phelps K, Taylor DL, Lanni F. 1986. Probing the structure of cytoplasm. *J. Cell Biol.* 102:2015–22

64. Lukacs GL, Haggie P, Seksek O, Lechardeur D, Freedman N, Verkman AS. 2000. Size-dependent DNA mobility in cytoplasm and nucleus. *J. Biol. Chem.* 275:1625–29

65. Mason TG, Ganesan K, van Zanten JV, Wirtz D, Kuo SC. 1997. Particle-tracking microrheology of complex fluids. *Phys. Rev. Lett.* 79:3282–85

66. Mason TG, Weitz D. 1995. Optical measurements of frequency-dependent linear viscoelastic moduli of complex fluids. *Phys. Rev. Lett.* 74:1254–56

67. Massiera G, Van Citters KM, Biancaniello PL, Crocker JC. 2007. Mechanics of single cells: rheology, time dependence, and fluctuations. *Biophys. J.* 93:3703–13

68. Minin AA, Kulik AV, Gyoeva FK, Li Y, Goshima G, Gelfand VI. 2006. Regulation of mitochondria distribution by RhoA and formins. *J. Cell Sci.* 119:659–70

69. Mizuno D, Tardin C, Schmidt CF, Mackintosh FC. 2007. Nonequilibrium mechanics of active cytoskeletal networks. *Science* 315:370–73

70. Morse DC. 1998. Viscoelasticity of concentrated isotropic solutions of semiflexible polymers. 1. Model and stress tensor. *Macromolecules* 31:7030–43

71. Mounkes L, Kozlov S, Burke B, Stewart CL. 2003. The laminopathies: nuclear structure meets disease. *Curr. Opin. Genet. Dev.* 13:223–30

72. O'Toole M, Lamoureux P, Miller KE. 2008. A physical model of axonal elongation: force, viscosity, and adhesions govern the mode of outgrowth. *Biophys. J.* 94:2610–20

73. Pai A, Sundd P, Tees DF. 2008. In situ microrheological determination of neutrophil stiffening following adhesion in a model capillary. *Ann. Biomed. Eng.* 36:596–603

74. Palmer A, Xu J, Kuo SC, Wirtz D. 1999. Diffusing wave spectroscopy microrheology of actin filament networks. *Biophys. J.* 76:1063–71

75. Palmer A, Xu J, Wirtz D. 1998. High-frequency rheology of crosslinked actin networks measured by diffusing wave spectroscopy. *Rheol. Acta* 37:97–108

76. Panorchan P, Lee JS, Daniels BR, Kole TP, Tseng Y, Wirtz D. 2007. Probing cellular mechanical responses to stimuli using ballistic intracellular nanorheology. *Methods Cell Biol.* 83:115–40

77. Panorchan P, Lee JS, Kole TP, Tseng Y, Wirtz D. 2006. Microrheology and ROCK signaling of human endothelial cells embedded in a 3D matrix. *Biophys. J.* 91:3499–507

78. Pelham RJ Jr, Wang Y. 1997. Cell locomotion and focal adhesions are regulated by substrate flexibility. *Proc. Natl. Acad. Sci. USA* 94:13661–65

79. Pollard TD. 2003. The cytoskeleton, cellular motility and the reductionist agenda. *Nature* 422:741–45

80. Pollard TD, Borisy GG. 2003. Cellular motility driven by assembly and disassembly of actin filaments. *Cell* 112:453–65

81. Qian H, Sheetz MP, Elson EL. 1991. Single particle tracking. Analysis of diffusion and flow in two-dimensional systems. *Biophys. J.* 60:910–21

82. Radmacher M. 2007. Studying the mechanics of cellular processes by atomic force microscopy. *Methods Cell Biol.* 83:347–72

83. Radmacher M, Cleveland JP, Fritz M, Hansma HG, Hansma PK. 1994. Mapping interaction forces with the atomic force microscope. *Biophys. J.* 66:2159–65

84. Radmacher M, Fritz M, Kacher CM, Cleveland JP, Hansma PK. 1996. Measuring the viscoelastic properties of human platelets with the atomic force microscope. *Biophys. J.* 70:556–67

85. Rogers SS, Waigh TA, Lu JR. 2008. Intracellular microrheology of motile *Amoeba proteus. Biophys. J.* 94:3313–22

86. Rogers SS, Waigh TA, Zhao X, Lu JR. 2007. Precise particle tracking against a complicated background: polynomial fitting with Gaussian weight. *Phys. Biol.* 4:220–27

87. Rudnick J, Gaspari G. 1987. The shapes of random walks. *Science* 237:384–89

88. Savin T, Doyle PS. 2005. Role of a finite exposure time on measuring an elastic modulus using microrheology. *Phys. Rev. E* 71:041106

89. Savin T, Doyle PS. 2005. Static and dynamic errors in particle tracking microrheology. *Biophys. J.* 88:623–38

90. Schnurr B, Gittes F, MacKintosh FC, Schmidt CF. 1997. Determining microscopic viscoelasticity in flexible and semiflexible polymer networks from thermal fluctuations. *Macromolecules* 30:7781–92

91. Seksek O, Biwersi J, Verkman AS. 1997. Translational diffusion of macromolecule-sized solutes in cytoplasm and nucleus. *J. Cell Biol.* 138:131–42

92. Sivaramakrishnan S, DeGiulio JV, Lorand L, Goldman RD, Ridge KM. 2008. Micromechanical properties of keratin intermediate filament networks. *Proc. Natl. Acad. Sci. USA* 105:889–94

93. Smith SB, Finzi L, Bustamante C. 1992. Direct mechanical measurements of the elasticity of single DNA molecules by using magnetic beads. *Science* 258:1122–26

94. Stamenovic D, Rosenblatt N, Montoya-Zavala M, Matthews BD, Hu S, et al. 2007. Rheological behavior of living cells is timescale-dependent. *Biophys. J.* 93:L39–41

95. Stewart-Hutchinson PJ, Hale CM, Wirtz D, Hodzic D. 2008. Structural requirements for the assembly of LINC complexes and their function in cellular mechanical stiffness. *Exp. Cell Res.* 314:1892–905

96. Suh J, Wirtz D, Hanes J. 2003. Efficient active transport of gene nanocarriers to the cell nucleus. *Proc. Natl. Acad. Sci. USA* 100:3878–82

97. Svitkina TM, Verkhovsky AB, McQuade KM, Borisy GG. 1997. Analysis of the actin-myosin II system in fish epidermal keratocytes: mechanism of cell body translocation. *J. Cell Biol.* 139:397–414

98. Tapon N, Hall A. 1997. Rho, Rac and Cdc42 GTPases regulate the organization of the actin cytoskeleton. *Curr. Opin. Cell Biol.* 9:86–92

99. Tseng Y, An KM, Esue O, Wirtz D. 2004. The bimodal role of filamin in controlling the architecture and mechanics of F-actin networks. *J. Biol. Chem.* 279:1819–26

100. Tseng Y, An KM, Wirtz D. 2002. Microheterogeneity controls the rate of gelation of actin filament networks. *J. Biol. Chem.* 277:18143–50

101. Tseng Y, Fedorov E, McCaffery JM, Almo SC, Wirtz D. 2001. Micromechanics and microstructure of actin filament networks in the presence of the actin-bundling protein human fascin: a comparison with α-actinin. *J. Mol. Biol.* 310:351–66

102. Tseng Y, Kole TP, Wirtz D. 2002. Micromechanical mapping of live cells by multiple-particle-tracking microrheology. *Biophys. J.* 83:3162–76

103. Tseng Y, Lee JS, Kole TP, Jiang I, Wirtz D. 2004. Micro-organization and visco-elasticity of the interphase nucleus revealed by particle nanotracking. *J. Cell Sci.* 117:2159–67

104. Tseng Y, Wirtz D. 2001. Mechanics and multiple-particle tracking microheterogeneity of alpha-actinin-cross-linked actin filament networks. *Biophys. J.* 81:1643–56

105. Tseng Y, Wirtz D. 2004. Dendritic branching and homogenization of actin networks mediated by Arp2/3 complex. *Phys. Rev. Lett.* 93:258104

106. Valberg PA, Albertini DF. 1985. Cytoplasmic motions, rheology, and structure probed by a novel magnetic particle method. *J. Cell Biol.* 101:130–40

107. Van Citters KM, Hoffman BD, Massiera G, Crocker JC. 2006. The role of F-actin and myosin in epithelial cell rheology. *Biophys. J.* 91:3946–56

108. Verkhovsky AB, Svitkina TM, Borisy GG. 1995. Myosin II filament assemblies in the active lamella of fibroblasts: their morphogenesis and role in the formation of actin filament bundles. *J. Cell Biol.* 131:989–1002

109. Wachsstock DH, Schwartz WH, Pollard TD. 1993. Affinity of alpha-actinin for actin determines the structure and mechanical properties of actin filament gels. *Biophys. J.* 65:205–14

110. Wachsstock DH, Schwartz WH, Pollard TD. 1994. Crosslinker dynamics determine the mechanical properties of actin gels. *Biophys. J.* 66:801–9

111. Wang N, Butler JP, Ingber DE. 1993. Mechanotransduction across the cell surface and through the cytoskeleton. *Science* 260:1124–27

112. Xu J, Palmer A, Wirtz D. 1998. Rheology and microrheology of semiflexible polymer solutions: actin filament networks. *Macromolecules* 31:6486–92

113. Xu J, Tseng Y, Wirtz D. 2000. Strain-hardening of actin filament networks—regulation by the dynamic crosslinking protein α-actinin. *J. Biol. Chem.* 275:35886–92

114. Xu J, Viasnoff V, Wirtz D. 1998. Compliance of actin filament networks measured by particle-tracking microrheology and diffusing wave spectroscopy. *Rheol. Acta* 37:387–98

115. Xu J, Wirtz D, Pollard TD. 1998. Dynamic cross-linking by alpha-actinin determines the mechanical properties of actin filament networks. *J. Biol. Chem.* 273:9570–76

116. Yamada S, Wirtz D, Kuo SC. 2000. Mechanics of living cells measured by laser tracking microrheology. *Biophys. J.* 78:1736–47

117. Zhou X, Rowe RG, Hiraoka N, George JP, Wirtz D, et al. 2008. Fibronectin fibrillogenesis regulates three-dimensional neovessel formation. *Genes Dev.* 22:1231–43

Bioimage Informatics for Experimental Biology*

Jason R. Swedlow,[1] Ilya G. Goldberg,[2]
Kevin W. Eliceiri,[3] and the OME Consortium[4]

[1]Wellcome Trust Centre for Gene Regulation and Expression, College of Life Sciences, University of Dundee, Dundee DD1 5EH, Scotland, United Kingdom; email: jason@lifesci.dundee.ac.uk

[2]Image Informatics and Computational Biology Unit, Laboratory of Genetics, National Institute on Aging, IRP, NIH Biomedical Research Center, Baltimore MD 21224; email: igg@nih.gov

[3]Laboratory for Optical and Computational Instrumentation, University of Wisconsin at Madison, Madison, Wisconsin 53706; email: eliceiri@wisc.edu

[4]**http://openmicroscopy.org/site/about/development-teams**

Annu. Rev. Biophys. 2009. 38:327–46

The *Annual Review of Biophysics* is online at biophys.annualreviews.org

This article's doi:
10.1146/annurev.biophys.050708.133641

Key Words

microscopy, file formats, image management, image analysis, image processing

Abstract

Over the past twenty years there have been great advances in light microscopy with the result that multidimensional imaging has driven a revolution in modern biology. The development of new approaches of data acquisition is reported frequently, and yet the significant data management and analysis challenges presented by these new complex datasets remain largely unsolved. As in the well-developed field of genome bioinformatics, central repositories are and will be key resources, but there is a critical need for informatics tools in individual laboratories to help manage, share, visualize, and analyze image data. In this article we present the recent efforts by the bioimage informatics community to tackle these challenges, and discuss our own vision for future development of bioimage informatics solutions.

Contents

INTRODUCTION

Modern imaging systems have enabled a new kind of discovery in cellular and developmental biology. With spatial resolutions running from millimeters to nanometers, analysis of cell and molecular structure and dynamics is now routinely possible across a range of biological systems. The development of fluorescent reporters, most notably in the form of genetically encoded fluorescent proteins (FPs), combined with increasingly sophisticated imaging systems has enabled direct study of molecular structure and dynamics (6, 52). Cell and tissue imaging assays have scaled to include all three spatial dimensions, a temporal component, and the use of spectral separation to measure multiple molecules such that a single image is now

a five-dimensional structure—space, time, and channel. High content screening (HCS) and fluorescence lifetime, polarization, and correlation are all examples of new modalities that further increase the complexity of the modern microscopy dataset. However, multidimensional data acquisition generates a significant data problem: A typical four-year project generates hundreds of gigabytes of images, perhaps on many different proprietary data acquisition systems, making hypothesis-driven research dependent on data management, visualization, and analysis.

Bioinformatics is a mature science that forms the cornerstone of much of modern biology. Modern biologists routinely use genomic databases to inform their experiments.

HCS: high content screening

In fact these databases are well-crafted multi-layered applications that include defined data structures, application programming interfaces (APIs), and use standardized user interfaces to enable querying, browsing, and visualization of the underlying genome sequences. These facilities serve as a great model of the sophistication necessary to deliver complex, heterogeneous datasets to bench biologists. However, most genomic resources work on the basis of defined data structures with defined formats and known identifiers that all applications can access (they also employ expert staff to monitor systems and databases, a resource that is rarely available in individual laboratories). There is no single agreed data format, but a defined number are used in various applications, depending on the exact application (e.g., FASTA and EMBL files). These files are accessed through a number of defined software libraries that translate data into defined data structures that can be used for further analysis and visualization. Because a relatively small number of sequence data generation and collation centers exist, standards have been relatively easy to declare and support. Nonetheless, a key to the successful use of these data was the development of software applications, designed for use by bench biologists as well as specialist bioinformaticists, that enabled querying and discovery based on genomic data held by and served from central data resources.

Given this paradigm, the same facility should in principle be available for all biological imaging data (as well as proteomics and, soon, deep sequencing). In contrast to centralized genomics resources, in most cases, these methods are used for defined experiments in individual laboratories or facilities, and the number of image datasets recorded by a single postdoctoral fellow (hundreds to thousands) can easily rival the number of genomes that have been sequenced to date. For the continued development and application of experimental biology imaging methods, it will be necessary to invest in and develop informatics resources that provide solutions for individual laboratories and departmental facilities. Is it possible to deliver flexible, powerful, and usable informatics tools

to manage a single laboratory's data that are comparable to that used to deliver genomic sequence applications and databases to the whole community? Why can't the tools used in genomics be immediately adapted to imaging? Are image informatics tools from other fields appropriate for biological microscopy? In this article, we address these questions, discuss the requirements for successful image informatics solutions for biological microscopy, and consider the future directions that these applications must take to deliver effective solutions for biological microscopy.

FLEXIBLE INFORMATICS FOR EXPERIMENTAL BIOLOGY

Experimental imaging data are by their very nature heterogeneous and dynamic. The challenge is to capture the evolving nature of an experiment in data structures that by their very nature are specifically typed and static, for later recall, analysis, and comparison. Achieving this goal in imaging applications means solving a number of problems.

Proprietary File Formats

There are over 50 different proprietary file formats (PFFs) used in commercial and academic image acquisition software packages for light microscopy (34). This number significantly increases if electron microscopy, new HCS systems, tissue imaging systems, and other new modes of imaging modalities are included. Regardless of the specific application, almost all store data in their own PFFs. Each of these formats includes the binary data (i.e., the values in the pixels) and the metadata (i.e., the data that describes the binary data). Metadata include physical pixel sizes, time stamps, spectral ranges, and any other measurements or values required to fully define the binary data. Because of the heterogeneity of microscope imaging experiments, there is no agreed upon community specification for a minimal set of metadata (see below). Regardless, the binary data and metadata

Application programming interface (API): an interface providing one software program or library easy access to its functionality with full knowledge of the underlying code or data structures

combined form the full output of the micro-scope imaging system, and each software appli-cation must contend with the diversity of PFFs and continually update its support for changing formats.

Experimental Protocols

Sample preparation, data acquisition meth-ods and parameters, and analysis workflow all evolve during the course of a project, and there are invariably differences in approach even be-tween laboratories doing similar work. This evolution reflects the natural progression of sci-entific discovery. Recording this evolution (e.g., "What exposure time did I use in the exper-iment last Wednesday?") and providing flexi-bility for changing metadata, especially when new metadata must be supported, are critical requirements for any experimental data man-agement system.

Image Result Management

Many experiments only use a single micro-scope, but the visualization and analysis of im-age data associated with a single experiment can generate many additional derived files of varying formats. Typically, these are stored on a hard disk using arbitrary directory struc-tures. Thus an experimental result typically reflects the compilation of many different im-ages, recorded across multiple runs of an exper-iment and associated processed images, analysis outputs, and result spreadsheets. Keeping these disparate data linked so that they can be recalled and examined at a later time is a critical require-ment and a significant challenge.

Remote Image Access

Image visualization requires significant com-putational resources. Many commercial image-processing tools use specific graphics CPU hardware (and thus depend on the accompa-nying driver libraries). Moreover, they often do not work well when analyzing data across a net-work connection to data stored on a remote

file system. As work patterns move to wire-less connections and more types of portable devices, remote access to image visualization tools, coupled with the ability to access and run powerful analysis and processing, will be required.

Image Processing and Analysis

Substantial effort has gone into the develop-ment of sophisticated image processing and analysis tools. In genome informatics, the link-age of related but distinct resources [e.g., WormBase (48) and FlyBase (13)] is possi-ble due to the availability of defined inter-faces that different resources use to provide access to underlying data. This facility is crit-ical to enable discovery and collaboration—any algorithm developed to ask a specific ques-tion should address all available data. This is especially critical as new image methods are developed—an existing analysis tool should not be made obsolete just because a new file format has been developed that it does not read. When scaled across the large number of analysis tool developers, this is an unacceptable code main-tenance burden.

Distributed Processing

As the sizes and numbers of images increase, access to larger computing facilities will be rou-tinely required by all investigators. Grid-based data processing is now available for specific analyses of genomic data, but the burden of moving many gigabytes of data even for a single experiment means that distributed computing must also be made locally available, at least in a form that allows laboratories and facilities to access their local clusters or to leverage an in-vestment in multi-CPU, multi-core machines.

Image Data and Interoperability

Strategic collaboration is one of the cornerstones of modern science and fun-damentally consists of scientists sharing resources and data with each other. Biological imaging is composed of several specialized subdisciplines—experimental image acquisition, image processing, and image data mining. Each requires its own domain of expertise and specialization, which is justified because each presents unsolved technical challenges as well as ongoing scientific research. For a group specializing in image analysis to make the best use of its expertise, it needs to have access to image data from groups specializing in acquisition. Ideally, this data should comprise current research questions and not historical image repositories that may no longer be scientifically relevant. Similarly, groups specializing in data mining and modeling must have access to image data and to results produced by image processing groups. Ultimately, this drives the development of useful tools for the community and certainly results in synergistic collaborations that enhance each group's advances.

TOWARDS BIOIMAGE INFORMATICS

The delivery of solutions for the problems detailed above requires the development of a new emerging field known as bioimage informatics (45), which includes the infrastructure and applications that enable discovery of insight using systematic annotation, visualization, and analysis of large sets of images of biological samples. For applications of bioimage informatics in microscopy, we include HCS, in which images are collected from arrayed samples and treated with large sets of siRNAs or small molecules (46), as well as large sets of time-lapse images (26), collections of fixed and stained cells or tissues (10, 18), and even sets of generated localization patterns (59) that define specific collections of localization for reference or for analysis. The development and implementation of successful bioimage informatics tools provide enabling technology for biological discovery in several different ways:

- Management: keeping track of data from large numbers of experiments
- Sharing with defined collaborators: allowing groups of scientists to compare

images and analytic tools with one another
- Remote access: ability to query, analyze, and visualize without having to connect to a specific file system or use specific video hardware on the user's computer or mobile device
- Interoperability: interfacing of visualization and analysis programs with any set of data, without concern for file format
- Integration of heterogeneous data types: collection of raw data files, analysis results, annotations, and derived figures into a single resource that is easily searchable and browsable.

BUILDING BY AND FOR THE COMMUNITY

Given these requirements, how should an image informatics solution be developed and delivered? It certainly will involve the development, distribution, and support of software tools that must be acceptable to bench biologists and must work with all the existing commercial and academic data acquisition, visualization, and analysis tools. Moreover, it must support a broad range of imaging approaches and, if at all possible, include the newest modalities in light and electron microscopy, support extensions into clinical research familiar with microscopy (e.g., histology and pathology), and provide the possibility of extension into modalities that do not use visible light (MRI, CT, ultrasound). Because many commercial image acquisition and analysis software packages are already established as critical research tools, all design, development, and testing must assume and expect integration and interoperability. It therefore seems prudent to avoid a traditional commercial approach and make this type of effort community led, using open source models that are now well defined. This does not exclude the possibility of successful commercial ventures being formed to provide bioimage informatics solutions to the experimental biological community, but a community-led, open source approach will be best placed to provide

Bioimage informatics: infrastructure including specifications, software, and interfaces to support experimental biological imaging

interfaces between all existing academic and commercial applications.

DELIVERING ON THE PROMISE: STANDARDIZED FILE FORMATS VERSUS "JUST PUT IT IN A DATABASE"

In our experience, there are a few commonly suggested solutions for biological imaging. The first is a common, open file format for microscope imaging. A number of specifications for file formats have been presented, including our own (2, 14). Widespread adoption of standardized image data formats has been successful in astronomy (FITS), crystallography (PDB), and clinical imaging (DICOM), where either most of the acquisition software is developed by scientists or a small number of commercial manufacturers adopt a standard defined by the imaging community. Biological microscopy is a highly fractured market, with at least 40 independent commercial providers. This combined with rapidly developing technology platforms acquiring new kinds of data has stymied efforts at establishing a commonly used data standard.

Against this background, it is worth asking whether defining a standardized format for imaging is at all useful and practical. Standardized file formats and minimum data specifications have the advantage of providing a single or, perhaps more realistically, a small number of data structures for the community to contend with. These facilitate interoperability—visualization and analysis tools developed by one lab may be easily used by another. This is an important step for collaboration and allows data exchange—moving a large multidimensional file from one software application to another, or from one lab or center to another. However, standardized formats only satisfy some of the requirements defined above and provide none of the search, query, remote access, or collaboration facilities discussed above, and thus are only a partial solution. However, the expression of a data model in a standardized file format, and especially the development of software that reads and writes that format, is a useful exercise. It tests the modeling concepts, relationships, and requirements (e.g., "If an objective lens is specified, should the numerical aperture be mandatory?") and provides a relatively easy way for the community to access, use, and comment on the data relationships defined by the project. This is an important component of data modeling and standardization and should not be minimized. Moreover, while not providing most of the functionality defined above, standardized formats have the practical value of providing a medium for the publishing and release of data to the scientific community. Unlike gene sequence and protein structure data, there is no requirement for release of images associated with published results, but the availability of standardized formats may facilitate this.

To provide some of the data management features described above, labs might use any number of commercial database products (e.g., Microsoft Access®, FileMaker Pro®) to build customized local databases on commercial foundations. This is certainly a potential solution for individual laboratories, but to date, these local database efforts have not simultaneously dedicated themselves to addressing interoperability, allowing broad support for alternative analysis and visualization tools that were not specifically supported when the database was built. Perhaps most importantly, single lab efforts often emphasize specific aspects of their own research (e.g., the data model supports individual cell lines, but not yeast or worm strains), and the adaptability necessary to support a range of disciplines across biological research, or even their own evolving repertoire of methods and experimental systems, is not included.

In no way does this preclude the development of local or targeted bioimage informatics solutions. In genomics, several community-initiated informatics projects focused on specific resources support the various biological model systems (13, 50, 57). It seems likely that similar projects will grow up around specific bioimage informatics projects, following the models of the Allen Brain Atlas, E-MAGE, and the Cell Centered Database (CCDB) (10, 18, 21). In genomics, there is underlying interoperability between specialized sources—ultimately all of the sequence data as well as the specialized annotation exist in common repositories and formats (e.g., GenBank). Common repositories are not yet feasible for multidimensional image data, but there will be value in linking through the gene identifiers themselves, ontological annotations, or perhaps, localization maps or sets of phenotypic features, once these are standardized (59). Once these links are made to images stored in common formats, distributed storage may effectively accomplish the same thing as centralized storage.

Several large-scale bioinformatics projects related to interoperability between large biological information datasets have emerged, including caBIG, which focuses on cancer research (7); BIRN, which focuses on neurobiology with a substantial imaging component (4); BioSig (44), which provides tools for large-scale data analysis; and myGrid, which focuses on simulation, workflows, and in silico experiments (25). Projects specifically involved in large-scale imaging infrastructure include the Protein Subcellular Location Image Database (PSLID) (16, 24), Bisque (5), CCDB (9, 21), and our own, the Open Microscopy Environment (OME) (31, 55). All these projects were initiated to support the specific needs of the biological systems and experiments in each of the labs driving the development of each project. For example, studies in neuroscience depend on a proper specification for neuroanatomy so that any image and resulting analysis can be properly oriented with respect to the physiological source. In this case, an ontological framework for neuroanatomy is then needed to support and compare the results from many different laboratories (21). A natural progression is a resource that enables sharing of specific images, across many different resolution scales, that are as well defined as possible (9). PSLID is an alternative repository that provides a well-annotated resource for subcellular localization by fluorescence microscopy. In all cases these projects are the result of dedicated, long-term collaboration

between computer scientists and biologists, indicating that the challenges presented by this infrastructure development represent the state of the art not only in biology but in computing as well. Many if not most of these projects make use of at least some common software and data models, and although full interoperability is not something that can be claimed today, key members of these projects regularly participate in the same meetings and working groups. In the future, it should be possible for these projects to interoperate to enable, for example, OME software to upload to PSLID or CCDB.

CCDB: Cell Centered Database

PSLID: Protein Subcellular Location Image Database

OME: Open Microscopy Environment

OME: A COMMUNITY-BASED EFFORT TO DEVELOP IMAGE INFORMATICS TOOLS

Since 2000, the Open Microscopy Consortium has been working to deliver tools for image informatics for biological microscopy. Our original vision (55), to provide software tools to enable interoperability between as many image data storage, analysis, and visualization applications as possible, remains unchanged. However, the project has evolved and grown since its founding to encompass a much broader effort and now includes subprojects dedicated to data modeling (37), file format specification and conversion (34, 35), data management (27), and image-based machine learning (29). The Consortium (28) also maintains links with many academic and commercial partners (32). While the challenges of running and maintaining a larger Consortium are real, the major benefits are synergies and feedback that develop when our own project has to use its own updates to data models and file formats. Within the Consortium, there is substantial expertise in data modeling and software development, and we have adopted a series of project management tools and practices to make the project as professional as possible, within the limits of working within academic laboratories. Moreover, our efforts occur within the context of our own image-based research activities. We make no pretense that this samples the full range of potential applications for our specifications and software, just that our

XML: extensible
markup language

ideas and work are actively tested and refined before release to the community. Most importantly, the large Consortium means that we can interact with a larger community, gathering requirements and assessing acceptance and new directions from a broad range of scientific applications.

THE FOUNDATION: THE OME DATA MODEL

Since its inception in 2000, the OME Consortium has dedicated itself to developing a specification for the metadata associated with the acquisition of a microscope image. Initially, our goal was to specify a single data structure that would contain spatial, temporal, and spectral components [often referred to as Z, C, T, which together form a 5D image (1)]. This has evolved into specifications for the other elements of the digital microscope system including objective lenses, fluorescence filter sets, illumination systems, and detectors. This effort has been greatly aided by many discussions about configurations and specifications with commercial imaging device manufacturers (32). This work is ongoing, with our current focus being the delivery of specifications for regions-of-interest [based on existing specifications from the geospatial community (42)] and a clear understanding of what data elements are required to properly define a digital microscope image. This process is most efficient when users or developers request updates to the OME data model—the project's Web site (37) accepts requests for new or modified features and fixes.

OME FILE FORMATS

The specification of an open, flexible file format for microscope imaging provides a tool for data exchange between distinct software applications. It is certainly the lowest level of interoperability, but for many situations it suffices in its provision of readable, defined structured image metadata. OME's first specification cast a full 5D image—binary and metadata—in an XML (extensible markup language) file (14).

Although conceptually sound, a more pragmatic approach is to store binary data as TIFF images and then link image metadata represented as OME-XML by including it within the TIFF image header or as a separate file (35). To ensure that these formats are in fact defined, we have delivered an OME-XML and OME-TIFF file validator (36) that can be used by developers to ensure files follow the OME-XML specification. As of this writing five commercial companies support these file formats in their software with a "Save as. . ." option, thus enabling export of image data and metadata to a vendor-neutral format.

SUPPORT FOR DATA TRANSLATION—BIO-FORMATS

PFFs are perhaps the most common informatics challenge faced by bench biologists. Despite the OME-XML and OME-TIFF specifications, PFFs will continue to be the dominant source of raw image for visualization and analysis applications for some time. Because all software must contend with PFFs, the OME Consortium has dedicated its resources to developing a software library that can convert PFFs to a vendor-neutral data structure—OME-XML. This led to the development, release, and continued maintenance of Bio-Formats, a standalone Java library for reading and writing life sciences image formats. The library is general, modular, flexible, extensible, and accessible. The project originally grew out of efforts to add support for file formats to the LOCI VisBio software (40, 49) for visualization and analysis of multidimensional image data, when we realized that the community was in acute need of a broader solution to the problems created by myriad incompatible microscopy formats.

Utility

Over the years we have repeatedly observed software packages reimplement support for the same microscopy formats [i.e., ImageJ (17), MIPAV (22), BioImageXD (3), and many commercial packages]. The vast majority of these efforts focus exclusively on adaptation of formats into each program's specific internal data model; Bio-Formats (34), in contrast, unites popular life sciences file formats under a broad, evolving data specification provided by the OME data model. This distinction is critical: Bio-Formats does not adapt data into structures designed for any specific visualization or analysis agenda, but rather expresses each format's metadata in an accessible data model built from the ground up to encapsulate a wide range of scientifically relevant information. We know of no other effort within the life sciences with as broad a scope as Bio-Formats and dedicated toward delivering the following features.

Modularity

The architecture of the Bio-Formats library is split into discrete, reusable components that work together but are fundamentally separable. Each file format reader is implemented as a separate module extending a common IFormatReader interface; similarly, each file format writer module extends a common IFormatWriter interface. Both reader and writer modules utilize the Bio-Formats MetadataStore API to work with metadata fields in the OME Data Model. Shared logic for encoding and decoding schemes (e.g., JPEG and LZW) are structured as part of the Bio-Formats codec package, so that future readers and writers that need those same algorithms can leverage them without reimplementing similar logic or duplicating any code.

When reading data from a dataset, Bio-Formats provides a tiered collection of reader modules for extracting or restructuring various types of information from the dataset. For example, a client application can instruct Bio-Formats to compute minimum and maximum pixel values using a MinMaxCalculator, combine channels with a ChannelMerger, split them with a ChannelSeparator, or reorder dimensional axes with a DimensionSwapper. Performing several such operations can be accomplished merely by stacking the relevant reader modules one on top of the other. Several auxiliary components are also provided; the most significant are a caching package for intelligent management of image planes in memory when storage requirements for the entire dataset would be too great, and a suite of graphical components for common tasks such as presenting the user with a file chooser dialog box or visualizing hierarchical metadata in a tree structure.

Flexibility

Bio-Formats has a flexible metadata API, built in layers over the OME Data Model itself. At the lowest level, the OME Data Model is expressed as an XML schema, called OME-XML, that is continually revised and expanded to support additional metadata fields. An intermediate layer known as the OME-XML Java library is produced using code generation techniques, which provides direct access to individual metadata fields in the OME-XML hierarchy. The Bio-Formats metadata API, which provides a simplified, flattened version of the OME Data Model for flexible implementation by the developer, leverages the OME-XML Java library layer and is also generated automatically from underlying documents to reduce errors in the implementation.

Extensibility

Adding a new metadata field to the data model is done at the lowest level, to the data model itself via the OME-XML schema. The supporting code layers—both the OME-XML Java library and the Bio-Formats metadata API—are programmatically regenerated to include the addition. The only remaining task is to add a small amount of code to each file format reader mapping the original data field into the appropriate location within the standardized OME Data Model.

Although the OME Data Model specifically targets microscopy data, in general, the Bio-Formats model of metadata extensibility is ideal for adaptation to alternative data models unrelated to microscopy. By adopting a similar pattern for the new data model, and introducing code generation layers corresponding to the new model, the Bio-Formats infrastructure could easily support additional branches of multidimensional scientific imaging data. In the future the Bio-Formats infrastructure will provide significant interoperability between the multiple established data models at points where they overlap by establishing a common base layer between them.

Bio-Formats is written in Java so that the code can execute on a wide variety of target platforms, and code and documentation for interfacing Bio-Formats with a number of different tools including ImageJ, MATLAB, and IDL are available (34). We provide documentation on how to use Bio-Formats both as an end user and as a software developer, including hints on leveraging Bio-Formats from other programming environments such as C++, Python, or a command shell. We have successfully integrated Bio-Formats with native acquisition software written in C++ using ICE middleware (58).

DATA MANAGEMENT APPLICATIONS: OME AND OMERO

Data management is a critical application for modern biological discovery, and in particular necessary for biological imaging because of the large heterogeneous datasets generated during data acquisition and analysis. We define data management as the collation, integration, annotation, and presentation of heterogeneous experimental and analytic data in ways that enable the physical, temporal, and conceptual relationships in experimental data to be captured and represented to users. The OME Consortium has built two data management tools—the original OME Server (29) and the recently released OME Remote Objects (OMERO; a port of the basic image data management functionality to a Java enterprise application) application platform (30). Both applications are now heavily used worldwide, but our development focus has

shifted from the OME Server toward OMERO, and that is where most future advances will occur.

The OME data management applications are specifically designed to meet the requirements and challenges described above, enabling the storage, management, visualization, and analysis of digital microscope image data and metadata. The major focus of this work is not on creating novel analysis algorithms, but instead on development of a structure that ultimately allows any application to read and use any data associated with or generated from digital microscope images.

A fundamental design concept in the OME data management applications is the separation of image storage, management, analysis, and visualization functions between a lab's or imaging facility's server and a client application (e.g., Web browser or Java user interface). This concept mandates the development of two facilities: a server that provides all data management, access control, and storage, and a client that runs on a user's desktop workstation or laptop and that provides access to the server and the data via a standard Internet connection. The key to making this strategy work is judicious choice of the functionality placed on client and server to ensure maximal performance.

The technical design details and principles of both systems have recently been described (23) and are available online (39). In brief, both the OME Server and the OMERO platform (**Figure 1**) use a relational database management system (RDMS) [PostgreSQL (47)] to provide all aspects of metadata management and an image repository to house all the binary pixel data. Both systems then use a middleware application to interact with the RDMS and read and write data from the image repository. The middleware applications include a rendering engine that reads binary data from the image repository and renders it for display by the client, and if necessary, compresses the image to reduce the bandwidth requirements for transferring across a network connection to a client. The result is access to high-performance data visualization, management,

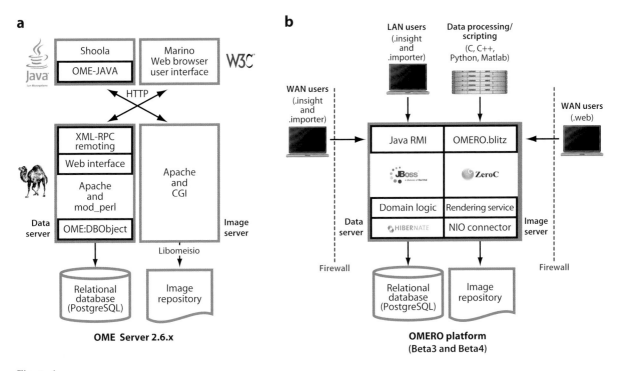

a

Shoola
OME-JAVA

Marino
Web browser
user interface

W3C

Java

HTTP

XML-RPC
remoting

Web interface

Apache
and
mod_perl

OME:DBObject

Apache
and
CGI

Data
server

Image
server

Libomeisio

Relational
database
(PostgreSQL)

Image
repository

OME Server 2.6.x

b

LAN users
(.insight
and
.importer)

Data processing/
scripting
(C, C++,
Python, Matlab)

WAN users
(.insight
and
.importer)

WAN users
(.web)

Java RMI

OMERO.blitz

JBoss

ZeroC

Domain logic

Rendering service

HIBERNATE

NIO connector

Data
server

Image
server

Firewall

Firewall

Relational
database
(PostgreSQL)

Image
repository

OMERO platform
(Beta3 and Beta4)

Figure 1

Architecture of the OME and OMERO servers and client applications. (*a*) Architecture of the OME Server, built using Perl for most of the software code and an Apache Web server. The main client application for the server is a Web browser-based interface. (*b*) The architecture of the OMERO platform, including OMERO.server and the OMERO clients. OMERO is based on the JBOSS JavaEE framework, but it also includes an alternative remoting architecture called ICE (58). For more details, see Reference 39.

and analysis in a remote setting. Both the OME Server (**Figure 1***a*) and OMERO (**Figure 1***b*) also provide well-developed data querying facilities to access metadata, annotations, and analytics from the RDMS. For user interfaces, the OME Server includes a Web browser-based interface that provides access to image, annotation, analytics, and visualization and also a Java interface (OME-JAVA) and remote client (Shoola) to support access from remote client applications. OMERO includes separate Java-based applications for uploading data to an OMERO server (OMERO.importer), for visualizing and managing data (OMERO.insight), and for Web browser-based server administration (OMERO.webadmin).

The OME Server has been installed in hundreds of facilities worldwide; however, after significant development effort it became clear

that the application, which we worked on for five years (2000–2005), had three major flaws. (*a*) The installation was too complex, and too prone to failure. (*b*) Our object-relational mapping library ("DBObject") was all custom code, developed by OME, and required significant code maintenance effort to maintain compatibility with new versions of Linux and Perl (**Figure 1***a*). Support for alternative RDMSs (e.g., Oracle®) was possible in principle but required significant work. (*c*) The data transport mechanisms available to us in a Perl-based architecture amounted to XML-RPC and SOAP. Although totally standardized and promoting interoperability, this mechanism, with its requirement for serialization/deserialization of large data objects, was too slow for working with remote client applications—simple queries with well-populated databases could take minutes to transfer from server to client.

With work, problem *a* became less of an issue, but problems *b* and *c* remained significant fundamental barriers to delivery of a great image informatics application to end users. For these reasons, we initiated work on OMERO. In taking on this project, it was clear that the code maintenance burden needed to be substantially reduced, the system must be simple to install, and the performance of the remoting system must be significantly improved. A major design goal was the reduction of self-written code through the reuse of existing middleware and tools where possible. In addition, OMERO must support as broad a range of client applications as possible, enabling the development of new user interfaces, as well as a wide range of data analysis applications.

We based the initial implementation of OMERO's architecture (**Figure 1b**) on the JavaEE5 specification, as it appeared to have wide uptake, clear specifications, and high performance libraries in active development from a number of projects. A full specification and description of OMERO.server is available (23). The architecture follows accepted standards and consists of services implemented as EJB3 session beans (53) that make use of Hibernate (15), a high-performance object-relational mapping solution, for metadata retrieval from the RDMS. Connection to clients is via Java Remote Method Invocation (Java RMI) (54). All released OMERO remote applications are written in Java and cross-platform. OMERO.importer uses the Bio-Formats library to read a range of file formats and load the data into an OMERO.server, along with simple annotations and assignment to the OME Project-Dataset-Image experimental model (for demonstrations, see Reference 33). OMERO.insight includes facilities for managing, annotating, searching, and visualizing data stored in an installation of OMERO.server. OMERO.insight also includes simple line and region-of-interest measurements and thus supports the simplest forms of image analysis.

OMERO ENHANCEMENTS: BETA3 AND BETA4

Through 2007, the focus of the OMERO project has been on data visualization and management, all the while laying the infrastructure for data analysis. With the release of OMERO3-Beta2, we began adding functionality that has the foundation for delivering a fully developed image informatics framework. In this section, we summarize the major functional enhancements that are being delivered in OMERO-Beta3 (released June 2008) and OMERO-Beta4 (released February 2009). Further information on all the items described below is available at the OMERO documentation portal (39).

OMERO.blitz

Starting with OMERO-Beta3, we provided interoperability with many different programming environments. We chose an ICE-based framework (58) rather than the more popular Web services–based GRID approaches because of the absolute performance requirements we had for the passage of large binary objects (image data) and large data graphs (metadata trees) between server and client. Our experience using Web services and XML-based protocols with the Shoola remote client and the OME Server showed that Web services, while standardized in most genomic applications, were inappropriate for client-server transfer of the much larger data graphs we required. Most importantly, the ICE framework provided immediate support for multiple programming environments (C, C++, and Python are critical for our purposes) and a built-in distribution mechanism [IceGRID (58)] that we have adapted to deliver OMERO.grid (39), a process distribution system. OMERO.blitz is three to four times faster than Java RMI and we are currently examining migrating our Java API and the OMERO clients from JBOSS to OMERO.blitz. This framework provides substantial flexibility—interacting with data in OMERO can be as simple as starting the Python interpreter and interacting with OMERO via the console. Most importantly, this strategy forms the foundation for our future work as we can now leverage the advantages and existing functionality in cross-platform Java, native C and C++, and scripted Python for rapidly expanding the functionality in OMERO.

Structured Annotations

Beginning with OMERO-Beta3, users can attach any type of data to an image or other OMERO data container—text, URL, or other data files (e.g., .doc, .pdf, .xls, .xml) providing essentially the same flexibility as email attachments. The installation of this facility followed feedback from users and developers concerning the strategy for analysis management built into the OME Server. The underlying data model supported hard semantic typing in which each analysis result was stored in relational tables with names that could be defined by the user (23, 55). This approach, although conceptually desirable, proved too complex and burdensome. As an alternative, OMERO uses Structured Annotations to store any kind of analysis result as untyped data, defined only by a unique name to ensure that multiple annotations are easily distinguished. The data are not queryable by standard SQL, but any text-based file can be indexed and therefore found by users. Interestingly, Bisque has implemented a similar approach (5), enabling tags with defined structures that are otherwise completely customized by the user. In both cases, whether this flexible strategy provides enough structure to manage large sets of analysis results will have to be assessed.

OMERO.search

As of OMERO-Beta3, OMERO includes a text-indexing engine based on Lucene (19), which can be used to provide indexed-based searches for all text-based metadata in an OMERO database. This includes metadata and annotations stored within the OMERO database and also any text-based documents or results stored as Structured Annotations.

OMERO.java

As of OMERO-Beta3, we have released OMERO.java, which provides access for all external Java applications via the OMERO.blitz interface. As a first test of this facility, we are using analysis applications written in MATLAB as client applications to read from and write to OMERO.server. As a demonstration of the utility of this library, we have adapted the popular open source MATLAB-based image analysis tool CellProfiler (8) to work as a client of OMERO, using the MATLAB Java interface.

OMERO.editor

In OMERO-Beta3, we also released OMERO.editor, a tool to help experimental biologists define their own experimental data models and, if desired, use other specified data models in their work. It allows users to create a protocol template and to populate this with experimental parameters. This creates a complete experimental record in one XML file, which can be used to annotate a microscope image or exchanged with other scientists. OMERO.editor supports the definition of completely customized experimental protocols but also includes facilities to easily import defined data models [e.g., MAGE-ML (56) and OME-XML (14)] and support for all ontologies included in the Ontology Lookup Service (11).

OMERO.web

Staring with OMERO-Beta4, we will release a Web browser-based client for OMERO.server. This new client is targeted specifically to truly remote access (different country, limited bandwidth connections), especially where collaboration with other users is concerned. OMERO.web includes all the standard functions for importing, managing, viewing, and annotating image data. However, a new function is the ability to share specific sets of data with another user on the system—this allows password-protected access to a specific set of data that can initiate or continue data sharing between two lab members or two collaborating scientists. OMERO.web also supports a publish function, in which a defined set of data is published to the world via a public URL. OMERO.web uses the Python API in OMERO.blitz for access to OMERO.server using the Django framework (12).

OMERO.scripts

In OMERO-Beta4, we will extend the analysis facility provided by OMERO.java to provide a scripting engine, based on Python scripts and the OMERO.blitz interface. OMERO.scripts is a scripting engine that reads and executes functions cast in Python scripts. Scripts are passed to processors specified by OMERO.grid that can be on the local server or on networked computing facilities. This is the facility that will provide support for analysis of large image sets or of calculations that require simple linear or branched workflows.

OMERO.fs

Finally, a fundamental design principle of OMERO.server is the presence of a single image repository for storing binary image data that is tightly integrated with the server application. This is the basis of the import model, which is the only way to get image data into an OMERO.server installation—data are uploaded to the server, and binary data are stored in the single image repository. In many cases, as the storage space required expands, multiple repositories must be supported. Moreover, data import takes significant time and, especially with large datasets, can be prohibitive. A solution to this involves using the OMERO.blitz Python API to access the file system search and notification facilities that are now provided as part of the Windows, Linux, and OS X operating systems. In this scenario, an OMERO client application, OMERO.fs, sits between the file system and OMERO.blitz and provides a metadata service that scans user-specified image folders or file systems and reads image metadata into an OMERO relational database using PFF translation provided by Bio-Formats. As the coverage of Bio-Formats expands, this approach means that essentially any data can be loaded into an OMERO.server instance.

WORKFLOW-BASED DATA ANALYSIS: WND-CHARM

WND-CHARM is an image analysis algorithm based on pattern recognition (43). It relies on supervised machine learning to solve image analysis problems by example rather than by using a preconceived perceptual model of what is being imaged. An advantage of this approach is its generality. Because the algorithms used to process images are not task specific, they can be used to process any image regardless of the imaging modality or the image's subject. Similar to other pattern recognition algorithms, WND-CHARM first decomposes each image to a set of predefined numeric image descriptors. Image descriptors include measures of texture, factors in polynomial decompositions, and various statistics of the image as a whole, as well as measurements and distribution of high-contrast objects in the image. The algorithms that extract these descriptors (features) operate on both the original image pixels as well as transforms of the original pixels (Fourier, wavelet, etc). Together, there are 17 independent algorithms comprising 53 computational nodes (algorithms used along specific upstream data flows), with 189 links (data flows) producing 1025 numeric values (**Figure 2**). Although the entire set of features can be modeled as a single algorithm, this set is by no means complete and will grow to include other algorithms that extract both more specific and more general image content. The advantage of modeling this complex workflow as independently functional units is that new units can be easily added to the existing ones. This workflow model is therefore more useful to groups specializing in pattern recognition. Conversely, a monolithic

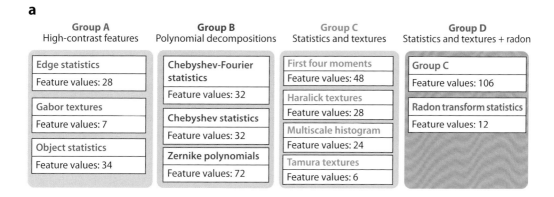

a

Group A High-contrast features	Group B Polynomial decompositions	Group C Statistics and textures	Group D Statistics and textures + radon
Edge statistics Feature values: 28	Chebyshev-Fourier statistics Feature values: 32	First four moments Feature values: 48	Group C Feature values: 106
Gabor textures Feature values: 7	Chebyshev statistics Feature values: 32	Haralick textures Feature values: 28	Radon transform statistics Feature values: 12
Object statistics Feature values: 34	Zernike polynomials Feature values: 72	Multiscale histogram Feature values: 24	
		Tamura textures Feature values: 6	

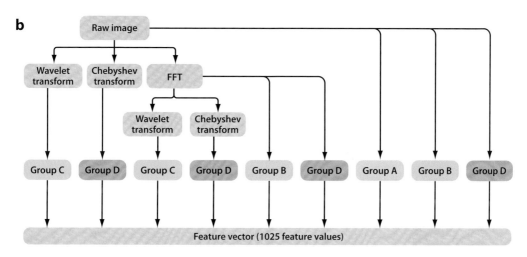

b

Raw image → Wavelet transform → Chebyshev transform → FFT

FFT → Wavelet transform → Chebyshev transform

Group C | Group D | Group C | Group D | Group B | Group D | Group A | Group B | Group D

Feature vector (1025 feature values)

Figure 2

Workflows in WND-CHARM. (*a*) List of feature types calculated by WND-CHARM. (*b*) Workflow of feature calculations in WND-CHARM. Note that different feature groups use different sets of processing tools.

representation of this workflow is probably more practical when implemented in a biology lab that would use a standard set of image descriptors applied to various imaging experiments. In neither case, however, should anyone be particularly concerned with what format was used to capture these images, or how they are represented in a practical imaging system. WND-CHARM is an example of a highly complex image-processing workflow and as such represents an important application for any system capable of managing workflows and distributed processing for image analysis. Currently, the fully modularized version of

WND-CHARM runs only on the OME Server. In the near future, the monolithic version of WND-CHARM (51) will be implemented using OMERO.blitz.

The raw results from a pattern recognition application are annotations assigned to whole images or image regions. These annotations are probabilities (or simply scores) that the image or region-of-interest belongs to a previously defined training set. In a dose-response experiment, for example, the training set may consist of control doses defining a standard curve, and the experimental images would be assigned an equivalent dose by the pattern recognition algorithm. Whereas the original experiment may be concerned with characterizing a collection of chemical compounds, the same image data could be analyzed in the context of a different set of training images—one defined by RNA interference, for example. When using these algorithms our group has found that performing these in silico experiments to reprocess existing image data in different contexts can be fruitful.

USABILITY

All the functionality discussed above must be built into OMERO.server and then delivered in a functional, usable fashion within the OMERO client applications OMERO.importer, OMERO.insight, and OMERO.web. This development effort is achieved by the OMERO development team and is invariably an iterative process that requires testing by our local community, as well as sampling feedback from the broader community of users. Therefore, the OMERO project has made software usability a priority throughout the project. A key challenge for the OME Consortium has been to improve the quality of the end user (i.e., the life scientist at their bench) experience. The first versions of OME software, the OME Server, provided substantial functionality but never received wide acceptance, despite dedicated work, mostly because its user interfaces were too complicated and the developed code, while open and available, was too complex for other developers to adopt and extend. In response to this failure, we initiated the Usable Image project (41) to apply established methods from the wider software design community, such as user-centered design and design ethnography (20), to the OME development process. Our goals were to initially improve the usability and accessibility of the OMERO client software and to provide a paradigm useful for the broader e-science and bioinformatics communities. The result of this combined usability and development effort has been a high level of success and acceptance of OMERO software. A wholly

unanticipated outcome has been the commitment to the user-centered design process by both users and developers. The investment in iterative, agile development practice has produced rapid, substantial improvements that the users appreciate, which in turn makes them more enthusiastic about the software. On the other hand, the developers have reliable, well-articulated requirements that, when implemented in software, are rewarded with more frequent use. This positive-feedback loop has transformed our development process and made usability analysis a core part of our development cycles. It has also forced a commitment to the development of usable code—readable, well-documented, tested, and continuously integrated—and the provision of up-to-date resources defining architecture and code documentation (38, 39).

SUMMARY AND FUTURE IMPACT

In this article we have focused on the OME Consortium's efforts (namely OME-XML, Bio-Formats, OMERO, and WND-CHARM), as we feel they are representative of the community-wide attempts to address many of the most pressing challenges in bioimaging informatics. While OME is committed to developing and releasing a complete image informatics infrastructure focused on the needs of the end user bench biologist, we are at least equally committed to the concept that beyond our software, our approach is part of a critical shift in how the challenges of data analysis, management, sharing, and visualization have been traditionally addressed in biology. In particular the OME Consortium has put an emphasis on flexibility, modularity, and inclusiveness that targets not only the bench biologist but also importantly the informatics developer to help ensure maximum implementation of and penetration into the bioimaging community. Key to this has been a dedication to allowing the biologist to retain and capture all available metadata and binary data from disparate sources, including proprietary ones, to map these data to a flexible data model, and to analyze these data in whatever environment he or she chooses. This ongoing effort requires an interdisciplinary approach that combines concepts from traditional bioinformatics, ethnography, computer science, and data visualization. It is our intent and hope that the bioimage informatics infrastructure that is developed by the OME Consortium will continue to have utility for its principal target community of experimental bench biologists, and also serve as a collaborative framework for developers and researchers from other closely related fields who might want to adopt the methodologies and code-based approaches for informatics challenges that exist in other communities. Interdisciplinary collaboration between biologists, physicists, engineers, computer scientists, ethnographers, and software developers is absolutely necessary for the successful maturation of the bioimage informatics community, and it will play an even larger role as this field evolves to fully support the continued evolution of imaging in modern experimental biology.

SUMMARY POINTS

1. Advances in digital microscopy have driven the development of a new field, bioimage informatics. This field encompasses the storage, querying, management, analysis, and visualization of complex image data from digital imaging systems used in biology.

2. Although standardized file formats have often been proposed to be sufficient to provide the foundation for bioimage informatics, the prevalence of PFFs and the rapidly evolving data structures needed to support new developments in imaging make this impractical.

3. Standardized APIs and software libraries enable interoperability, which is a critical unmet need in cell and developmental biology.

4. A community-driven development project is best placed to define, develop, release, and support these tools.

5. A number of bioimage informatics initiatives are underway, and collaboration and interaction are developing.

6. The OME Consortium has released specifications and software tools to support bioimage informatics in the cell and developmental biology community.

7. The next steps in software development will deliver increasingly sophisticated infrastructure applications and should deliver powerful data management and analysis tools to experimental biologists.

FUTURE ISSUES

1. Further development of the OME Data Model must keep pace with and include advances in biological imaging, with a particular emphasis on improving support for image analysis metadata and enabling local extension of the OME Data Model to satisfy experimental requirements with good documentation and examples.

2. Development of Bio-Formats to include as many biological image file formats as possible and extension to include data from non-image-based biological data.

3. Continue OMERO development as an image management system with a particular emphasis on ensuring client application usability and the provision of sophisticated image visualization and analysis tools.

4. Support both simple and complex analysis workflow as a foundation for common use of data analysis and regression in biological imaging.

5. Drive links between the different bioimage informatics enabling transfer of data between instances of the systems so that users can make use of the best advantages of each.

DISCLOSURE STATEMENT

The authors are not aware of any biases that might be perceived as affecting the objectivity of this review.

ACKNOWLEDGMENTS

JRS is a Wellcome Trust Senior Research Fellow and work in his lab on OME is supported by the Wellcome Trust (Ref 080087 and 085982), BBSRC (BB/D00151X/1), and EPSRC (EP/D050014/1). Work in IGG's lab is supported by the National Institutes of Health. The OME work in KWE's lab is supported by NIH grants R03EB008516 and R01EB005157.

LITERATURE CITED

1. Andrews PD, Harper IS, Swedlow JR. 2002. To 5D and beyond: quantitative fluorescence microscopy in the postgenomic era. *Traffic* 3:29–36
2. Berman J, Edgerton M, Friedman B. 2003. The tissue microarray data exchange specification: a community-based, open source tool for sharing tissue microarray data. *BMC Med. Inform. Decis. Making* 3:5
3. BioImageXD. 2008. *BioImageXD—open source software for analysis and visualization of multidimensional biomedical data*. **http://www.bioimagexd.net/**
4. BIRN. 2008. **http://www.nbirn.net/**
5. Centre for Bio-Image Informatics. 2008. *Bisque database*. **http://www.bioimage.ucsb.edu/downloads/ Bisque%20Database**
6. Bullen A. 2008. Microscopic imaging techniques for drug discovery. *Nat. Rev. Drug Discov.* 7:54–67
7. caBIG Community Website. **http://cabig.nci.nih.gov/**
8. Carpenter A, Jones T, Lamprecht M, Clarke C, Kang I, et al. 2006. CellProfiler: image analysis software for identifying and quantifying cell phenotypes. *Genome Biol.* 7:R100
9. Cell Centered Database. 2008. **http://ccdb.ucsd.edu/CCDBWebSite/**
10. Christiansen JH, Yang Y, Venkataraman S, Richardson L, Stevenson P, et al. 2006. EMAGE: a spatial database of gene expression patterns during mouse embryo development. *Nucleic Acids Res.* 34:D637–41
11. Cote RG, Jones P, Martens L, Apweiler R, Hermjakob H. 2008. The Ontology Lookup Service: more data and better tools for controlled vocabulary queries. *Nucleic Acids Res.* 36:W372–76
12. Django Project. 2008. **http://www.djangoproject.com/**
13. Drysdale R, FlyBase Consortium. 2008. FlyBase: a database for the *Drosophila* research community. *Methods Mol. Biol.* 420:45–59
14. Goldberg IG, Allan C, Burel J-M, Creager D, Falconi A, et al. 2005. The Open Microscopy Environment (OME) data model and XML file: open tools for informatics and quantitative analysis in biological imaging. *Genome Biol.* 6:R47
15. hibernate.org. 2008. **http://www.hibernate.org/**

16. Huang K, Lin J, Gajnak JA, Murphy RF. 2002. Image content-based retrieval and automated interpretation of fluorescence microscope images via the protein subcellular location image database. *Proc. 2002 IEEE Int. Symp. Biomed. Imaging* 325–28

17. Image J. 2008. **http://rsbweb.nih.gov/ij/**

18. Lein ES, Hawrylycz MJ, Ao N, Ayres M, Bensinger A, et al. 2007. Genome-wide atlas of gene expression in the adult mouse brain. *Nature* 445:168–76

19. Apache Lucene. 2008. *Overview.* **http://lucene.apache.org/java/docs/**

20. Macaulay C, Benyon D, Crerar A. 2000. Ethnography, theory and systems design: from intuition to insight. *Int. J. Hum. Comput. Stud.* 53:35–60

21. Martone ME, Tran J, Wong WW, Sargis J, Fong L, et al. 2008. The Cell Centered Database project: an update on building community resources for managing and sharing 3D imaging data. *J. Struct. Biol.* 161:220–31

22. MIPAV. 2008. **http://mipav.cit.nih.gov/**

23. Moore J, Allan C, Burel J-M, Loranger B, MacDonald D, et al. 2008. Open tools for storage and management of quantitative image data. *Methods Cell Biol.* 85:555–70

24. Murphy RF. 2008. *Murphy lab—Imaging services—PSLID.* **http://murphylab.web.cmu.edu/services/PSLID/**

25. myGrid. 2008. **http://www.mygrid.org.uk/**

26. Neumann B, Held M, Liebel U, Erfle H, Rogers P, et al. 2006. High-throughput RNAi screening by time-lapse imaging of live human cells. *Nat. Methods* 3:385–90

27. OME Consortium. 2008. *Data management.* **http://www.openmicroscopy.org/site/documents/data-management**

28. OME Consortium. 2008. *Development teams.* **http://openmicroscopy.org/site/about/development-teams**

29. OME Consortium. 2008. *OME server.* **http://openmicroscopy.org/site/documents/data-management/ome-server**

30. OME Consortium. 2008. *OMERO platform.* **http://openmicroscopy.org/site/documents/data-management/omero**

31. OME Consortium. 2008. **http://openmicroscopy.org/site**

32. OME Consortium. 2008. *Partners.* **http://openmicroscopy.org/site/about/partners**

33. OME Consortium. 2008. *Videos for Quicktime.* **http://openmicroscopy.org/site/videos**

34. OME Consortium. 2008. *OME at LOCI—software—Bio-formats library.* **http://www.loci.wisc.edu/ome/formats.html**

35. OME Consortium. 2008. *OME at LOCI—OME-TIFF—overview and rationale.* **http://www.loci.wisc.edu/ome/ome-tiff.html**

36. OME Consortium. 2008. *OME validator.* **http://validator.openmicroscopy.org.uk/**

37. OME Consortium. 2008. **http://ome-xml.org/**

38. OME Consortium. 2008. *Continuous build & integration system for the Open Microscopy Environment project.* **http://hudson.openmicroscopy.org.uk**

39. OME Consortium. 2008. *ServerDesign.* **http://trac.openmicroscopy.org.uk/omero/wiki/ServerDesign**

40. OME Consortium. 2008. *VisBio—introduction.* **http://www.loci.wisc.edu/visbio/**

41. OME Consortium. 2008. *The usable image project.* **http://usableimage.org**

42. OpenGIS®. 2008. *Standards and specifications.* **http://www.opengeospatial.org/standards**

43. Orlov N, Shamir L, Macura T, Johnston J, Eckley DM, Goldberg IG. 2008. WND-CHARM: multi-purpose image classification using compound image transforms. *Pattern Recognit. Lett.* 29:1684–93

44. Parvin B, Yang Q, Fontenay G, Barcellos-Hoff MH. 2003. BioSig: An imaging and informatic system for phenotypic studies. *IEEE Trans. Syst. Man Cybern. B* 33:814–24

45. Peng H. 2008. Bioimage informatics: a new area of engineering biology. *Bioinformatics* 24:1827–36

46. Pepperkok R, Ellenberg J. 2006. High-throughput fluorescence microscopy for systems biology. *Nat. Rev. Mol. Cell Biol.* 7:690–96

47. PostgreSQL. 2008. **http://www.postgresql.org/**

48. Rogers A, Antoshechkin I, Bieri T, Blasiar D, Bastiani C, et al. 2008. WormBase 2007. *Nucleic Acids Res.* 36:D612–17

49. Rueden C, Eliceiri KW, White JG. 2004. VisBio: a computational tool for visualization of multidimensional biological image data. *Traffic* 5:411–17

50. *Saccharomyces* Genome Database. 2008. **http://www.yeastgenome.org/**

51. Shamir L, Orlov N, Eckley DM, Macura T, Johnston J, Goldberg IG. 2008. Wndcharm—an open source utility for biological image analysis. *Source Code Biol. Med.* 3:13

52. Shaner NC, Campbell RE, Steinbach PA, Giepmans BN, Palmer AE, Tsien RY. 2004. Improved monomeric red, orange and yellow fluorescent proteins derived from *Discosoma* sp. red fluorescent protein. *Nat. Biotechnol.* 22:1567–72

53. Sun Microsystems. 2008. *Enterprise Javabeans Technology.* **http://java.sun.com/products/ejb**

54. Sun Microsystems. 2008. *Remote method invocation home.* **http://java.sun.com/javase/technologies/core/basic/rmi/index.jsp**

55. Swedlow JR, Goldberg I, Brauner E, Sorger PK. 2003. Informatics and quantitative analysis in biological imaging. *Science* 300:100–2

56. Whetzel PL, Parkinson H, Causton HC, Fan L, Fostel J, et al. 2006. The MGED Ontology: a resource for semantics-based description of microarray experiments. *Bioinformatics* 22:866–73

57. WormBase. 2008. **http://www.wormbase.org/**

58. ZeroC^TM. 2008. **http://www.zeroc.com/**

59. Zhao T, Murphy RF. 2007. Automated learning of generative models for subcellular location: Building blocks for systems biology. *Cytometry A* 71:978–90

Site-Directed Spectroscopic Probes of Actomyosin Structural Dynamics

David D. Thomas, David Kast, and Vicci L. Korman

Department of Biochemistry, Molecular Biology, and Biophysics, University of Minnesota, Minneapolis, Minnesota 55455; email: ddt@umn.edu, kast0040@umn.edu, korma002@umn.edu

Annu. Rev. Biophys. 2009. 38:347–69

First published online as a Review in Advance on February 5, 2009

The *Annual Review of Biophysics* is online at biophys.annualreviews.org

This article's doi:
10.1146/annurev.biophys.35.040405.102118

Key Words

EPR, spin label, fluorescence, luminescence, disorder, mobility

Abstract

Spectroscopy of myosin and actin has entered a golden age. High-resolution crystal structures of isolated actin and myosin have been used to construct detailed models for the dynamic actomyosin interactions that move muscle. Improved protein mutagenesis and expression technologies have facilitated site-directed labeling with fluorescent and spin probes. Spectroscopic instrumentation has achieved impressive advances in sensitivity and resolution. Here we highlight the contributions of site-directed spectroscopic probes to understanding the structural dynamics of myosin II and its actin complexes in solution and muscle fibers. We emphasize studies that probe directly the movements of structural elements within the myosin catalytic and light-chain domains, and changes in the dynamics of both actin and myosin due to their alternating strong and weak interactions in the ATPase cycle. A moving picture emerges in which single biochemical states produce multiple structural states, and transitions between states of order and dynamic disorder power the actomyosin engine.

Contents

INTRODUCTION

Since this subject was last covered in an *Annual Review* (74), two major advances outside the realm of spectroscopy have helped produce a renaissance in the spectroscopic study of muscle. First, high-resolution structures of the actin protomer (36) and the myosin head (58) were solved in the early 1990s (**Figure 1a**), and considerable further progress has been made in crystallography, coordinated with cryo-electron microscopy, of myosin and actin (30). This greatly increased the molecular detail and plausibility of models explaining how actin and myosin transform the chemical energy from ATP hydrolysis to the mechanical work of muscle contraction.

Figure 1b illustrates some of the principal hypotheses that developed from these advances. The best-characterized actin-myosin complex is the post-powerstroke rigor state, in which ADP-bound myosin binds strongly to actin at a ~45° angle. It was proposed that the actin-binding cleft is closed in this complex, while the ATP-binding pocket is open. This complex is short-lived, since ATP binds quickly and dissociates myosin from actin when the nucleotide

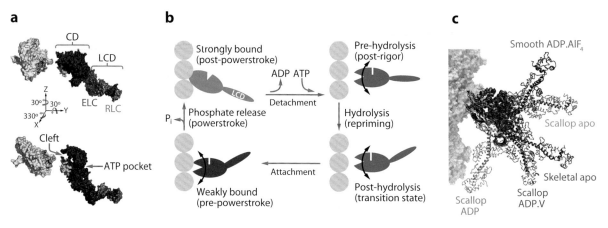

Figure 1

(*a*) Crystal structures of actin monomer (36) and myosin head (S1) (58). (*Top*) Oriented as in rigor, muscle fiber axis vertical, showing catalytic domain (CD), light-chain domain (LCD), essential light chain (ELC, *dark gray*) and regulatory light chain (RLC, *light gray*). (*Bottom*) Rotated to show the ATP-binding pocket and the actin-binding cleft. Residues proposed to form the actomyosin interface are cyan on myosin, tan on actin. (*b*) Proposed structural changes within S1 during muscle contraction based on crystal structures (19, 57) and on EPR data showing that a disorder-to-order transition occurs when myosin isomerizes from weak to strong binding (9). (*c*) Variable angle between CD [all shown bound to actin in the rigor orientation (*red*)] and LCD from crystal structures of S1 from scallop (33), chicken smooth muscle (19), and chicken skeletal muscle (58).

pocket closes and the actin-binding cleft opens. According to this model, ATP hydrolysis induces a large rotation of the light-chain domain (LCD), acting as a lever arm. A weakly bound pre-powerstroke complex is formed, characterized by dynamic disorder. In the proposed force-generating powerstroke, the lever arm rotates, coupled to closing of the cleft, opening of the pocket, and rigid ordering of the strongly bound myosin head. **Figure 1b** illustrates three key areas in which crystallographic results are ambiguous. (*a*) In myosin crystallography, which relies on the use of nucleotide analogs to trap stable complexes that mimic intermediates in the ATPase cycle, the correlation between structural state and biochemical state (defined by the bound nucleotide analog) is not consistent (**Figure 1c**) (19, 25, 28), probably because crystal packing perturbs the distribution between structural states, compared with solution, in which multiple structural states can populate a single biochemical state (51). (*b*) The key structures to be determined are complexes of actin and myosin, but there are no crystal structures containing both proteins. (*c*) Some of the key structural transitions involve dynamically disordered states, which cannot be defined by static crystal structures (20, 38, 76, 77). Spectroscopic methods have the potential to overcome these shortcomings and provide the means for direct testing and refining of these models.

In order for spectroscopic methods to test mechanisms in molecular detail, they must be site-directed—certain locations within the actin-myosin complex must be probed specifically. This has been made possible by a second major advance: technology for site-directed mutagenesis and expression of mutant muscle proteins, which now makes it possible to label virtually any residue by site-directed labeling (SDL). As discussed below, for myosin light chains this has been achieved in *Escherichia coli* (66), but it has not been achieved for skeletal muscle myosin heavy chain. Therefore, most SDL of myosin II heavy chain has been done either in *Dictyostelium discoideum* [starting from the native myosin gene (70)] or in insect

(*Spodoptera frugiperda*) cells [for smooth muscle myosin fragments (92)]. Similarly, SDL of actin has been done mainly on yeast actin (39), although recent progress has been reported in expression of muscle actin in insect cells (2). SDL usually entails first producing a "Cys-lite" version of the protein, in which all the reactive Cys have been replaced by a nonreactive amino acid. A specific Cys labeling site is then introduced by mutagenesis, followed by labeling with a Cys-reactive probe (39, 70). Alternatively, spectroscopic signals can be introduced by the use of genetically encoded probes (73, 95) or nucleotide analogs (49). All these labeling methods are applicable to actin-myosin complexes, and some can be used in active muscle fibers. SDL has allowed researchers to probe previously unexplored domains of muscle proteins, and to test and refine new models suggested by crystal structures.

In spectroscopy, there have also been significant technological advances, such as time-resolved luminescence resonance energy transfer (LRET) (43, 84, 85), single-molecule fluorescence (93), total internal reflection fluorescence (TIRF) detection (26), time-domain electron paramagnetic resonance (EPR) (37), multifrequency EPR (51), and electrostatic potential measured by EPR (72). These breakthroughs have allowed spectroscopy to test previously untestable molecular models of muscle structure and function. The present review surveys the recent use of spectroscopic probes to study myosin and actin structure and dynamics, focusing on site-specific probes that are applicable not only to isolated proteins in solution, but also to protein complexes and muscle fibers. To keep this review relevant to muscle, this review focuses on myosin II, the isoform found in muscle. Although primary emphasis is placed on probes of myosin, we also highlight some exciting recent probes of actin structural dynamics.

There have been recent reviews on site-directed spin labeling (34), distance measurement by EPR (35, 71), single-molecule fluorescence (63), and resonance energy transfer (RET) (64, 68) that discuss these spectroscopic methods in greater detail than this review.

Actin-binding cleft: gap between upper and lower 50-kDa domains of myosin that may close to allow strong binding to actin

LCD: light-chain domain

SDL: site-directed labeling

LRET: luminescence resonance energy transfer

Fluorescence: a type of luminescence in which light is emitted from the singlet excited state, which typically has a lifetime in the nanosecond range

Electron paramagnetic resonance (EPR): spectroscopic detection of microwave absorption by paramagnetic molecules in a magnetic field, providing high-resolution information about the probe's environment, orientation, and motion

Resonance energy transfer (RET): transfer of excited state energy from a luminescent donor to an acceptor, with a rate proportional to r^{-6}

There are also more detailed reviews on actomyosin dynamics (76, 94), actomyosin interactions (87, 88), muscle mechanics (44), and muscle regulation (23) that have focused on spectroscopic data.

LUMINESCENCE AND ELECTRON PARAMAGNETIC RESONANCE

This review focuses on optical emission (luminescence) and nitroxide EPR, because these are the only techniques that have the combination of sensitivity and resolution needed to analyze purified proteins as well as large complexes and muscle fibers. These methods are capable of measuring both probe orientation (due to anisotropic interactions of the probe with polarized light or magnetic field) and interprobe distances in the 2–10 nm range (due to dipole-dipole interactions) with high resolution, quantitate both static and dynamic disorder of these distances and angles, and detect solvent accessibility. Because these techniques differ in the type of probe used and the physical parameters observed, they are complementary. EPR has the advantage that most spin labels are smaller than most optical labels and thus usually cause smaller steric perturbations. Another advantage is that a single nitroxide probe and a single EPR spectrometer can be used to detect rotational dynamics in the nanosecond (by conventional EPR) and microsecond (by saturation transfer EPR, ST-EPR) time ranges, whereas optical spectroscopy typically requires two different probes and two different spectrometers, fluorescence and phosphorescence (74). EPR also has superior resolution in both orientation and probe-probe distance and can thus provide more quantitative and definitive information about the presence of disorder and multiple structural states. On the other hand, because the energy of an optical photon is 10^5 times that of a microwave photon, fluorescence is much more sensitive. Whereas EPR typically requires nanomoles of probe, fluorescence can be measured even at the single-molecule level, which can remove some sources of heterogeneity

within an ensemble and thus compensate for the relatively low spectroscopic resolution of fluorescence (63). When both techniques are performed at micromolar concentrations, fluorescence data can be acquired much more rapidly, even within the millisecond timescale of biochemical kinetics (47, 91).

It is particularly in the measurement of interprobe distances that EPR and luminescence reveal their complementary natures. Conventional, continuous wave EPR is sensitive to distances from 0.8 to 2.5 nm, almost exactly complementing the range of 2 to 10 nm that is accessible to RET. The recently introduced pulsed EPR technique called DEER (double-electron-electron resonance) is sensitive to distances from 1.5 to 6 nm (24, 37), allowing EPR to take the place of RET in many cases. On the other hand, DEER requires frozen samples, whereas RET does not. Thus when constructing Cys mutants for labeling, it is not predictable whether EPR or luminescence will prove more useful or valid, and the wisest course usually is to do both.

MYOSIN LIGHT-CHAIN DOMAIN
Global LCD Dynamics

The lever arm hypothesis suggests that the LCD behaves as a rigid body that amplifies and transmits ATP-driven conformational changes in the myosin catalytic domain (CD) to the thick filament core (58). This hypothesis predicts a large force-generating rotation of the LCD relative to the CD (**Figure 1b**), and there have been many attempts to test this model using orientation-sensitive probes by EPR or luminescence. The first direct evidence in support of this model in contracting muscle was obtained from a study in which spin-labeled regulatory light chain (RLC) was exchanged for endogenous RLC in skinned scallop muscle fibers, and then analyzed with the high orientational resolution of EPR (4). These fibers were perfused with solutions to induce the three physiological states of muscle, and revealed two populations of LCD angles that varied only in their mole fraction (**Figure 2**).

In rigor (no ATP, heads strongly bound to actin), a 38° population is predominant, with a much broader orientational distribution than observed previously for the CD (75), suggesting that there is some flexibility between the CD and LCD. In relaxation (ATP, no Ca^{2+}), 38° and 74° angles were equally populated, and this distribution shifted toward the 38° angle in contraction (ATP + Ca^{2+}) by 17%. In other words, both pre- and post-powerstroke structural states are populated in both relaxation and contraction, but only 17% of the heads change orientation upon activation of muscle, consistent with the 20% of heads previously found to be strongly attached to actin in active muscle, using probes on the CD (17). This 36° rotation is only about half that predicted from the most popular models, indicating that the actual lever arm swing in contraction is less than that suggested by crystal structures (**Figure 1**). Without the high resolution of EPR, even this 36° rotation would not be resolved, because the mean rotation between contraction and relaxation is only 3° (**Figure 2**), explaining why fluorescence polarization is insensitive to this rotation (discussed below). Transient EPR following the photolysis of caged ATP shows that this LCD rotation lags behind the initial phase of force generation (42). This observation suggests that the powerstroke has two roughly equal components—a disorder-to-order transition of the CD followed by a lever arm rotation (**Figure 1**).

The quantitative resolution by EPR of the pre- and post-powerstroke structural states of myosin in active muscle (4), coupled with simultaneous measurement of muscle force, enabled analysis of the coupling between thermodynamics and structural mechanics within the muscle filament lattice (5). This work showed that inorganic phosphate decreases active isometric muscle force by decreasing the average force per strongly attached myosin head, without changing the number of strongly attached heads as classically assumed. Thus force is limited by the free energy of the weak-to-strong transition within the coupled ensemble of actomyosin interactions (6).

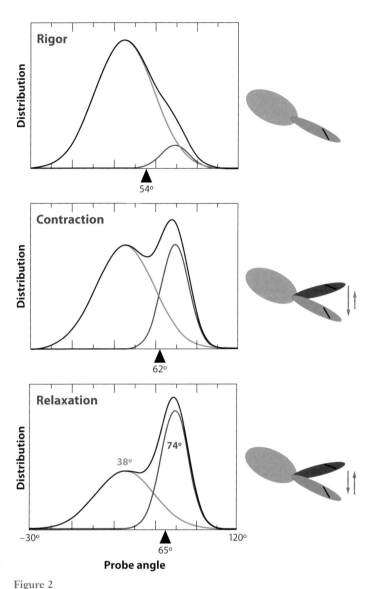

Figure 2

Distributions of regulatory light chain (RLC)-bound probe angle (relative to fiber axis) detected by EPR in muscle fibers (4). The total distribution (*black*; mean indicated by *black arrowhead*) is a linear combination of pre- (*red*, 74° mean) and post-powerstroke (*green*, 38° mean) distributions.

LCD orientation was also measured from the polarized emission of a fluorescent probe on RLC (18, 31, 65). To determine accurately the orientation of the LCD, a bifunctional rhodamine dye was reacted with a series of Cys pairs engineered in a Cys-lite RLC construct (18). This labeled RLC was reconstituted into skinned rabbit muscle fibers, and fluorescence polarization was then measured as a function of physiological state. The mean tilt of the LCD changed by only a few degrees from rigor to relaxation, and the change from relaxation to contraction was not significant (18). As explained above, this result is consistent with that obtained using EPR (4), which had the necessary resolution to detect the small population of actively rotating heads. Subsequent studies employing rapid length steps did show significant LCD rotation that correlated with force, supporting the lever arm model (32). By studying several probes attached at different orientations, researchers found that both tilting and twisting of the LCD occurs upon rapid length changes in rigor (18, 32).

Fluorescence polarization was used to detect LCD orientation at the single-molecule level, with the potential of circumventing some sources of ensemble heterogeneity. In smooth muscle myosin, the RLC adopts two orientational states on actin during force generation (81), attributed to a disordered weak-binding state followed by an ordered strong-binding state, consistent with the disorder-to-order model of force generation (**Figure 1b**). In a single-molecule study of labeled RLC on skeletal myosin in rigor, using bifunctional rhodamine and TIRF polarization microscopy, a more heterogeneous orientational distribution was observed for the two-headed heavy

meromyosin (HMM) than for the single-headed S1, suggesting a flexible connection between the CD and LCD (55).

Clear resolution by fluorescence of transitions between two distinct myosin head orientations on actin was finally obtained by combining time resolution, single-molecule detection, and myosin V, a processive motor in which most heads bind to actin even in the presence of ATP (27). This study resolved two interchanging orientations of the LCD (with bifunctional rhodamine attached to a calmodulin light chain) during active sliding on actin. These two LCD angles were tilted by about 42° (**Figure 3**) (27) in remarkable agreement with the 36° tilt resolved previously by EPR in contracting muscle (**Figure 3**) (4). Owing to the $\cos^2\theta$ dependence of the observed signals, it is equally likely that the tilt is on the order of 70° in both cases.

The above methods measure the orientational distribution of the LCD but do not reveal whether disorder is static or dynamic. Microsecond rotational dynamics of the LCD has been measured by phosphorescence (56) and ST-EPR (61), using RLC-bound probes. Both studies showed microsecond rotational motions in relaxation, slower and more restricted motions in contraction, and much more restricted motions in rigor. Thus the LCD undergoes a disorder-to-order transition as it transitions from relaxation to contraction to rigor, as observed for the CD (9). However, the LCD is (a) less dynamic (more ordered) than the CD in relaxation and (b) more dynamic (less ordered) than the CD in rigor, adding to the evidence of a flexible linkage between the CD and LCD.

While changes in probe orientation and rotational motion dominated early work in testing the lever arm rotation model, distance measurement by RET between the CD and LCD has become more important in recent years. Initial LRET studies were limited to native labeling sites, which were not strategically placed to detect this predicted interdomain rotation (14, 52, 84, 86). Real progress required SDL. A breakthrough was achieved by fusing green and blue fluorescent proteins (GFP and BFP) to the N and C termini of *D. discoideum*

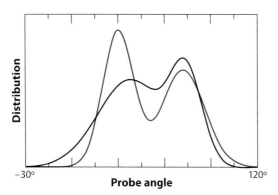

Figure 3

LCD probe angular distributions from EPR on contracting scallop muscle fibers (*black*) (4) and single-molecule fluorescence polarization on myosin V (*red*) (27).

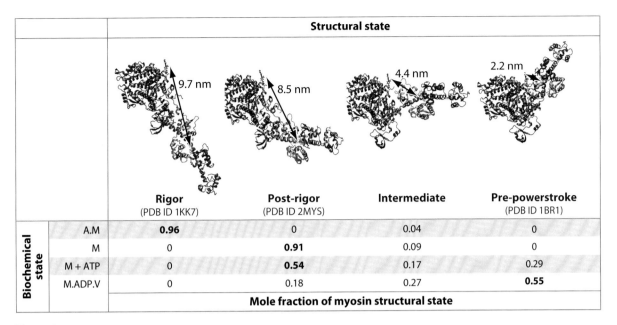

		Structural state			
		Rigor (PDB ID 1KK7)	**Post-rigor** (PDB ID 2MYS)	**Intermediate**	**Pre-powerstroke** (PDB ID 1BR1)
Biochemical state	A.M	**0.96**	0	0.04	0
	M	0	**0.91**	0.09	0
	M + ATP	0	**0.54**	0.17	0.29
	M.ADP.V	0	0.18	0.27	**0.55**
		Mole fraction of myosin structural state			

Figure 4

Position of the light chain domain relative to the catalytic domain (CD) measured by time-resolved fluorescence resonance energy transfer (FRET) within *Dictyostelium discoideum* S1. (*Top*) Four structural states resolved by FRET, with structural models constructed as described in Reference 70, by adding the converter domain and lever arm of the indicated PDBs to the *D. discoideum* CD. The structural states illustrate agreement with the FRET-measured interprobe distances. (*Bottom*) Table showing the distribution of these structural states within each of the four biochemical states. Fractions for M, M + ATP, and M.ADP.V are calculated from data provided in Reference 70 by normalizing to these three structural states; A.M data and mole fractions are from D. Kast (personal communication).

CD (S1dC) (73). Fluorescence resonance energy transfer (FRET) showed these fluorescence proteins move apart 15 to 18 Å when ATP or a post-hydrolysis analog (ADP.AlF$_4$ or ADP.V) is bound, but not when ADP or a pre-hydrolysis analog (ATPγS) is bound, supporting the proposal that the lever arm (attached to the C terminus) rotates, producing a severely bent myosin head, only in the post-hydrolysis (pre-powerstroke) state (**Figure 1b**). However, this construct did not even contain a significant portion of the lever arm, so this experiment did not provide a quantitative test of the model.

RET studies discussed thus far could not match the orientational resolution of EPR, which resolved two distinct LCD states (4). Equivalent resolution with FRET was finally achieved by combining time-resolved fluorescence detection with SDL of Cys-lite myosin (**Figure 4**). A labeling site, C250, was engineered in the CD of Cys-lite *D. discoideum* myosin II full-length S1 that was optimally placed to detect movement of the LCD (70). FRET was measured to this site, labeled with tetramethylrhodamine (acceptor), from Oregon Green 488 (donor) on RLC (C114 or C116). Large interdomain movements were detected. Nucleotides decreased the distances between probes, indicating more acute angles, as predicted. However, time resolution revealed that, as was discovered previously with EPR in muscle fibers (4), multiple structural states were populated in a single biochemical state. Each of the rigor (A.M) and apo (M) biochemical states is populated predominantly by a single structural state, but actin perturbs the structure of apo-S1, making it even straighter. With the addition of ATP, three structural states are populated, but the major one is still post-rigor. With the addition of the post-hydrolysis analog ADP.V, the predominant state is the compact

S1dC: catalytic domain of *Dictyostelium discoideum* myosin II

FRET: fluorescence resonance energy transfer

pre-powerstroke state as predicted (19), but two other structural states are also populated, including an intermediate structure not previously observed. Thus these FRET measurements confirm some of the predictions from crystallography, but they describe a heterogeneous structural distribution in solution, even in a well-defined biochemical state, and they suggest that at least one myosin structural state has not been revealed by crystallography.

Starting from a similar Cys-lite construct of smooth muscle myosin S1, LRET from a chelated Tb^{3+} donor (attached to C210, essentially the same site labeled in Reference 70) to an acceptor on RLC (85) confirmed that a post-hydrolysis nucleotide analog (in this case $ADP.AlF_4$) makes S1 more compact and found that actin makes S1.ADP slightly more compact, consistent with previous electron microscopic results.

LRET, using Tb^{3+} chelates bound to single-Cys RLC mutants, was used to measure distances between LCDs on the two heads of myosin in scallop muscle fibers (43) and in skeletal muscle HMM (15). In the muscle fiber study, the head-head separation was greatest in relaxation, least in rigor, and intermediate in contraction. Using four different labeling sites on RLC, it was found that the relative tilt of the two LCDs decreases by 30° in the force-generating weak-to-strong transition (43). The HMM study concluded that the two heads of myosin bind to actin in rigor in a strained configuration that is distinct from that of single-headed binding (15).

Flexibility Within the LCD

Spectroscopic measurements with probes on both essential light chain (ELC) and RLC have been compared to determine whether the LCD acts as a rigid body (lever arm). ST-EPR (8) and time-resolved phosphorescence anisotropy (TPA) (13) in myosin filaments showed that the microsecond rotational mobility of ELC is only slightly greater than that of RLC and substantially less than that of the CD. These results indicate that most of the flexibility within S1

is between the CD and LCD, not between the ELC and RLC. Similarly, RET between ELC and RLC probes showed little or no effect of nucleotides (52). These results are consistent with the proposition that the LCD acts as a semirigid lever arm.

The fluorescence polarization of bifunctional rhodamine, previously applied to the RLC as discussed above, was recently extended to the ELC in skinned rabbit psoas fibers (38). The data obtained in rigor, when combined with previous RLC data (12), could be reconciled with the nucleotide-free myosin crystal structure (**Figure 5a**) only by introducing both a 45° tilt (β) and a 25° twist (φ) between the ELC and RLC (**Figure 5b**). In rigor, the ELC had a narrower orientational distribution than the RLC, consistent with its proximity to the rigid actin bond. In relaxation, the RLC was more ordered, consistent with its proximity to the thick filament backbone. In contraction, the ELC orientational distribution was slightly more ordered than in relaxation, consistent with previous observations of RLC orientation by fluorescence (65) and EPR (4). These results, together with the ELC/RLC probe results discussed above, indicate that the principal joint of flexibility (compliance) within the myosin head is between the CD and LCD, but there is also some flexibility within the LCD (**Figure 5**).

RLC Structure and Dynamics

Phosphorylation of S19 on RLC activates the actin-activated ATPase of smooth muscle myosin, inducing contraction. Less dramatic activation occurs in cardiac and skeletal muscle. The structural basis of this regulatory mechanism remains unknown, primarily because there is no crystal structure of the N-terminal 24 amino acids of any RLC, presumably because this segment is disordered and/or susceptible to proteolysis. This structural problem has been solved by site-directed spectroscopy. In a pioneering study, FRET was measured between probes attached to an engineered Cys at position 2 on skeletal RLC and three other sites

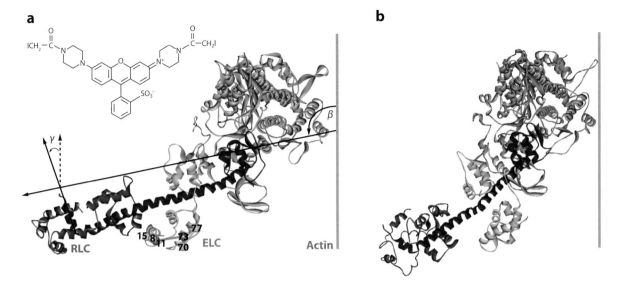

Figure 5

Conventional model for the rigor acto-S1 complex (*a*) compared with revised structural model (*b*), based on fluorescence polarization data using bifunctional rhodamine. Adapted with permission from Reference 38. ELC, essential light chain; RLC, regulatory light chain.

within the LCD (66, 83). Although these results were not enough to construct a structural model, the measured distances later proved useful as constraints for modeling.

A more complete structural analysis was later achieved by site-directed spin labeling (50). Single-Cys mutagenesis was performed throughout smooth muscle RLC expressed in *E. coli*, and then EPR experiments were performed on spin-labeled RLC bound to intact and functional HMM to determine side chain dynamics and solvent accessibility as affected by phosphorylation at S19. A significant phosphorylation effect was only observed in the N-terminal 25 amino acids, showing that this region can properly be designated the phosphorylation domain. Cys-scanning SDL was performed on this domain, and the pattern of oxygen accessibility along the sequence was analyzed by EPR (**Figure 6a**). In the absence of phosphorylation, little or no periodicity was observed, suggesting a lack of secondary structural order in this region. However, phosphorylation induced a strong helical pattern in the

first 17 residues, while increasing accessibility throughout the first 24 residues. The results support a model in which this disorder-to-order transition within the phosphorylation domain results in decreased head-head interactions, activating myosin in smooth muscle (50). Molecular dynamics simulations confirmed this result and produced the first atomic-resolution structural models of the phosphorylation domain in the presence and absence of phosphorylation (20) (**Figure 6b**). In these models, phosphorylation stabilizes a salt bridge between pS19 and R16, which in turn stabilizes the N-terminal helix. In silico mutation of this arginine to alanine or glutamate destabilizes this helix (21). Although the structural mechanism of communication between the phosphorylation domain and CD is not yet known, a fluorescence quenching study with acrylodan demonstrated that the phosphorylation domain is sensitive to the nucleotide state of myosin (48). Because this sensitivity depends on phosphorylation, it is likely that the more ordered form of the N-terminal helix reverses inhibitory head-head

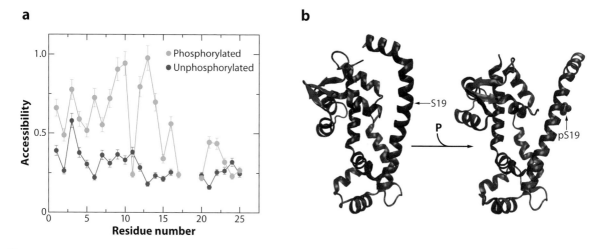

a

Accessibility

- Phosphorylated
- Unphosphorylated

Residue number

b

←S19

P

pS19

Figure 6

Effect of phosphorylation on smooth muscle myosin regulatory light chain (RLC) structure. (*a*) EPR from site-directed spin labeling; the sequence dependence of accessibility changes dramatically, implying a transition from disorder to helical order (50). (*b*) Structural model. Blue, myosin heavy chain. Red, RLC (25–168) from crystal structure (PDB ID 1SEM). Purple, RLC (1–24), based on FRET (66), EPR (50), and molecular dynamics simulation (20, 21), showing that phosphorylation induces a disorder-to-order transition in the N-terminal domain of RLC.

or head-tail interactions (50). Thermodynamic analysis of simulations revealed that phosphorylation of S19 in RLC is accompanied by a large, unfavorable decrease in entropy, which is offset by a slightly larger, favorable decrease in enthalpy (21). The small size of the total free energy change (6 kcal mol⁻¹) helps ensure that this phosphorylation-dependent disorder-to-order transition is a highly reversible regulatory switch.

MYOSIN CATALYTIC DOMAIN

Nucleotide Pocket

Based on X-ray crystal structures, it has been hypothesized that the nucleotide-binding pocket closes in weak-binding states and opens in strong-binding states (**Figure 1***b*). FRET measurement between W512- and MANT-labeled nucleotide analogues was used to test and refine this hypothesis. When MANT-ADP is bound (strong binding), FRET efficiency is 13 ± 1.3% compared to 35.4 ± 5.5% for MANT-ATP (weak binding), corresponding to an increase of 6.7 Å in probe separation,

suggesting that the nucleotide pocket is more open in the strong-binding state (90). Solvent accessibility measurements were consistent with these results. More recently, FRET was measured from an engineered W344 (upper 50-kDa domain) to MANT nucleotides (60). Again, MANT-ADP produced less FRET than MANT-ATP did, indicating that the two sites become closer in the weak (ATP) to strong (ADP) transition by 8.5 Å. This provides compelling evidence of a nucleotide-dependent structural change within the upper 50-kDa domain, becoming more compact in the weak-to-strong transition.

Nucleotide spin labels have provided a direct probe of opening and closing of the nucleotide pocket. An EPR study (49) using spin-labeled nucleotides and analogues [SL-ADP (**Figure 7**), SL-AMPPNP, SL-ADP·BeF₃, or SL-ADP·AlF₄] showed that the spin label is restricted (**Figure 7***a*) in all cases when myosin is free in solution or weakly bound to actin, indicating a closed pocket, as predicted by crystal structures (**Figure 1***b*). However, when SL-ADP was bound to myosin, the formation of a strongly bound complex (A.M.D) with actin revealed a new spectral component implying greatly increased mobility (**Figure 7a**) and an open nucleotide pocket, which had been predicted (**Figure 1a**) but never observed. Unexpectedly, it was observed that both the open and closed structural states of the pocket are populated (**Figure 7a**), indicating that there are two distinct structural states in the A.M.D biochemical state (**Figure 7a**). EPR was used to measure the K_{eq} value for this transition. The thermodynamics of this structural equilibrium was determined from a van't Hoff plot (**Figure 7b**), revealing that the large positive (unfavorable) enthalpy change in the pocket-opening reaction is offset almost precisely by a favorable entropy increase (order-to-disorder transition), thus balancing the reaction.

Actin-Binding Cleft

One hypothesis to arise from the first actomyosin model (57) with support from more recent models (29, 79) is that the actin-binding cleft between the upper and lower 50-kDa domains closes upon rigor binding to actin (**Figure 1b**). Several spectroscopic studies have been carried out to test this hypothesis in solution. Starting with a Trp-lite construct of the smooth myosin CD, Trp was engineered at position 425 in the actin-binding cleft (91). Solvent exposure of W425 was decreased by actin binding and increased by ATP, with the same kinetics as pyrene-actin fluorescence, suggesting that cleft closure coincides with the weak-to-strong actomyosin transition. Subsequently, a pair of Cys residues (537 and 416), one on each side of the cleft, was introduced into a Cys-lite mutant of S1dC (16). Pyrene fluorescence increased upon ATP binding and decreased upon actin binding. Because the opposite behavior had been expected, it was suggested that the pyrene probes are too large to fit inside the closed cleft. When spin labels were attached to the same sites, and spin-spin distances were measured by both continuous wave EPR and DEER, the results were more consistent with the conventional models (22). A recent EPR study on S1dC employed five different pairs of spin-labeling sites

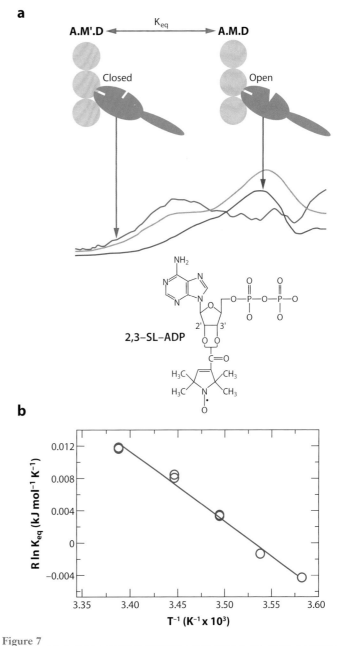

Figure 7

Structure and thermodynamics of the nucleotide pocket deduced from EPR (49). (*a*) Low-field portion of EPR spectrum of SL-ADP bound to myosin in rabbit psoas muscle (*orange*), deconvoluted into its two components (*blue and red*). (*b*) van't Hoff plot of K_{eq} for opening of the pocket, determined from EPR.

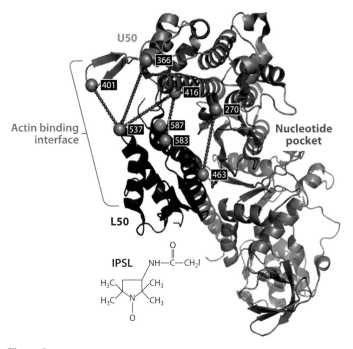

Figure 8

Pairs of spin-labeled Cys residues used to probe cleft closure in S1dC (37).

(**Figure 8**) and thus provided a more comprehensive picture (37). The data were consistent with actin-induced cleft closure, but most labeling sites gave evidence of two conformations of the cleft. These two structural states (open and closed clefts) were populated in each of the biochemical states (pre-hydrolysis, post-hydrolysis, and rigor), with the closed state most highly populated in rigor. Thus actin binding is not required for cleft closure, nor is actin dissociation required for cleft opening. These results suggest that the conformational distribution of the cleft in solution is more complex and dynamic than in the crystal. This conclusion is consistent with previous studies of probe dynamics and accessibility (39).

Force-Generating Region

Early spectroscopic studies of myosin showed that ATP binding enhances Trp fluorescence (82), and there is a further enhancement upon ATP hydrolysis (3). Spectral and structural

analysis of Trp residues in the CD of skeletal myosin suggested that these changes arise primarily from W510 (59) [W512 in smooth muscle (90) and W501 in *D. discoideum* (7)], which is located on the rigid relay loop (45), just N-terminal from the relay helix that directly links the nucleotide-binding site to the SH1-SH2 helix and the converter domain and thus to the LCD. In particular, the relay loop and helix are coupled directly to the nucleotide-binding pocket through switch 2, which responds directly to both nucleotide binding and hydrolysis, so the fluorescence of this conserved Trp has proven to be a powerful tool in the study of myosin's transient ATPase kinetics (46, 78). However, Trp fluorescence intensity does not provide well-defined structural information, as can be obtained from EPR or FRET.

Crystal structures suggest that the power-stroke, a transition from bent to straight structural states (**Figure 9a**), is driven by a similar bent-to-straight transition in the relay helix (**Figure 9b**). The reversal of this step, the recovery stroke, should be observable in isolated myosin upon the hydrolysis of ATP or with the binding of post-hydrolysis nucleotide analogs such as ADP.V or ADP.AlF$_4$. This was tested in S1dC, by measuring the distance between two probes at engineered Cys within the CD, one at the N terminus of the relay helix (C498) and another at 515 in the upper 50-kDa domain. Both DEER and FRET showed the predicted result: Addition of ATP, ADP.V, or ADP.AlF$_4$ decreased the interprobe distance, demonstrating that the relay helix bends during the recovery stroke. Unexpectedly, both bent (pre-powerstroke) and straight (post-powerstroke) structural states (**Figure 9**) were populated simultaneously in a single biochemical state, trapped by a pre-hydrolysis analog (ADP.BeF$_3$). This was confirmed by transient FRET after rapid ATP addition, which showed that the bent structural state is populated before ATP is hydrolyzed. This direct structural observation is consistent with previous kinetic measurements of Trp fluorescence in S1dC (46, 47).

The SH1-SH2 helix was among the first sites labeled covalently with probes on myosin, because SH1 (Cys 707) is the most reactive Cys residue and can therefore be labeled specifically with a wide range of thiol reagents. This is a sensitive region, so most probes result in partial inhibition of ATPase activity. The mobility of an iodoacetamide spin label (IASL), attached to SH1 in rabbit skeletal myosin, is sensitive to ATP binding and hydrolysis; its EPR spectra resolve three distinct structural states of this region (reviewed in References 74 and 76). This resolution is even greater when the microwave frequency is increased by a factor of 10 from the conventional 9.4 GHz (X-band) to 94 GHz (W band). When EPR spectra at both frequencies were recorded for IASL-labeled S1, it was possible not only to resolve the three structural states, but also to define the spin label's amplitude and rate of motion in each state, giving detailed insight into the environment of the SH1-SH2 helix, which is so central to the force-generating function of myosin (51). Spectra obtained with a series of nucleotides permitted assignment of these structural states to crystal structures.

To extend this analysis to *D. discoideum*, which lacks SH1, a single-Cys mutant (T688C) was engineered to have Cys at the equivalent site (1). Spectra were identical for the two myosins in the apo and ADP-bound states, in which single (but distinct) structural states (probe mobilities) were observed. However, with bound ADP and phosphate analogs, (*a*) both proteins exhibit two resolved structural states (pre-powerstroke and post-powerstroke) in a single biochemical state (defined by the bound nucleotide); (*b*) in the two myosins, these structural states are essentially identical but are occupied to different extents as a function of the biochemical state. It is clear that myosin structural and biochemical states do not have a one-to-one correspondence, and that *D. discoideum* myosin II differs from muscle myosin in the coupling between biochemical and structural states. As in the case of the nucleotide spin labels, the quantitative resolution of EPR permitted the thermodynamic analysis of the powerstroke structural transition, revealing an increase in entropy (51), consistent with the hy-

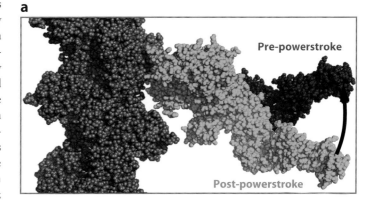

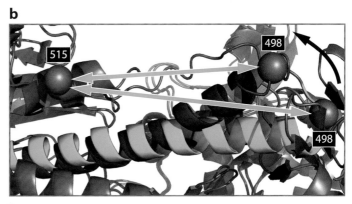

Figure 9

(*a*) Reversal of powerstroke (recovery stroke) detected by EPR (4). (*b*) Bending of relay helix in myosin recovery stroke, predicted by crystal structures and detected by DEER and FRET (R. Agafanov & Y. Nesmelov, personal communication).

pothesis that the powerstroke is a disorder-to-order transition in which enthalpic and entropic changes are both large and compensatory.

Global CD Dynamics

Cross-linking Cys 707 (SH1) and Cys 697 (SH2) in the CD of myosin weakens the actin-myosin bond, producing a biochemical state with properties expected for the post-hydrolysis complex A.M.ADP.P (reviewed in Reference 77). When these two thiols were cross-linked with a bifunctional spin label (BSL), this weak-binding state was trapped and the spin label was completely immobilized (**Figure 10**), reflecting the global orientation and dynamics of the myosin CD (77). Conventional EPR of

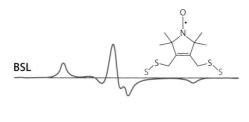

BSL

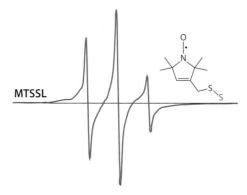

MTSSL

Figure 10

EPR spectra of myosin S1 labeled with a bifunctional methanethiosulfonate spin label (BSL), which is strongly immobilized, compared with the monofunctional methanethiosulfonate spin label (MTSSL), which undergoes almost complete nanosecond rotational mobility (77).

BSL-S1 on oriented actin showed that the CD is partially disordered, as previously observed for ATP-induced weak binding. However, ST-EPR showed that this disorder is slow on the microsecond timescale, as previously observed for strong-binding states, suggesting that the trapped state is a missing link between weak and strong binding.

Results on myosin-bound spectroscopic probes, discussed above, were used to construct the model in **Figure 11**, showing that force and actin movement are generated in two steps, first a disorder-to-order transition in the CD and then rotation of the LCD (77). Of course, these and other transitions in **Figure 11** are probably not strictly sequential, since in several biochemical states evidence is described above for a structural equilibrium among structural states.

ACTIN DYNAMICS

Spectroscopy has shown that the changes in structural dynamics caused by the formation of the actin-myosin complex and the weak-to-strong transition are not limited to myosin. Indeed, one of the most widely exploited muscle probes is pyrene iodoacetamide attached to

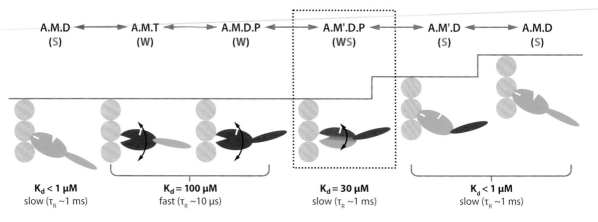

Figure 11

Updated model for the coupling of actomyosin ATP hydrolysis to force and movement, summarizing key results discussed in this review. Only actin-bound states are shown. Red and green signify weak (pre-force) and strong (force-generating) complexes, curved arrows signify orientational disorder, and upward steps indicate stages in the cycle where force and movement are likely to be imparted to actin. A, actin; M, myosin; T, ATP; D, ADP; P, inorganic phosphate. Text under each state indicates distinguishing properties of the catalytic domain (77). Prime symbol (′) indicates a second structural state corresponding to the same biochemical state (defined by the active-site ligand). Although these primed and unprimed states are shown in a sequence, they are presumably interconverting in solution or muscle fibers.

C374 on actin, which is strongly quenched by strong, but not weak, myosin binding (40). Previous spectroscopic studies have shown that the unbound actin filament exhibits intermonomer torsional and bending motion, as well as internal dynamics within the actin monomer (reviewed in Reference 76). Both optical (mainly TPA) and EPR (mainly ST-EPR) experiments showed that strong myosin binding constrains the dynamics of the actin filament at substoichiometric levels, indicating that the effects are propagated cooperatively along the filament. These observations suggest that as myosin isomerizes from the weak to strong binding state, in a disorder-to-order transition, the actin filament also undergoes a transition from dynamic disorder to order. There is a correlation between inhibited actomyosin interactions due to actin modification (e.g., covalent modification, isoform differences, or mutation) and alterations in actin filament structural dynamics (76).

How does actin structural dynamics change in the transition between weak and strong binding? This was analyzed by performing TPA experiments in the absence and presence of saturating ATP, in the brief steady state before ATP is depleted (54). In the absence of ATP, the final anisotropy rises nonlinearly with myosin binding, indicating that each myosin cooperatively restricts the dynamics of about three actin protomers (**Figure 12a**). At saturation, weakly bound myosin has almost as much effect as strongly bound myosin, but a much smaller effect is seen at lower binding levels. Strikingly, the dependence is precisely linear—the cooperativity is completely lost. Thus (*a*) actin undergoes a partial ordering transition upon weak binding and a more profound one upon strong binding (**Figure 12b**), and (*b*) in the weak-to-strong transition both myosin and actin undergo disorder-to-order transitions. The functional significance of this finding was recently underscored by the finding that oxidative stress inhibits muscle contraction by perturbing actin dynamics in the weak complex but not in the strong complex,

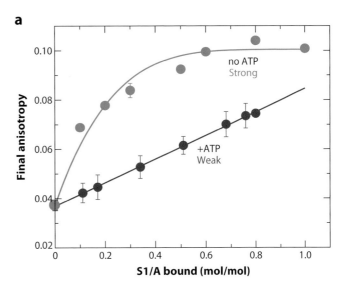

a

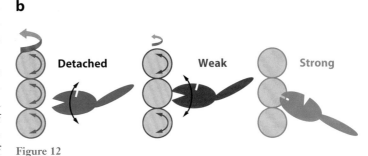

b

Figure 12

(*a*) The effect of S1 on TPA of actin in the absence (strong binding) and the presence (weak binding) of saturating ATP (54). (*b*) Both myosin and actin undergo disorder-to-order transitions upon attachment and upon the weak-to-strong transition.

thus decreasing the magnitude of the weak-to-strong transition (53).

Further support for actin's active role in the actomyosin interaction came from a single-molecule FRET study, which measured structural changes within the actin protomer, using rhodamine on N41 as donor and IC5 on C374 as acceptor (41). Actin switches between two states, defined by low- and high-FRET efficiencies, on a timescale of seconds. Active interaction with myosin favors the high-FRET conformation, but cross-linked actin (which does not interact actively with myosin) favors the low-FRET conformation.

Anisotropy (*r*): luminescence polarization signal used to measure rotational dynamics in the time scale of excited-state lifetime

Recent studies have probed actin's structural response to myosin in skinned muscle fibers. Actin was labeled with either rhodamine-phalloidin or Alexa-ATP, and polarization showed a myosin-dependent biphasic response to an ATP pulse (from photolysis of caged ATP), indicating that actin rotates during contraction (10). A similar approach was used to detect orientational signals simultaneously from actin (labeled on phalloidin), myosin (labeled on RLC), and Alexa-ADP (bound to the active site of myosin) (69). Following caged ATP photolysis, the initial phases of RLC rotation (presumably due to cross-bridge detachment), actin rotation, and ADP release (which is presumably simultaneous with ATP binding) were essentially simultaneous, consistent with the model that actin is responding to ATP-induced dissociation of myosin. Another study measured the effects of weak and strong myosin binding on the orientation and dynamics of actin in muscle (11). Actin monomers were labeled either in the nucleotide-binding pocket or on one of five actin residues and then incorporated into muscle fibers. Strongly and weakly bound

myosin produced opposite effects on probe orientation, confirming similar conclusions from solution studies (54) (**Figure 12**) and indicating that actin is an active participant in muscle contraction.

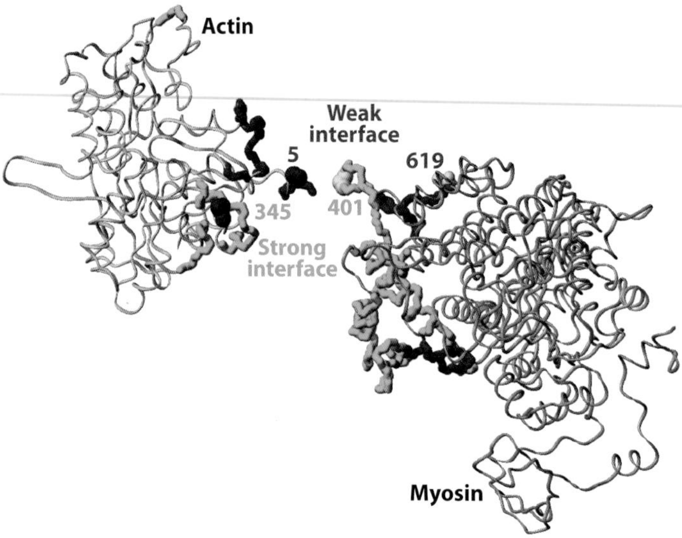

ACTOMYOSIN INTERFACE

There are no high-resolution structures of actin-myosin cocrystals; however, several models have been constructed for these complexes by docking crystal structures into electron microscopy images (29, 57, 67, 80). Although these actomyosin structural models are only for the strongly bound complexes and they do not include dynamics of the interaction, they provide a testable hypothesis for spectroscopic measurements in solution.

Before the development of SDL, spectroscopic studies on the actomyosin complex were limited to placing probes on residues located outside of the proposed interface (reviewed in Reference 76). A recent study used yeast actin and *D. discoideum* myosin to introduce single Cys mutations into four labeling sites located in the proposed weak-binding and strong-binding interfaces of each protein, and both fluorescent and spin labels were attached to each (**Figure 13**) (39). While formation of the rigor complex resulted in decreased probe mobility at all four sites, both myosin sites and one of the actin sites showed remarkably high probe mobility even after complex formation. Complex formation decreased solvent accessibility for both actin-bound probes but increased it for the myosin-bound probes. These results are not consistent with a simple model in which there are discrete weak and strong interfaces with only the strong interface forming under strong-binding conditions, nor are they consistent with a model in which surface residues become rigid and inaccessible upon complex formation. In fact, the actin-myosin interface remains dynamic, especially on the surface of myosin, even under rigor binding conditions.

RET has been used to measure distances between labels on actin and myosin, with the objective of testing models for the actin-myosin

Figure 13

Proposed actin-myosin interface: Yeast actin (*left*) and S1dC (*right*) are shown in their rigor orientation (filament axis vertical) but separated horizontally to reveal the proposed interface involved in strong binding (*green*) and weak binding (*red*). The four labeling sites used in Reference 39 are indicated.

complex. In a study on smooth muscle myosin, two distances within the rigor complex were consistent with most models, but no effect of ADP was observed, indicating that the known effect of ADP on the LCD orientation of actin-bound smooth muscle S1 is not detectable with the chosen probe in the CD (89). Further studies of this kind, with several pairs of probes in well-defined sites, coupled with computational analysis incorporating data from crystallography, electron microscopy, and FRET or DEER, are needed to provide more rigorous tests of actin-myosin complex structures (62).

CONCLUSIONS AND PERSPECTIVES

Structural dynamics, disorder, and conformational heterogeneity are essential features of the actomyosin machine, especially involving myosin II, which is a low-duty-cycle motor. Therefore, crystal structures serve as the building blocks (models for structural states) that the spectroscopist uses to determine the actual mechanisms (i.e., the coupling between biochemical and structural states in solution and in muscle). EPR and luminescence have the required properties—sensitivity and resolution—to lead this effort, and the two approaches are complementary. Spectroscopic resolution, in both frequency and time, has allowed these methods to demonstrate that a single biochemical state of myosin or actin is populated by two or more structural states. Recent results show that both myosin and actin undergo coupled weak-to-strong (disorder-to-order) transitions that are crucial to muscle contraction and regulation. Future advances will continue to depend on advances in SDL, computational simulation, and spectroscopic instrumentation, correlating structural changes in multiple domains with functional measurements of force, movement, and biochemical kinetics. These advances will lead to more direct tests of complex molecular models, connecting the separate parts of actomyosin in space and time, and defining which of these relationships and events are most crucial for the function of muscle contraction.

SUMMARY POINTS

1. The discovery of high-resolution structures for myosin and actin along with the development of expression systems for these proteins has helped to produce a renaissance in the spectroscopic study of muscle. Crystal structures are the building blocks (structural states) that the spectroscopist uses to determine the actual mechanisms (i.e., the coupling between biochemical and structural states).

2. Unlike crystallography, spectroscopy can resolve structural features of (*a*) proteins and protein complexes that are not crystallizable and (*b*) partially disordered regions of proteins.

3. EPR and luminescence are complementary methods and are most effectively used in tandem.

4. Although luminescence (fluorescence) is much more sensitive, EPR has smaller probes and more intrinsic resolution.

5. The weak-to-strong transition in myosin, which generates force on actin, occurs in at least two structural steps, first a dynamic disorder-to-order transition of the catalytic domain and then a tilting of the LCD.

6. Spectroscopic resolution—in both frequency and time—allows both EPR and fluorescence to show that a single biochemical state of myosin or actin is populated by two or more structural states.

7. Quantitation of structural states' populations gives insight into thermodynamics as well as structure.

8. Myosin and actin both undergo coupled weak-to-strong (disorder-to-order) transitions that are crucial to muscle contraction and its regulation. Dynamics and disorder are critical aspects of myosin and actin structure and function.

FUTURE ISSUES

1. To assess the relationships of actomyosin dynamics to biochemical kinetics, site-directed spectroscopy should be measured in the transient phase of the ATPase reaction. This will require further development of transient spectroscopic methods.

2. To assess the coordination among structural elements of the actomyosin machine, simultaneous spectroscopic measurements should be made with probes on two or more sites.

3. To assess the functional relevance of detected structural changes within actomyosin, spectroscopic measurements should be done in the presence and absence of known functional mutations.

4. This field will continue to be driven by advancements in technology: multifrequency EPR, high-throughput time-resolved spectroscopy, single-molecule detection of fluorescence, and solid-state NMR.

5. Because myosin II is a low-duty-cycle motor, insight into its most important structural transition—the transition from weak binding (dynamic disorder) to strong binding (order)—requires more measurements capable of resolving weak (brief) interactions and dynamic disorder.

6. The recent development of in vivo labeling techniques is likely to have a profound impact on this field in the next decade.

7. An increasing number of spectroscopic studies will continue to be done in relationship to muscle disorders. This will depend to some extent on biological technology (e.g., generation of transgenic animals and expression systems) and spectroscopic developments that increase sensitivity for studying small and unstable samples.

DISCLOSURE STATEMENT

The authors are not aware of any biases that might be perceived as affecting the objectivity of this review.

ACKNOWLEDGMENTS

We thank Octavian Cornea for expert assistance in preparation of the manuscript. We obtained helpful comments from many colleagues, but we especially thank Josh Baker, Roger Cooke, Nariman Naber, Malcolm Irving, Yuri Nesmelov, and Ewa Prochniewicz. This work was supported by NIH grants AR32961 and AG26160 to DDT. DK was supported by NIH training grant AR07612.

LITERATURE CITED

1. Agafonov RV, Titus MA, Nesmelov YE, Thomas DD. 2008. Muscle and nonmuscle myosins probed by a spin label at equivalent sites in the force-generating domain. *Proc. Natl. Acad. Sci. USA* 105:13397–402

2. Anthony Akkari P, Nowak KJ, Beckman K, Walker KR, Schachat F, Laing NG. 2003. Production of human skeletal alpha-actin proteins by the baculovirus expression system. *Biochem. Biophys. Res. Commun.* 307:74–79

3. Bagshaw CR, Eccleston JF, Eckstein F, Goody RS, Gutfreund H, Trentham DR. 1974. The magnesium ion-dependent adenosine triphosphatase of myosin. Two-step processes of adenosine triphosphate association and adenosine diphosphate dissociation. *Biochem. J.* 141:351–64

4. Baker JE, Brust-Mascher I, Ramachandran S, LaConte LE, Thomas DD. 1998. A large and distinct rotation of the myosin light chain domain occurs upon muscle contraction. *Proc. Natl. Acad. Sci. USA* 95:2944–49

5. Baker JE, LaConte LE, Brust-Mascher II, Thomas DD. 1999. Mechanochemical coupling in spin-labeled, active, isometric muscle. *Biophys. J.* 77:2657–64

6. Baker JE, Thomas DD. 2000. A thermodynamic muscle model and a chemical basis for A.V. Hill's muscle equation. *J. Muscle Res. Cell Motil.* 21:335–44

7. Batra R, Manstein DJ. 1999. Functional characterisation of *Dictyostelium* myosin II with conserved tryptophanyl residue 501 mutated to tyrosine. *Biol. Chem.* 380:1017–23

8. Baumann BA, Hambly BD, Hideg K, Fajer PG. 2001. The regulatory domain of the myosin head behaves as a rigid lever. *Biochemistry* 40:7868–73

9. Berger CL, Thomas DD. 1994. Rotational dynamics of actin-bound intermediates of the myosin adenosine triphosphatase cycle in myofibrils. *Biophys. J.* 67:250–61

10. Borejdo J, Shepard A, Dumka D, Akopova I, Talent J, et al. 2004. Changes in orientation of actin during contraction of muscle. *Biophys. J.* 86:2308–17

11. Borovikov YS, Dedova IV, dos Remedios CG, Vikhoreva NN, Vikhorev PG, et al. 2004. Fluorescence depolarization of actin filaments in reconstructed myofibers: the effect of S1 or pPDM-S1 on movements of distinct areas of actin. *Biophys. J.* 86:3020–29

12. Brack AS, Brandmeier BD, Ferguson RE, Criddle S, Dale RE, Irving M. 2004. Bifunctional rhodamine probes of myosin regulatory light chain orientation in relaxed skeletal muscle fibers. *Biophys. J.* 86:2329–41

13. Brown LJ, Klonis N, Sawyer WH, Fajer PG, Hambly BD. 2001. Independent movement of the regulatory and catalytic domains of myosin heads revealed by phosphorescence anisotropy. *Biochemistry* 40:8283–91

14. Burmeister Getz E, Cooke R, Selvin PR. 1998. Luminescence resonance energy transfer measurements in myosin. *Biophys. J.* 74:2451–58

15. Chakrabarty T, Xiao M, Cooke R, Selvin PR. 2002. Holding two heads together: stability of the myosin II rod measured by resonance energy transfer between the heads. *Proc. Natl. Acad. Sci. USA* 99:6011–16

16. Conibear PB, Bagshaw CR, Fajer PG, Kovacs M, Malnasi-Csizmadia A. 2003. Myosin cleft movement and its coupling to actomyosin dissociation. *Nat. Struct. Biol.* 10:831–35

17. Cooke R, Crowder MS, Thomas DD. 1982. Orientation of spin labels attached to cross-bridges in contracting muscle fibres. *Nature* 300:776–78

18. Corrie JE, Brandmeier BD, Ferguson RE, Trentham DR, Kendrick-Jones J, et al. 1999. Dynamic measurement of myosin light-chain-domain tilt and twist in muscle contraction. *Nature* 400:425–30

19. Dominguez R, Freyzon Y, Trybus KM, Cohen C. 1998. Crystal structure of a vertebrate smooth muscle myosin motor domain and its complex with the essential light chain: visualization of the prepower stroke state. *Cell* 94:559–71

20. Espinoza-Fonseca LM, Kast D, Thomas DD. 2007. Molecular dynamics simulations reveal a disorder-to-order transition upon phosphorylation of smooth muscle myosin. *Biophys. J.* 93:2083–90

21. Espinoza-Fonseca LM, Kast D, Thomas DD. 2008. Thermodynamic and structural basis of phosphorylation-induced disorder-to-order transition in the regulatory light chain of smooth muscle myosin. *J. Am. Chem. Soc.* 130:12208–9

22. Fajer MI, Li H, Yang W, Fajer PG. 2007. Mapping electron paramagnetic resonance spin label conformations by the simulated scaling method. *J. Am. Chem. Soc.* 129:13840–46

23. Fajer P. 2004. Conformational switching in muscle. *Adv. Exp. Med. Biol.* 547:61–80

4. EPR study, first experimental evidence demonstrating lever arm rotation in contraction.

24. Fajer PG, Gyimesi M, Málnási-Csizmadia A, Bagshaw CR, Sen KI, Song L. 2007. Myosin cleft closure by double electron-electron resonance and dipolar EPR. *J. Phys. Condens. Matter* 19:285208 (10pp.)

25. Fisher AJ, Smith CA, Thoden JB, Smith R, Sutoh K, et al. 1995. X-ray structures of the myosin motor domain of *Dictyostelium discoideum* complexed with MgADP.BeFx and MgADP.AlF4. *Biochemistry* 34:8960–72

26. Forkey JN, Quinlan ME, Goldman YE. 2005. Measurement of single macromolecule orientation by total internal reflection fluorescence polarization microscopy. *Biophys. J.* 89:1261–71

27. Forkey JN, Quinlan ME, Shaw MA, Corrie JE, Goldman YE. 2003. Three-dimensional structural dynamics of myosin V by single-molecule fluorescence polarization. *Nature* 422:399–404

28. Holmes KC. 1998. Muscle contraction. *Novartis Found. Symp.* 213:76–92

29. Holmes KC, Angert I, Kull FJ, Jahn W, Schroder RR. 2003. Electron cryo-microscopy shows how strong binding of myosin to actin releases nucleotide. *Nature* 425:423–27

30. Holmes KC, Schroder RR, Sweeney HL, Houdusse A. 2004. The structure of the rigor complex and its implications for the power stroke. *Philos. Trans. R. Soc. London B Biol. Sci.* 359:1819–28

31. Hopkins SC, Sabido-David C, Corrie JE, Irving M, Goldman YE. 1998. Fluorescence polarization transients from rhodamine isomers on the myosin regulatory light chain in skeletal muscle fibers. *Biophys. J.* 74:3093–110

32. Hopkins SC, Sabido-David C, Van Der Heide UA, Ferguson RE, Brandmeier BD, et al. 2002. Orientation changes of the myosin light chain domain during filament sliding in active and rigor muscle. *J. Mol. Biol.* 318:1275–91

33. Houdusse A, Szent-Gyorgyi AG, Cohen C. 2000. Three conformational states of scallop myosin S1. *Proc. Natl. Acad. Sci. USA* 97:11238–43

34. Hubbell WL, Cafiso DS, Altenbach C. 2000. Identifying conformational changes with site-directed spin labeling. *Nat. Struct. Biol.* 7:735–39

35. Jeschke G. 2002. Distance measurements in the nanometer range by pulse EPR. *Chemphyschem* 3:927–32

36. Kabsch W, Mannherz HG, Suck D, Pai EF, Holmes KC. 1990. Atomic structure of the actin:DNase I complex. *Nature* 347:37–44

37. Klein JC, Burr AR, Svensson B, Kennedy DJ, Allingham J, et al. 2008. Actin-binding cleft closure in myosin II probed by site-directed spin labeling and pulsed EPR. *Proc. Natl. Acad. Sci. USA* 105:12867–72

38. Knowles AC, Ferguson RE, Brandmeier BD, Sun YB, Trentham DR, Irving M. 2008. Orientation of the essential light chain region of myosin in relaxed, active and rigor muscle. *Biophys. J.* 95:3882–91

39. Korman VL, Anderson SE, Prochniewicz E, Titus MA, Thomas DD. 2006. Structural dynamics of the actin-myosin interface by site-directed spectroscopy. *J. Mol. Biol.* 356:1107–17

40. Kouyama T, Mihashi K. 1981. Fluorimetry study of N-(1-pyrenyl)iodoacetamide-labelled F-actin. Local structural change of actin protomer both on polymerization and on binding of heavy meromyosin. *Eur. J. Biochem.* 114:33-38

41. Kozuka J, Yokota H, Arai Y, Ishii Y, Yanagida T. 2006. Dynamic polymorphism of single actin molecules in the actin filament. *Nat. Chem. Biol.* 2:83–86

42. LaConte LE, Baker JE, Thomas DD. 2003. Transient kinetics and mechanics of myosin's force-generating rotation in muscle: resolution of millisecond rotational transitions in the spin-labeled myosin light-chain domain. *Biochemistry* 42:9797–803

43. Lidke DS, Thomas DD. 2002. Coordination of the two heads of myosin during muscle contraction. *Proc. Natl. Acad. Sci. USA* 99:14801–6

44. Lombardi V, Piazzesi G, Reconditi M, Linari M, Lucii L, et al. 2004. X-ray diffraction studies of the contractile mechanism in single muscle fibres. *Philos. Trans. R. Soc. London B Biol. Sci.* 359:1883–93

45. Malnasi-Csizmadia A, Kovacs M, Woolley RJ, Botchway SW, Bagshaw CR. 2001. The dynamics of the relay loop tryptophan residue in the *Dictyostelium* myosin motor domain and the origin of spectroscopic signals. *J. Biol. Chem.* 276:19483–90

46. Malnasi-Csizmadia A, Pearson DS, Kovacs M, Woolley RJ, Geeves MA, Bagshaw CR. 2001. Kinetic resolution of a conformational transition and the ATP hydrolysis step using relaxation methods with a *Dictyostelium* myosin II mutant containing a single tryptophan residue. *Biochemistry* 40:12727–37

47. Malnasi-Csizmadia A, Woolley RJ, Bagshaw CR. 2000. Resolution of conformational states of *Dictyostelium* myosin II motor domain using tryptophan (W501) mutants: implications for the open-closed transition identified by crystallography. *Biochemistry* 39:16135–46

48. Mazhari SM, Selser CT, Cremo CR. 2004. Novel sensors of the regulatory switch on the regulatory light chain of smooth muscle myosin. *J. Biol. Chem.* 279:39905–14

49. Naber N, Purcell TJ, Pate E, Cooke R. 2007. Dynamics of the nucleotide pocket of myosin measured by spin-labeled nucleotides. *Biophys. J.* 92:172–84

50. Nelson WD, Blakely SE, Nesmelov YE, Thomas DD. 2005. Site-directed spin labeling reveals a conformational switch in the phosphorylation domain of smooth muscle myosin. *Proc. Natl. Acad. Sci. USA* 102:4000–5

51. Nesmelov YE, Agafonov RV, Burr AR, Weber RT, Thomas DD. 2008. Structure and dynamics of the force-generating domain of myosin probed by multifrequency electron paramagnetic resonance. *Biophys. J.* 95:247–56

52. Palm T, Sale K, Brown L, Li H, Hambly B, Fajer PG. 1999. Intradomain distances in the regulatory domain of the myosin head in prepower and postpower stroke states: fluorescence energy transfer. *Biochemistry* 38:13026–34

53. Prochniewicz E, Spakowicz DJ, Thomas D. 2008. Changes in actin structural transitions associated with oxidative inhibition of muscle contraction. *Biochemistry* 47:11811–17

54. Prochniewicz E, Walseth TF, Thomas DD. 2004. Structural dynamics of actin during active interaction with myosin: different effects of weakly and strongly bound myosin heads. *Biochemistry* 43:10642–52

55. Quinlan ME, Forkey JN, Goldman YE. 2005. Orientation of the myosin light chain region by single molecule total internal reflection fluorescence polarization microscopy. *Biophys. J.* 89:1132–42

56. Ramachandran S, Thomas DD. 1999. Rotational dynamics of the regulatory light chain in scallop muscle detected by time-resolved phosphorescence anisotropy. *Biochemistry* 38:9097–104

57. Rayment I, Holden HM, Whittaker M, Yohn CB, Lorenz M, et al. 1993. Structure of the actin-myosin complex and its implications for muscle contraction. *Science* 261:58–65

58. Rayment I, Rypniewski WR, Schmidt-Base K, Smith R, Tomchick DR, et al. 1993. Three-dimensional structure of myosin subfragment-1: a molecular motor. *Science* 261:50–58

59. Reshetnyak YK, Andreev OA, Borejdo J, Toptygin DD, Brand L, Burstein EA. 2000. The identification of tryptophan residues responsible for ATP-induced increase in intrinsic fluorescence of myosin subfragment 1. *J. Biomol. Struct. Dyn.* 18:113–25

60. Robertson CI, Gaffney DP 2nd, Chrin LR, Berger CL. 2005. Structural rearrangements in the active site of smooth-muscle myosin. *Biophys. J.* 89:1882–92

61. Roopnarine O, Szent-Gyorgyi AG, Thomas DD. 1998. Microsecond rotational dynamics of spin-labeled myosin regulatory light chain induced by relaxation and contraction of scallop muscle. *Biochemistry* 37:14428–36

62. Root DD, Stewart S, Xu J. 2002. Dynamic docking of myosin and actin observed with resonance energy transfer. *Biochemistry* 41:1786–94

63. Rosenberg SA, Quinlan ME, Forkey JN, Goldman YE. 2005. Rotational motions of macromolecules by single-molecule fluorescence microscopy. *Acc. Chem. Res.* 38:583–93

64. Roy R, Hohng S, Ha T. 2008. A practical guide to single-molecule FRET. *Nat. Methods* 5:507–16

65. Sabido-David C, Hopkins SC, Saraswat LD, Lowey S, Goldman YE, Irving M. 1998. Orientation changes of fluorescent probes at five sites on the myosin regulatory light chain during contraction of single skeletal muscle fibres. *J. Mol. Biol.* 279:387–402

66. Saraswat LD, Lowey S. 1998. Subunit interactions within an expressed regulatory domain of chicken skeletal myosin. Location of the NH2 terminus of the regulatory light chain by fluorescence resonance energy transfer. *J. Biol. Chem.* 273:17671–79

67. Schroder RR, Manstein DJ, Jahn W, Holden H, Rayment I, et al. 1993. Three-dimensional atomic model of F-actin decorated with *Dictyostelium* myosin S1. *Nature* 364:171–74

68. Selvin PR. 2002. Principles and biophysical applications of lanthanide-based probes. *Annu. Rev. Biophys. Biomol. Struct.* 31:275–302

49. Spin-labeled nucleotide analogs demonstrate that open and closed states of the myosin nucleotide pocket are populated in actin-bound myosin.ADP.

50. Site-directed spin labeling shows that phosphorylation induces a disorder-to-order transition in the N-terminal domain of the RLC.

54. Phosphorescence shows that weak and strong binding of myosin have profoundly different effects on actin filament dynamics.

69. Shepard AA, Dumka D, Akopova I, Talent J, Borejdo J. 2004. Simultaneous measurement of rotations of myosin, actin and ADP in a contracting skeletal muscle fiber. *J. Muscle Res. Cell Motil.* 25:549–57

70. Shih WM, Gryczynski Z, Lakowicz JR, Spudich JA. 2000. A FRET-based sensor reveals large ATP hydrolysis-induced conformational changes and three distinct states of the molecular motor myosin. *Cell* 102:683–94

71. Steinhoff HJ. 2004. Inter- and intramolecular distances determined by EPR spectroscopy and site-directed spin labeling reveal protein-protein and protein-oligonucleotide interaction. *Biol. Chem.* 385:913–20

72. Surek JT, Thomas DD. 2007. A paramagnetic molecular voltmeter. *J. Magn. Reson.* 190:7–25

73. Suzuki Y, Yasunaga T, Ohkura R, Wakabayashi T, Sutoh K. 1998. Swing of the lever arm of a myosin motor at the isomerization and phosphate-release steps. *Nature* 396:380–83

74. Thomas DD. 1987. Spectroscopic probes of muscle cross-bridge rotation. *Annu. Rev. Physiol.* 49:691–709

75. Thomas DD, Cooke R. 1980. Orientation of spin-labeled myosin heads in glycerinated muscle fibers. *Biophys. J.* 32:891–906

76. Thomas DD, Prochniewicz E, Roopnarine O. 2002. Changes in actin and myosin structural dynamics due to their weak and strong interactions. *Results Probl. Cell Differ.* 36:7–19

77. Thompson AR, Naber N, Wilson C, Cooke R, Thomas DD. 2008. Structural dynamics of the Actomyosin complex probed by a bifunctional spin label that crosslinks SH1 and SH2. *Biophys. J.* 95:5238–46

78. van Duffelen M, Chrin LR, Berger CL. 2005. Kinetics of structural changes in the relay loop and SH3 domain of myosin. *Biochem. Biophys. Res. Commun.* 329:563–72

79. Volkmann N, Hanein D, Ouyang G, Trybus KM, DeRosier DJ, Lowey S. 2000. Evidence for cleft closure in actomyosin upon ADP release. *Nat. Struct. Biol.* 7:1147–55

80. Volkmann N, Ouyang G, Trybus KM, DeRosier DJ, Lowey S, Hanein D. 2003. Myosin isoforms show unique conformations in the actin-bound state. *Proc. Natl. Acad. Sci. USA* 100:3227–32

81. Warshaw DM, Hayes E, Gaffney D, Lauzon AM, Wu J, et al. 1998. Myosin conformational states determined by single fluorophore polarization. *Proc. Natl. Acad. Sci. USA* 95:8034–39

82. Werber MM, Szent-Gyorgyi AG, Fasman GD. 1972. Fluorescence studies on heavy meromyosin-substrate interaction. *Biochemistry* 11:2872–83

83. Wolff-Long VL, Tao T, Lowey S. 1995. Proximity relationships between engineered cysteine residues in chicken skeletal myosin regulatory light chain. A resonance energy transfer study. *J. Biol. Chem.* 270:31111–18

84. Xiao M, Li H, Snyder GE, Cooke R, Yount RG, Selvin PR. 1998. Conformational changes between the active-site and regulatory light chain of myosin as determined by luminescence resonance energy transfer: the effect of nucleotides and actin. *Proc. Natl. Acad. Sci. USA* 95:15309–14

85. Xiao M, Reifenberger JG, Wells AL, Baldacchino C, Chen LQ, et al. 2003. An actin-dependent conformational change in myosin. *Nat. Struct. Biol.* 10:402–8

86. Xu J, Root DD. 1998. Domain motion between the regulatory light chain and the nucleotide site in skeletal myosin. *J. Struct. Biol.* 123:150–61

87. Yanagida T. 2007. Muscle contraction mechanism based on actin filament rotation. *Adv. Exp. Med. Biol.* 592:359–67

88. Yengo CM, Berger CL. 2002. Fluorescence resonance energy transfer in acto-myosin complexes. *Results Probl. Cell Differ.* 36:21–30

89. Yengo CM, Chrin LR, Berger CL. 2000. Interaction of myosin LYS-553 with the C-terminus and DNase I-binding loop of actin examined by fluorescence resonance energy transfer. *J. Struct. Biol.* 131:187–96

90. Yengo CM, Chrin LR, Rovner AS, Berger CL. 2000. Tryptophan 512 is sensitive to conformational changes in the rigid relay loop of smooth muscle myosin during the MgATPase cycle. *J. Biol. Chem.* 275:25481–87

91. Yengo CM, De La Cruz EM, Chrin LR, Gaffney DP 2nd, Berger CL. 2002. Actin-induced closure of the actin-binding cleft of smooth muscle myosin. *J. Biol. Chem.* 277:24114–19

92. Yengo CM, Fagnant PM, Chrin L, Rovner AS, Berger CL. 1998. Smooth muscle myosin mutants containing a single tryptophan reveal molecular interactions at the actin-binding interface. *Proc. Natl. Acad. Sci. USA* 95:12944–49

70. Frequency-domain FRET between CD and LCD of myosin S1 resolves multiple lever arm angles.

84. Developed LRET to detect lever arm motion in skeletal myosin.

90. Engineered single tryptophan mutant in smooth muscle myosin.

93. Yildiz A, Forkey JN, McKinney SA, Ha T, Goldman YE, Selvin PR. 2003. Myosin V walks hand-overhand: single fluorophore imaging with 1.5-nm localization. *Science* 300:2061–65
94. Zeng W, Conibear PB, Dickens JL, Cowie RA, Wakelin S, et al. 2004. Dynamics of actomyosin interactions in relation to the cross-bridge cycle. *Philos. Trans. R. Soc. London B Biol. Sci.* 359:1843–55
95. Zhang J, Campbell RE, Ting AY, Tsien RY. 2002. Creating new fluorescent probes for cell biology. *Nat. Rev. Mol. Cell Biol.* 3:906–18

Lessons from Structural Genomics*

Thomas C. Terwilliger,[1] David Stuart,[2] and Shigeyuki Yokoyama[3]

[1] Los Alamos National Laboratory, Los Alamos, New Mexico 87545; email: terwilliger@lanl.gov

[2] Division of Structural Biology, The Wellcome Trust Center for Human Genetics, University of Oxford, Headington, Oxford OX3 7BN, United Kingdom

[3] RIKEN Systems Structural Biology Center, Yokohama 230-0045, Japan

Annu. Rev. Biophys. 2009. 38:371–83

First published online as a Review in Advance on February 5, 2009

The *Annual Review of Biophysics* is online at biophys.annualreviews.org

This article's doi: 10.1146/annurev.biophys.050708.133740

Copyright © 2009 by Annual Reviews. All rights reserved

1936-122X/09/0609-0371$20.00

*The U.S. Government has the right to retain a nonexclusive, royalty-free license in and to any copyright covering this paper.

Key Words

international cooperation, protein structure, X-ray crystallography, nuclear magnetic resonance

Abstract

A decade of structural genomics, the large-scale determination of protein structures, has generated a wealth of data and many important lessons for structural biology and for future large-scale projects. These lessons include a confirmation that it is possible to construct large-scale facilities that can determine the structures of a hundred or more proteins per year, that these structures can be of high quality, and that these structures can have an important impact. Technology development has played a critical role in structural genomics, the difficulties at each step of determining a structure of a particular protein can be quantified, and validation of technologies is nearly as important as the technologies themselves. Finally, rapid deposition of data in public databases has increased the impact and usefulness of the data and international cooperation has advanced the field and improved data sharing.

Contents

INTRODUCTION

What is Structural Genomics?

Structural genomics is the large-scale determination of protein structures. In the late 1990s many structural biologists realized that major breakthroughs in technologies for structure determination, combined with the success of genome projects, laid a foundation for a systematic worldwide effort to determine the structures of proteins (12, 22, 42, 63, 66). Public and private funding agencies in the United States, Europe, and Japan sponsored workshops on what this new field might accomplish and how international cooperation could help (50, 53). This led to support for major structural genomics efforts in the United States [the NIH Protein Structure Initiative (PSI), **http://www.nigms.nih.gov/Initiatives/PSI/**], Europe (The Protein Structure Factory, **http://www.proteinstrukturfabrik.de/**, and SPINE, **http://www.spineurope.org/**), and Japan [The National Project on Protein Structural and Functional Analyses (Protein 3000) and RIKEN Structural Genomics/Proteomics Initiative; **http://protein.gsc.riken.jp/**]. Over the next decade many additional structural genomics efforts were started around the world, two of the largest of which are the Canadian-U.K.-Swedish Structural Genomics Consortium (**http://www.sgc.utoronto.ca/**) and the Japanese Targeted Proteins Research Program (**http://www.tanpaku.org/**). The TargetDB database at **http://sg.pdb.org/target_centers.html** has a current list of structural genomics efforts and their status, and a volume and a review devoted to structural genomics have recently been published (19, 59).

Technologies Forming a Foundation for Structural Genomics

Structural genomics is a technology-driven field. Key foundations for structural genomics included developments in macromolecular X-ray crystallography and NMR along with general systems for cloning and expression of proteins. In X-ray crystallography the increasing use of tunable synchrotron radiation in collecting anomalous diffraction data on protein crystals, the use of selenomethionine in crystallographic phase determination (68), automation of nanodroplet crystallization (11,

52), cryo-cooling techniques that reduced radiation damage to those crystals (30, 49), and automation of the structure determination process for macromolecular crystallography (47, 62) suggested that determining a protein structure once crystals were available would be increasingly straightforward. In the NMR field, the availability of cryo-probe high-field NMR devices (32), the development of multidimensional techniques for data acquisition (60), and the introduction of automated procedures for data analysis (8, 28, 43) similarly suggested that NMR structure determination was headed toward ever-higher throughput. Generalized systems for cloning and expression of proteins, particularly the addition of histidine tags at the N or C termini of recombinant proteins (48), were becoming increasingly used and allowed the use of generic techniques such as nickel-affinity chromatography, suggesting that standardized protocols might be developed that could be used for protein production in a factory-like setting.

Genomic Sequences and Structural Genomics

In parallel with these major technical developments, the availability of genomic sequences from dozens of organisms and the promise of sequences of many more over the following years were major stimulants for the idea of structural genomics. It was widely recognized that these genomic sequences could be used to identify sets of protein structures that would be highly informative, such as all the structures from a particular organism, representatives of all unique protein folds, representatives of all unique protein families, or all members of a biochemical pathway (12, 22, 35, 42, 63, 69) (**http://www.thermus.org/e_index.htm**). Combined with the technological advances, this led to a great deal of excitement about the idea of large-scale protein structure determination, or "structural genomics" or "structural proteomics" as it has variously come to be called.

Challenges in Structural Genomics

It was recognized early on that genomic sequencing was fundamentally different from structure determination because the diversity among proteins and their solubility and physical properties is much greater than that of fragments of DNA, so that the challenge in determining all protein structures from an organism was vastly more difficult than that of determining the genome sequence for that organism. Nevertheless, the idea of completeness and of structural coverage of all proteins was an important part of much of the thinking in structural genomics (50, 53).

The biggest challenges in structural genomics have largely been clear from the early days of the field: Production of proteins in a soluble form and crystallization (for X-ray crystallography) or suitability for production of high-quality spectra (for NMR) were bottlenecks in all structural biology laboratories. Moreover, only a few structures of membrane proteins had been determined by that time.

What was not known at the start of structural genomics was whether it would be possible to overcome bottlenecks in protein expression and structure determination for various classes of proteins, and whether large-scale centers could build pipelines that could routinely determine the structures of targeted proteins. The cost of structure determination by conventional means was generally estimated to be in the range of $100,000–$300,000 per protein structure, and it was anticipated that cost savings could be obtained by large-scale structure determination, but the extent of possible savings was unclear (20). Similarly, the importance of structures of proteins of unknown function was not known. The role of rapid deposition of data was not known but was generally expected to be a key one. It was generally expected that technology development would be a critical part of structural genomics (58), but it was not clear at the start of structural genomics how much development in this area would actually occur. Finally, at the start of structural genomics the importance of international collaboration was

Structural genomics: large-scale determination of protein structures, typically using robotics in many steps of the process, often carried out by consortia of research efforts; also known as structural proteomics

Structure determination: identifying the shape of a protein, normally represented by the coordinates of the nonhydrogen atoms in a model of the protein

U.S. National Institutes of Health Protein Structure Initiative (U.S. NIH PSI): PSI-1 was the first five-year program, devoted to pilot-scale initiatives, and PSI-2 included large-scale structural genomics efforts

SPINE and SPINE-2: Structural Proteomics in Europe

Crystallization: proteins can be crystallized by adding salts or other compounds to a solution containing the purified protein

Production of proteins in a soluble form: to crystallize a protein or to obtain an NMR spectrum of a protein, usually the protein must first be purified, separating it from other proteins, and remain soluble, retaining its 3D shape and not aggregating

Table 1 Lessons from structural genomics

Lessons
1. It is possible to construct large-scale facilities that can determine the structures of a hundred or more proteins per year.
2. The difficulties at each step of determining a structure of a particular protein can be quantified.
3. Structures from structural genomics can have an important impact on scientific research.
4. Rapid deposition of data in public databases increases the impact and usefulness of the data.
5. Technology development has played a critical role in structural genomics.
6. Validation of technologies is nearly as important as the technologies themselves.
7. Structures from structural genomics are of high quality.
8. International cooperation advances the field and improves data sharing.

unclear, although experience with the genome projects suggested that cooperation would be highly beneficial.

Protein Data Bank (PDB): an open repository of structural information on proteins

WHAT CAN BE ACCOMPLISHED BY STRUCTURAL GENOMICS?

Table 1 lists key lessons from a decade of structural genomics. One of the most important of these is that it is indeed possible to construct large-scale facilities that can determine the

structures of a hundred or more proteins per year. This has been accomplished in efforts around the world, including each of the current four U.S. PSI Large-Scale Centers [Midwest Center for Structural Genomics, **http://www.mcsg.anl.gov/**; Joint Center for Structural Genomics (JCSG), **http://www.jcsg.org/**; Northeast Structural Genomics Consortium, **http://www.nesg.org**; and the New York SGX Research Center for Structural Genomics, **http://www.nysgrc.org/nysgrc**], the Japanese RIKEN structural genomics effort and the Canadian-UK-Swedish Structural Genomics Consortium.

The U.S. PSI-2 Large-Scale Centers provide a good set of examples of what is possible when a concerted effort is made to produce a large number of structures from prokaryotic organisms (13). Each of the four Centers has been operating since 2001. The rates of depositing new protein structures into the Protein Data Bank (PDB) (4, 5) for these Centers have increased dramatically during this period, from an average of about 25 structures per year each in 2002 to an average of 155 each per year in 2007 (**Figure 1**). The corresponding cost per structure has decreased from about $500,000 to about $70,000, as calculated by dividing the total cost of the projects by the number of structures produced (13, 16).

Similar rates of structure determination and costs have been achieved by RIKEN and by the Structural Genomics Consortium (SGC). During the fiscal years 2002–2006 RIKEN deposited 2675 structures into the PDB, with an average cost per structure of approximately

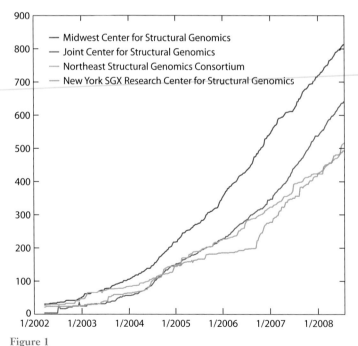

— Midwest Center for Structural Genomics
— Joint Center for Structural Genomics
— Northeast Structural Genomics Consortium
— New York SGX Research Center for Structural Genomics

Figure 1

PDB depositions by NIH Protein Structure Initiative Large-Scale Centers (see **http://www.mcsg.anl.gov** for a current plot).

$55,000. These structures included many small protein structures determined by NMR, as well as many eukaryotic proteins. Focusing on eukaryotic proteins, the SGC has deposited 750 structures into the PDB since 2003 (**http://www.sgc.utoronto.ca/structures/target_progress.php**), a particularly notable accomplishment as more than 600 of these structures were from human cells and included two membrane proteins. The cost per structure was approximately $135,000. Together RIKEN and the SGC account for over 50% of newly determined structures of human proteins.

Together, these structural genomics efforts have deposited (as of August 2008) some 6048 structures into the PDB (see **http://targetdb.pdb.org/**). These 6048 structures represent about 11.5% of all structures in the PDB. The successes of these large structural genomics projects show that it is indeed possible to create a pipeline for large-scale protein structure determination.

SUCCESS RATES: WHICH STRUCTURES WILL BE EASY TO DETERMINE AND WHICH WILL BE DIFFICULT?

A second major lesson from structural genomics is that the difficulties at each step of determining a structure of a particular protein can be quantified. There are two aspects of this type of analysis. One aspect is that the probability of overall success for a particular step or for obtaining a structure can be estimated. The second and more important aspect is that the relative probabilities of success for trying various approaches at a particular stage can be

estimated. The development of ways to estimate the effectiveness of various approaches toward successful structure determination is a major step forward for structural biology because it allows a rational approach to choosing what methods to try for any particular protein.

Success Rates for Individual Steps in Structure Determination

One of the earliest impacts of structural genomics on the structural biology field was the ability to identify overall success rates for the major steps in structure determination. Early on, structural genomics efforts around the world agreed to post the status of their targeted structures on the TargetDB Web site maintained by the PDB (**http://sg.pdb.org/**). This allowed anyone to count, for example, the number of targets that were cloned, purified, crystallized, had NMR data collected, or were deposited into the PDB at any time. For the first time a quantitative measure of success rates for each step in structure determination was available. **Table 2** shows the success rates for the major steps in structure determination as of July 2008. Each of these steps is (on average) successful from about one-third to two-thirds of the time, with the lowest success rate at the steps between expressing a protein and purifying it (which include obtaining the protein in a soluble form and successfully isolating the pure protein) and between purified protein and obtaining useful crystals or NMR spectra. This type of analysis has been important because it shows which steps need to be improved most (obtaining soluble protein, crystallization, and obtaining samples suitable for

NMR spectra: used to determine structures of proteins by identifying pairs of atoms that are close together in the structure

Table 2 Success rates for major steps in structure determination

Status	Total number of targets	% Success (step)	% Success (overall)
Cloned	125,316	100	100
Expressed	83,115	66.3	66.3
Purified	29,409	35.4	23.5
Diffraction-quality crystals or NMR spectrum	8,690	29.5	6.9
In PDB	5,811	66.9	4.6

high-quality NMR spectra) and which are relatively successful (protein expression and structure determination).

A more sophisticated approach has been developed as well in which success rates, both overall and for individual steps, are analyzed as functions of the physical properties of proteins (where these physical properties are inferred from their amino acid sequences). Success rates are highly dependent on the isoelectric point, hydrophobicity, and propensity for disordered structure of a protein (17, 24, 55). Owing to this, protein sequences have been classified as optimal, suboptimal, average, difficult, and very difficult to get to the stage of crystallization (55). Such a classification can be used to estimate how difficult a particular structure would be to determine, but more importantly it can be used to decide the relative allocation effort of various proteins. For example, if an investigator is not concerned about the amino acid differences between a pair of proteins, then work can be focused on the one with the higher probability of success. Alternatively, a large-scale effort can continually reprioritize entire targeted sets of proteins on the basis of their current relative probabilities of success.

Identifying the Best Approach to Take at Each Step in Structure Determination

Prior to structural genomics efforts, every structural biology laboratory had accumulated in-house a set of techniques that could be applied to obtain soluble protein, crystals and X-ray data, or NMR data. The order in which these techniques might be applied, and the point at which a project might be abandoned, would largely depend on the personal experience of the investigator and anecdotal evidence gathered from the experiences of other investigators. Success in structure determination depended strongly on the ability of the investigator to synthesize this evidence and identify the best approaches to apply to the problem at hand. This paradigm is now changing because structural genomics provides the opportunity for a systematic evaluation of success rates for various approaches to overcoming bottlenecks.

One of the earliest applications of this approach was the identification of conditions likely to lead to protein crystallization. The thousands of proteins produced by structural genomics efforts allowed the evaluation of relative effectiveness of many precipitants and additives, leading to new standardized sets of screens for protein crystallization (14, 45, 46). Similar crystallization data, combined with information on the physical properties of proteins obtained from their amino acid sequences, also led to the ability to predict the overall probability of crystallization success depending on the isoelectric point or the hydrophobicity of a protein (14, 56).

An important use of systematic information about relative success rates for different approaches is in the choice of which approach to apply next after failure at some step in structure determination (36). For example, if soluble protein is not obtained upon expression of a cloned gene, possible choices might include changing the expression vector or host, expressing the protein in the presence of chaperones or cofactors, expressing domains, or engineering the protein sequence to increase solubility. A good way to decide which of these to try would be to balance the probability of success of each possible approach with the cost in time or effort of applying that approach, weighting the benefits and cost according to how important they are to the circumstance at hand. This type of approach is made possible by detailed analyses of the chances of success for a wide variety of approaches at each step of structure determination.

Synthesis of Experience from Structural Genomics Laboratories Around the World

The systematic analysis of chances of success at each step in structure determination has had qualitative practical outcomes in addition to the quantitative ones described above. The combined experience of a large group of structural genomics efforts has led to many protocols for carrying out these steps, including general protocols with recommendations for each step in structure determination (2, 21, 27).

VALUE OF STRUCTURES FROM STRUCTURAL GENOMICS

The next major lesson from structural genomics is that the structures that result from such a worldwide effort can have an important impact. From the start of structural genomics efforts it has been recognized that the choice of which protein structures to determine is a critical one. Many approaches have been suggested for choosing which proteins to target, and the structural genomics efforts around the world have had a range of emphases. The most fundamental choice has been whether to target proteins with identified biochemical or biological importance to provide structural information directly applicable to their functions, or to target proteins representative of large families of related proteins to provide a coarse level of structural information for the entire families.

The U.S. PSI-1 and PSI-2 have been largely targeted at determining structures of representative proteins from protein families with many members (**http://www.nigms. nih.gov/Initiatives/PSI**). Centers supported by PSI-1 also targeted proteins from *Mycobacterium tuberculosis* (the TB Structural Genomics Consortium, **http://www.doe-mbi.ucla.edu/ TB/**) and from human parasites (the Structural Genomics of Pathogenic Protozoa Consortium; **http://www.sgpp.org/**) as well as complete coverage of the *Thermotoga maritima* genome (JCSG, **http://www.jcsg.org**). Two of the PSI-2 Specialized Centers are targeting membrane proteins (the Center for Structures of Membrane Proteins, **http://csmp. ucsf.edu**; and the New York Consortium on Membrane Protein Structure, **http://www. nycomps.org**) and one is focused on structures of eukaryotic proteins (the Center for Eukaryotic Structural Genomics, **http://www. uwstructuralgenomics.org**).

The SPINE-1 and SPINE-2 efforts targeted eukaryotic macromolecular structures and complexes relevant to human disease (**http://www.spineurope.org; http://www. spine2.eu/SPINE2/**), and the Protein Structure Factory effort targeted eukaryotic proteins. The RIKEN structural genomics efforts targeted coverage of the complete genome of *Thermus thermophilus* as well as structure determination of eukaryotic proteins with identified function. The SGC has targeted human proteins of therapeutic interest, and the Japanese Targeted Proteins Research Program has targeted protein structures for understanding fundamental biology, medical importance, food, and the environment.

New large-scale efforts in the United States are targeting proteins from a variety of human pathogens (the Center for Structural Genomics of Infectious Diseases, **http://www.csgid.org**; and the Seattle Structural Genomics Center for Infectious Diseases, **http://ssgcid.org**). Other efforts around the world have targeted proteins to identify their functions (the Structure 2 Function project, **http://s2f.umbi.umd.edu/**, and the Yeast Structural Genomics pilot project, **http://genomics.eu.org/spip/**).

Some of the structures determined by structural genomics projects were targeted on the basis of their importance and have had obvious individual impacts. These include, for example, structures of a protein secretion apparatus from *Pseudomonas aeruginosa* (44), structures of an aquaporin (64), and structures of hormone receptor-ligand interactions (9). Proteins from *M. tuberculosis*, targeted by the TB Structural Genomics Consortium, are of clear interest because of their potential as targets for antituberculosis therapeutics. This consortium has determined structures of 116 different proteins from *M. tuberculosis* (3), including at least 9 that are active targets for therapeutics in pharmaceutical companies (J. Sacchettini, personal communication).

Other proteins have been targeted in structural genomics efforts simply because they were proteins of unknown function and the investigators hoped that determination of their structures would hint at their biochemical and cellular functions (23, 54). In some cases, particularly those in which a ligand or cofactor has been discovered bound to a protein in crystals, this has been successful (38). In other cases the structures were similar to those of characterized proteins of known function, allowing information from the characterized proteins to be transferred (69) to the targeted proteins (26, 61). In still other cases, a combination of methods had to be applied to identify the function of a protein (51).

Efforts to determine the structures of all proteins from a microorganism, while not yielding completeness, have nevertheless given a picture of the range of structures involved in supporting a living organism (13, 31, 34, 35, 54) (**http://www.thermus.org/e_index.htm**). More importantly, they form a foundation for future efforts to develop detailed models of each step in metabolism, regulation, and other cellular functions. The availability of structural information is a prerequisite for atomic-level simulations of these processes and makes it possible to think about developing such comprehensive models.

Another important avenue through which structures from structural genomics efforts will have increasing impacts is by adding to the database of structures in the PDB. A useful way to look at this is that these structures are being determined now so that they will be available when they are needed. An example is the structure of TM0936 from *T. maritima*, determined by the JCSG, which provided key information for identification of the function of this protein by docking potential intermediates some five years later (29). A different type of example is that structures from structural genomics are being used as molecular replacement models for the determination of new proteins of identified interest. The contribution of structural genomics to the novel structures in the PDB has been increasing, with structural genomics efforts now determining about 50% of the structures that represent new protein families (15, 37, 40).

IMPORTANCE OF RAPID DEPOSITION OF DATA

The experience from structural genomics strengthens another important lesson made clear earlier by the genomic sequencing efforts. Rapid deposition of data in public databases vastly increases the impact and usefulness of the data. In the case of structural genomics, the international community agreed at the outset (see **http://www.isgo.org**) to deposit the identities of structures that were targeted on the TargetDB Web site (**http://sg.pdb.org/**) and to continually update this target list with the status of each target. As discussed above, this has made analyses of success rates possible. Additionally, this target information has facilitated efforts to avoid duplication of effort, as a number of structural genomics groups have had a policy of scanning TargetDB on a regular basis and discontinuing work on structures that are solved or nearly solved (16).

The structural genomics community also agreed at the outset to deposit structures and raw data (e.g., structure factors for crystallographic data) into the PDB (**http://www.pdb.org**) promptly upon completion of the structures (with up to six months delay in exceptional cases). This rapid deposition of structural information has made the results of structural genomics efforts rapidly accessible to the broader community.

Beyond the simple listing of target status and deposition of the final structures and data, structural genomics efforts have made a systematic effort to deposit and make publicly accessible the methods used to produce proteins and determine their structures as well as to make materials such as expression clones generally available. For example, the U.S. PSI has developed a KnowledgeBase portal (**http://kb.psi-structuralgenomics.org/KB/**) that is intended to allow general access to data, flow charts of structure determination, methods, intermediate data files, structures, and interpretations of structures. Additionally, the JCSG and SGC (**http://www.thesgconline.org/**) have extensive annotations of structures available online. For SGC targets that have been deprioritized, the SGC Web site contains the methods used to determine structures and the availability of clones.

IMPORTANCE OF TECHNOLOGY DEVELOPMENT

It has been clear from the beginning of structural genomics efforts that technology development made structural genomics possible, that continued development would be necessary for the success of high-throughput structure determination, and that developments in structural genomics would be applicable to other areas of structural and general biology. One lesson from structural genomics is that technology development is indeed important. Another lesson is that the systematic validation of the utilities of new technologies is almost as important as the technologies themselves.

The high-throughput needs of structural genomics have spurred efforts to develop diverse sets of technologies. These include the development of high-throughput cell-free systems for protein expression (33, 65), methods for improving solubilities of proteins (67), highly parallel, small-volume screening systems for protein crystallization (6, 52), automation of X-ray data collection (18, 25, 57), and automated macromolecular structure determination procedures for X-ray crystallography (1, 41) and NMR (39). These developments have had an impact on all of structural biology. Because of the influence of structural genomics, a user of most beamlines for macromolecular X-ray crystal structure determination can now expect to have available a highly automated system that allows robotic mounting of crystals on the X-ray beam and automated screening of a set of crystals to find those that show the best diffraction. This has completely changed the strategy of data collection and has vastly increased the potential throughput of structure determination at X-ray beamlines.

QUALITY OF STRUCTURES FROM STRUCTURAL GENOMICS

At the start of structural genomics there was some concern that the quality of structures obtained from high-throughput efforts might be lower than that for structures determined in individual researchers' laboratories. A powerful lesson from structural genomics is that in general the structures determined in structural genomics pipelines are similar in quality compared with those determined by nonstructural genomics research (10). For X-ray structures the quality is typically higher for structural genomics, whereas for NMR structures the quality is somewhat lower for structural genomics (7). In retrospect the high quality of structures from structural genomics is not surprising, as structural genomics projects can devote the necessary effort to develop standardized procedures for structure determination and quality control.

ROLE OF INTERNATIONAL COOPERATION

A final lesson from structural genomics is the importance of international cooperation, particularly during the early stages of the field. During the first several years of structural genomics efforts, a series of workshops were held, first on the feasibility and potential impact of structural genomics and later on data sharing and collaboration. The Argonne workshop in 1998 was a defining workshop for the field (53). It was attended by structural biologists, bioinformaticians, and representatives of funding agencies, and it helped identify the possible approaches to targeting proteins for structure determination along with the tools that were available and those that would be needed.

A set of meetings sponsored by the Wellcome Trust, the U.S. NIH, and the Japanese Ministry of Science was critical in defining the environment in which structural genomics was to be carried out. At these meetings, held in 2000 and 2001, a charter for the International Structural Genomics Organization (ISGO; http://www.isgo.org) was drafted. The participants agreed on the guidelines and principles of ISGO and that structural genomics efforts around the world would follow them. These guidelines included the rapid

Structural genomics pipeline: an integrated procedure for determining the structures of proteins

ISGO: International
Structural Genomics
Organization

deposition of structural information with the raw data supporting them. The international nature of these agreements was critical in this process. Structural biologists in individual countries used it to convince their respective governments that as the rest of the world was going to agree to rapid deposition, so should they (31). The resulting agreements have been a key motivating force for target status reporting and rapid deposition for structural genomics efforts worldwide.

The atmosphere of collaboration fostered by the structural genomics community and the ISGO has led to international meetings on structural genomics (ICSG 2000 in Yokohama, Japan; ICSG 2002 in Berlin; ICSG 2004 in Washington, DC; ICSG 2006 in Beijing; ICSG 2008 in Oxford, U.K.; see **http://www.isgo.org**) and to workshops designed to share technologies such as cell-free expression and new methods in NMR structure determination.

PERSPECTIVES

Structural genomics is now a mature field, with highly successful large-scale centers around the world and thousands of structures determined. The lessons learned from structural genomics are important not only for this field but also for other fields. The fact that high-throughput centers have been developed that can successfully carry out the complicated and difficult task of macromolecular structure determination has implications for future efforts in other challenging fields, from proteomics to cell biology. The developments in methods for the identification of which structures will be feasible, the recognition of the importance of systematic validation of methods, and approaches for estimating the relative efficacy of different procedures are applicable to almost any field. The important roles of international cooperation, data and method sharing, and rapid data deposition are key lessons as well for all disciplines generating large amounts of data. Efforts worldwide have now shown that structural genomics is possible and practical and how it can be carried out. The future for structural genomics is to continue to apply this powerful approach to determine structures that are of both current and long-term interest, providing a foundation for understanding macromolecules whose biological roles are known now and for those whose roles will be identified in the future.

DISCLOSURE STATEMENT

The authors are not aware of any biases that might be perceived as affecting the objectivity of this review.

ACKNOWLEDGMENTS

The authors are grateful to the many colleagues around the world who initiated and developed the field of structural genomics. Los Alamos National Laboratory requests that the publisher identify this article as work performed under the auspices of the U.S. Department of Energy. Los Alamos National Laboratory strongly supports academic freedom and a researcher's right to publish; as an institution, however, the Laboratory does not endorse the viewpoint of a publication or guarantee its technical correctness.

LITERATURE CITED

1. Adams PD, Grosse-Kunstleve RW, Hung L-W, Ioerger TR, McCoy AJ, et al. 2002. PHENIX: building new software for automated crystallographic structure determination. *Acta Crystallogr. D* 58:1948–54

2. Alzari PM, Berglund H, Berrow NS, Blagova E, Busso D, et al. 2006. Implementation of semiautomated cloning and prokaryotic expression screening: the impact of SPINE. *Acta Crystallogr. D* 62:1103–13

3. Baker EN. 2007. Structural genomics as an approach towards understanding the biology of tuberculosis. *J. Struct. Funct. Genomics* 8:57–65

4. Berman HM, Battistuz T, Bhat TN, Bluhm WF, Bourne PE, et al. 2002. The Protein Data Bank. *Acta Crystallogr. D* 58:899–907

5. Bernstein FC, Koetzle TF, Williams GJB, Meyer EF, Brice MD, et al. 1977. Protein Data Bank—computer-based archival file for macromolecular structures. *J. Mol. Biol.* 112:535–42

6. Berry IM, Dym O, Esnouf RM, Harlos K, Meged R, et al. 2006. SPINE high-throughput crystallization, crystal imaging and recognition techniques: current state, performance analysis, new technologies and future aspects. *Acta Crystallogr. D* 62:1137–49

7. Bhattacharya A, Tejero R, Montelione G. 2007. Evaluating protein structures determined by structural genomics consortia. *Proteins* 66:778–95

8. Bhavesh N, Panchal S, Hosur R. 2001. An efficient high-throughput resonance assignment procedure for structural genomics and protein folding research by NMR. *Biochemistry* 40:14727–35

9. Billas IML, Iwema T, Garnier JM, Mitschler A, Rochel N, Moras D. 2003. Structural adaptability in the ligand-binding pocket of the ecdysone hormone receptor. *Nature* 426:91–96

10. Brown EN, Ramaswamy S. 2007. Quality of protein crystal structures. *Acta Crystallogr. D* 63:941–50

11. Brown J, Walter TS, Carter L, Abrescia NGA, Aricescu AR, et al. 2003. A procedure for setting up high-throughput nanolitre crystallization experiments. II. Crystallization results. *J. Appl. Crystallogr.* 36:315–18

12. Burley SK, Almo SC, Bonanno JB, Capel M, Chance MR, et al. 1999. Structural genomics: beyond the Human Genome Project. *Nat. Genet.* 23:151–57

13. Burley SK, Joachimiak A, Montelione GT, Wilson IA. 2008. Contributions to the NIH-NIGMS Protein Structure Initiative from the PSI Production Centers. *Structure* 16:5–11

14. Canaves JM, Page R, Wilson IA, Stevens RC. 2004. Protein biophysical properties that correlate with crystallization success in *Thermotoga maritima*: maximum clustering strategy for structural genomics. *J. Mol. Biol.* 344:977–91

15. Chandonia JM, Brenner SE. 2006. The impact of structural genomics: expectations and outcomes. *Science* 311:347–51

16. Chandonia JM, Kim S-H, Brenner SE. 2006. Target selection and deselection at the Berkeley Structural Genomics Center. *Proteins Struct. Funct. Bioinform.* 62:356–70

17. Christendat D, Yee A, Dharamsi A, Kluger Y, Savchenko A, et al. 2000. Structural proteomics of an archaeon. *Nat. Struct. Mol. Biol.* 7:903–9

18. Cohen AE, Ellis PJ, Miller MD, Deacon AM, Phizackerley RP. 2002. An automated system to mount cryo-cooled protein crystals on a synchrotron beamline, using compact sample cassettes and a small-scale robot. *J. Appl. Crystallogr.* 35:720–26

19. Edwards A. 2009. Large-scale structural biology of the human proteome. *Annu. Rev. Biochem.* 78:In press

20. Fletcher L. 2000. Efforts to commercialize structural genomics may be limited. *Nat. Biotechnol.* 18:1036–36

21. Fox BG, Goulding C, Malkowski MG, Stewart L, Deacon A. 2008. Structural genomics: from genes to structures with valuable materials and many questions in between. *Nat. Methods* 5:129–32

22. Gaasterland T. 1998. Structural genomics: bioinformatics in the driver's seat. *Nat. Biotechnol.* 16:625–27

23. Gilliland GL, Teplyakov A, Obmolova G, Tordova M, Thanki N, et al. 2002. Assisting functional assignment for hypothetical *Haemophilus influenzae* gene products through structural genomics. *Curr. Drug Targets Infect. Disord.* 2:339–53

24. Goh C-S, Lan N, Douglas SM, Wu B, Echols N, et al. 2004. Mining the structural genomics pipeline: identification of protein properties that affect high-throughput experimental analysis. *J. Mol. Biol.* 336:115–30

25. Gonzalez A, Moorhead P, McPhillips SE, Song J, Sharp K, et al. 2008. Web-Ice: integrated data collection and analysis for macromolecular crystallography. *J. Appl. Crystallogr.* 41:176–84

26. Graille M, Quevillon Cheruel S, Leulliot N, Zhou C, de La Sierra Gallay I, et al. 2004. Crystal structure of the YDR533c *S. cerevisiae* protein, a class II member of the Hsp31 family. *Structure* 12:839–47

27. Graslund S, Nordlund P, Weigelt J, Bray J, Gileadi O, et al. 2008. Protein production and purification. *Nat. Methods* 5:135–46

28. Guntert P. 2003. Automated NMR protein structure calculation. *Prog. NMR Spectrosc.* 43:105–25
29. Hermann JC, Marti-Arbona R, Fedorov AA, Fedorov E, Almo SC, et al. 2007. Structure-based activity prediction for an enzyme of unknown function. *Nature* 448:775–79
30. Hope H. 1988. Cryocrystallography of biological macromolecules—a generally applicable method. *Acta Crystallogr. B* 44:22–26
31. Iino H, Naitow H, Nakamura Y, Nakagawa N, Agari Y, et al. 2008. Crystallization screening test for the whole-cell project on *Thermus thermophilus* HB8. *Acta Crystallogr. F* 64:487–91
32. Kennedy MA, Montelione GT, Arrowsmith CH, Markley JL. 2002. Role for NMR in structural genomics. *J. Struct. Funct. Genomics* 2:155–69
33. Kigawa T, Yabuki T, Matsuda N, Matsuda T, Nakajima R, et al. 2004. Preparation of *Escherichia coli* cell extract for highly productive cell-free protein expression. *J. Struct. Funct. Genomics* 5:63–68
34. Kim S-H, Shin D-H, Kim R, Adams P, Chandonia J-M. 2008. Structural genomics of minimal organisms: pipeline and results. *Methods Mol. Biol.* 426:475–96
35. Kuramitsu S, Kawaguchi S, Hiramatsu Y. 1995. Database of heat-stable proteins from *Thermus thermophilus* HB8. *Protein Eng.* 8:964
36. Lesley SA, Wilson IA. 2005. Protein production and crystallization at the Joint Center for Structural Genomics. *J. Struct. Funct. Genomics* 6:71–79
37. Levitt M. 2007. Growth of novel protein structural data. *Proc. Natl. Acad. Sci. USA* 104:3183–88
38. Liger D, Graille M, Zhou C-Z, Leulliot N, Quevillon-Cheruel S, et al. 2004. Crystal structure and functional characterization of yeast YLR011wp, an enzyme with NAD(P)H-FMN and ferric iron reductase activities. *J. Biol. Chem.* 279:34890–97
39. Liu GH, Shen Y, Atreya HS, Parish D, Shao Y, et al. 2005. NMR data collection and analysis protocol for high-throughput protein structure determination. *Proc. Natl. Acad. Sci. USA* 102:10487–92
40. Liu J, Montelione GT, Rost B. 2007. Novel leverage of structural genomics. *Nat. Biotechnol.* 25:849–51
41. Minor W, Cymborowski M, Otwinowski Z, Chruszcz M. 2006. HKL-3000: the integration of data reduction and structure solution—from diffraction images to an initial model in minutes. *Acta Crystallogr. D* 62:859–66
42. Montelione GT, Anderson S. 1999. Structural genomics: keystone for a human proteome project. *Nat. Struct. Biol.* 6:11–12
43. Moseley HN, Monleon D, Montelione GT. 2001. Automatic determination of protein backbone resonance assignments from triple resonance nuclear magnetic resonance data. *Methods Enzymol.* 339:91–108
44. Mougous JD, Cuff ME, Raunser S, Shen A, Zhou M, et al. 2006. A virulence locus of *Pseudomonas aeruginosa* encodes a protein secretion apparatus. *Science* 312:1526–30
45. Newman J, Egan D, Walter TS, Meged R, Berry I, et al. 2005. Towards rationalization of crystallization screening for small- to medium-sized academic laboratories: the PACT/JCSG plus strategy. *Acta Crystallogr. D* 61:1426–31
46. Page R, Grzechnik SK, Canaves JM, Spraggon G, Kreusch A, et al. 2003. Shotgun crystallization strategy for structural genomics: an optimized two-tiered crystallization screen against the *Thermotoga maritima* proteome. *Acta Crystallogr. D* 59:1028–37
47. Perrakis A, Morris R, Lamzin VS. 1999. Automated protein model building combined with iterative structure refinement. *Nat. Struct. Biol.* 6:458–63
48. Porath J, Carlsson J, Olsson I, Belfrage G. 1975. Metal chelate affinity chromatography, a new approach to protein fractionation. *Nature* 258:598–99
49. Rodgers DW. 1994. Cryocrystallography. *Structure* 2:1135–40
50. Sali A. 1998. Meeting on 100,000 protein structures for the biologist (Avalon, New Jersey, USA). *Nat. Struct. Biol.* 5:1029–32
51. Sanishvili R, Yakunin AF, Laskowski RA, Skarina T, Evdokimova E, et al. 2003. Integrating structure, bioinformatics, and enzymology to discover function: BioH, a new carboxylesterase from *Escherichia coli*. *J. Biol. Chem.* 278:26039–45
52. Santarsiero BD, Yegian DT, Lee CC, Spraggon G, Gu J, et al. 2002. An approach to rapid protein crystallization using nanodroplets. *J. Appl. Crystallogr.* 35:278–81
53. Shapiro L, Lima CD. 1998. The Argonne Structural Genomics Workshop: Lamaze class for the birth of a new science. *Structure* 6:265–67

54. Shin DH, Hou J, Chandonia J-M, Das D, Choi I-G, et al. 2007. Structure-based inference of molecular functions of proteins of unknown function from Berkeley Structural Genomics Center. *J. Struct. Funct. Genomics* 8:99–105

55. Slabinski L, Jaroszewski L, Rodrigues APC, Rychlewski L, Wilson IA, et al. 2007. The challenge of protein structure determination–lessons from structural genomics. *Protein Sci.* 16:2472–82

56. Smialowski P, Schmidt T, Cox J, Kirschner A, Frishman D. 2006. Will my protein crystallize? A sequence-based predictor. *Proteins* 62:343–55

57. Snell G, Cork C, Nordmeyer R, Cornell E, Meigs G, et al. 2004. Automated sample mounting and alignment system for biological crystallography at a synchrotron source. *Structure* 12:537–45

58. Stevens RC, Yokoyama S, Wilson IA. 2001. Global efforts in structural genomics. *Science* 294:89–92

59. Sussman JL, Silman I. 2008. *Structural Proteomics and Its Impact on the Life Sciences*. Hackensack, NJ: World Sci.

60. Szyperski T, Braun D, Banecki B, Wuthrich K. 1996. Useful information from axial peak magnetization in projected NMR experiments. *J. Am. Chem. Soc.* 118:8146–47

61. Teplyakov A, Pullalarevu S, Obmolova G, Doseeva V, Galkin A, et al. 2004. Crystal structure of the YffB protein from *Pseudomonas aeruginosa* suggests a glutathione-dependent thiol reductase function. *BMC Struct. Biol.* 4:5

62. Terwilliger TC, Berendzen J. 1999. Automated MAD and MIR structure solution. *Acta Crystallogr. D* 55:849–61

63. Terwilliger TC, Waldo G, Peat TS, Newman JM, Chu K, Berendzen J. 1998. Class-directed structure determination: foundation for a Protein Structure Initiative. *Protein Sci.* 7:1851–56

64. Tornroth-Horsefield S, Wang Y, Hedfalk K, Johanson U, Karlsson M, et al. 2006. Structural mechanism of plant aquaporin gating. *Nature* 439:688–94

65. Vinarov DA, Lytle BL, Peterson FC, Tyler EM, Volkman BF, Markley JL. 2004. Cell-free protein production and labeling protocol for NMR-based structural proteomics. *Nat. Methods* 1:149–53

66. Vitkup D, Melamud E, Moult J, Sander C. 2001. Completeness in structural genomics. *Nat. Struct. Biol.* 8:559–66

67. Waldo GS, Standish BM, Berendzen J, Terwilliger TC. 1999. Rapid protein-folding assay using green fluorescent protein. *Nat. Biotechnol.* 17:691–95

68. Yang W, Hendrickson WA, Crouch RJ, Satow Y. 1990. Structure of RNase H phased at 2 Å resolution by MAD analysis of the selenomethionyl protein. *Science* 249:1398–405

69. Yokoyama S, Hirota H, Kigawa T, Yabuki T, Shirouzu M, et al. 2000. Structural genomics in Japan. *Nat. Struct. Biol.* 7(Suppl.):943–45

Structure and Dynamics of Membrane Proteins by Magic Angle Spinning Solid-State NMR

Ann McDermott

Department of Chemistry, Columbia University, New York, New York 10027;
email: aem5@columbia.edu

Annu. Rev. Biophys. 2009. 38:385–403

First published online as a Review in Advance on February 26, 2009

The *Annual Review of Biophysics* is online at biophys.annualreviews.org

This article's doi:
10.1146/annurev.biophys.050708.133719

Key Words

biological NMR, dynamics, membrane proteins, solid-state NMR

Abstract

Membrane proteins remain difficult to study by traditional methods. Magic angle spinning solid-state NMR (MAS SSNMR) methods present an important approach for studying membrane proteins of moderate size. Emerging MAS SSNMR methods are based on extensive assignments of the nuclei as a basis for structure determination and characterization of function. These methods have already been used to characterize fibrils and globular proteins and are being increasingly used to study membrane proteins embedded in lipids. This review highlights recent applications to intrinsic membrane proteins and summarizes recent technical advances that will enable these methods to be utilized for more complex membrane protein systems in the near future.

Contents

INTRODUCTION: DEFINING THE OPPORTUNITIES

Proteins that sit in membranes have myriad crucial roles. They supervise the traffic of information and materials across the cell membrane. They defend against invading pathogens, maintain the energy supply, and deliver complex instructions for the cell cycle. With such important roles to play, they have been squarely in our sights as structural biology targets, encouraged by the first structure in the mid 1980s (19). Membrane proteins are still strongly underrepresented in the publically accessible collection of structures, comprising 20–30% of the proteins in most genomes yet accounting for less than 1% of the proteins whose structure have been determined and deposited in the Protein Data Bank (7). Although this observation was made a decade ago (126) and has been repeated frequently since, it remains valid and important. 3D structures of many intrinsic membrane targets of interest to drug discovery efforts are often not available (58) because they are difficult to prepare and crystallize. Although homology models remain an essential emerging area of membrane structural biology, these tools are not routinely applicable to key novel problems in membrane protein structure either (24). Each new target for X-ray analysis in a structural genomics effort is truly formidable, and each success is still monumental.

Monotopic: intrinsic membrane proteins residing on one side of the membrane, involving membrane-embedded but "reentrant" segment(s), in contrast with transmembrane proteins

It is nevertheless impressive that 150+ unique membrane protein structures have been determined as of 2008, mainly by X-ray crystallography, although many smaller systems were characterized through NMR studies and a handful from electron microscopy and diffraction studies. This collection contains some real gems. As made clear in the White group's online compendium (**http://blanco.biomol.uci.edu/Membrane_Proteins_xtal.html**), many proteins come from disease vectors and several come from mammals. There are, to mention only a select few, proteins key to the signaling event across a membrane, for example, seven transmembrane structures. Membrane-bound enzymes such as intrinsic membrane serine proteases and monotopic proteins are crucial in hormone biochemistry. There are key players in transport across the membranes such as ABC transporters and prokaryotic and eukaryotic channels, including some channel structures at high resolution. These proteins offer a first view into the exciting world of the cell membrane. Among the broad lessons that can be learned from our present database of structures are issues of typical architecture: Nearly all are helical folds, with the exception of prokaryotic outer membrane proteins (barrels made typically from an even number of antiparallel, tilted β-strands). Hydrophobic nonpolar side chains tend to face the lipid interior, aromatic side chains tend to be at the headgroup and interface region, positively charged residues tend to be in the inside (nontranslocated) loops, and lipid molecules often copurify and cocrystallize with the protein. Beyond these interesting generalities, the structures are varied, with many different packing motifs for the transmembrane helices.

Membrane proteins, especially transporters and signaling transducers, can be expected to attain a range of conformations to fulfill their function. It is assumed that a static protein with static partners could not carry out the key functions of transporting information and materials across a cell membrane. To capture the key functional features of intrinsic membrane enzymes, transporters, and signaling players, a collection of structures and their interacting partners is needed. Such a powerful collection of structures in a variety of states has not been forthcoming generally for membrane systems. Beyond the issue of trapping the various states in a collection of still life poses, the issue of their dynamic interconversion is important: On what timescales are the transitions made? How are these dynamic excursions integrated into the whole kinetic picture for function? The paradigm that proteins are inherently flexible and that their flexibility is in some sense optimized for function has been an important intellectual centerpiece in many NMR studies of soluble proteins (41, 53, 105, 111), and recent methods allow characterization of otherwise elusive minor conformers in exchange with the major conformers (56, 93). For the cell's gatekeepers, the intrinsic membrane proteins, insight of this kind into conformational dynamics and inherent flexibility will have even more significant impact, and solid-state NMR (SSNMR) promises to have unique capabilities to probe these phenomena.

Often the native lipid bilayer environment is critically important in supporting function and the various conformations and functionally important states. The lipid composition is crucial for function of membrane proteins, presumably both in terms of the chemical features dictating specific binding of lipids and in terms of physical or mechanical aspects of the bilayer environment (89). The dual challenge of characterizing membrane protein structure and dynamics in a bilayer of relevant composition and accessing the variety of functionally important states and dynamical properties of the system by controlling sample conditions constitutes wonderful opportunities for NMR, as both enterprises challenge established methods exactly within the skill set of the growing powerful tools of magic angle spinning (MAS) SSNMR.

Over the past decade, considerable progress has been made in studies of proteins by emerging high-resolution SSNMR methods. SSNMR is remarkable in its richness of information and capability of providing highly detailed information on structure, conformational

dynamics, chemical state and chemical dynamics, and local nonbonded interactions. To carry out SSNMR measurements, there is no in principle requirement that the system be either soluble or crystalline. Thus for biological macromolecules, many kinds of systems can be studied that are not currently amenable to X-ray crystallography or solution NMR. Moreover, that crystals are not required implies that a breadth of conditions can be applied in a given problem to achieve a range of functional substates, including those that require low temperature to be kinetically trapped. Because there are clear indications that the native bilayer environment is crucial for supporting function, the availability of a site-specific structural tool that can be applied in native or near-native bilayers is critical. Recently, MAS-based methods have been utilized to provide extensive site-specific information on whole domains or whole proteins in the context of microcrystals, fibrils, native assemblies, and near-native bilayer environments. This breakthrough, the culmination of many years of technical development in SSNMR, has been rapidly validated and extended by studies in more than a dozen laboratories, so that although the first assignments occurred at the change of the millennium, many systems are currently under study through these methods. Thus it is an exciting time to be in this field and a useful time to describe key progress.

HIGH-RESOLUTION NMR STUDIES OF COMPLEX NONSOLUBLE BIOPOLYMERS: SPECTRAL ASSIGNMENTS

NMR measurements carried out in the solid-state differ fundamentally from those more commonly carried out in solution: The absence of rapid isotropic or near-isotropic molecular tumbling results in anisotropic interactions in the detected lines. Rather than isotropic values for the chemical shift and the dipolar couplings, solid-state samples exhibit couplings and shifts that depend explicitly on the rotational orientation of the functional group with respect to the magnetic field. This effect is a rich

Solid-state NMR (SSNMR): methods developed to study the NMR spectra of samples that are not in solution and hence do not exhibit spectral simplification due to rapid Brownian tumbling

Magic angle spinning (MAS): kHz frequency rotation of samples about an angle Arccos(1/sqrt(3)) with respect to the applied magnetic field, for the purpose of narrowing NMR lines

Dipolar couplings: magnetic, through-space interactions between nuclear spins that give rise to splittings, whose strength is dependent on structure

source of structural and dynamic information. SSNMR studies of static samples, with macroscopic alignment (for example, of the bilayers) relative to the applied magnetic field, have had a particular impact on intrinsic membrane proteins and have been the subject of an excellent recent review (90) in which the importance of anisotropic interactions is highlighted.

Much of the power of solution NMR for determining 3D structures of proteins derives from the use of isotropic shifts of organic materials and their narrow lines in solution. Obtaining NMR spectra of solid biopolymers with narrow lines is one key ingredient for efficient studies of structure and function. Narrow lines at the isotropic frequency for solid samples can be achieved using MAS, in which the sample is rotated about an axis of approximately 54° relative to the applied field, as initially demonstrated by Andrew et al. (2), Lowe (74), and Waugh and coworkers (97). These groups showed that during MAS of a solid, the anisotropic interactions for nuclei with spin 1/2 are averaged over time to produce narrow NMR lines at the isotropic shift. This approach has been combined with decoupling of the spins and is a good approach for obtaining resolved spectra for an impressive range of naturally occurring complex biopolymers, systems not at all amenable to crystallography or solution NMR, including, for example, whole biological tissues (106). Even early on, MAS-based studies of the structure and function of intrinsic membrane proteins (37) illustrated that these methods were complementary to and useful for closure on diffraction studies.

The narrowed lines also need to be correlated to the lines corresponding to the neighboring spins to achieve site-specific assignment of the spectra. This is complicated by the fact that MAS averages not only the chemical shift anisotropy (CSA), but also the useful dipolar interactions that serve as the basis for correlation spectra and distance measurements. With modern MAS-based techniques, however, it is possible to make use of or reintroduce dipolar interactions if appropriate rotor synchronized radio frequency (RF) pulse sequences are

applied. This approach has allowed for the use of heteronuclear (107) or homonuclear (100) interactions during spinning. Soon after development, these recoupling approaches were utilized to accomplish structural characterization, including precise and accurate internuclear distances in a chromophore in a membrane protein, and correlations between neighboring atoms (18, 79). In addition, these methods were used early on for dynamic characterization of proteins (108), providing evidence for fast limit averaging of dipolar interactions. These two aspects, narrowing the NMR lines to the extent possible and recoupling the spins during spinning, remain active areas of technical development.

Site-specific assignment of the sometimes congested groups of lines in a protein acts as a gatekeeper for extensive structure-function and biophysical studies. The challenge of this objective is illustrated in **Figures 1** and **2**, which show typical 2D homonuclear ^{13}C spectra and 3D ^{15}N,^{13}C spectra of an intrinsic membrane protein, which for many sites the third dimension is essential for obtaining resolved lines and indications of sequential contacts.

Many kinds of insights into the system follow directly from having the MAS spectrum assigned. When shifts are assigned for the heteroatoms, secondary structure analysis can be performed based on the database of known solution NMR (17, 131). Such an analysis of the shifts and of the secondary structure is a key step toward structural studies, allowing comparisons to predicted structural models or to other conformations of the same system. MAS SSNMR experiments can be used to identify tertiary contacts and thereby define the 3D structure. Moreover, the assigned shifts, if they are extensive, serve as invaluable markers to identify surfaces where ligands or other macromolecular partners bind to the protein, and therefore could be used for ligand screening (137). The assigned shifts also have provided the basis for inclusive, site-specific studies of conformational dynamics. Thus assigning a system paves the way for high-level structural and functional studies.

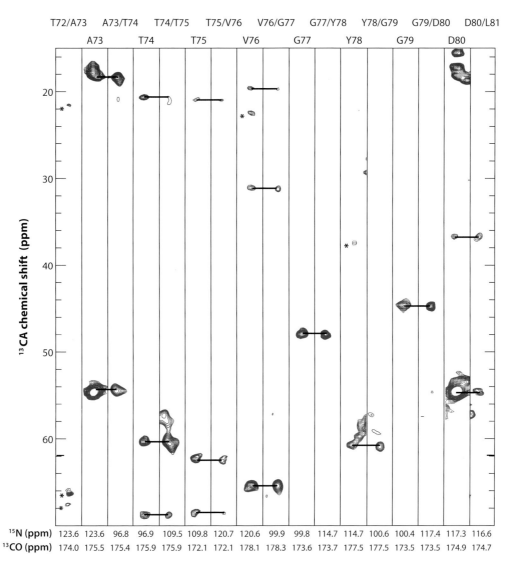

Figure 1

Sequential assignments are illustrated using the example residues Thr72-Leu81 of KcsA, the prokaryotic potassium channel, prepared in lipid bilayers. Magic angle spinning NMR studies correlating amide ^{15}N shifts (values below plot strips) to neighboring ^{13}C backbone and side chain shifts. Intraresidue correlations are in red and sequential correlations are in blue.

Prescient ideas about efficient assignments of NMR shifts measured using MAS-based methods were reported in the past few decades (16, 113), and relatively soon afterward came the first reports of nearly complete assignments on uniformly enriched globular proteins (8, 47, 48, 81, 83, 95). The first handful of globular proteins that were assigned using MAS SSNMR methods involved uniform ^{13}C^{15}N enrichment and 2D or 3D heteronuclear correlations between well-resolved amide ^{15}N shifts to neighboring ^{13}C side chain shifts via selective recoupling sequences (8, 47, 48, 81, 83, 95). At the time of this manuscript, 11 different proteins have fairly complete assignments of heteroatoms deposited into the Biological Magnetic Resonance Bank (**http://www.bmrb.wisc.edu/**)

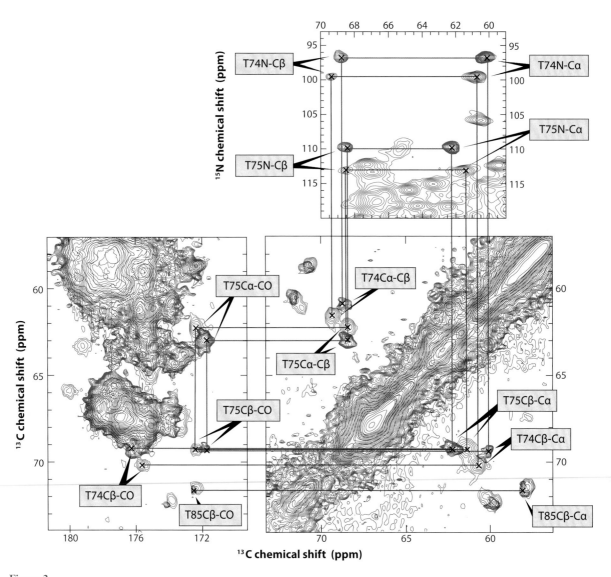

Figure 2

Homonuclear ^{13}C correlations for KcsA in lipid bilayers contrasting sample conditions, including <3 mM (*red*) versus 50 mM (*blue*) K$^+$. The spectra show pronounced changes in the selectivity filter region (Thr74 and Thr75) but lesser or no detectible changes in a variety of control sites (Thr85). Similar results are seen for other residues.

(22): mastoparan 6214, GB1 15156, kaliotoxin 6351, ubiquitin 7111, HET-s 11028, Crh 5757, dsbb 15546, c subunit of ATP synthase 10021, Pf1 coat 15138, LH2 6348, and GB3 15283. Several others, including thioredoxin (80), a mutant of human prion protein (40), α-synuclein in the fibrillar (Parkinson's disease related) form (38), Cu II-containing superoxide dismutase (98), and matrix metalloproteinase (4), have been reported in the literature as being mainly or extensively assigned. The general approach to assigning small- or medium-sized proteins (ca. 100 amino acids) by correlations of heteroatoms using moderate- to high-magnetic-field instruments (400–800 MHz) is robust and has breadth of scope.

Some of these proteins are intrinsic membrane proteins, some are fibrils, and some contain paramagnetic centers. Remarkably, many of these studies were carried out by first-time investigators and by solution-state NMR practitioners performing their first SSNMR study. These ventures are solid indications of the future for SSNMR methods.

A particular impediment for studying or even assigning membrane proteins is the severe spectral congestion that occurs for the ^{13}C spectra with the repetitive use of hydrophobic amino acids and the unrelenting helical secondary structure. These and other challenging protein systems continue to push the field to identify profound improvements on the experimental approach to assignments. In working with such congested spectra, it is important to emphasize the use of high magnetic fields, higher-dimensional spectra, ^{15}N shifts, which tend to be better dispersed than ^{13}C shifts. 4D correlations of chemical shifts have advantages in congested systems (29). One particularly important advance involves protocols based upon J couplings (12, 13, 23, 67, 86, 87), especially methods involving coherence selection, resulting in particularly narrow lines. These through bond methods can be expected to have great importance for systems in which flexibility sometimes thwarts the use of dipolar based correlation protocols. Whereas both X-ray studies and NMR studies of membrane proteins often disappoint, with missing bits of the protein, there is no reason to think this problem is unsolvable by NMR methods; the missing or mobile fragments can be targeted with specific labeling schemes and pulse sequences. In this regard, the J-based methods mentioned above are likely to be an important piece.

In the pursuit of more biologically relevant samples, it is important to use a free range of sample conditions without concern for the instrumental limitations. Probe design can be important in these experiments, not only for maximizing sensitivity but also for controlling the sample temperature during high-power RF irradiation. Recent advances in probe design, involving low inductance coils that result in less electric field penetration in the sample and therefore less sample heating, allow the study of a broader range of biological samples (34, 114).

In studying samples in their native conditions, sensitivity of detection will arise as a key impediment, particularly if 4D and higher-dimensional experiments are applied (29). There have been many efforts to improve detection sensitivity. Naturally, the use of the highest applied magnetic field strengths possible is important both in relieving congestion and in improving sensitivity. The use of directly detected protons rather than the heteronuclei (15, 49, 84, 96, 102, 139) improves detection sensitivity considerably.

Traditionally, the overall throughput of data collection has been constrained by the need for high-power proton decoupling. Typically, 100–300 W proton irradiation is used to decouple protons from heteronuclei and achieve narrow NMR lines in the solid state. This irradiation must be run at a relatively low duty cycle for many reasons, including the requirements of the amplifiers, the vulnerability of the RF elements in the NMR probe, and most fundamentally the deleterious effects on the sample from associated electrical fields from RF irradiation. Milder proton decoupling powers can be utilized if the proton homonuclear couplings are otherwise attenuated, either because of deuteration of the sample or through rapid MAS (21, 25, 57). Moreover, pulse sequences for coherent transfer of polarization among heteroatoms without proton irradiation have been recently reported (5, 78). These protocols will lead to more rapid data collection and hence dramatic overall improvements in sensitivity.

Dynamic nuclear polarization, the transfer of polarization from unpaired electrons to nearby weakly coupled nuclei, is a useful tool in dramatically enhancing signal detection sensitivity in high-field, high-resolution MAS-based NMR studies (45, 75). These studies are typically carried out at low temperatures, at which the electron spin characteristics are favorable. The use of low temperatures presents considerable challenges, both instrumental and in terms of the sample conditions, but recent reports

Heteronuclei: refers to spins other than protons in the study of NMR of proteins

Dynamic nuclear polarization: the use of electron nuclear methods to transfer polarization from unpaired electron spin reservoirs to nuclear spin reservoirs indicate a robust method for carrying out the low-temperature measurements (118). The low sample and coil temperatures offer an inherent advantage to achieving both lower thermal noise and stronger signals, which could have considerable impact on overall sensitivity as well. These approaches for improving sensitivity (high magnetic fields, improved probe designs, low-temperature measurements, dynamic nuclear polarization, proton detection, J-based transfers with selective coherence transfer, high-duty-cycle measurements) can in principle be combined to achieve a remarkable boost in sensitivity, and allow this field to address ever more complex membrane targets.

The success of site-specific assignments depends most critically on the quality and quantity of the sample and on the ability to control the NMR-active isotopes used in MAS-based experiments. Although important efforts have been made to use other expression systems such as *Pichia*, cell-free systems, insect cell lines, and even mammalian cell culture, MAS SSNMR studies are typically carried out on proteins that express well in *Escherichia coli*, a host in which isotopic composition is most conveniently controlled. Currently, it remains challenging to generate large quantities of functional key mammalian membrane proteins, and the use of alternative expression systems is likely to become important.

To some extent, the approach to spectral analysis of more complex systems involves issues of judgment and overall project vision. For success in achieving site-specific assignments, the resolution and therefore the linewidths and inherent dispersion of the protein's NMR peaks are also important. These properties in turn reflect the sample conditions, specifically the homogeneity in chemical composition and in conformation, and the conformational dynamics. This is no trivial concern at this stage. Although small globular proteins can be assigned with less than a year's effort, larger more complex systems still tend to be a major undertaking. The success of these major undertakings is linked to the choice of target and sample conditions. In this choice there is an apparent inherent tension. On the one hand, there is a fairly unique opportunity for SSNMR to perform site-specific measurements on key proteins in near physiological conditions, in which activity is maximal or deliberately controlled. On the other hand, there is a strategic advantage if sample conditions and protein constructs are screened to obtain benchmark spectra, regardless to some degree of the biological relevance of the conditions. The field as a whole has advanced through both kinds of projects. Only if samples are prepared to retain maximal biological interest can SSNMR methods resolve issues of debate in biology and medical chemistry. But, if samples are screened for spectra with ultimate resolution, which could be time and material intensive, then daring technique-development projects will be predisposed to success. The vision of the field as a whole, and of our future impact, will be to develop a disciplined approach that in some measure addresses both of these crucial concerns.

A New Dawn Rising: Structural Studies

We have deposited entries into the Protein Data Bank (7) for about a dozen proteins, including the 62-residue globular SH3 domain of α-spectrin, 1m8m (10); the 93-residue globular catabolite repressor protein Crh, 2rlz (68); the 76-residue globular protein ubiquitin, 2jzz (77); the 38-residue toxin kaliotoxin, 2uvs (55); the 56-residue globular B1 immunoglobulin-binding domain of protein G, 2jsv (30); the 22-residue β-2 microglobulin, 2e8d (50); the 18-residue antibiotic membrane-embedded peptide protegrin-1, 1zy6 (76); the 11-residue fibril-forming transthyretin peptide, 1rvs (52); the 15-residue membrane peptide mastoparan, 2czp (120); the 79-residue fungal protein in fibril form, HET-s, 2rnm (127). Other fibrillar systems of keen interest to human health have been assigned and structurally characterized by MAS-based methods, including a key fragment of the Alzheimer's Aβ protein (122). Although many of these studies served primarily to demonstrate methods, because the structure itself was known at high precision in advance by X-ray crystallography, a few concern outstanding unsolved structural problems of key biological interest and moderate complexity, particularly the fibrillar systems (40, 122, 127). These studies can serve as the basis of important structural mechanistic, binding, and dynamical studies in intrinsic membrane systems.

Studies of a group of structurally characterized, microcrystalline, thermostable, small (<100 amino acids) globular proteins with well-dispersed NMR shifts have provided the basis for developing protocols for structure determination by SSNMR. For example, in several of these systems, homonuclear ^{13}C spin diffusion has provided the basis for identifying tertiary contacts. Combined with chemical shift analysis for secondary structure by database methods (17, 130), the spin diffusion measurements have allowed researchers to compute structures at moderate resolution. Some variations on this approach have been reported, including passive spin diffusion (in absence of proton irradiation), or so-called proton-driven spin diffusion (PDSD) (10); recoupling of the ^{13}C nuclei through proton irradiation at a rotary resonance condition ($\omega_1 = \omega_r$) (also known as dipolar assisted rotational recoupling, DARR) (138); and the use of conveniently introduced, somewhat sparse isotopic enrichment (44) or, alternatively, uniform ^{13}C enrichment (77, 82, 134). Finally, related methods involve transfer to directly bonded protons that then serve as a spin diffusion reservoir (C–H...H–C). This approach is intended to enhance the detection of important tertiary contacts at the hydrophobic interior of the protein relative to intraresidue correlations, although at some cost in terms of overall transfer efficiency. The heteronuclear transfers can be carried out selectively for methyl groups with J-based transfers (69) or nonselectively based on heteronuclear dipolar correlations (62), as used in the study of kaliotoxin (55) and Crh (68).

Because these spin-diffusion-based approaches are technically not burdensome and relatively time efficient, and because the calculation protocols are analogous to solution NMR procedures, there has been relatively good dissemination of these approaches. Several laboratories have validated the general approach (with various elaborations) on several protein systems, all of which are in some measure successful, although the precision and accuracy of the structures cannot be said to be atomic. In the respect that these methods result in a large number of mainly ambiguous tertiary contacts, and in the respect that relayed transfers can be expected to be an important issue, these experiments are close analogs to the NOE in solution and will be subject to many of the same data analysis challenges.

A few approaches to achieving high-resolution structures have been developed. Measurements of correlated dipolar vectors from neighboring spin pairs (103) provide powerful constraints on torsion angles. Recent approaches actually retrieve internuclear distances that are remarkably accurate in extensively or uniformly isotopically enriched solids. Accurate measurements of internuclear distances pertaining to tertiary contacts distances by MAS SSNMR methods were originally predicated on the introduction of selective pairs of labels. One approach to measuring higher-precision distances in extensively isotopically enriched samples was based on TEDOR, a heteronuclear recoupling method (51). For homonuclear chemical shift correlation, broadband recoupling schemes, preferably those that are not demanding with respect to RF irradiation, can be important (5, 115). On the other hand, for determining specific homonuclear internuclear distances with precision (59, 125), spectrally selective recoupling sequences are preferable (20, 36, 51, 59). Even when few in number, accurate distances can be expected to have considerable power in refining structures determined by SSNMR. These ideas (the development of which is still ongoing) constitute the final breakthrough in the field, finally allowing for a variety of structural studies. These methods collectively provide the centerpiece of an approach to determine structures at atomic or subangstrom accuracy (30, 52, 104). These efforts toward high-resolution structures delineate an important future challenge, namely the correction for conformational dynamics.

Moving forward to structural studies of membrane proteins will also present considerable challenges. Many of the key players are homo-oligomeric or hetero-oligomeric structures, and most show severe spectral congestion. Therefore, advances in selective labeling schemes will continue to be important, as has arguably been the case for amyloid structures. Tertiary contact information will be more ambiguous. Ambiguous restraints have been discussed in the context of nearly every structure solved by SSNMR methods, and the available tools for use of ambiguous restraints have played a large role in shaping the thinking of the first structures calculated (28) and have continued to be an important aspect in almost all the structures reported to date (68, 77, 138).

Methods to identify intermolecular versus intramolecular contacts through mixed labeling strategies will be important for the oligomeric structures. Such an approach has been used in studies of thioredoxin (135), and related approaches were demonstrated in the case of Crh (26) and HET-s (127).

A more nuanced and expanded role for chemical shift analysis is likely to be significant as this field progresses toward more complex systems as well. Knowledge-based or bioinformatics-based methods have figured prominently in the advances in solution NMR (11, 112) and will undoubtedly in solid-state studies as well. Recent work has paved the way for the use of CSA as a constraint (132, 133).

The use of chemical tags introduced into the protein is also an option for solving problems in spectrally congested systems. As has been the case for solution NMR studies of intrinsic membrane proteins, the use of paramagnetic tags will presumably become important as well. Paramagnetic centers cause a range of useful effects in spectra of complex materials (63, 88, 128), and there is strong potential for the use of paramagnetic tags to probe internuclear distances and solvent depth (4, 63, 88, 98). Solvent accessibility and membrane depth have been probed by a variety of methods (31, 64, 109). Another promising method for characterizing complex systems is the dephasing of spins by irradiation of selectively introduced rare spins with high gamma, e.g., ^{19}F or ^{31}P (35).

For the next set of advances in structure and function studies by MAS methods, an approach for integrating dynamical information into solving structures is needed. Established methods allow for probing motions on a variety of timescales, but many of them rely on isolated spins and/or static samples. Recent developments have allowed for the characterization of a variety of dynamic variables by MAS-based methods for extensively isotopically enriched samples in a site-specific fashion, including order parameters (71–73) and spin lattice relaxation rates (14, 32). The challenge of optimizing these methods and interpreting these dynamic measurements and integrating them into structural studies will encompass exciting biochemical and spectroscopic opportunities for future scientists. These methods uniquely promise to provide atomic level description of conformational dynamics in intrinsic membrane proteins.

Insights Regarding Membrane Proteins in Action: A View from 30,000 Pulses

SSNMR studies of membrane proteins ligands have provided crucial insights into their structure and function. Frequently, the structural data available from diffraction studies on membrane proteins have omitted loops and more importantly binding surfaces for partners and the ligands themselves, or the ligands are at too low resolution to obtain details of chemical interest. Recent work highlights the importance of studying membrane protein ligands by SSNMR. Using MAS SSNMR, Williamson et al. (129) showed that the conformation of acetylcholine in its binding site on the membrane-embedded nicotinic acetylcholine receptor was extended. The coupling of retinal isomerization to activation of rhodopsin has been studied in atomic detail (94), and a clear picture for energy transduction in the early bacteriorhodopsin intermediates has emerged from SSNMR measurements (75). Neurotensin, a peptide that binds with high affinity to a G-protein-coupled receptor, was contrasted in a lipid model phase versus a receptor-bound form (39).

Numerous NMR studies characterize relatively small membrane peptides that play key roles in shaping the properties of bilayers and have exciting biological consequences. For example, bioactive peptides can disrupt membranes, sometimes acting as antibiotics, and they can cause tightly timed fusion of membranes. Many of these systems have been structurally and dynamically characterized in biologically relevant conditions using solution NMR, static SSNMR, and MAS-based NMR methods. SSNMR experiments on specifically ^{15}N-labeled antibiotic peptides PGLa and magainin in phospholipid bilayer samples show that the helix axis is parallel to the plane of the bilayers, with some residues being highly mobile (6). Channel-forming colicins, bacterial toxins that spontaneously insert into the inner cell membrane of sensitive bacteria, were investigated by SSNMR to elucidate their topology and segmental motion (44). Protegrin-1, a β-hairpin antimicrobial peptide, forms a multimer involving a membrane-inserted β-barrel surrounded by disordered lipids (76). Membrane perturbation of a synthetic antibiotic amphipathic peptide is caused by induction of positive curvature strain, induced when the peptide associates laterally with the bilayer surface (92). Similarly, two other amphipathic antimicrobial peptides derived from magainin-2 and melittin also repress the lamellar-to-inverted hexagonal phase transition by inducing positive curvature strain while associating with the helix oriented nearly perpendicular to the bilayer normal (101). A structural model for a functional mimic of lung surfactant protein B, which lowers surface tension in the alveoli, was derived recently from SSNMR data (85). SSNMR studies of an HIV fusion peptide associated with native-like membranes have demonstrated the presence of a fully extended conformation possibly with antiparallel strand registries (99). Many of these systems are engaged in rapid rotational diffusion under the conditions of the MAS experiments, a situation that results in considerable narrowing of the lines and can allow for simplified analysis of angular long-range constraints (9). Constraints on the structural relation between lipids and lipid-bound peptides have been obtained from MAS-based methods (116, 117).

The coat protein of a filamentous bacteriophage, Pf1, has been studied by MAS-based methods in the context of a fully infectious viral particle, and full assignments of heteroatoms were determined (33). These measurements allowed for analysis of the likely asymmetric unit in the viral packing as well as the secondary structure. Moreover, a dynamical analysis of the backbone and side chains was carried out, indicating both static and fast limit dynamic heterogeneity in the vicinity of the DNA (72). This system and related systems have been extensively studied by static SSNMR methods (91), and the two methods generally agree that all peptides are chemically equivalent in the assembly and that the coat protein is helical from residue 6 to the C terminus.

The progress in assigning resonances using extensive enrichment in heteroatoms in more complex proteins in the membrane indicates that structural insights into these systems will be forthcoming. Nearly complete assignments of the light-harvesting complex II from *Rhodobacter*, LH2, were reported. These studies illustrate creative and painstaking use of selective isotopic labels to relieve the inherent congestion in intrinsic membrane protein NMR spectra (123). The analogous but more structurally elusive LH1 has also been partially assigned using mainly uniformly isotopically enriched materials (46). Both studies noted that it was difficult to detect the solvent-exposed loops by using these methods. NMR data were reported on a uniformly ^{13}C ^{15}N-enriched variant of the 52-residue phospholamban, which is involved in regulating the flow of ions in cardiac muscle cells. Recent studies raise issues regarding the secondary structure of the cytoplasmic portion and its associations with lipids.

One SSNMR study supported a model with an α-helical transmembrane segment and a high degree of disorder in the cytoplasmic domain (3). Another study from a different group supports a pentameric pinwheel-like geometry, in which the cytoplasmic helix is perpendicular to the membrane surface (121); this conclusion was also supported by static SSNMR measurements in lipids carefully selected to ensure function (1).

Assignments have also been reported for sensory rhodopsin, a prokaryotic seven-transmembrane receptor protein (27). Optimization of sample conditions has been reported for the intrinsic membrane protein diacyl glycerol kinase (70), in which crystalline samples produce better-resolved spectra than do samples reconstituted in proteoliposomes. In this system, ^{31}P NMR spectra allowed identification of bound lipids. The NMR spectra of the ATP synthase subunit have been analyzed in dried films, an intermediate stage in an organic solvent-based purification protocol (54). Several residues could be assigned site specifically. Significant progress toward the assignments of the resonances of an outer membrane pore protein, OmpG, has been reported (42, 43), illustrating the importance of selective isotopic enrichment protocols. Recent studies of DsbB, a 20-kDa redox-active intrinsic membrane protein involved in disulfide bond formation in periplasmic proteins or prokaryotes, report sequential chemical shift assignments for most of the residues in the transmembrane helices, based on 3D correlation experiments on a

uniformly ^{13}C,^{15}N-labeled sample (carried out at 750 MHz). An interesting aspect was the clarification of congested regions by supplementing the 3D spectra with 4D correlation spectra. Note that here again the loops were difficult to detect by MAS-based methods (65, 66).

The prokaryotic potassium channel, KcsA, has been studied by a number of NMR groups. The structure of this fascinating molecule was reported a decade ago, and it exhibits a number of novel features. The conformational dynamics and range of alternative conformers remain an open research topic ripe for further discovery. Partial assignments of the channel based on 3D correlation experiments (124) have been reported, encompassing the selectivity filter and portions of the inner helix. More complete assignments (110) were used to address the interactions with a group of selective toxin blockers and the induced structural changes in the channel upon toxin binding (60, 136). The effect of ion concentration has also been reported (61, 119). To illustrate data that form the basis of sequential assignments and functional binding assays, **Figures 1** and **2** show strip plots for the assignment of the residues in the selectivity filter and the dependence of side chain shifts on K$^+$ ion concentration, respectively.

Looking forward, extra challenges can be expected in studies of more complex membrane proteins, and there is good cause for optimism in these studies. Many satisfying puzzles lie ahead for spectroscopists and biochemists in this burgeoning field.

SUMMARY POINTS

1. As demonstrated by studies on a variety of globular proteins, MAS SSNMR provides a powerful method for obtaining highly detailed structural and dynamic information regarding proteins in a site-specific fashion.

2. Studies of several small membrane proteins, such as antibiotic peptides, and recent spectral assignments of a handful of medium-sized prokaryotic membrane proteins, including membrane enzymes and ion channels, indicate that MAS SSNMR methods can be applied to research the structure and dynamics of intrinsic membrane proteins under near-native conditions and without need for crystals.

FUTURE ISSUES

1. Technical developments from the past few years include a variety of approaches that improve sensitivity of detection, provide more powerful spectral assignment protocols, sharpen the precision of structural constraints, and simplify the procedure for moderate-resolution structures.

2. These recent methodological improvements are likely to allow for an enlarged scope of applicability of these methods, for example, for studies of more complex intrinsic membrane protein systems.

DISCLOSURE STATEMENT

The author is not aware of any biases that might be perceived as affecting the objectivity of this review.

ACKNOWLEDGMENTS

A grant from the NIH GM75026 supported this research. The author thanks current and prior students and postdocs for useful discussions, especially Yisong Tao, Lin Tian, Ansgar Siemer, Ben Wylie, and Tatyana Polenova.

LITERATURE CITED

1. Abu-Baker S, Lu JX, Chu S, Shetty KK, Gor'kov PL, Lorigan GA. 2007. The structural topology of wild-type phospholamban in oriented lipid bilayers using N-15 solid-state NMR spectroscopy. *Protein Sci.* 16:2345–49

2. Andrew ER, Bradbury A, Eades RG. 1958. Nuclear magnetic resonance spectra from a crystal rotated at high speed. *Nature* 182:1659

3. Andronesi OC, Becker S, Seidel K, Heise H, Young HS, Baldus M. 2005. Determination of membrane protein structure and dynamics by magic-angle-spinning solid-state NMR spectroscopy. *J. Am. Chem. Soc.* 127:12965–74

4. Balayssac S, Bertini I, Falber K, Fragai M, Jehle S, et al. 2007. Solid-state NMR of matrix metalloproteinase 12: an approach complementary to solution NMR. *ChemBioChem* 8:486–89

5. Bayro MJ, Ramachandran R, Caporini MA, Eddy MT, Griffin RG. 2008. Radio frequency-driven recoupling at high magic-angle spinning frequencies: homonuclear recoupling sans heteronuclear decoupling. *J. Chem. Phys.* 128:052321

6. Bechinger B, Zasloff M, Opella SJ. 1998. Structure and dynamics of the antibiotic peptide PGLa in membranes by solution and solid-state nuclear magnetic resonance spectroscopy. *Biophys. J.* 74:981–87

7. Berman HM, Westbrook J, Feng Z, Gilliland G, Bhat TN, et al. 2000. The Protein Data Bank. *Nucleic Acids Res.* 28:235–42

8. Bockmann A, Lange A, Galinier A, Luca S, Giraud N, et al. 2003. Solid state NMR sequential resonance assignments and conformational analysis of the 2 × 10.4 kDa dimeric form of the Bacillus subtilis protein Crh. *J. Biomol. NMR* 27:323–39

9. Cady SD, Goodman C, Tatko CD, DeGrado WF, Hong M. 2007. Determining the orientation of uniaxially rotating membrane proteins using unoriented samples: a H-2°C-13, and N-15 solid-state NMR investigation of the dynamics and orientation of a transmembrane helical bundle. *J. Am. Chem. Soc.* 129:5719–29

10. Castellani F, van Rossum B, Diehl A, Schubert M, Rehbein K, Oschkinat H. 2002. Structure of a protein determined by solid-state magic-angle-spinning NMR spectroscopy. *Nature* 420:98–102

11. Cavalli A, Salvatella X, Dobson CM, Vendruscolo M. 2007. Protein structure determination from NMR chemical shifts. *Proc. Natl. Acad. Sci. USA* 104:9615–20

12. Chen L, Kaiser JM, Polenova T, Yang J, Rienstra CM, Mueller LJ. 2007. Backbone assignments in solid-state proteins using J-based 3D heteronuclear correlation spectroscopy. *J. Am. Chem. Soc.* 129:10650–52

13. Chen LL, Kaiser JM, Lai JF, Polenova T, Yang J, et al. 2007. J-based 2D homonuclear and heteronuclear correlation in solid-state proteins. *Magn. Reson. Chem.* 45:S84–S92

14. Chevelkov V, Diehl A, Reif B. 2008. Measurement of N-15-T-1 relaxation rates in a perdeuterated protein by magic angle spinning solid-state nuclear magnetic resonance spectroscopy. *J. Chem. Phys.* 128:052316

15. Chevelkov V, Reif B. 2008. TROSY effects in MAS solid-state NMR. *Concepts Magn. Reson. Pt. A* 32:143–56

16. Cole HBR, Sparks SW, Torchia DA. 1988. Comparison of the solution and crystal-structures of staphylococcal nuclease with C-13 and N-15 chemical-shifts used as structural fingerprints. *Proc. Natl. Acad. Sci. USA* 85:6362–65

17. Cornilescu G, Delaglio F, Bax A. 1999. Protein backbone angle restraints from searching a database for chemical shift and sequence homology. *J. Biomol. NMR* 13:289–302

18. Creuzet F, McDermott A, Gebhard R, Vanderhoef K, Spijkerassink MB, et al. 1991. Determination of membrane-protein structure by rotational resonance NMR—bacteriorhodopsin. *Science* 251:783–86

19. Deisenhofer J, Epp O, Miki K, Huber R, Michel H. 1985. Structure of the protein subunits in the photosynthetic reaction center of Rhodopseudomonas viridis at 3A resolution. *Nature* 318:618–24

20. De Paepe G, Lewandowski JR, Griffin RG. 2008. Spin dynamics in the modulation frame: application to homonuclear recoupling in magic angle spinning solid-state NMR. *J. Chem. Phys.* 128:124503

21. Detken A, Hardy EH, Ernst M, Meier BH. 2002. Simple and efficient decoupling in magic-angle spinning solid-state NMR: the XiX scheme. *Chem. Phys. Lett.* 356:298–304

22. Doreleijers JF, Mading S, Maziuk D, Sojourner K, Yin L, et al. 2003. BioMagResBank database with sets of experimental NMR constraints corresponding to the structures of over 1400 biomolecules deposited in the Protein Data Bank. *J. Biomol. NMR* 26:139–46

23. Elena B, Lesage A, Steuernagel S, Bockmann A, Emsley L. 2005. Proton to carbon-13 INEPT in solid-state NMR spectroscopy. *J. Am. Chem. Soc.* 127:17296–302

24. Elofsson A, von Heijne G. 2007. Membrane protein structure: Prediction versus reality. *Annu. Rev. Biochem.* 76:125–40

25. Ernst M, Samoson A, Meier BH. 2003. Low-power XiX decoupling in MAS NMR experiments. *J. Magn. Reson.* 163:332–39

26. Etzkorn M, Bockmann A, Lange A, Baldus M. 2004. Probing molecular interfaces using 2D magic-angle-spinning NMR on protein mixtures with different uniform labeling. *J. Am. Chem. Soc.* 126:14746–51

27. Etzkorn M, Martell S, Andronesi OC, Seidel K, Engelhard M, Baldus M. 2007. Secondary structure, dynamics, and topology of a seven-helix receptor in native membranes, studied by solid-state NMR spectroscopy. *Angew. Chem. Int. Ed.* 46:459–62

28. Fossi M, Castellani T, Nilges M, Oschkinat H, van Rossum BJ. 2005. SOLARIA: a protocol for automated cross-peak assignment and structure calculation for solid-state magic-angle spinning NMR spectroscopy. *Angew. Chem. Int. Ed.* 44:6151–54

29. Franks WT, Kloepper KD, Wylie BJ, Rienstra CM. 2007. Four-dimensional heteronuclear correlation experiments for chemical shift assignment of solid proteins. *J. Biomol. NMR* 39:107–31

30. Franks WT, Wylie BJ, Schmidt HLF, Nieuwkoop AJ, Mayrhofer RM, et al. 2008. Dipole tensor-based atomic-resolution structure determination of a nanocrystalline protein by solid-state NMR. *Proc. Natl. Acad. Sci. USA* 105:4621–26

31. Gallagher GJ, Hong M, Thompson LK. 2004. Solid-state NMR spin diffusion for measurement of membrane-bound peptide structure: gramicidin A. *Biochemistry* 43:7899–906

32. Giraud N, Blackledge M, Bockmann A, Emsley L. 2007. The influence of nitrogen-15 proton-driven spin diffusion on the measurement of nitrogen-15 longitudinal relaxation times. *J. Magn. Reson.* 184:51–61

33. Goldbourt A, Gross BJ, Day LA, McDermott AE. 2007. Filamentous phage studied by magic-angle spinning NMR: resonance assignment and secondary structure of the coat protein in Pf1. *J. Am. Chem. Soc.* 129:2338–44

34. Gor'kov PL, Witter R, Chekmenev EY, Nozirov F, Fu R, Brey WW. 2007. Low-E probe for F-19-H-1 NMR of dilute biological solids. *J. Magn. Reson.* 189:182–89

35. Graesser DT, Wylie BJ, Nieuwkoop AJ, Franks WT, Rienstra CM. 2007. Long-range F-19-N-15 distance measurements in highly-C-13°N-15-enriched solid proteins with F-19-dephased REDOR shift (FRESH) spectroscopy. *Magn. Reson. Chem.* 45:S129–S34

36. Griffin RG. 1998. Dipolar recoupling in MAS spectra of biological solids. *Nat. Struct. Biol.* 5:508–12

37. Harbison GS, Herzfeld J, Griffin RG. 1983. Solid-state N-15 nuclear magnetic-resonance study of the Schiff-base in bacteriorhodopsin. *Biochemistry* 22:1–5

38. Heise H, Celej MS, Becker S, Riede D, Pelah A, et al. 2008. Solid-state NMR reveals structural differences between fibrils of wild-type and disease-related A53T mutant alpha-synuclein. *J. Mol. Biol.* 380:444–50

39. Heise H, Luca S, de Groot BL, Grubmuller H, Baldus M. 2005. Probing conformational disorder in neurotensin by two-dimensional solid-state NMR and comparison to molecular dynamics simulations. *Biophys. J.* 89:2113–20

40. Helmus JJ, Surewicz K, Nadaud PS, Surewicz WK, Jaroniec CP. 2008. Molecular conformation and dynamics of the Y145Stop variant of human prion protein. *Proc. Natl. Acad. Sci. USA* 105:6284–89

41. Henzler-Wildman K, Kern D. 2007. Dynamic personalities of proteins. *Nature* 450:964–72

42. Hiller M, Higman VA, Jehle S, van Rossum BJ, Kuhlbrandt W, Oschkinat H. 2008. [2,3-C-13]-labeling of aromatic residues—getting a head start in the magic-angle-spinning NMR assignment of membrane proteins. *J. Am. Chem. Soc.* 130:408–9

43. Hiller M, Krabben L, Vinothkumar KR, Castellani F, van Rossum BJ, et al. 2005. Solid-state magic-angle spinning NMR of outer-membrane protein G from Escherichia coli. *ChemBioChem* 6:1679–84

44. Hong M, Jakes K. 1999. Selective and extensive C-13 labeling of a membrane protein for solid-state NMR investigations. *J. Biomol. NMR* 14:71–74

45. Hu KN, Song C, Yu HH, Swager TM, Griffin RG. 2008. High-frequency dynamic nuclear polarization using biradicals: a multifrequency EPR lineshape analysis. *J. Chem. Phys.* 128:052302

46. Huang L, McDermott A. 2008. Partial site-specific assignment of a uniformly (13)C, (15)N enriched membrane protein, light-harvesting complex 1 (LH1), by solid state NMR. *Biochim. Biophys. Acta* 1777:1098–108

47. Igumenova TI, McDermott AE, Zilm KW, Martin RW, Paulson EK, Wand AJ. 2004. Assignments of carbon NMR resonances for microcrystalline ubiquitin. *J. Am. Chem. Soc.* 126:6720–27

48. Igumenova TI, Wand AJ, McDermott AE. 2004. Assignment of the backbone resonances for microcrystalline ubiquitin. *J. Am. Chem. Soc.* 126:5323–31

49. Ishii Y, Tycko R. 2000. Sensitivity enhancement in solid state N-15 NMR by indirect detection with high-speed magic angle spinning. *J. Magn. Reson.* 142:199–204

50. Iwata K, Fujiwara T, Matsuki Y, Akutsu H, Takahashi S, et al. 2006. 3D structure of amyloid protofilaments of beta(2)-microglobulin fragment probed by solid-state NMR. *Proc. Natl. Acad. Sci. USA* 103:18119–24

51. Jaroniec CP, Filip C, Griffin RG. 2002. 3D TEDOR NMR experiments for the simultaneous measurement of multiple carbon-nitrogen distances in uniformly C-13°N-15-labeled solids. *J. Am. Chem. Soc.* 124:10728–42

52. Jaroniec CP, MacPhee CE, Bajaj VS, McMahon MT, Dobson CM, Griffin RG. 2004. High-resolution molecular structure of a peptide in an amyloid fibril determined by magic angle spinning NMR spectroscopy. *Proc. Natl. Acad. Sci. USA* 101:711–26

53. Kempf JG, Jung JY, Ragain C, Sampson NS, Loria JP. 2007. Dynamic requirements for a functional protein hinge. *J. Mol. Biol.* 368:131–49

54. Kobayashi M, Matsuki Y, Yumen I, Fujiwara T, Akutsu H. 2006. Signal assignment and secondary structure analysis of a uniformly [C-13°N-15]-labeled membrane protein, H+-ATP synthase subunit c, by magic-angle spinning solid-state NMR. *J. Biomol. NMR* 36:279–93

55. Korukottu J, Schneider R, Vijayan V, Lange A, Pongs O, et al. 2008. High-resolution 3D structure determination of kaliotoxin by solid-state NMR spectroscopy. *PLoS ONE* 3:e2359

56. Korzhnev DM, Kay LE. 2008. Probing invisible, low-populated states of protein molecules by relaxation dispersion NMR spectroscopy: an application to protein folding. *Acc. Chem. Res.* 41:442–51

57. Kotecha M, Wickramasinghe NP, Ishii Y. 2007. Efficient low-power heteronuclear decoupling in C-13 high-resolution solid-state NMR under fast magic angle spinning. *Magn. Reson. Chem.* 45:S221–S30

58. Lacapere JJ, Pebay-Peyroula E, Neumann JM, Etchebest C. 2007. Determining membrane protein structures: still a challenge! *Trends Biochem. Sci.* 32:259–70

59. Ladizhansky V, Griffin RG. 2004. Band-selective carbonyl to aliphatic side chain C-13-C-13 distance measurements in U-C-13,N-15-labeled solid peptides by magic angle spinning NMR. *J. Am. Chem. Soc.* 126:948–58

60. Lange A, Giller K, Hornig S, Martin-Eauclaire MF, Pongs O, et al. 2006. Toxin-induced conformational changes in a potassium channel revealed by solid-state NMR. *Nature* 440:959–62

61. Lange A, Giller K, Pongs O, Becker S, Baldus M. 2006. Two-dimensional solid-state NMR applied to a chimeric potassium channel. *J. Recept. Signal Transduct.* 26:379–93

62. Lange A, Seidel K, Verdier L, Luca S, Baldus M. 2003. Analysis of proton-proton transfer dynamics in rotating solids and their use for 3D structure determination. *J. Am. Chem. Soc.* 125:12640–48

63. Lee H, Polenova T, Beer RH, McDermott AE. 1999. Lineshape fitting of deuterium magic angle shinning spectra of paramagnetic compounds in slow and fast limit motion regimes. *J. Am. Chem. Soc.* 121:6884–94

64. Lesage A, Gardiennet C, Loquet A, Verel R, Pintacuda G, et al. 2008. Polarization transfer over the water-protein interface in solids. *Angew. Chem. Int. Ed.* 47:5851–54

65. Li Y, Berthold DA, Frericks HL, Gennis RB, Rienstra CM. 2007. Partial C-13 and N-15 chemical-shift assignments of the disulfide-bond-forming enzyme DsbB by 3D magic-angle spinning NMR spectroscopy. *ChemBioChem* 8:434–42

66. Li Y, Berthold DA, Gennis RB, Rienstra CM. 2008. Chemical shift assignment of the transmembrane helices of DsbB, a 20-kDa integral membrane enzyme, by 3D magic-angle spinning NMR spectroscopy. *Protein Sci.* 17:199–204

67. Linser R, Fink U, Reif B. 2008. Proton-detected scalar coupling based assignment strategies in MAS solid-state NMR spectroscopy applied to perdeuterated proteins. *J. Magn. Reson.* 193:89–93

68. Loquet A, Bardiaux B, Gardiennet C, Blanchet C, Baldus M, et al. 2008. 3D structure determination of the Crh protein from highly ambiguous solid-state NMR restraints. *J. Am. Chem. Soc.* 130:3579–89

69. Loquet A, Laage S, Gardiennet C, Elena B, Emsley L, et al. 2008. Methyl proton contacts obtained using heteronuclear through-bond transfers in solid-state NMR spectroscopy. *J. Am. Chem. Soc.* 130:10625–32

70. Lorch M, Fahem S, Kaiser C, Weber I, Mason AJ, et al. 2005. How to prepare membrane proteins for solid-state NMR: a case study on the α-helical integral membrane protein diacylglycerol kinase from E. coli. *ChemBioChem* 6:1693–700

71. Lorieau J, McDermott AE. 2006. Order parameters based on (CH)-C-13-H-1, (CH2)-C-13-H-1 and (CH3)-C-13-H-1 heteronuclear dipolar powder patterns: a comparison of MAS-based solid-state NMR sequences. *Magn. Reson. Chem.* 44:334–47

72. Lorieau JL, Day LA, McDermott AE. 2008. Conformational dynamics of an intact virus: order parameters for the coat protein of Pf1 bacteriophage. *Proc. Natl. Acad. Sci. USA* 105:10366–71

73. Lorieau JL, McDermott AE. 2006. Conformational flexibility of a microcrystalline globular protein: order parameters by solid-state NMR spectroscopy. *J. Am. Chem. Soc.* 128:11505–12

74. Lowe IJ. 1959. Free induction decays of rotating solids. *Phys. Rev. Lett.* 2:285–87

75. Mak-Jurkauskas ML, Bajaj VS, Hornstein MK, Belenky M, Griffin RG, Herzfeld J. 2008. Energy transformations early in the bacteriorhodopsin photocycle revealed by DNP-enhanced solid-state NMR. *Proc. Natl. Acad. Sci. USA* 105:883–85

76. Mani R, Cady SD, Tang M, Waring AJ, Lehrert RI, Hong M. 2006. Membrane-dependent oligomeric structure and pore formation of beta-hairpin antimicrobial peptide in lipid bilayers from solid-state NMR. *Proc. Natl. Acad. Sci. USA* 103:16242–47

77. Manolikas T, Herrmann T, Meier BH. 2008. Protein structure determination from C-13 spin-diffusion solid-state NMR spectroscopy. *J. Am. Chem. Soc.* 130:3959–66

78. Marin-Montesinos I, Brouwer DH, Antonioli G, Lai WC, Brinkmann A, Levitt MH. 2005. Heteronuclear decoupling interference during symmetry-based homonuclear recoupling in solid-state NMR. *J. Magn. Reson.* 177:307–17

79. Marshall GR, Beusen DD, Kociolek K, Redlinski AS, Leplawy MT, et al. 1990. Determination of a precise interatomic distance in a helical peptide by REDOR NMR. *J. Am. Chem. Soc.* 112:963–66

80. Marulanda D, Tasayco ML, Cataldi M, Arriaran V, Polenova T. 2005. Resonance assignments and secondary structure analysis of E. coli thioredoxin by magic angle spinning solid-state NMR spectroscopy. *J. Phys. Chem. B* 109:18135–45

81. Marulanda D, Tasayco ML, McDermott A, Cataldi M, Arriaran V, Polenova T. 2004. Magic angle spinning solid-state NMR spectroscopy for structural studies of protein interfaces. Resonance assignments of differentially enriched *Escherichia coli* thioredoxin reassembled by fragment complementation. *J. Am. Chem. Soc.* 126:16608–20

82. Matsuki Y, Akutsu H, Fujiwara T. 2007. Spectral fitting for signal assignment and structural analysis of uniformly C-13-labeled solid proteins by simulated annealing based on chemical shifts and spin dynamics. *J. Biomol. NMR* 38:325–39

83. McDermott A, Polenova T, Bockmann A, Zilm KW, Paulson EK, et al. 2000. Partial NMR assignments for uniformly (C-13°N-15)-enriched BPTI in the solid state. *J. Biomol. NMR* 16:209–19

84. McDermott AE, Creuzet FJ, Kolbert AC, Griffin RG. 1992. High-resolution magic-angle-spinning NMR-spectra of protons in deuterated solids. *J. Magn. Reson.* 98:408–13

85. Mills FD, Antharam VC, Ganesh OK, Elliott DW, McNeill SA, Long JR. 2008. The helical structure of surfactant peptide KL4 when bound to POPC: POPG lipid vesicles. *Biochemistry* 47:8292–300

86. Mueller LJ, Elliott DW, Kim KC, Reed CA, Boyd PDW. 2002. Establishing through-bond connectivity in solids with NMR: structure and dynamics in HC60+. *J. Am. Chem. Soc.* 124:9360

87. Mueller LJ, Elliott DW, Leskowitz GM, Struppe J, Olsen RA, et al. 2004. Uniform-sign cross-peak double-quantum-filtered correlation spectroscopy. *J. Magn. Reson.* 168:327–35

88. Nadaud PS, Helmus JJ, Hofer N, Jaroniec CP. 2007. Long-range structural restraints in spin-labeled proteins probed by solid-state nuclear magnetic resonance spectroscopy. *J. Am. Chem. Soc.* 129:7502–3

89. Opekarova M, Tanner W. 2003. Specific lipid requirements of membrane proteins—a putative bottleneck in heterologous expression. *Biochim. Biophys. Acta Biomembr.* 1610:11–22

90. Opella SJ, Marassi FM. 2004. Structure determination of membrane proteins by NMR spectroscopy. *Chem. Rev.* 104:3587–606

91. Opella SJ, Zeri AC, Park SH. 2008. Structure, dynamics, and assembly of filamentous bacteriophages by nuclear magnetic resonance spectroscopy. *Annu. Rev. Phys. Chem.* 59:635–57

92. Ouellet M, Doucet JD, Voyer N, Auger M. 2007. Membrane topology of a 14-mer model amphipathic peptide: a solid-state NMR spectroscopy study. *Biochemistry* 46:6597–606

93. Palmer AG, Massi F. 2006. Characterization of the dynamics of biomacromolecules using rotating-frame spin relaxation NMR spectroscopy. *Chem. Rev.* 106:1700–19

94. Patel AB, Crocker E, Eilers M, Hirshfeld A, Sheves M, Smith SO. 2004. Coupling of retinal isomerization to the activation of rhodopsin. *Proc. Natl. Acad. Sci. USA* 101:10048–53

95. Pauli J, Baldus M, van Rossum B, de Groot H, Oschkinat H. 2001. Backbone and side-chain C-13 and N-15 signal assignments of the alpha-spectrin SH3 domain by magic angle spinning solid-state NMR at 17.6 tesla. *ChemBioChem* 2:272–81

96. Paulson EK, Morcombe CR, Gaponenko V, Dancheck B, Byrd RA, Zilm KW. 2003. Sensitive high resolution inverse detection NMR spectroscopy of proteins in the solid state. *J. Am. Chem. Soc.* 125:15831–36

97. Pines A, Gibby MG, Waugh JS. 1973. Proton-enhanced NMR of dilute spins in solids. *J. Chem. Phys.* 59:569–90

98. Pintacuda G, Giraud N, Pierattelli R, Bockmann A, Bertini I, Emsley L. 2007. Solid-state NMR spectroscopy of a paramagnetic protein: assignment and study of human dimeric oxidized Cu-II-Zn-II superoxide dismutase (SOD). *Angew. Chem. Int. Ed.* 46:1079–82

99. Qiang W, Bodner ML, Weliky DP. 2008. Solid-state NMR spectroscopy of human immunodeficiency virus fusion peptides associated with host-cell-like membranes: 2D correlation spectra and distance measurements support a fully extended conformation and models for specific antiparallel strand registries. *J. Am. Chem. Soc.* 130:5459–71

100. Raleigh DP, Levitt MH, Griffin RG. 1988. Rotational resonance in solid-state NMR. *Chem. Phys. Lett.* 146:71–76

101. Ramamoorthy A, Thennarasu S, Lee DK, Tan AM, Maloy L. 2006. Solid-state NMR investigation of the membrane-disrupting mechanism of antimicrobial peptides MSI-78 and MSI-594 derived from magainin 2 and melittin. *Biophys. J.* 91:206–16

102. Reif B, Jaroniec CP, Rienstra CM, Hohwy M, Griffin RG. 2001. H-1-H-1 MAS correlation spectroscopy and distance measurements in a deuterated peptide. *J. Magn. Reson.* 151:320–27

103. Rienstra CM, Hohwy M, Mueller LJ, Jaroniec CP, Reif B, Griffin RG. 2002. Determination of multiple torsion-angle constraints in U-C-13,N-15-labeled peptides: 3D H-1-N-15-C-13-H-1 dipolar chemical shift NMR spectroscopy in rotating solids. *J. Am. Chem. Soc.* 124:11908–22

104. Rienstra CM, Tucker-Kellogg L, Jaroniec CP, Hohwy M, Reif B, et al. 2002. De novo determination of peptide structure with solid-state magic-angle spinning NMR spectroscopy. *Proc. Natl. Acad. Sci. USA* 99:10260–65

105. Rozovsky S, McDermott AE. 2001. The time scale of the catalytic loop motion in triosephosphate isomerase. *J. Mol. Biol.* 310:259–70

106. Schaefer J, Kramer KJ, Garbow JR, Jacob GS, Stejskal EO, et al. 1987. Aromatic cross-links in insect cuticle—detection by solid-state C-13 and N-15 NMR. *Science* 235:1200

107. Schaefer J, McKay RA, Stejskal EO. 1979. Double-cross-polarization NMR of solids. *J. Magn. Reson.* 34:443

108. Schaefer J, Stejskal EO, Buchdahl R. 1977. Magic-angle C-13 NMR analysis of motion in solid glassy polymers. *Macromolecules* 10:384–405

109. Scheidt HA, Huster D. 2008. The interaction of small molecules with phospholipid membranes studied by H-1 NOESY NMR under magic-angle spinning. *Acta Pharmacol. Sinica* 29:35–49

110. Schneider R, Ader C, Lange A, Giller K, Hornig S, et al. 2008. Solid-state NMR spectroscopy applied to a chimeric potassium channel in lipid bilayers. *J. Am. Chem. Soc.* 130:7427–35

111. Schnell JR, Dyson HJ, Wright PE. 2004. Structure, dynamics, and catalytic function of dihydrofolate reductase. *Annu. Rev. Biophys. Biomol. Struct.* 33:119–40

112. Shen Y, Lange O, Delaglio F, Rossi P, Aramini JM, et al. 2008. Consistent blind protein structure generation from NMR chemical shift data. *Proc. Natl. Acad. Sci. USA* 105:4685–90

113. Straus SK, Bremi T, Ernst RR. 1998. Experiments and strategies for the assignment of fully C–13/N-15-labelled polypeptides by solid state NMR. *J. Biomol. NMR* 12:39–50

114. Stringer JA, Bronnimann CE, Mullen CG, Zhou DHH, Stellfox SA, et al. 2005. Reduction of RF-induced sample heating with a scroll coil resonator structure for solid-state NMR probes. *J. Magn. Reson.* 173:40–48

115. Takegoshi K, Nakamura S, Terao T. 2003. C-13-H-1 dipolar-driven C-13-C-13 recoupling without C-13 rf irradiation in nuclear magnetic resonance of rotating solids. *J. Chem. Phys.* 118:2325–41

116. Tang M, Waring AJ, Hong M. 2007. Phosphate-mediated arginine insertion into lipid membranes and pore formation by a cationic membrane peptide from solid-state NMR. *J. Am. Chem. Soc.* 129:11438–46

117. Tang M, Waring AJ, Lehrer RL, Hong M. 2008. Effects of guanidinium-phosphate hydrogen bonding on the membrane-bound structure and activity of an arginine-rich membrane peptide from solid-state NMR spectroscopy. *Angew. Chem. Int. Ed.* 47:3202–5

118. Thurber KR, Tycko R. 2008. Biomolecular solid state NMR with magic-angle spinning at 25 K. *J. Magn. Reson.* 195:179–86

119. Tian L. 2007. *Functional studies of KcsA, a potassium channel, by solid state NMR.* PhD thesis. Columbia Univ.

120. Todokoro Y, Yumen I, Fukushima K, Kang SW, Park JS, et al. 2006. Structure of tightly membrane-bound mastoparan-X, a G-protein-activating peptide, determined by solid-state NMR. *Biophys. J.* 91:1368–79

121. Traaseth NJ, Verardi R, Torgersen KD, Karim CB, Thomas DD, Veglia G. 2007. Spectroscopic validation of the pentameric structure of phospholarnban. *Proc. Natl. Acad. Sci. USA* 104:14676–81

122. Tycko R. 2006. Molecular structure of amyloid fibrils: insights from solid-state NMR. *Q. Rev. Biophys.* 39:1–55

123. van Gammeren AJ, Buda F, Hulsbergen FB, Kiihne S, Hollander JG, et al. 2005. Selective chemical shift assignment of B800 and B850 bacteriochlorophylls in uniformly [C-13,N-15]-labeled light-harvesting complexes by solid-state NMR spectroscopy at ultrahigh magnetic field. *J. Am. Chem. Soc.* 127:3213–19

124. Varga K, Tian L, McDermott AE. 2007. Solid-state NMR study and assignments of the KcsA potassium ion channel of S. lividans. *Biochim. Biophys. Acta* 1774:1604–13

125. Verel R, Ernst M, Meier BH. 2001. Adiabatic dipolar recoupling in solid-state NMR: the DREAM scheme. *J. Magn. Reson.* 150:81–99

126. Wallin E, von Heijne G. 1998. Genome-wide analysis of integral membrane proteins from eubacterial, archaean, and eukaryotic organisms. *Protein Sci.* 7:1029–38

127. Wasmer C, Lange A, Van Melckebeke H, Siemer AB, Riek R, Meier BH. 2008. Amyloid fibrils of the HET-s(218–289) prion form a beta solenoid with a triangular hydrophobic core. *Science* 319:1523–26

128. Wickramasinghe NP, Shaibat MA, Jones CR, Casabianca LB, de Dios AC, et al. 2008. Progress in C-13 and H-1 solid-state nuclear magnetic resonance for paramagnetic systems under very fast magic angle spinning. *J. Chem. Phys.* 128:052210

129. Williamson PTF, Verhoeven A, Miller KW, Meier BH, Watts A. 2007. The conformation of acetylcholine at its target site in the membrane-embedded nicotinic acetylcholine receptor. *Proc. Natl. Acad. Sci. USA* 104:18031–36

130. Wishart DS, Sykes BD. 1994. The C-13 chemical-shift index—a simple method for the identification of protein secondary structure using C-13 chemical-shift data. *J. Biomol. NMR* 4:171–80

131. Wishart DS, Sykes BD, Richards FM. 1992. The chemical-shift index—a fast and simple method for the assignment of protein secondary structure through NMR-spectroscopy. *Biochemistry* 31:1647–51

132. Wylie BJ, Rienstra CM. 2008. Multidimensional solid state NMR of anisotropic interactions in peptides and proteins. *J. Chem. Phys.* 128:052207

133. Wylie BJ, Sperling LJ, Frericks HL, Shah GJ, Franks WT, Rienstra CM. 2007. Chemical-shift anisotropy measurements of amide and carbonyl resonances in a microcrystalline protein with slow magic-angle spinning NMR spectroscopy. *J. Am. Chem. Soc.* 129:5318–19

134. Yang J, Paramasivan S, Marulanda D, Cataidi M, Tasayco ML, Polenova T. 2007. Magic angle spinning NMR spectroscopy of thioredoxin reassemblies. *Magn. Reson. Chem.* 45:S73–S83

135. Yang J, Tasayco ML, Polenova T. 2008. Magic angle spinning NMR experiments for structural studies of differentially enriched protein interfaces and protein assemblies. *J. Am. Chem. Soc.* 130:5798–807

136. Zachariae U, Schneider R, Velisetty P, Lange A, Seeliger D, et al. 2008. The molecular mechanism of toxin-induced conformational changes in a potassium channel: relation to C-type inactivation. *Structure* 16:747–54

137. Zech SG, Olejniczak E, Hajduk P, Mack J, McDermott AE. 2004. Characterization of protein-ligand interactions by high-resolution solid-state NMR spectroscopy. *J. Am. Chem. Soc.* 126:13948–53

138. Zech SG, Wand AJ, McDermott AE. 2005. Protein structure determination by high-resolution solid-state NMR spectroscopy: application to microcrystalline ubiquitin. *J. Am. Chem. Soc.* 127:8618–26

139. Zhou DH, Shea JJ, Nieuwkoop AJ, Franks WT, Wylie BJ, et al. 2007. Solid-rate protein-structure determination with proton-detected triple-resonance 3D magic-angle-spinning NMR spectroscopy. *Angew. Chem. Int. Ed.* 46:8380–83

Cumulative Indexes

Contributing Authors, Volumes 34–38

H

Haseltine EL, 36:1–19
Haustein E, 36:151–69
Hebda JA, 38:125–52
Hille B, 37:175–95
Hilvert D, 37:153–73
Holbrook SR, 37:445–64
Holowka D, 37:265–88
Howard J, 38:217–34
Hribar-Lee B, 34:173–99
Hud NV, 34:295–318
Hurley JH, 35:277–98

I

Iyengar R, 34:319–49

J

Jäckel C, 37:153–73
Jackson MB, 35:135–60
Janovjak H, 36:233–60
Jasiak AJ, 37:337–52
Jawhari A, 37:337–52
Jennebach S, 37:337–52

K

Kaback H, 35:67–91
Kahn R, 38:153–71
Kamenski T, 37:337–52
Karplus M, 35:1–47
Kasai RS, 34:351–78
Kast D, 38:347–69
Kast P, 37:153–73
Kedrov A, 36:233–60
Kerppola TK, 37:465–87
Kettenberger H, 37:337–52
Koeppe RE II, 36:107–30
Kondo J, 34:351–78
Korman VL, 38:347–69
Kono H, 34:379–98
Kraft ML, 38:53–74
Kuhlman B, 35:49–65
Kuhn C-D, 37:337–52
Kusumi A, 34:351–78

L

Larijani B, 38:107–24
Larson DR, 38:173–96

Lee NK, 37:417–44
Lehman E, 37:337–52
Leike K, 37:337–52
Leschziner AE, 36:43–62
Levy Y, 35:389–415
Li G-W, 37:417–44
Lia G, 37: 417–44
Lim C, 37:97–116
Lionnet T, 38:173–96
Lipfert J, 36:307–27
Lührmann R, 35:435–57
Lukavsky PJ, 35:319–42

M

Ma'ayan A, 34:319–49
Maheshri N, 36:413–34
Malmberg NJ, 34:71–90
Maritan A, 36:261–80
McDermott A, 38:385–403
McIntosh TJ, 35:177–98
McMahon HT, 37:65–95
Melin J, 36:213–31
Minton AP, 37:375–97
Miranker AD, 38:125–52
Mitra K, 35:299–317
Montange RK, 37:117–33
Müller DJ, 36:233–60
Muñoz V, 36:395–412
Murakoshi H, 34:351–78
Murase K, 34:351–78
Muthukumar M, 36:435–50

N

Nakada C, 34:351–78
Nilsson BL, 34:91–118
Nogales E, 36:43–62

O

Oldfield CJ, 37:215–46
Onuchic JN, 35:389–415
O'Shea EK, 36:413–34
Ozkan SB, 37:289–316

P

Pallan PS, 36:281–305
Pan T, 35:161–75
Pande VS, 34:43–69

Perkins GA, 35:199–224
Peters R, 36:371–94
Poccia DL, 38:107–24
Pollard TD, 36:449–75
Pyle AM, 37:317–36

Q

Quake SR, 36:213–31

R

Raines RT, 34:91–118
Raj A, 38:255–70
Raunser S, 38:89–105
Rhee YM, 34:43–69
Riek R, 35:319–42
Ritchie K, 34:351–78
Rivas G, 37:375–97
Roux B, 34:153–71
Ruda VM, 38:173–96

S

Sable JE, 35:417–34
Santangelo TJ, 35:343–60
Sapra KT, 36:233–60
Sarai A, 34:379–98
Schulman BA, 36:131–50
Schultz PG, 35:225–49
Schwille P, 36:151–69
Selvin PR, 36:349–69
Sheetz MP, 35:417–34
Shell MS, 37:289–316
Simon SA, 35:177–98
Singer RH, 38:173–96
Snow CD, 34:43–69
Soellner MB, 34:91–118
Sokurenko E, 37:399–416
Sorin EJ, 34:43–69
Sosnick T, 35:161–75
Soto A, 34:221–43
Stahelin RV, 34:119–51
Stark H, 35:435–57
Strick TR, 34:201–19
Stuart D, 38:371–83
Suh B-C, 37:175–95
Suzuki K, 34:351–78
Swedlow JR, 38:327–46
Sydow J, 37:337–52
Sykes MT, 38:197–215

T

Takamoto K, 35:251–76
Tama F, 35:115–33
Terwilliger TC, 38:371–83
Thomas DD, 38:347–69
Thomas WE, 37:399–416
Tjian R, 38:173–96
Toprak E, 36:349–69
Torres AJ, 37:265–88
Truskett TM, 34:173–99
Tzakos AG, 35:319–42

U

Udgaonkar JB, 37:489–510
Uhlenbeck OC, 34:415–40
Uversky VN, 37:215–46

V

Vannini A, 37:337–52

van Oudenaarden A, 38:255–70
Verkman AS, 37:247–63
Vilfan ID, 34:295–318
Vlachy V, 34:173–99
Vogel V, 35:459–88; 37:399–416
Völker J, 34:21–42
von Ballmoos C, 37:43–64
von Heijne G, 37:23–42
von Hippel PH, 36:79–105

W

Walz T, 38:89–105
Wang L, 35:225–49
Wang MD, 35:343–60
Ward LD, 36:329–47
Waterman-Storer CM,
   35:361–87
Weber PK, 38:53–74
Weikl TR, 37:289–316
White SH, 37:23–42
Williamson JR, 38:197–215

Wirtz D, 38:301–26
Woodside MT, 36:171–90
Wu M, 37:265–88

Y

Yao J, 38:173–96
Yokoyama S, 38:371–83

X

Xie J, 35:225–49
Xie XS, 37:417–44

Z

Zenklusen D, 38:173–96
Zhang X, 34:267–94
Zhou H-X, 36:21–42; 37:375–97
Zhuang X, 34:399–414
Zocchi G, 38:75–88